D0119679

Understanding the Earth System

Global Change Science for Application

Earth system science has been described as 'a science struggling with problems too large for its participants, but too important to ignore'. This exciting new book provides an overview of this enormous, rapidly growing field, tackling current scientific debates and policy-relevant questions on the global environment.

The multi-disciplinary author team explains the what, the how and the why of climate science, providing a review of research from the last decade, illustrated with cutting-edge data and observations. A key focus is the development of analysis tools that can be used to demonstrate society's options for mitigating and adapting to increasing climate risks. Emphasis is given to the importance of Earth system feedback mechanisms and the role of the biosphere. The book explains advances in modelling, process understanding and observations, and the development of consistent and coherent studies of past, present and 'possible' climates.

This highly illustrated, data-rich book is written both for those who use Earth system model outputs and need to know more about the science behind them, and those who develop Earth system models and need to know more about the broader context of their research. The author team is made up of leading scientists involved in QUEST, a major, recently completed, UK-led research programme. The book forms a concise and up-to-date reference for academic researchers or students in the fields of climatology, Earth system science and ecology. By highlighting the application of scientific results to contemporary policy issues, the book is also a vital resource for professionals and policy-makers working on any aspect of global change.

'This beautifully organized and written book connects the fundamental natural sciences – meteorology, oceanography, ecology and many others – to provide the most complete understanding yet of how our planet works. But it doesn't stop there. It lays out a seamless storyline from the deep past through the present and into the future that contextualises the current phenomenon of global change. Critically, the book brings humanity fully into the picture, from the impacts of environmental change to potential stewardship of the planet, while always maintaining the rigour that good Earth System research demands.'

– **Professor Will Steffen**, *Executive Director, ANU Climate Change Institute, Australian National University*

'With a scope extending across paleoclimate, current climate, feedbacks, human dimensions impacts, adaptation, and mitigation, this ambitious book succeeds in providing a deep yet comprehensive view of the earth system in all its facets. Particularly impressive and novel is its rich set of clear and original figures to illustrate each issue in vibrant ways that will be especially useful for educators and students.'

– **Professor Susan Solomon**, *Ellen Swallow Richards Professor of Atmospheric Chemistry and Climate Science, Massachusetts Institute of Technology*

Understanding the Earth System

Global Change Science for Application

Edited by
Sarah E. Cornell
Stockholm Resilience Centre
I. Colin Prentice
Macquarie University
Joanna I. House
University of Bristol
Catherine J. Downy
European Space Agency Climate Office, Harwell

CAMBRIDGE UNIVERSITY PRESS
Cambridge, New York, Melbourne, Madrid, Cape Town,
Singapore, São Paulo, Delhi, Mexico City

Cambridge University Press
The Edinburgh Building, Cambridge CB2 8RU, UK

Published in the United States of America by Cambridge University Press, New York

www.cambridge.org
Information on this title: www.cambridge.org/9781107009363

First published 2012

Printed and bound in the United Kingdom by the MPG Books Group

A catalogue record for this publication is available from the British Library

Library of Congress Cataloging in Publication Data
Understanding the earth system : global change science for application /
[edited by] Sarah Cornell and I. Colin Prentice ; with Joanna House and Catherine Downy.
pages cm
Includes bibliographical references and index.
ISBN 978-1-107-00936-3
1. Dynamic climatology. 2. Dynamic meteorology. I. Cornell, Sarah, 1969- II. Prentice, I. Colin, 1952- III. House, Joanna. IV. Downy, Catherine.
QC981.7.D94U53 2012
551.6–dc23 2012018849

ISBN 978-1-107-00936-3 Hardback

Additional resources for this publication at www.cambridge.org/QUEST

Contents

Editors

Sarah E. Cornell (executive editor) works on integrative socio-environmental research at the Stockholm Resilience Centre. In her previous role at the University of Bristol, she was responsible for the science management and the synthesis phase of QUEST, the UK Natural Environment Research Council's research programme for Earth system science. Her research background is in marine and atmospheric chemistry, and her interdisciplinary interests are in the anthropogenic changes in global biogeochemistry, socio-economics, environmental management, and the philosophy and methodology of integrative research. Dr Cornell was a founding member of the UK Human Dimensions Committee on Global Change. She established the University of Bristol's MSc in Earth System Science, the UK's first post-graduate programme on this topic, and is active in promoting education for sustainability. In recent years, she has become more engaged in use-oriented transdisciplinary research, with a particular focus on conceptualizations of humans in the Earth system.

I. Colin Prentice (senior editor and chair of the scientific editorial team) served as the scientific leader and chair of the QUEST research programme. He is now a professor at both Macquarie University, Sydney and Imperial College, London, where he and his collaborators are developing a 'next-generation' biosphere model combining Earth system observations with developments in plant functional ecology. His overall research goal is to understand the interplay of the biosphere and its physical environment, including the causes and consequences of natural and human-caused changes in climate and the global carbon and nitrogen cycles. He was awarded the Milutin Milankovitch medal in 2002, and shared in the award of the Nobel Peace Prize to the Intergovernmental Panel on Climate Change in 2007. Professor Prentice co-chaired the International Geosphere–Biosphere Programme's (IGBP's) Analysis, Integration and Modelling of the Earth System (AIMES) project until 2010, and now directs the Australian Terrestrial Ecosystem Research Network's modelling facility. He jointly co-ordinates the biodiversity theme for Macquarie's Climate Futures Centre, and is affiliated to the Grantham Institute for Climate Change at Imperial College, London.

Joanna I. House (policy editor) has been working in the field of climate science for over 20 years. As science and policy officer for QUEST, her role was scientific analysis, synthesis and coordination of the liaison between researchers and the policy/practice user communities. Now a Leverhulme Research Fellow in the Cabot Institute at the University of Bristol, her research focuses on the carbon cycle; land-use change and greenhouse-gas emissions; mitigation of climate change through avoided deforestation, forestry and bioenergy; emissions scenarios; and policy implications. Dr House was a contributing author for the Nobel Prize-winning IPCC Third Assessment Report, and a convening lead author on the Millennium Ecosystem Assessment, which received the Zayeed Prize for Environmental Science. She has also contributed her expertise to reports such as the *Stern Review on the Economics of Climate Change* (2006) and the *Eliasch Review of Forests and Climate Change* (2008).

Catherine J. Downy (technical editor) has a background in science management and liaison, working to bring scientists together under the growing banner of 'Earth system science'. She ran the UK's first open conference in this field, Earth System Science 2010, in Edinburgh, successfully uniting natural scientists from a range of disciplines, together with those from the social sciences and humanities. She now holds the post of liaison officer for the IGBP and European Space Agency (ESA), based at the Climate Office in ESA Harwell.

Scientific editorial team members

Eric Wolff is the science leader of the Chemistry and Past Climate programme, at the British Antarctic Survey. He also holds an honorary visiting professorship at Southampton University's School of Ocean and Earth Science. Professor Wolff's career has centred on investigations of Quaternary and recent climate using ice cores. His areas of interest bridge the physical and chemical characteristics of ice, informing understanding of palaeoclimate and polar atmospheric chemistry. Professor Wolff has served on many polar science field seasons, both in Antarctica and Greenland. He was the science chair of the European Project for Ice Coring in Antarctica (EPICA), a major research initiative that has produced detailed records of climate and atmospheric composition spanning the last 800,000 years, providing unprecedented insights into the workings of Earth's climate system. He now co-chairs the international ice-core co-ordinating group IPICS, and its European counterpart, EuroPICS. He is a member of the steering committee of the IGBP's Past Global Changes (PAGES) project, and a former committee member for the International Global Atmospheric Chemistry (IGAC) project. He was the 2009 Agassiz medallist of the European Geosciences Union (EGU) cryosphere division, and was elected Fellow of the Royal Society in 2010.

Pierre Friedlingstein is Professor of Mathematical Modelling of Climate Systems at Exeter University. His research interests are in the field of global carbon cycle and global biogeochemical cycles, with particular interest in the interactions between climate and biogeochemistry over timescales ranging from the glacial cycles to future IPCC-like projections. His research contributed to the identification of the positive feedback between climate change and carbon cycle, a process that profoundly alters our perspectives on future climate change. This work drives his interest in the evaluation of land surface models and Earth system models over the last millennium. He coordinates the IGBP/WCRP's Coupled Climate Carbon Cycle Intercomparison Project (C^4MIP). He is a member of the science steering committee of AIMES and of the Earth System Science Partnership's Global Carbon Project. Since 1994, he has participated in the IPCC's climate assessments, and is currently a lead author in Working Group I of the IPCC Fifth Assessment Report.

Marko Scholze is a Natural Environment Research Council (NERC) Advanced Research Fellow in the School of Earth Sciences, University of Bristol, and a visiting fellow at the University of Hamburg. His research focuses on the dynamics of terrestrial ecosystems and their interactions with the climate system. The responses of the terrestrial biosphere and carbon fluxes to changes in climate can be seen over various timescales. Dr Scholze's research therefore spans palaeo glacial–interglacial transitions, interannual variability and future global warming. His research goal is to quantify the anthropogenic perturbation of the carbon cycle, including the detection of measures to mitigate climate change. He uses observational data and data assimilation techniques to optimize models of the Earth system. His research also addresses the pressing question of what constitutes 'dangerous climate change' and how to assess the risks of climate change on ecosystems. Dr Scholze contributed to the Millennium Ecosystem Assessment.

Richard Betts is Head of Climate Impacts at the Met Office Hadley Centre. His research specialism is in ecosystem–hydrology–climate interactions, extending to climate impacts on urbanization, health, industry and finance. He is particularly interested in climate-change impacts on water resources and the role of vegetation in modifying this impact, and in climate-change impacts on large-scale ecosystems and interactions with deforestation (with a particular interest in Amazonia). He leads the Met Office's climate consultancy area, which works directly with end-users in a wide range of sectors, to ensure climate-change information is used effectively for decision-making. Dr Betts leads the impacts theme of the Joint UK Land Environment Simulator (JULES) community land surface modelling programme. He was a lead author for Working Group I and a contributing author for Working Group II of the IPCC's Fourth Assessment Report, and is a Working Group II lead author for the IPCC Fifth Assessment Report. He was also a lead author in the Millennium Ecosystem Assessment, and a reviewer for the UK Government-commissioned Stern Review on Economics of Climate Change.

Contributing authors

Icarus Allen is Head of Science (Today's Models, Tomorrow's Futures) at the Plymouth Marine Laboratory. He specialises in the numerical modelling and operational forecasting of marine systems from individual cells to shelf-wide ecosystems.

Nigel Arnell is the director of the Walker Institute for Climate Change Research and Professor of Climate System Science at Reading University. His research focuses on the impacts of climate change on river flows, water resources and their management.

Peter Baines is a professorial fellow in the Department of Infrastructure Engineering at the University of Melbourne. He is a meteorologist and oceanographer interested in climate dynamics on decadal timescales and on the effects of topography on atmospheric flows.

Jessica Bellarby is a research fellow at the University of Aberdeen. Her background is in bioremediation and environmental microbiology. She now works on improving process understanding of soil carbon and organic matter.

Richard Betts leads the Climate Impacts strategic area of the Met Office Hadley Centre. His interests are in large-scale modelling of ecosystem-hydrology-climate interactions and integrated impacts modelling, with particular interest in land use change and deforestation.

Eleanor Blyth leads the Land Surface Processes modelling group at the NERC Centre for Ecology and Hydrology. Her research involves representing the land surface in meteorological and hydrological models.

Penelope Boorman is a scientist in the Climate Impacts strategic area of the UK Met Office Hadley Centre. Her work focuses on improving the understanding and assessment of dangerous regional impacts of global climate change.

Hannes Böttcher is a research scholar with the Ecosystem Services and Management Program at the International Institute for Applied Systems Analysis. He is a forest scientist with particular interest in systems analysis and integrated modelling, specializing on the impact of forestry on the carbon cycle.

Emily Boyd is a reader in environmental change and human communities at the University of Reading, also affiliated to the Stockholm Resilience Centre. Her research focus is on the governance and politics of natural resource management in the context of global environmental change.

Matthew Brander is a senior analyst at Ecometrica, providing expert advice on greenhouse-gas accounting and ecosystem services, with particular interests in bioenergy generation.

Bill Collins leads research into the interactions between atmospheric composition and climate at the UK Met Office, where he led the development of the Hadley Centre Earth system model HadGEM2.

Sarah Cornell coordinates the Planetary Boundaries research initiative at the Stockholm Resilience Centre. Her research background is in global biogeochemistry and in interdisciplinary approaches to environmental management.

Anne-Laure Daniau is a palaeoclimatologist, exploring the palaeorecord to understand the nature of the relationships between fire, climate, land and ocean changes, and human perturbation. Formerly at the School of Geographical Sciences, University of Bristol, she now works at University of Bordeaux.

Cat Downy works at the European Space Agency's Climate Office in Harwell. She has a background in science management and liaison, working to bring scientists together under the growing banner of Earth system science.

Evan Fraser holds the Canada Research Chair in Global Human Security and is an associate professor in geography at the University of Guelph. He is also an author writing on sustainability and food. His research interests are in climate change, food security, farmer responses to environmental change, landscape and history.

Pierre Friedlingstein is Professor of Mathematical Modelling of Climate Systems at the University of Exeter. He is interested in the interactions and feedbacks of the climate system and global biogeochemical cycles.

Angela Gallego-Sala is a research scientist at the University of Exetes and the University of Bristol. Her background is in microbial biogeochemistry. She investigates the effects of climate change on carbon storage in high-latitude peatlands.

Sandy Harrison is Professor of Ecology and Evolution at Macquarie University, Sydney. She is a palaeoclimatologist, reconstructing past climates and landscapes using regional to global syntheses of palaeoenvironmental observations and dynamic models of the land biosphere.

Fiona Hewer has a background in meteorology and consultancy, and now leads her own environmental consultancy, Fiona's Red Kite, working on climate change, knowledge exchange and business development.

Joanna House is a Leverhulme research fellow in the School of Geographical Sciences, University of Bristol. Her research addresses land-use change, greenhouse-gas emissions and climate mitigation.

Chris Huntingford is a climate modeller at NERC's Centre for Ecology and Hydrology. His research interests include detection and attribution of climate change, climate impacts in Amazonia, large-scale land–atmosphere interactions and characterizing uncertainty in future predictions for differing emissions scenarios.

Manoj Joshi is a lecturer in climate dynamics at the University of East Anglia. In addition to his research into various mechanisms of forcing and response of the climate system he project-managed the development of the QUEST Earth system model.

Nicole Kalas is a researcher at the Centre for Environmental Policy at Imperial College, London. Her interests include land-use change and climate, life-cycle analysis for biofuels, and mitigation practice and policy.

Neil Kaye is a GIS specialist working at the UK Met Office Hadley Centre, where he has developed a range of techniques to visualize and analyse climate-model output, with the aim of facilitating the assessment of climate impacts and improving the communication of climate model uncertainty.

Reto Knutti is a professor at the Institute for Atmospheric and Climate Science at ETH Zurich. He is a climate modeller, interested in scenarios for anthropogenic climate change and using observations to improve climate models.

Jason Lowe is Head of Climate Knowledge Integration and Mitigation Advice at the UK Met Office Hadley Centre. He is the chief scientist for the Avoiding Dangerous Climate Change programme, providing key advice to the UK government.

Valérie Masson-Delmotte is a senior scientist at the Commissariat à l'Energie Atomique et aux Energies Alternatives (LSCE–CEA), where she heads the Climate Dynamics and Archives team, researching climate variability using palaeoclimate reconstructions and model simulations.

Mark McCarthy is a climate scientist at the UK Met Office Hadley Centre, where he is currently working on improving understanding of urban microclimates and their interactions with regional and global climate change.

Doug McNeall is a physical scientist at the UK Met Office Hadley Centre, developing and applying statistical and analytical techniques for use with model output and observational data, to improve understanding of climate impacts.

Colin Prentice is a biologist and a pioneer of global ecosystem modelling, with interests in the carbon cycle and biosphere–climate interactions. He served as the scientific leader of QUEST. He is now Professor in Ecology and Evolution at Macquarie University, Sydney and Professor in Biosphere and Atmosphere Interactions at Imperial College, London.

Maria Fernanda Sanchez Goñi is Professor of Palaeoclimatology at the École Pratique des Hautes Études (EPHE, Paris-Bordeaux) with expertise in vegetation–climate interactions and, in particular, in tracing abrupt climate changes using pollen and charcoal preserved in marine records.

Michael Sanderson works in the Climate Impacts group of the Met Office Hadley Centre. His background is in atmospheric chemistry, and his current research interests are in future changes in extreme temperature and precipitation, and urban climates.

Marko Scholze is a research fellow at the University of Bristol, also affiliated to the University of Hamburg. His interests are in data assimilation techniques for climate–carbon-cycle models, and climate risk.

Sonia Seneviratne is Professor of Land-Climate Interactions at ETH Zurich, where she leads a research group focusing on quantifying land–climate feedbacks, modelling extreme events and developing reference data sets for benchmarking models.

Pete Smith is Professor of Soils and Global Change at the University of Aberdeen. He models greenhouse-gas emissions and mitigation, bioenergy for fossil fuel offsets and biological carbon sequestration. He is also the science director of Scotland's Climate Change Centre of Expertise.

Robin Smith is a research fellow at NCAS-Climate, Reading University. He develops and works with Earth system models, addressing large-scale coupled climate processes and also exploring all kinds of imagined Earth-like worlds.

Renato Spahni is a climatologist at the Physics Institute at the University of Bern. He researches past and future climate change, investigating the behaviour of the greenhouse gases methane and nitrous oxide trapped in polar ice cores.

Allan Spessa is a research scientist at the Max Planck Institute for Chemistry in Mainz, where he is working on a new Earth system model. His research focuses on vegetation–climate–fire interactions. He is the joint developer of the SPITFIRE (Spread and InTensity of FIRes and Emissions) model.

Parv Suntharalingam is Research Council UK Academic Fellow at the University of East Anglia, researching the biogeochemical cycles of climatically important species (carbon, nitrogen and sulfur) in the atmosphere and ocean.

Richard Tipper is the managing director of Ecometrica, a company developing and applying standards for greenhouse-gas accounting for climate mitigation and the assessment of ecosystem services.

Oliver Wild, a reader in atmospheric science at Lancaster University, develops and applies models of atmospheric composition and chemistry, with a particular focus on the distribution of ozone in the troposphere and the role of tropospheric chemistry in the Earth System.

Andrew Wiltshire is a climate impacts scientist at the Met Office Hadley Centre, where his interests include glacier melting, changing water resources, and impacts on agriculture and forests. He is also a science communicator.

Eric Wolff leads the British Antarctic Survey's scientific research on chemistry and past climate. He was the science leader of the EPICA ice-core drilling project at Dome C in Antarctica, which produced detailed records of eight glacial cycles in the past 800,000 years of Earth's climate history.

Jeremy Woods, a lecturer in bioenergy at Imperial College, London, researches the links between climate mitigation, development and land use, with particular interests in advanced biorenewables and their implications for food security.

Martin Wooster is Professor of Earth Observation Science at Kings College, London, heading the Environmental Monitoring and Modelling research group, which develops climate-relevant products from satellite data. He applies his remote-sensing approaches to research as diverse as fires, lake and marine monitoring, vegetation and food supply.

Foreword

In 1999, as the new chief executive of the UK Natural Environment Research Council (NERC), I was struggling to find a simple, high-level description of what the NERC did, to use in discussions with our political masters. A conversation with Professor Chris Rapley (then the director of the British Antarctic Survey, part of the NERC) struck a chord; the Medical Research Council is about understanding how the human body works, NERC is about understanding how the planet works. The NERC does Earth system science. Simple, but was it actually true? Certainly, NERC-funded scientists studied all the components that make up the Earth system, but by and large what we were not doing was studying the interactions *between* these key components – between the biosphere and the atmosphere for example – and whilst "NERC strives to understand how the planet works" was a convincing argument to use in negotiations with government, in reality we needed to do better. So QUEST was born.

I don't remember who first suggested a NERC-directed programme in Earth system science (it wasn't me), but once the idea was floated it seemed blindingly obvious, and *Quantifying and Understanding the Earth System* emerged from some intense and scientifically exciting discussions among the members of the research community in 2002. I had the privilege of appointing Colin Prentice to be the scientific leader and chair of the QUEST research programme, and the programme itself started in 2003. I retired from NERC in 2005, and lost touch with what the NERC was doing in general, and with QUEST in particular. So it was with pleasure and surprise that I received Sarah Cornell's e-mail out of the blue in December last year, asking me to write the foreword to this book.

It is an amazing piece of work. When I was running the NERC I was in the fortunate position of being surrounded by colleagues working on all aspects of environmental science. If I wanted to know something, all I had to do was ask the experts, which I frequently did. One of the things I missed most when I retired was losing touch with major developments in environmental research outside my own immediate areas of expertise. Earth system science is a good example. But now here is the next best thing! A truly wonderful, up-to-date synthesis of the major links that exist between our climate system and the biosphere. The contributing authors and the scientific editorial team read like Europe's 'who's who' of Earth system science, writing with great clarity about not only what we know about the scale of the impacts of the human enterprise on our planet, but also what we don't know.

There can be no more important scientific endeavour than to show policy-makers what we are doing to the planet, and what it means for our future. I am under no illusion that explaining to policy-makers what the science says will necessarily get them to 'do the right thing'; the translation of science into policy is a messy, iterative, slow process, fraught with difficulties and frustrations. But a work of this quality must make a difference, and I feel privileged to have been in at the beginning.

Professor Sir John Lawton
York
February 2012

Preface

Why have we written this book?

In 2001, the former chief executive of the UK Natural Environment Research Council (NERC), Sir John Lawton (Lawton, 2001), wrote:

> One of the great scientific challenges of the 21st century is to forecast the future of planet Earth. ...We find ourselves, literally, in uncharted territory, performing an uncontrolled experiment with planet Earth that is terrifying in its scale and complexity.

In the year that followed, the research council consulted widely among its scientists, policy stakeholders and the international research community about how to address that challenge. By the autumn of 2002, a plan of action was in place. The research council had earmarked a very substantial research budget for 'Quantifying and Understanding the Earth System', matched by an ambitious vision for the science that this research programme – QUEST – would address:

> QUEST will seek to provide a more robust understanding of the global carbon cycle. QUEST will require partnerships, both within the UK, and between colleagues in Europe and the USA. NERC's planned investment in QUEST is substantial. It has to be if we are really to make a difference. It is difficult to think of a more important thing to search for (NERC, 2002).

We, the authors of this book, have worked together over several years under the auspices of QUEST (Box 1). QUEST ran from 2003 to 2011, as one of several initiatives worldwide aligned with the internationally developed Earth system science agenda for collaborative research. The research programme sought to do more than 'just' provide a more robust understanding of the global carbon cycle, although interactions between biogeochemical cycles and climate have been at the heart of the programme.

Box 1 – About QUEST

Quantifying and Understanding the Earth System (QUEST), the UK Natural Environment Research Council's directed research programme for Earth system science, ran from 2003 until 2011. Nearly 300 scientists from over 50 institutions were involved in QUEST through its collaborative research projects and related activities. Their mission was to quantify Earth system processes and feedbacks for better-informed assessments of alternative futures of the global environment. The programme's research objectives were to make substantial progress in resolving the following scientific questions:

How important are biotic feedbacks to contemporary climate change?

QUEST investigated the contemporary carbon cycle and its interactions with climate and atmospheric chemistry. New modelling and data analysis tools were developed, incorporating a broader suite of the biogeochemical processes that occur on land and in the oceans. This new breadth enables Earth system models to be used to assess the feedbacks among physical, chemical and biological processes that have contributed to determining the contemporary atmospheric greenhouse-gas content.

How are climate and atmospheric composition naturally regulated?

Palaeoclimate records compiled from diverse marine and terrestrial data sets give a richer picture of landscapes, ecosystems and environmental changes in the past. These reconstructions of past climates have been compared with simulations made using climate and biogeochemistry models, to improve understanding of the interactions between atmospheric composition and climate, primarily on timescales of up to a million years.

How much climate change is 'dangerous'? And how much difference could managing the biosphere make?

Climate change is likely to have serious consequences for ecosystems, and for the societies that depend on them. QUEST carried out interdisciplinary, multi-sectoral studies that provide information about potential global and regional impacts of different degrees of environmental change. QUEST has also assessed the potential for biosphere management to mitigate climate change in a way that accounts for the constraints of land availability and environmental change.

Addressing these 'big-picture' scientific questions about the Earth system presents new operational challenges for research. QUEST explicitly set out to tackle the problematic interfaces between diverse science areas, notably those between the land, atmospheric and marine domains; modelling and observations; palaeoclimate and the contemporary Earth; and the natural and human sciences. Making advances towards integrating knowledge across these component areas of the Earth system requires sustained cooperation by scientists from different specialist backgrounds, institutions and cultures. QUEST was set up explicitly to encourage such cooperation, operating a programme of collaborative interdisciplinary research, investing in carefully designed multi-institution consortium projects, and supporting cross-cutting synthesis activities to respond to today's scientific challenges. QUEST scientists were also active in international collaborative networks and agenda-shaping assessments. Notably, QUEST included a very strong international programme of collaboration, supported by the CNRS of France. Also, through the life of the programme, project scientists maintained active engagement with policy stakeholders, recognizing the societal significance of the research.

One major theme QUEST addressed was feedbacks between climate and the biosphere (see Box 2), with the development of new dynamic global models of land and marine ecosystems and atmospheric chemistry for coupling into climate models. Together with improved approaches to link them with observational data obtained from satellite remote sensing and field and laboratory experimental and empirical studies, these are the tools that allow Earth system interactions to be identified and explored in the contemporary world.

Another key research question is: 'What are the natural controls that regulate Earth's climate and atmospheric composition over much longer timescales?' Society is concerned about human-induced, or anthropogenic, changes to climate, associated with accelerated emissions of greenhouse gases into the atmosphere and changes to the land surface. These changes are taking place in the context of a great deal of natural variability, particularly when set into the context of geological timescales. Understanding the drivers of these changes, and assessing the consequences to ecosystems and landscapes of warming and cooling events is a major challenge that QUEST has sought to address.

Improving our understanding of the interconnected physical, ecological and biogeochemical dimensions of climate change opens opportunities to 'forecast' planet Earth, or rather, to make robust predictions of what consequences would be likely to result from different climatic conditions. This understanding has implications for human society, which is increasingly recognizing the vital need to adapt to climate change and mitigate by reducing emissions of carbon dioxide (CO_2) and other greenhouse gases. Earth system science can be deployed in assessing the potential impact of different degrees of climate warming on key socio-economic sectors (such as agriculture, fisheries, water resources, biodiversity and human health). Improved understanding of the role of the biosphere in climate also offers the potential for managing land ecosystems (forests, farmlands and biomass for energy use) differently to optimize their mitigation potential.

Sir John Lawton was right in that this research effort has been an international enterprise. Nations recognize the strategic importance of research into the dynamic processes of planet Earth and, as concern about global change grows, more effort is being directed in many countries towards research that integrates knowledge about all the sub-systems of our world. With the UK investment in this area, QUEST was able to build strong collaborative relationships, particularly with a large team of Earth system scientists supported by the French Centre national de la recherche scientifique (CNRS), Swiss scientists at the University of Bern and Zürich's Eidgenössische Technische Hochschule (ETH Zürich) and with scientists from many nations who participated in QUEST's extensive programme of working groups. Overall, QUEST has been a major initiative in global-change research, so we draw substantially on our collective experience in this book.

Our motivation in writing this book is to provide an overview of the current state of knowledge of Earth system science, a very rapidly developing field of research. We have highlighted the areas where there is scientific consensus, but perhaps the more important motivation for us is our desire to explain the many areas where there are misconceptions, active debates and open questions still to be addressed. We are not making a detailed scientific assessment of global change; that enormous synthesis process is the domain of the Intergovernmental Panel on Climate Change (although several members of the author team are involved in the IPCC as authors or reviewers). The IPCC's in-depth worldwide peer-review and political-approval process means that it tends not to highlight debated and contested areas of the science. These contested areas are of particular interest to us, in fact, because a clear understanding of the issues, including the arguments and uncertainties, is needed by policy- and decision-makers.

The origins and evolution of Earth system science

The concept of Earth – the whole planet – as a complex interacting system was set out in the late 1960s by James Lovelock whose provocative ideas about planetary-scale feedbacks developed into the Gaia hypothesis (Lovelock, 1979). Although the idea of Earth operating like a living organism, with its dynamic processes acting in concert for self-stabilization, was and still is controversial, it has also been a fruitful source of new thinking.

The period from the late 1960s onwards saw remarkable technological developments in space exploration, new Earth observation technologies, and the development of computers capable of handling and storing very large data sets and running calculations at unprecedented rates. The Bretherton Report (Earth System Sciences Committee, 1988) was a landmark document setting out the scope for an Earth system research agenda that would make the most of these opportunities. This new field of study would integrate studies of Earth's 'spheres' (Box 2) – the atmosphere, oceans, ice sheets, land surfaces, and marine and terrestrial ecosystems, explicitly looking at the interactions between the spheres at various timescales.

Box 2– The 'spheres' of the Earth system

When we conceptualize Earth as a system, our analysis focuses on the characteristics and interactions of a set of interdependent components. These components are often referred to as 'spheres' (see Figure P1), which have different intrinsic rates and patterns of dynamic change:

- ***Atmosphere*** – the fluid layer of air that surrounds Earth, bound by gravity. The atmosphere plays a vital role in Earth's energy balance. It receives solar radiation, the energy that drives life on Earth. It is characterized by relatively rapid physical and chemical processes that shape heat transfer, weather patterns and the long-range transport of dust and aerosols on timescales of hours to seasons.
- ***Hydrosphere*** – all water on Earth, including the oceans, freshwater and groundwater as well as the water cycling through the atmosphere. Over 95% of the volume of water on Earth is found in the oceans, while the rest is split between ice, groundwater, lakes, soil moisture, the atmosphere, rivers and the biosphere. The hydrological cycle describes the processes associated with the movement of water through these different reservoirs, powered by solar radiation. Processes that affect the hydrosphere are generally slower than those of the atmosphere, but vary greatly in time and scale; the residence time of water can be thousands of years in the oceans but only 10 days for water vapour in the atmosphere. Because the heat capacity of water is greater than that of air, the hydrosphere is particularly important for the global redistribution of heat.
- ***Cryosphere*** – although often classed as part of the hydrosphere, Earth's areas of ice cover are climatically important for other reasons. Residence times for ice and snow vary much more than the hydrosphere, from seasonal up to one million years for the ancient ice in eastern Antarctica. The reflection of incoming solar radiation off ice surfaces has a cooling effect, so their high albedo makes them an important factor in geophysical feedback processes. Another key process that determines the extent of the cryosphere and its interaction with other spheres is the thermal diffusivity of snow and ice. This is much lower than air, meaning that it cannot transfer heat very quickly, and therefore any land or ocean covered with snow and ice are

decoupled from their normal interactions with the atmosphere.

- ***Lithosphere*** – the rocky component of Earth, usually extending from the crust to the upper mantle, although in the context of Earth system science it can also include the mantle and core. This provides the ultimate control on supply of essential elements to the biosphere, and the variations in topography and landscape are key physical controls on climate. With exceptions such as earthquakes and volcanic eruptions, the Earth system processes of the lithosphere are generally slow. Processes like erosion, tectonics and geological uplift operate on timescales of centuries to millions of years.

Together, these components comprise the ***geosphere***, or the physical climate system. Earth system science also is concerned with the processes of life itself:

- ***Biosphere*** – comprised of all of Earth's living organisms, on land and in water bodies, the ecosystem processes of this sphere operate at multiple timeframes, but diurnal and seasonal cyclic processes are particularly important features for the Earth system. Energy is obtained mainly through photosynthesis and is used by the organisms on Earth up through the trophic levels. Important processes include the cycling of nutrients such as carbon, nitrogen and phosphorus.

The interdependence of these spheres means that an event can trigger consequences through the whole system. Earth system analysis explores these causal chains of interactions, the patterns of effects in the different spheres, and the explanatory mechanisms for changes.

In an extension of the concept of Earth's spheres, the ***anthroposphere*** is seen as a distinctive subset of the biosphere, capable of disproportionate impacts on all the other spheres of the Earth system through the deployment of technology. The expansion of communication technologies has driven a much more rapid connectivity on a global scale. In earlier phases of human history, new discoveries or inventions that led to greater efficiency in human transformation of landscapes or appropriation of natural resources were spatially independent. Now, ideas can have global consequences within very short timescales, wherever they arise. Proponents of the idea of the anthroposphere, and the related concept of the cybersphere, argue that incorporating understanding of the dynamics of human activities – including information flows – is essential for explaining and predicting the behaviour of the Earth system.

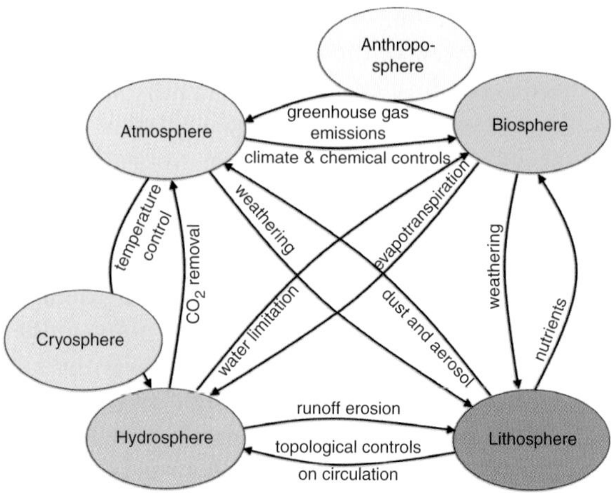

Figure P1 – The 'spheres' of the Earth system and examples of their exchanges.

The two vital strands to this work – Earth observation and global modelling – are both enormous enterprises, and right from the start it was recognized that international efforts would be needed to pull together the fundamental knowledge of Earth's processes and the infrastructure to support an Earth system research effort. Because these research networks, institutional structures and policy-dialogue mechanisms feature throughout this book, we describe them briefly here. Box 3 sets these developments in their historic context.

Scientific institutions in Earth system research

Scientific institutional developments to support the global Earth system research effort included the establishment by the International Council of Science Unions (ICSU) of a set of international collaborative global-change programmes. These have the remit to agree on scientific priorities and help coordinate collaborative international activities funded by national research agencies. The World Climate Research Programme was the first of these programmes to be set up by ICSU, in partnership with the World Meteorological Organisation (WMO) and later also with support from UNESCO's International Oceanographic Commission. The International Geosphere–Biosphere Programme (IGBP) followed in 1987, to address biogeochemical interactions on a global scale. Diversitas was set up in 1991 to address biodiversity and ecosystem change, with the support of UNESCO, the International Union of Biological Sciences and the Scientific Committee on Problems of the Environment. The International Human Dimensions Programme on Global Change was established in 1996 as a partnership of ICSU and the International Social Sciences Council. (It is now also sponsored by the United Nations University.) Together, these four programmes cover the key fields of global-change research, from the physical climate system, through biogeochemical cycles, ecosystems and human society. But even this level of integration of knowledge reached its limits. The challenges of bridging all these fields of knowledge are huge. In 2001, the four programmes collectively formed the Earth System Science Partnership,[1] to provide another level of international coordination and cooperation in research and societal engagement. This structure is still evolving[2] as society's needs, and the perceived opportunities for technological innovation, change.

It was evident from the start of these programmes that although there was excellent scientific understanding of environmental processes at a small scale, there was a dearth of basic information at the global scale. There was a need for long-term time-series studies so that the dynamics of key processes could be observed. Global budgets were needed for the major biogeochemical cycles of water, carbon, nitrogen, sulfur and other elements, with a better understanding of their environmental (and human) sources and sinks. Together with partner organizations at the national level, the programmes coordinated studies linking space and detailed in-situ process studies in land and marine environments, and of the atmosphere. Early examples include NASA's 'Mission to Planet Earth' (Wickland, 1991); 'Global Change and Terrestrial Ecosystems' (Walker and Steffen, 1996); 'Past Global Changes' (Alverson, 1998); and the 'Joint Global Ocean Flux Study' (Fasham *et al.*, 2001). As each project reached completion, the new data and models opened new vistas and raised new questions, so the process of international observation and collaborative networking has continued. QUEST was actively engaged in this process from its inception. Together, all this effort has combined to give unprecedented insight into workings of Earth as a system.

[1] www.essp.org
[2] www.icsu.org/future-earth

Institutions at the science–policy interface in Earth system research

Another area where institutional change has been required is in the interface between science and policy. The most visible of these interfaces is IPCC,[3] established in 1989 by the UN Environment Programme and the WMO to provide the world's governments with a consensus scientific view of climate change, its economic and social impacts, and a scientifically and technologically informed assessment of possible responses. At the time of writing, the IPCC is now working on its fifth comprehensive assessment report. Since its creation, it has expanded the range of climate-linked topics it addresses through special reports and technical guidelines. Yet even this is not enough for today's challenges of environmental change. Recognizing the global scale of ecosystem change, and the importance for human society of biodiversity and ecosystem health, another international body is being created using similar scientific assessment and worldwide policy engagement mechanisms to the IPCC: the newly approved Intergovernmental Platform on Biodiversity and Ecosystem Services,[4] which will provide synthesis reports targeted for policy-users in the area of conservation and sustainable use of natural resources. Alongside these global-scale activities, there are many mechanisms and processes for information exchange at the science–policy interface at both the national level (such as research centres linked to government departments, and forums linked to national academies of science) and the international level (notably the technical and scientific advisory bodies associated with global environmental conventions and treaties).

Box 3 – A history of Earth system science

The foundations of Earth system science lie in physical climate science, but the field has developed as a result of its particular focus on the nature of the interactions between the many components of the Earth system, the human controls of system processes, and of course the global scale of inquiry. As in many other fields of study, Earth system science began with the collection of empirical or observational data, shifting from the anecdotal towards more coordinated, systematic inquiry, coupled with the development of explanatory theory. These developments in the field of knowledge

[3] www.ipcc.ch
[4] www.ipbes.net

have taken place in the context of increasingly formalized scientific cooperation, and the creation of intergovernmental organizations to support research and science policy dialogues. Fuller treatments of the history of climate and Earth system science are given in Weart (2008), Liverman *et al.* (2002) and Cornell (2010).

~1700s The foundations were set for climate science in its 'parent disciplines' of meteorology and oceanography. These foundations included the development of observational data sets from around the world, such as surface temperature, ocean currents and atmospheric dynamics; a systematization of study; and the development of theory to explain climatic phenomena such as the tides, winds and monsoon systems.

~1800s Many fundamentals of the climate system were determined: Earth's greenhouse effect was calculated by Joseph Fourier. The properties of greenhouse gases were identified by John Tyndall. Svante Arrhenius noted the anthropogenic greenhouse effect of CO_2, including projections for the extent of future warming.

Structures were established for formalized scientific cooperation, including major oceanographic expeditions and the creation of organizations to support climate research. The International Meteorological Organization was formed as a specialist intergovernmental agency for scientific exchange and cooperation between national research bodies.

Early 1900s The 'Great Acceleration' in human activity was enabled as oil fields were first exploited on a large – and rapidly expanding – scale. G. S. Callendar made early measurements of the warming trend, having calculated the 'artificial production of carbon dioxide' through fuel combustion. The natural controls were also being explained: Milutin Milanković explained the role of solar orbital changes on climate.

The USA led a major post-war research investment in climate science and Earth observation technology.

1950s Numerical modelling emerged as a powerful tool for global-change science, with the quantification of feedbacks, global atmospheric modelling (including the GCMs) and the explanation of the links between atmospheric structure, chemical composition and radiative forcing.

The first observations of Earth were made from space.

1960s Charles Keeling's high-precision measurements in Hawai'i showed that the atmospheric CO_2 concentration was increasing with time with a pronounced seasonal signal.

Key concepts relating to climate as a system were explained and accepted, including positive and negative feedbacks, the role of ice and albedo changes, and the evidence of orbital changes in palaeorecords. Syukuro Manabe and Richard Wetherald made the first model calculations of climate sensitivity to a doubling of CO_2.

The international Global Atmospheric Research Program (GARP) was established, linking governmental (UN) and scientific institutions in ways that set the pattern for subsequent international collaboration and science–policy interactions.

1970s Key Earth system insights were obtained into episodes of comparatively rapid climate changes in the past, including feedback processes; the biosphere's effects on climate; and the role of atmospheric aerosols and dust. Space exploration gave information about the 'system behaviour' (and past climates) of other planets. The first weather satellite (NASA GOES, launched in 1975) provided data that could be used for model validation.

The recognition that human-induced changes, including aerosol production and deforestation, have climate consequences fed into growing societal concern about the environment. A coherent environmental movement emerged and strengthened.

1980s Essentially, this was a decade of consolidation of evidence addressing debates that had begun earlier: Greenland ice cores showed that very rapid (century-scale) climatic changes were possible; the Vostok ice core showed the very strong coupling of CO_2 and temperature over several glacial cycles; the strong warming trend over the previous decade was observed, and the cooling effect of anthropogenic aerosol that partially disguised the warming was determined.

This was also a period of scientist mobilization, with the creation of the IPCC for the science–policy interface; and the evolution of GARP into the global-change programmes IGBP and WCRP, supporting strategic research cooperation. The Bretherton Diagram mapped out an international collaborative scientific agenda for Earth system science.

1990s The UN Framework Convention on Climate Change was agreed. The first IPCC Assessment Reports were published, refining research questions and driving progressive improvements in global-change models and process-based research.

Two new international, non-governmental global-change programmes were formed: Diversitas provided an umbrella programme for biodiversity research, and the IHDP addressed social science research on global environmental change. The Human Interaction Working Group's Social Process Diagram identified key research areas for the social sciences concerned with global environmental change.

2000s A strong scientific consensus now exists on the anthropogenic control on climate.

Stronger warming has been observed, increasing fairly steadily since the 1970s. The effects on ice sheets are a growing concern, because of sea-level rise and poorly understood physical feedbacks. Understanding biophysical and biogeochemical feedbacks is a research priority, because they shape the timeframes and scope for society's adaptive responses to climate change. A key focus remains in bridging and integrating the knowledge of the natural and social sciences in the Earth system context.

The aims of Earth system science

Earth's climate history has been subject to enormous changes, and our growing understanding of these changes indicates that the interactive Earth system is astonishingly complex and an exceedingly rich subject for research. By gaining a better understanding of the processes and control mechanisms that make up the climate system, the aim of today's research effort is to provide society with knowledge about how the system might respond to future perturbations.

Developing this kind of predictive power means:

- **Bringing together the best available understanding of Earth's living and non-living components**, or – using the 'spheres' terminology, linking the biosphere (including the anthroposphere) with the geosphere, atmosphere and hydrosphere. This requires much improved understanding of physical, chemical and biological processes, which we have summarized in Chapter 2. It also requires many key gaps in climate models to be filled. We address those gaps and progress in Earth system modelling in Chapter 5.
- **Finding independent ways to verify or constrain this dynamic process understanding**. Chapters 3 and 4 describe how we use the palaeorecord (fossils and other measures from the past) together with contemporary science, and combine Earth observation from space with ground-level observations. Very often, the picture has to be pieced together from many strands of evidence, so we give particular attention to data synthesis and comparisons of models and data involving multiple data sources.
- **Bringing computer modellers, 'ground-up' field and laboratory experimentalists and 'top-down' Earth observational researchers together**. This has proved to be a surprisingly big challenge. As science has become ever more specialized, bigger differences have developed between different knowledge communities, who may have very different ideas of what counts as scientific evidence, what are the best ways to present research findings, how best to share and store data, and so on (described humorously in Randall and Wielicki, 1997). Moss and Schneider (2000) suggested that increasing confidence in a scientific finding requires coherence in the quality and amount of theoretical underpinning, observations, model results and expert consensus. Achieving 'joined-up' Earth system science requires careful meshing of very different gears across different physical scales. This integrative effort is a key theme throughout the remainder of this book, but is addressed particularly in Chapter 5.

In turn, these new insights demand new dialogues that are currently developing worldwide (these are summarized briefly here; research in these areas is addressed in more detail in Chapter 1):

- **Improved academic interaction between the natural and social sciences**. Many 'environmental issues' of the last few decades are now seen as *socio*-environmental issues, such as air and water pollution, the over-exploitation of land and natural resources, and even wildlife and natural-habitat conservation. The cross-disciplinary challenges of delivering integrated knowledge in this area are evidently even greater than linking across the different knowledge communities in the biogeochemical and physical sciences, but the need to do so is no less pressing. Chapters 6 and 7 show how Earth system science offers important tools for assessing and mitigating the impacts of changing climate on ecosystems and the human societal systems that depend on them.

- **Deeper, more responsive interaction between science and policy communities.** The whole history of Earth system science as a field of study has taken place in a context of awareness of the societal importance of its research findings and engagement with policy in the co-development of research priorities. In common with previous global environmental challenges, the trans-boundary and multicultural issues in Earth system science require societal engagement and response mechanisms that operate across the science–policy interface. However, as our knowledge develops more explanatory and predictive power, the nature of this interface changes, with different policy areas becoming engaged, and at different levels of governance. In other words, the deeper the scientific understanding, the more complex the policy terrain for engagement. And the dialogue is emphatically not a one-way provision of science for decision-makers. Many argue for more 'democratization of science', both in terms of scientists engaging directly in informing society's responses to environmental change challenges, and in terms of opening up more transparent deliberative processes about the science that is used in the policy process. Our final chapter sets the scientific issues described in the earlier chapters back in their social context.

The Editors on behalf of the Author Team

References

Alverson, K. (1998). PAGES: Past Global Changes. *JOIDES Journal*, **24**(2), 34–35.

Cornell, S. (2010). Climate change: brokering interdisciplinarity across the physical and social sciences. In R. Bhaskar, C. Frank, K. G. Høyer, P. Næss and J. Parker, eds. *Interdisciplinarity and Climate Change: Transforming Knowledge and Practice*. London: Routledge, pp. 116–134.

Earth System Sciences Committee (The Bretherton Report) (1988). *Earth System Sciences: A Closer View*. Washington, DC: NASA.

Fasham, M. J., Balino, B. M. and Bowles, M. C., eds. (2001). A new vision of ocean biogeochemistry after a decade of the Joint Global Ocean Flux Study (JGOFS). *AMBIO*, Special Report 10.

Lawton, J. (2001). Earth System Science. *Science*, **292**(5524), 1965.

Liverman, D., Yarnal, B. and Turner II, B. L. (2002). Origins and institutional setting of human dimensions of global change research. In *Geography in America at the Dawn of the 21st Century: The Human Dimensions of Global Change*, eds. G. L. Gaile and C. J. Willmott. Oxford: Oxford University Press, pp. 267–282.

Lovelock, J. E. (1979). *Gaia: A New Look at Life on Earth*. Oxford: Oxford University Press.

Moss, R. H. and Schneider, S. H. (2000). Uncertainties in the IPCC TAR: Recommendations to Lead Authors for more consistent assessment and reporting. In *Guidance Papers on the Cross-Cutting Issues of the Third Assessment Report of the IPCC*, eds. R. Pachauri, T. Taniguchi and K. Tanaka. Geneva: World Meteorological Organization, pp. 33–51.

NERC (2002). Editorial: John Lawton. *Planet Earth*, Autumn edition. UK Natural Environment Research Council, Swindon, UK. Available at: www.nerc.ac.uk/research/programmes/quest/resources/editorial.asp.

Randall, D. A. and Wielicki, B. A. (1997). Measurements, models, and hypotheses in the atmospheric sciences. *Bulletin of the American Meteorological Society*, **78**, 399–406.

Walker, B. H. and Steffen, W.L., eds. (1996). *Global Change and Terrestrial Ecosystems*. IGBP Book Series No. 2, Cambridge: Cambridge University Press.

Weart, S. (2008). *The Discovery of Global warming*, 2nd edn. Boston, MA: Harvard University Press.

Wickland, D. E. (1991). Mission to Planet Earth: the ecological perspective. *Ecology*, **72**(6), 1923–1933.

Acknowledgements

We gratefully acknowledge the financial support of the UK Natural Environment Research Council (NERC), which has funded the QUEST programme from 2003–2011, and has generously supported the development of this book.

We are grateful to the extensive community of QUEST scientists, external advisors and policy colleagues whose contributions to the strategic development and the delivery of the programme's ambitious and innovative research have provided the vital scientific underpinning to this book.

Many individuals also made important contributions: Julie Shackleford helped QUEST run smoothly from its outset. Pru Foster and Wolfgang Knorr, our colleagues in QUEST's core science team, contributed scientific insights, model code and output, and good company. Matt Fortnam and Jenneth Parker provided research synthesis and editorial support at various stages in the development of the book. Alistair Seddon, Steve Murray and Helen Thomas combined scientific knowledge, technical expertise and an artistic eye in producing many of the graphics and images in this book. We are also grateful to the many people and organizations who gave consent for their images to be reproduced or data to be used in order to generate images published here.

The issues explored in Chapter 1 were informed by science carried out in QUEST's Research Theme 3 (How much climate change is dangerous?), discussions under the aegis of the QUEST Working Group on Humans and the Earth System, and by debates led by Diana Liverman at the post-graduate Earth System Science Summer School (ES4). Reina Mashimo contributed input to the discussions made in Chapters 1, 5 and 6 of uncertainty in socio-environmental systems.

Chapter 2 benefited from discussions with many climate scientists, especially those involved in the QUEST Working Group on the Hydrological Cycle (Dartington Hall, May 2007) including Mike Raupach, Julia Slingo and Graeme Stephens, and the QUEST project Climate-Carbon Modelling, Assimilation and Prediction (CCMAP). Figure 2.1 was provided by Andrew Manning. Figure 2.7(a) was supplied by Rick Lumpkin. Josh Fisher processed the data for Figure 2.12. Pru Foster developed the codes to produce Figures 2.12, 2.16 and 2.17. Wang Han processed the data for Figure 2.13.

Chapter 3 includes research carried out under QUEST's Theme 2 (How are climate and atmospheric composition regulated on timescales up to a million years?), particularly the project Dynamics of the Earth System and the Ice Core Record (DESIRE), the Working Groups on Abrupt Climate Change and Fire and the Pinatubo Interdisciplinary Multi-Model Study, and the Palaeoclimate Model Intercomparison Project, which received support from QUEST.

Chapter 4 draws on research and discussions supported by several activities in QUEST's Theme 1 (How important are biotic feedbacks for 21st century climate change?), notably the QUEST Earth System Model and Climate-Carbon Modelling, Assimilation and Prediction projects, and the joint QUEST/Environment Agency project on carbon in peatlands and uplands.

In Chapter 5, Figure 5.3 was provided by Chris Jones. The chapter draws on scientific discussions that informed QUEST's development of its Earth system model, QESM, and the research projects Marine Biogeochemistry and Ecosystem Initiative in QUEST; CCMAP; and the Working Group on Earth System Model Benchmarking.

Chapter 6 includes research findings from the QUEST project Global Scale Impacts of Climate Change, and collaborative activities with the UK Government funded project AVOID.

Chapter 7 includes research carried out under the QUEST projects QUATERMASS (QUAntifying the potential of TERrestrial bioMASS to mitigate climate change) and the Joint Implementation Initiative for Forestry-based Climate Mitigation (JIFor). We are very grateful for the enjoyable, challenging and constructive discussions with Robert Matthews from Forest Research, who was involved in both these projects. The chapter also includes work carried out as part of the UK Government funded project AVOID.

The inspiration for Chapter 8 came from our own internal discussions about what we do and what it is for – but of course it resonates with contemporary concerns that are being addressed much more widely as we collectively face the challenges of global change. We have already thanked NERC for their vision and support in this regard. We are also enormously grateful to the wider research community, and the collaborative international global-change projects IGBP, WCRP, IHDP and Diversitas for their support of these fields of inquiry and for continually pressing for new knowledge about the Earth system of which we are part.

Units

We have followed international scientifically agreed standards for the units in this book, most of which are defined by the International System of Units (SI). Many of the basic units are very familiar:

mass	gram	g
length	metre	m
temperature	kelvin	K
quantity of substance	mole	mol
time	second	s
	hour	h
	year	a

We also use scientific notation and exponents for quantities, so for example:

- cubic metres per second is denoted as $m^3\ s^{-1}$
- watts per square metre is denoted as $W\ m^{-2}$
- per year is denoted as a^{-1}

Many of these units require prefixes in order to avoid presenting very big numbers, which are taxing on the eye when reading! The prefixes we use are:

Name	Factor	Symbol
kilo	1000, or 10^3	k
mega	10^6	M
giga	10^9	G
tera	10^{12}	T
peta	10^{15}	P
exa	10^{18}	E

One petagram (Pg) is equivalent to a gigatonne, an alternative unit often used in global-change science because of the magnitudes of processes like the carbon cycle. A tonne is 1000 kg.

We also refer to atmospheric concentrations in terms of mixing ratios, as parts per million (ppm), and parts per billion (ppb, that is 10^{-9}).

Earth system science and society: a focus on the anthroposphere

Sarah E. Cornell, Catherine J. Downy, Evan D. G. Fraser and Emily Boyd

In this chapter, we explore the challenges that Earth system researchers face in addressing human-induced global environmental changes and the societal consequences of global change within their research toolkit. We focus on areas of research that have particular resonance with today's social and political demands.

1.1 The Earth system and the 'problematic human'[1]

1.1.1 The state of play and our position

The great scientific challenge faced by today's global change scientists is to understand the Earth system. Part of this is knowing that we ourselves, as human beings, are an influential component of that system and that the understanding we develop shapes our responses to the environmental changes we see around us. In scientific terms, most of the fundamental workings of our planet, including the processes that change climate and landscapes on short and long timescales, were already well understood by the end of the twentieth century. Earth system science is the field of study that has brought these areas of knowledge together. It has not just provided insight into the phenomena of global environmental change, but also explained the 'hows' and 'whys' behind them, bringing insights into the future prospects for our planet. The enormity of the challenge lies in the realization that we are seeking to understand and predict the properties of a complex adaptive system of which we are a part, recognizing that our choices and our agency as human beings are important controls on its workings. More than that, our ability to deploy our knowledge and make choices about our actions is an important facet, perhaps even a characterizing trait, of our existence.

For scientists in all the contributing fields of inquiry, this development marks a shift from the pursuit of knowledge largely for its own sake to robust predictive knowledge that is required – urgently, many argue – for application in the real world. The prediction of any system where humans play a part has long challenged both scientists and philosophers. Without venturing into those debates here, the prediction of socio-environmental systems nevertheless presents us with a very practical conundrum: our current understanding, even the knowledge codified in the most sophisticated models, is a partial and simplified picture of reality. Our predictions based on this understanding may be wrong and the unintended consequences of action based on those predictions may be severe. However, not to use the available understanding would be to take a perverse and unhelpful position.

In this chapter, we often take the historic view, looking at past scientific efforts in global change research, particularly those efforts framed in terms of global systems. Quantitative models of human dynamics, such as Thomas Malthus' eighteenth-century calculations of Earth's carrying capacity for human population, and the Club of Rome's efforts in the 1970s to measure the limits to socio-economic growth, have generally done a poor job. Nevertheless, it is important to learn from these efforts. One reason they failed is that they did not

[1] With thanks to Lesley Head, University of Wollongong, for this expression. See Head, L. (2007) Cultural ecology: the problematic human and the terms of engagement, *Progress in Human Geography*, **31**, 837.

Understanding the Earth System: Global Change Science for Application, eds. Sarah E. Cornell, I. Colin Prentice, Joanna I. House and Catherine J. Downy. Published by Cambridge University Press © Cambridge University Press 2012.

adequately take into account human agency. In short, we have continually underestimated the role of the social and economic context when we have tried to model the impact of global change. Even where features of socio-economic change evident at the global level allow for a measure of scientific generalization to be made, they are often not included nor examined in models. For example, below we mention the widely observed demographic transition in human population growth: this is just as good and precise a 'law of nature' as most in biology. At the opposite end of the spectrum, there are models of the Earth system that omit human activity altogether. Contemporary physical climate models work as effectively (which is to say, very effectively indeed in describing climate dynamics) for all planets with an atmosphere, ocean and land surface. They reach the limits of their predictive power when they need to bring people into the equation. One challenge we face is that the climate modelling enterprise has defined contemporary climate change in strictly physical terms, i.e. as a physical change driven by increasing amounts of greenhouse gases in the atmosphere. However, from a socio-environmental perspective, there are many ways of looking at and defining climate change. There is a risk that using one dominant way of looking at the problem can drive the policy agenda to the exclusion of other important approaches for finding practically useful solutions. Predictive power alone is not synonymous with usefulness.

What do we require from Earth system science, defined broadly as the science of both the climate system and the human dimension (or the 'anthroposphere')? Ideally, we would like to develop a science that addresses both the human and the natural-environmental components of the system, and that can tell us something about how this complex, coupled socio-ecological system works (Young *et al.*, 2006). Modelling is an essential part of this process. Much of the remainder of this book tries to describe the present state of the models (conceptual and numerical) that underpin both the science and the political decision-making process relating to global environmental change. Earth system science should also be informed by an effective understanding of how society steers itself, so that the scientific process can be more transparent and more responsive to the needs of society. In the context of the unprecedented magnitude and rate of global environmental changes, many people consider that major social, political and economic changes are needed to cope with, manage or avert the worst impacts. These changes should involve new structures and dialogues between scientists and other members of society, for participation in collective decision-making about the future.

1.1.2 The human dimensions of global environmental change: controls, consequences and context

The rapidly expanding science of Earth's climate, biogeochemical processes, and their interconnections is complex, yet it needs to be understood much more widely if society is to respond to current environmental pressures and projected future changes in an informed way. Although our main focus in this book is on summarizing and explaining the biophysical science of global change, a deep understanding of the Earth system will also include insights from the study of human behaviour.

First of all, human activities, more than ever before, are important ***controls*** on Earth's biophysical processes.

The trajectory of human population, especially since the Industrial Revolution, has been one of steady and rapid growth. In the early nineteenth century, the political economist Thomas Malthus famously argued that unconstrained population growth would naturally follow a 'geometric ratio', or exponential growth, while the supply of life-sustaining resources would not. The result would be an overshoot of the human population and a catastrophic check (famine, disease, conflict) on human numbers. Although Malthus was plausibly describing the trends he observed, by simply extending them into the future, his predictions were very wrong. Earth's population has shown some periods of exponential growth – for instance, doubling from 1.5 billion to 3 billion inhabitants between about 1880 and 1960 (80 years), and then again to 6 billion between 1960 and 2000 (40 years). But it is unlikely that the Earth's human population will double again, so both the unfolding of history and a more sophisticated understanding of the dynamics of population have now removed the spectre of catastrophic overpopulation (Cohen, 1998). In 2011, world population reached 7 billion inhabitants (United Nations, 2011), and current estimates suggest that our numbers will peak at approximately 9 billion sometime over the next century (Lutz *et al.*, 2008; United Nations, 2011). Malthus was wrong because he failed to predict the 'demographic transition', a widely observed phenomenon where birth *and* death rates both fall as economic development progresses (e.g. Chesnais, 1992), halting the exponential pattern of growth. Rates of growth of global population have declined in the last two or three decades (US Census Bureau, 2011). Malthus also failed to predict the Haber–Bosch process for nitrogenous fertilizer production and the hybridization of seeds (along with other technological innovations and transformations in agricultural systems), which have made it possible for the world to produce enough food for today's population. Today, the problem of hunger is primarily one of economics and politics affecting food supplies, not one of Earth's capacity for food production. So, while some argue that the world will need to produce considerably more food by mid-century (Bruinsma, 2009), others point out that using the food we currently have more efficiently should be enough to continue feeding the world (Smil, 2001).

Steffen *et al.* (2004) reviewed the impacts on Earth of this rapid population growth, finding similar rapid rises in natural resource use (Figure 1.1), bringing unintended consequences for ecosystems such as rising levels of air and water pollution and environmental degradation. However, the impact of human activity on the natural environment is far from being a simple linear function of population. The widely quoted 'IPAT equation' (Ehrlich and Holdren, 1971; Commoner, 1972; Box 1.1) frames the environmental impact of human activities as a function of technological change and economic growth as well as population: impact = *f*(population × affluence × technology). It highlights the fact that society's increasing technological capability (the *T* in the equation) means people can access more of Earth's natural resources and transform them for their use more effectively, and also that increased affluence (*A*) enables individuals in the population to use and consume more resources. The analysis conducted by Steffen *et al.* (2004) shows how affluence and technology are often much more influential on impact than population numbers.

Box 1.1 The IPAT equation

A simple formulation that has widely been used in Earth system science describes the *impact* of human activity on the natural environment as a function of *population*, *affluence* and *technology*. The interactions of these three terms have been explored empirically in a range of socio-environmental contexts, including resource consumption, food security and energy-systems analysis.

$$\text{Impact} = f(\text{Population, Affluence, Technology})$$

The relationship between impact and the PAT terms is not simple (Chertow, 2001). The original equation was simply multiplicative ($I = P \times A \times T$). This is a useful heuristic for linking impact and socio-economic development, and many nations in the world have shown increasing impacts as they developed in the past, but it is not a predictive law. Empirical studies show strongly non-linear relationships between the terms (Dietz and Rosa, 1997; Dietz *et al.*, 2007). Recent studies (e.g. Pitcher, 2009; Davis and Caldeira, 2010) highlight the fact that consumption levels (the *A* and *T* dimensions) are the 'thermostats' on impact, not population itself, warning against the errors of simple Malthusianism and using the formulation in 'green-revolution' arguments for technological innovation to reduce impact.

A close relative of the IPAT equation, the Kaya Identity (Kaya and Yokobori, 1993) has been developed in the context of energy and greenhouse-gas emissions. It has been applied in IPCC emission studies and scenario development (IPCC, 2001).

$$CO_2 \text{ emissions} = \text{Population} \times \underset{\text{(goods consumed per capita)}}{\text{Consumption intensity}} \times \underset{\text{(energy input per unit goods)}}{\text{Energy intensity}} \times \underset{(CO_2\text{ output per unit energy})}{\text{Carbon intensity}}$$

These formulations show that multi-pronged responses to environmental impact can – and should – be explored. Thus, for climate change, responses could include societal learning and change, economic incentives and instruments, improved efficiency, energy substitution, CO_2 sequestration.

The impact of the human endeavour is now manifest at the global scale (Figure 1.2). Only the most hostile environments – deserts, ice-covered lands and seas, and some of the densest areas of forest remote from population centres – can still be regarded as near-pristine. The rates of human-induced changes to land, marine and atmospheric environments and the fact that they have become discernible at the global scale, have prompted a proposal for the adoption of the term 'Anthropocene' as a geological period. This is not merely a light-hearted neologism to describe contemporary environmental change: stratigraphers are engaging in international discussions about the merit and feasibility of defining a period of human perturbation of the global environment, and designating the Anthropocene as a formal unit of geological time (Zalasiewicz *et al.*, 2010).

The second important reason that Earth system science needs to give attention to knowledge from

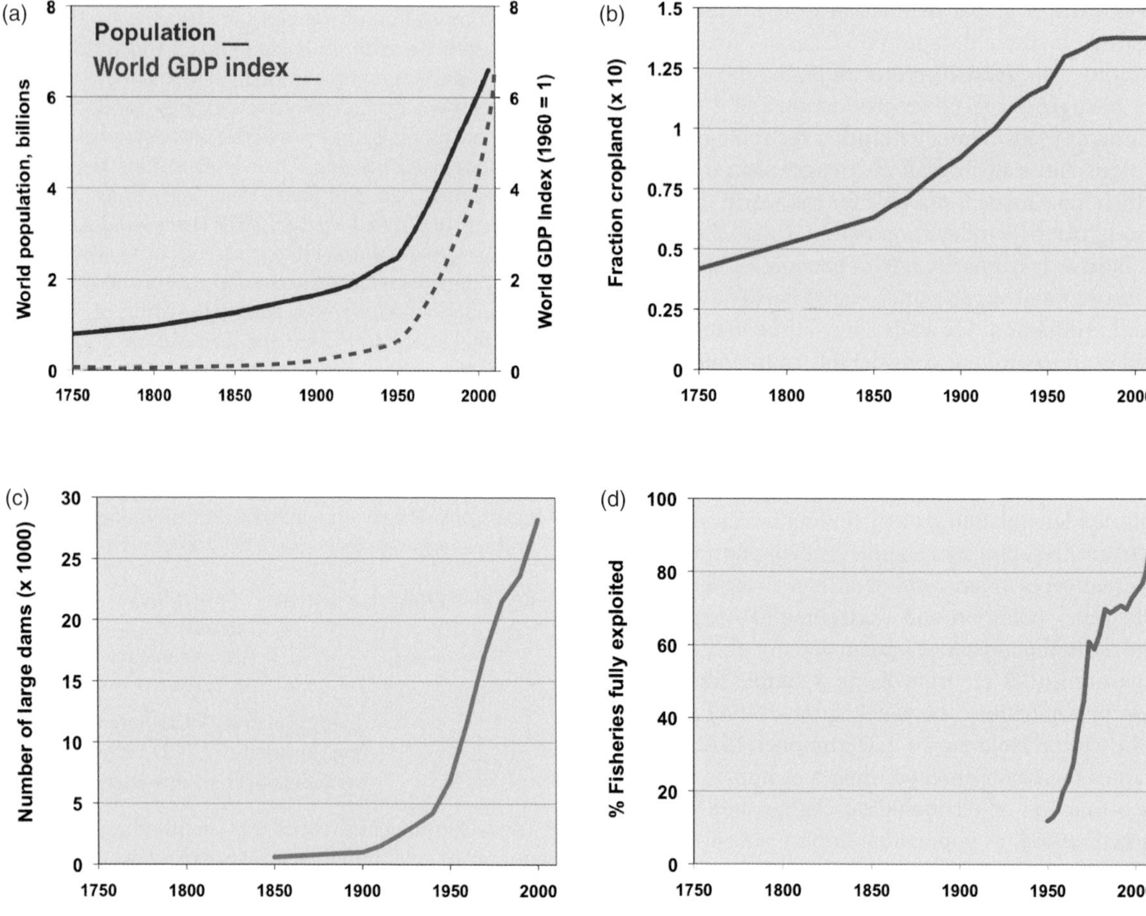

Figure 1.1 Some global socio-environmental trends since industrialization.

(a) World human population (US Bureau of the Census International Database, www.census.gov/ipc/www/worldpop.html; solid line) and the aggregate world gross domestic product indexed against 1960 world GDP (dashed line; data up to 2005 from the Earth Policy Institute, www.earth-policy.org/indicators/C53; updated to 2010 with data from the International Monetary Fund World Economic Outlook, www.imf.org/external/pubs/ft/weo).

(b) Global land use as cropland (Ramankutty and Foley, 1999; fraction of total land area multiplied by 10 for convenient scaling).

(c) Number of dams larger than 15 m on the world's rivers (data from World Commission on Dams (2000), including estimates for 2000 shown in the annex of that report).

(d) Percentage of global fisheries that are fully exploited, overfished or depleted (data from 1974 to present from FAO (2010); earlier data from FAOSTAT (2002) statistical databases, cited in Steffen *et al.*, 2004).

the social sciences is that humans experience the ***consequences*** of global environmental change.

The fact that the natural environment presents hazards to people is nothing new. Newspapers are full of reports of humanitarian disasters caused by droughts, floods, storms and other geological and climatic events. Similarly, it is widely recognized that human society has the capability to create serious environmental risks for itself. Communities, societies and, arguably, entire empires, have done so in the past with catastrophic consequences (Ponting, 2007). In this context, Earth system science provides powerful concepts and tools that can be used in assessing and predicting the risks to people and society of climate change and other global environmental changes, and in informing responses to these potential changes. Understanding human vulnerability in the context of these changes is as essential as understanding their biophysical dynamics.

Bringing a systems perspective to bear on these issues also allows the interplay of causes and consequences to be addressed. An iconic example of human-caused environmental disaster, often used as a warning metaphor for society's current unsustainable

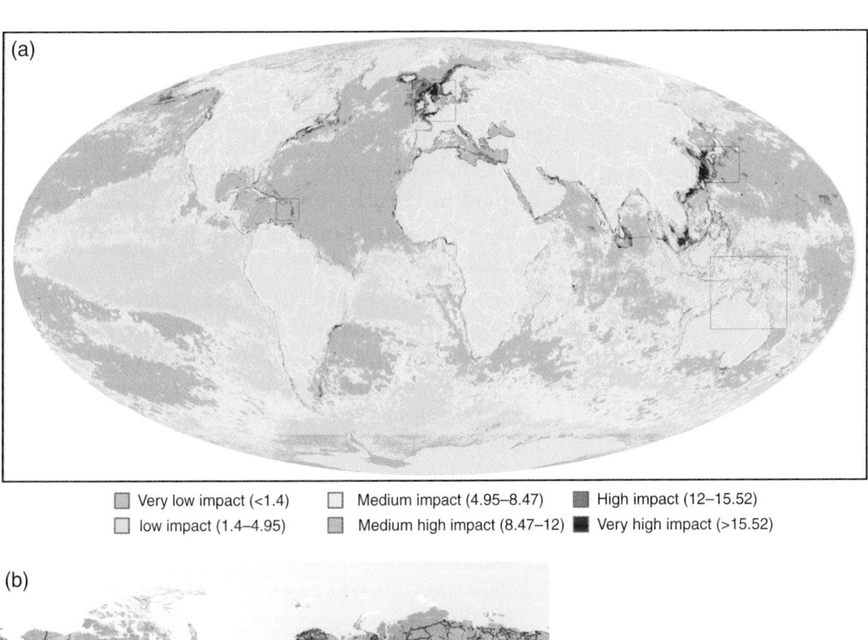

Figure 1.2 The scale of global anthropogenic impact.

(a) Human perturbation of marine ecosystems This map was constructed by overlaying 17 global data sets of drivers of ecosystem change, such as fishing activity, shipping, and riverine and long-range pollution. Reproduced with permission from Halpern *et al.*, 2008.

(b) Human impact on the land environment, modelled using GLOBIO-2 (image by Hugo Ahlenius, UNEP/GRID-Arendal, 2002; reproduced with permission). The GLOBIO model (www.globio.info) makes spatial assessments of the consequences for land biodiversity of human drivers like land use, pollution and infrastructure.

situation, is the total deforestation of Easter Island in the eighteenth and nineteenth centuries, and the linked socio-political turbulence. It is also an example of the need to take a deeper and more critical look at the human dimensions of change: the balance of social science research (summarized in Rainbird, 2002) indicates that, apart from tree-felling, other, yet very familiar, human factors played a major role in the degeneration of Easter Island's society and environment. Population collapsed, and social structures with them, following contact with European explorers who introduced new diseases and destructive animals, and raided repeatedly for slaves. In Section 1.3 below, we summarize areas of social science research that are actively exploring global environmental change issues. These fields of research enable a fuller perspective to be taken, necessary to prevent over-simplification or the proliferation of modern myths of environmental threats.

The third reason for explicitly addressing human society and its activities within the field of Earth system science is that society is the *context* in which Earth system research actually happens, and where the knowledge produced is directed towards practical action. The knowledge that scientists produce will go into the public and policy domains, where it faces many possible fates: this knowledge can be debated, reconfigured and developed in the context of other fields of knowledge, and naturally it can be used in decision-making processes. Earth system science is now deeply embedded in the processes and institutions that inform society's planning for climate change adaptation and mitigation, and for many other policy responses to global environmental change. This situation represents a marked shift in the way that science is done and, in our view, it brings new responsibilities and challenges along with new academic insights into our Earth.

1.2 Conceptualizing the 'human dimension' from an Earth system perspective

Given the importance of people in causing and being affected by global environmental change, the research community interested in these issues faces a serious challenge: how to conceptualize and embed human agency and socio-economic context in our understanding of the Earth system.

This challenge is highlighted by a debate that has been simmering between scholars for at least 350 years. To a very large extent, a primary activity in post-Enlightenment science has been *analysis* – the breaking up of the complex world into comprehensible pieces for in-depth investigation. Established by the likes of René Descartes, whose famous maxim *Cogito ergo sum* reflected an attempt to explain the human experience rationally, by reducing it to fundamental truths, analysis through reductionism has been the basis for many (if not most) of our scientific discoveries in the modern era. This point is highlighted by futurist and social commentator Alvin Toffler (1984) who wrote,

> One of the most highly developed skills in contemporary Western civilization is dissection: the split-up of problems into their smallest possible components. We are good at it. So good, we often forget to put the pieces back together again.

Without denying the immense value of reductionist research in both the biophysical and human domains, Toffler, and many others since, have argued that to address the global challenges that face us today, the balance needs to shift back towards *synthesis*: bringing our collective perspective back up to the better-rendered 'big picture' of our world. This world is a complex world, and it is inescapable that improving understanding requires the integration of multiple perspectives. If Earth system science is to be a part of this integration process, it requires a much fuller recognition of the interconnectivity of its social and environmental components.

There have been many calls for more such integrated knowledge from environmental research funders, government bodies and the research community itself (Box 1.2), in response to the challenges posed by global social and environmental changes. The International Human Dimensions Programme on Global Environmental Change came into existence in the mid 1990s, sponsored by both the International Council of Science Unions (ICSU) and the International Social Sciences Council (ISSC), to foster social-science research on global environmental change and to support collaborative research efforts across the social and natural sciences. In 2001, the four international global change research programmes jointly issued the Amsterdam Declaration on Global Change, which included a description of how this more integrative research should develop:

> The scientific communities of [the] international global change research programmes … recognise that, in addition to the threat of significant climate change, there is growing concern over the ever-increasing human modification of other aspects of the global environment and the consequent implications for human wellbeing. A new system of global environmental science is required. This is beginning to evolve from complementary approaches of the international global change research programmes and needs strengthening and further development. It will draw strongly on the existing and expanding disciplinary base of global change science; integrate across disciplines, environment and development issues and the natural and social sciences; collaborate across national boundaries on the basis of shared and secure infrastructure; intensify efforts to enable the full involvement of developing country scientists; and employ the complementary strengths of nations and regions to build an efficient international system of global environmental science'.[2]

Box 1.2 The current research and policy focus on 'understanding the Anthropocene'

Many organizations involved in the research process are currently orienting themselves towards better transdisciplinary integrated knowledge of global change, recognizing the importance of delivering this knowledge in a timely way to decision-makers in society in order to meet the growing sustainability challenge. This emphasis on better interaction between natural and human sciences, and on the urgency of the knowledge need, is evident at all levels – international and intergovernmental, regional and national:

- The United Nations Education, Scientific and Cultural Organisation (UNESCO) launched its Climate Change Initiative in 2009. It supports interdisciplinary integration of knowledge about climate, promoting the use of its biosphere reserves and World Heritage sites for research and implementation of climate risk management policies. A core programme focuses on ensuring

[2] Text in full available on the Earth System Science Partnership website, www.essp.org/index.php?id=41

that environmental ethics, social and human sciences are entrained in responding to climate. Also, UNESCO has designated 2005–2014 as the Decade for Education for Sustainable Development, in which climate change, biodiversity and sustainable lifestyles are key themes. See: www.unesco.org/new/en/natural-sciences/special-themes/global-climate-change and www.unesco.org/new/en/education/themes/leading-the-international-agenda/education-for-sustainable-development/about-us/.

- The ICSU and the ISSC identified a set of 'Grand Challenges' in Earth system science for global sustainability (Reid *et al.*, 2010). They acknowledge the huge advances made in understanding the functioning of the Earth system, but issue a call to action to researchers from the full range of sciences and humanities. The ICSU and the ISSC are also alert to the difficulties of this new mode of working, in terms of institutional structures, research methods and the incentives for participation in this evolving area. See: www.icsu-visioning.org/grand-challenges/
- The Belmont Forum is made up of representatives of funders of global change research from many countries around the world, many of which have supported socio-environmental research since questions of global environmental change first arose in the research agenda. As funders, they are influential in dealing practically with many of the difficulties that ICSU/ISSC identified in their Grand Challenges Visioning initiative. In 2011, the Belmont members made a shared commitment to supporting research that yields knowledge for action to avoid the detrimental impacts of climate change, explicitly requiring interaction of the natural and social sciences. See: www.igfagcr.org/index.php/challenge
- The IPCC itself, as the key organization providing scientific synthesis for decision-makers, has been criticized in the past for having a physical-sciences bias but, since it was formed, its reports have both reflected and shaped moves in the research community towards greater integration in order to better understand the human dimensions of global change. The Fifth Assessment Report currently being prepared puts more emphasis than any previous report on the interplay of socio-economics and biophysical changes, as well as on sustainability and risk management in the adaptation and mitigation responses. See the brochure on www.ipcc.ch/organization/organization_history.shtml
- Recognizing the strategic and practical challenges of integrated research on global change, the four global change programmes (IHDP, IGBP, WCRP and Diversitas) set up the Earth System Science Partnership to provide enabling mechanisms for the science community. It has supported joint projects on cross-cutting issues such as water, carbon, food and human health, as well as regional studies.
- In Europe, the European Commission's Research Advisory Board (EURAB) in 2004 spelled out some necessary improvements to enable Europe's research systems to better meet the transdisciplinary challenges presented by complex environmental systems (European Research Advisory Board, 2004). Its successor, the European Research Area Board envisions a 'New Renaissance', where new ways of thinking will emerge from better linkages between the natural and human sciences. See: http://ec.europa.eu/research/erab/pdf/erab-first-annual-report-06102009_en.pdf.
- The European Science Foundation (ESF) reported on its first Forward Look activity on Earth system science in 2003. That study focused primarily on biophysical changes, informing subsequent research, modelling and observation programmes (including the UK's QUEST programme). In 2009, the ESF launched a second Earth system science Forward Look, this time explicitly concerned with global change and the Anthropocene. This activity, *Responses to Environmental and Societal Challenges for our Unstable Earth*, also grappled less with the science base itself than with the need for new integrative structures and processes in research that are capable of addressing the socio-environmental issues of greatest concern. See: www.esf.org/index.php?id=6198.
- Many national projects and programmes have been developed recently to address the interlinked human and biophysical dimensions of global change. One example in the UK is *Living with Environmental Change* (www.lwec.org.uk), which is a multi-partner initiative involving research councils, government departments and agencies and businesses. *Nordic Strategic Adaptation Research* (www.nord-star.info) links interdisciplinary researchers across the Nordic nations with decision-makers in policy and business, and provides a model (and resources) for similar networks elsewhere in the world. *td-Net* (www.transdisciplinarity.ch) is a network for transdisciplinary research, which shares

information about methods and good practice in this frontier area of study. The *Grupo de Pesquisa em Mudanças Climáticas*, the climate change research group of Brazil's national space research institute INPE (mudancasclimaticas.cptec.inpe.br) is engaged in socio-environmental research involving multiple government, business and academic partners in Brazil and worldwide, but it also seeks to engage directly with journalism forums and organized civil society groups, and it provides a regional focus for other national research and societal engagement efforts across Latin America.

All these groups have recognized the evident need for new kinds of knowledge to equip society better for responding to the many linked challenges of global change. They have set out some institutional and operational principles for working. However, the nature of this newly integrated knowledge is still open to debate. Despite near-universal acknowledgement of the complexity of the Earth system, we still tend to deploy discipline-based approaches to the identification of the research problem. For instance, climate scientists define climate change as a physicochemical problem, the consequence of perturbed atmospheric chemistry, planetary albedo and the like. The Summary for Policy Makers of the IPCC's Fourth Assessment Report (IPCC, 2007) states, '*Causes of Change: Changes in atmospheric concentrations of greenhouse gases and aerosols, land cover and solar radiation alter the energy balance of the climate system.*' In contrast, a leading sociologist, Anthony Giddens, addressed climate change entirely as a political problem in his recent book on the topic (Giddens, 2009); while for economists, it is seen as the '*biggest market failure*' (e.g. van Ierland *et al.*, 2002; Stern, 2007). Of course, climate change is a consequence of all of these things together. The challenge remains in how we understand all of these aspects together, including their interactions.

Various tools have been proposed and tried out. Figure 1.3 shows one of the iconic conceptualizations of the Earth system, known as the Bretherton diagram. It was included in the NASA-sponsored Bretherton Report, '*Earth System Science: A Closer View*' (NASA Advisory Council, 1988), which set out a scientific research agenda for the emerging field of Earth system science. The Bretherton Report was profoundly influential. The development of Earth system models in the period since it was published can be seen as the progressive inclusion of sub-models representing the different boxes and arrows in the diagram, drawing on the findings of many international collaborative research programmes for biogeochemistry and climate science.

The figure shows how scientists viewed the Earth system as a set of interactions between the physical climate system and the biosphere mediated through various global biogeochemical cycles. The dynamics of the system are 'forced' by energy changes associated with natural variations in solar intensity and with the reductions of incoming solar radiation reaching Earth's surface caused by volcanic eruptions shooting ash and sulphate aerosol into the upper layers of the atmosphere. People feature in this diagram as a semi-external forcing on an intricately coupled biogeophysical system. Human activities cause land-use change and are a source of CO_2 and pollutants. These human activities are clearly affected by climate change and dependent on (land) ecosystems.

The Bretherton diagram was an important first step in demonstrating the links between human activity and environmental processes. Overall, however, people were not presented in this approach as being a fully endogenous part of the system. In this framework, the few elements representing all human activity contrast sharply with the more richly resolved processes of the natural world. This asymmetry has been reflected in research investments and international research infrastructure in the area of global environmental change until comparatively recently.

Social research at the global scale emerging at around the same period reflected these growing concerns about societal and environmental change linked to globalization and economic development. The priority research questions articulated in the late 1980s still look very topical today, but they do not fit easily into the Bretherton schema:

> What are the persistent, broad-scale social structures and processes that underlie these changes? In particular, what are the relative roles of the amount and concentration of human population, the character and use of technology, the changing relation between places of production and consumption, and the 'reach' and power of state and other institutional structures? How does the relative importance of these roles for environmental change vary across cultures, and through history? (SSRC, 1988, cited in Clark, 1988)

These questions involve structures, processes, changing dynamics, and causalities – all terms common to systems analysis and familiar to Earth system scientists, but they also address important social concepts, such as power, culture and institutions, that do not

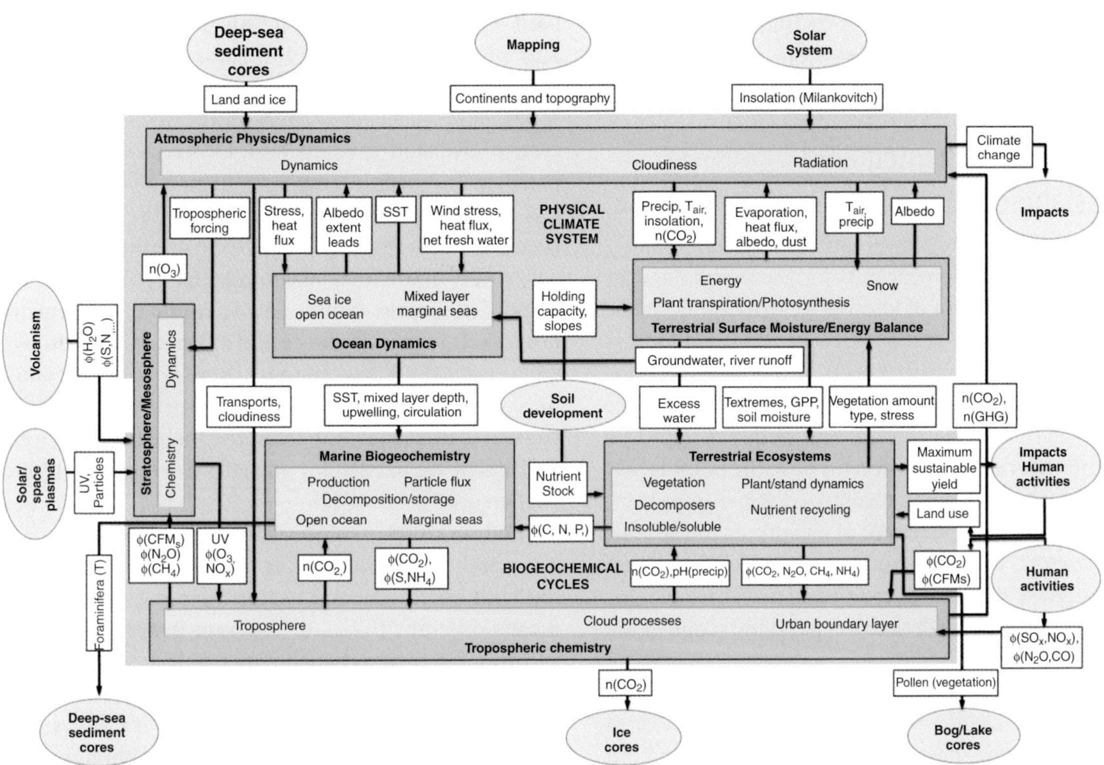

Figure 1.3 The Bretherton diagram (redrawn with permission from NASA; original figure published in *Earth System Science: A Closer View*, Report of the Earth System Science Committee of the NASA Advisory Council (1988), pp. 29–30).
This conceptual model, developed as part of a strategic research plan by the Earth System Science Committee of the NASA Advisory Council, represents Earth system processes occurring at timescales from decades to centuries. The ovals show exchanges with the external environment, or processes that operate over much longer timescales. The arrows connecting the sub-systems represent quantifiable measures that can be included in Earth system models. Contemporary Earth system models now include most of these processes.

translate well into quantitative measurements or computer models. Thus, research exploring these issues developed alongside the biophysical and climate studies, despite the recognition of the inextricable link between human development and the natural environment. Gro Harlem Brundtland, the chair of the World Commission on Environment and Development, wrote in the foreword of *Our Common Future*, the Brundtland Report (1987): 'Environment *is where we all live; and* development *is what we all do in attempting to improve our lot within that abode. The two are inseparable*'. Yet in terms of the research that has informed environmental and development policy, and the framing of the research questions, the social and natural sciences have largely followed separate paths.

Many still regard the complexity and the interdisciplinarity of the research effort now needed as an enormous challenge. The Bretherton diagram is a representation of the Earth system from a physical-sciences perspective. It has been a very important visualization of a set of priority areas for research into environmental and climate processes, but there is a growing discussion in the global change research community of the limitations of thinking of the Earth system in this particular way. Major questions for today's Earth system scientists are how far we have progressed from Bretherton's early conceptualization, and whether and how human activities might be better represented using the next generations of models. '*Structuring support for human dimensions research only around themes defined by natural science is inadequate*' (as stated by the Human Dimensions of Global Environmental Change committee, NRC, 1999; p. 62), but what might a consideration of the key processes and interconnections look like from other perspectives? In particular, what can now be gained in terms of understanding the Earth system if it is viewed from more than just the physical-sciences perspective? This effort must draw on the existing knowledge resources available in a wide range of disciplines, so the focus is increasingly on the 'integration' of knowledge, in ways that accommodate the multiple perspectives and insights from these different fields.

1.2.1 Towards an integrated understanding of the Earth system

The physical-sciences community has been making the most visible and vocal calls for the wider engagement and entrainment of knowledge from other fields, in large part because of the increasing demand for scientific insights to inform policy on climate and global environmental change. The changing interaction between Earth system science and policy is explored more fully later in this chapter; for now, the point is that Earth system science has recognized some important limitations in the deployment of its outputs in the real-world context, with many of these constraints relating to its interface with the human sciences. This recognition is what drives the desire to expand the scope of the field.

However, the methods and approaches of Earth system science are seen by many scholars in other fields as not entirely fit for these purposes, without some kind of transformation. In the view of many social scientists, Earth system science has followed a scientific tradition based on the search for universal laws and principles. In fields with more descriptive and interpretative traditions, there is a concern that the dynamics of the planet and its human inhabitants cannot be adequately described by reductive analysis of the components. Fortunately, many disciplines – and 'interdisciplines' – have well-established approaches to understanding the complex dynamics of a changing world that Earth system science can combine and draw upon.

Another well-debated theme from the social sciences that is beginning to resonate in contemporary Earth system science is their intensely critical concern with how the position of the researcher influences the outcome of the research.

For example, Demeritt (2001) explores the tacit social, cultural and political commitments of climate science, which shape the ways in which particular issues are defined as worthy of research, and determine the methods and techniques for the research inquiry itself. He argues that there is a tendency to '*concentrate upon the uses of scientific knowledge "downstream" in the political process*', and '*discount the ways in which a politics – involving particular cultural understandings, social commitments, and power relations – gets built "upstream" through the technical practices of science itself* (p. 306). The risks arising from this situation, where embedded assumptions and judgements are not acknowledged, include the messy battles of climate scepticism, opacity in what should be democratic processes of decision-making, and, Demeritt suggests, the problem that people simply are turned off by an overly technical and globally undifferentiated scientific line, precisely when there is a need to engage the global citizenry fully in the societal changes that are needed. The response to public doubts and scientific uncertainty is not merely to provide more and more technical knowledge about the Earth system. There is also a pressing need to recognize, reflect upon and work with the social context of this science, to build the social trust and solidarity that are needed for any effective response to the challenges.

Hulme (2008) explores a different perspective, but also one that is a major concern for social scientists dealing with environmental change: the fact that climate means different things to different people in diverse cultures. He argues that climate needs to be conceptualized and presented as a '*manifestation of both Nature and Culture*'. From this starting point, it follows that scientific insights about climate change, although it is a global phenomenon, need to be communicated at the level at which they are experienced: in terms of local weather, and also of how that weather relates to local environments and cultural practices. Like Demeritt, Hulme also gives pause for thought about the position of power – in academic and policy debates – of physical climate science. The failure consciously and deliberately to recognize the cultural context and dimensions of Earth system science means its research products can be appropriated by any of a growing range of ideologies when they are channelled into the policy process: '*Climate change becomes a malleable envoy enlisted in support of too many rulers*' (Hulme, 2008; p. 10). Hulme also points out that the language of the natural sciences, with their complex models, graphs, maps and so on, has more power – what he calls universality and authority – than the generally more context-dependent and context-specific findings of the social sciences, resulting in a narrowed climate policy agenda that excludes other approaches.

Both Hulme (2008) and Demeritt (2001) address the context in which more integrative research is needed, but the *content* of this research is also a focus of debate. At times, it can feel like an impasse between the different methods of the human and physical sciences. The debate is often framed rather bluntly in terms of the contrast between the physical science's focus on quantifiable and generalizable laws about objective phenomena, and the social sciences, where approaches are frequently less concerned with identifying general causalities, but instead can be narrative, interpretative or idiographic, seeking out the idiosyncrasies and specificities of a situation as a means of arriving at an in-depth understanding. In reality, the current situation in global change research is not so simply polarized: many areas of the social sciences have rich traditions of quantification, including the definition of precise formalisms, and of the identification of 'social mechanisms'. Causal relationships can be investigated 'scientifically' in the human domain, just as in the natural world. And Earth system science is not a rigid framework of the immutable laws of physics; its methods and approaches are supposed to allow the investigation and understanding of complex causalities, contingent behaviours, adaptive responses, and a wide range of other interdependencies, including those between the human and natural domains.

1.2.2 Current approaches to integration

We can now see a continuum of efforts for this integration in global change research, ranging from 'meeting in a conceptual middle', mediated by a shared set of tools and approaches in which modelling plays a key part, through to a 'rich-portfolio' approach that is more focused on the process of knowledge production and sharing, including dialogue and reflexivity. Before we move on to explore the key social, economic and policy concerns in contemporary global change research, we will describe some of these approaches to integration more fully.

Impacts modelling

One active area of development in Earth system science involves the inclusion of more processes and connections within existing Earth system models. This more comprehensive simulation, together with rapidly improving spatial resolution in the models and techniques for scaling down from global to regional and local scales, allows for a more robust assessment of the interactions among climate, vegetation and hydrology, and other domains. This improved understanding in turn provides invaluable insights to questions about future water supplies, crops, some environmental hazards – and hence projected impacts on human society. This area of research is a key theme in later chapters of this book.

The steadily expanding scope of mainstream Earth system modelling increasingly looks towards processes relating directly to human activities. Since the early years of global change research, human activities and impacts have been flagged as important Earth system processes, but they have also been comparatively under-specified. The Bretherton diagram of physical and biogeochemical processes maps conceptually and structurally onto the Earth system models and Earth observation programmes that have subsequently and progressively been developed. At the time that the Bretherton Report was published, the human dimensions research community noted the power and value of having such an over-arching representation of their research fields, '*to convey visually the interconnections among diverse cultural, economic, political, social and institutional phenomena and to begin to relate these theoretically*' (Balstad Miller and Jacobson, 1992; p. 175). Figure 1.4, known as the Social Process diagram (Kuhn *et al.*, 1992), was drawn up by the Human Interaction Working Group of the Consortium for International Earth Science Information Network (CIESIN), prompted by and developed in response to the Bretherton diagram, with these objectives in mind.

In the last two decades, the international human dimensions research agenda (as outlined, for example, by the IHDP (2007) and its previous strategic plans) has indeed been shaped broadly around the diagram's set of driving forces relating to the interactions between human activities and global environmental change. However, this diagram has not developed into a social science modelling and observation programme that slots seamlessly into the 'human activities' box of the Bretherton diagram. This is at least in part because of the very great diversity within the social sciences, in terms of concepts, theories and methodologies. The requirements in Earth system modelling for categorization and the fixed (quantitative, computational) representation of causal processes do not marry well with many of the fields of social inquiry. Another obstacle to wider social science engagement with this approach is that our impacts models make the same mistake as the tradition of environmental determinism, namely that they embed an assumption that all the world's societies react in a similar way to environmental conditions. Later in this chapter, we explore potential developments in this area.

Integrated assessment models

Modelling that combines scientific and economic aspects of global change is one area of integration that already has a very well-established track record. Integrated assessment models (IAMs; Figure 1.5) are

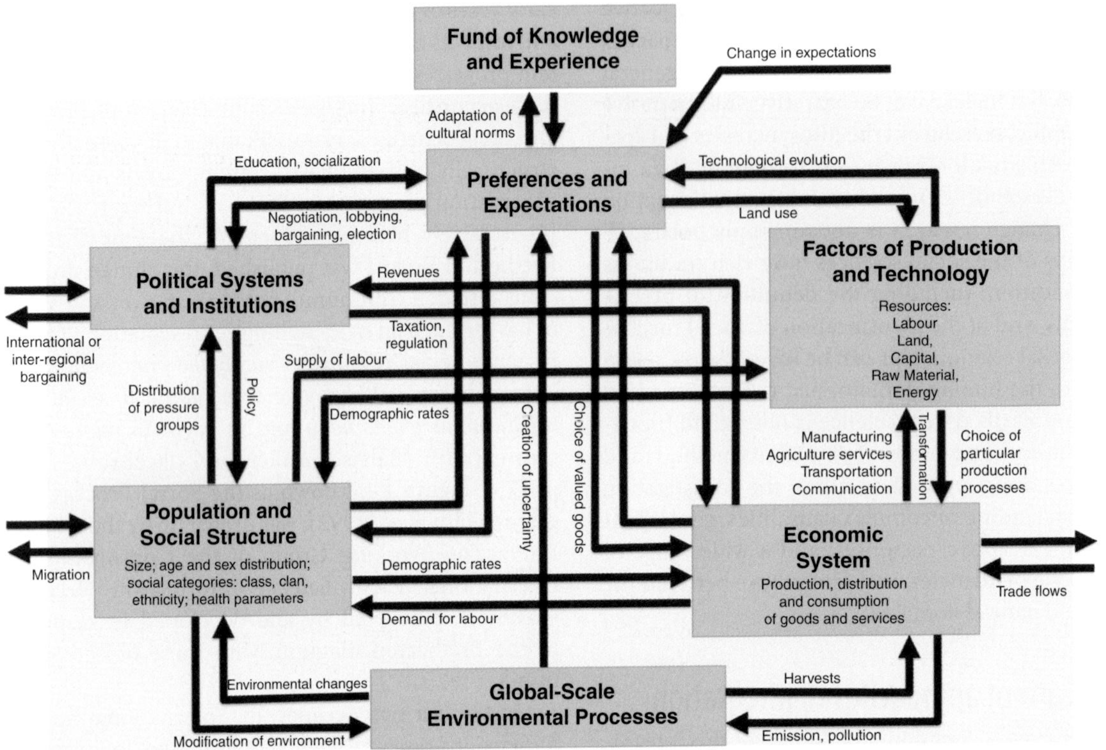

Figure 1.4 The Social Process diagram (redrawn with permission from CIESIN; original figure published in Kuhn *et al.* (1992) *Pathways of Understanding: The Interactions of Humanity and Global Environmental Change.* University Center, MI: The Consortium for International Earth Science Information Network, pp. 32–33).
This diagram developed from discussions of the Human Interactions Working Group of CIESIN, which had been funded by NASA to forge a link between the natural and social sciences in the context of global change research. The diagram represents the human system as a structure consisting of seven 'building blocks' connected by driving forces of change.

designed to explore or optimize policy options, and have been developed for a wide range of contexts where the interactions of technology, environment and the economy have societal consequences, including air pollution, land-use change, energy security and so on. They underpin contemporary climate change science and policy because of their application in the development of quantitative scenarios of potential future emissions of greenhouse gases (e.g. the SRES scenarios described in IPCC, 2000; and the new scenarios for the IPCC's Fifth Assessment Report based on Representative Concentration Pathways outlined in Moss *et al.*, 2010). Among the earliest integrative efforts of this kind were the Club of Rome's global models used in *World Dynamics* (Forrester, 1971) and *The Limits to Growth* (Meadows *et al.*, 1972), which linked representations of population, natural resources, the economy and pollution, with an explicit intention to '*clarify the course of human events in a way that can be transmitted to governments and peoples*' (Forrester, 1971, cited in Meadows *et al.*, 1982). By the 1980s, several different models of different types, complexity, underlying methodologies and purpose had been developed worldwide to address issues like energy, food security, and environmental change, and a long-standing series of symposia for information exchange on the developments in global modelling had been established by the International Institute of Applied Systems Analysis (Bruckmann, 1981).

In their delightful and thoughtful reflection on the first decade of global system modelling, Meadows *et al.* (1982) made the perhaps obvious point that '*as long as there are global problems, there will be a need for global models*' (p. xxiii). Nevertheless, they recognized the profound methodological challenges required to increase the understanding of the '*complex, interlocking, interdisciplinary sorts of issues*' that typify global environmental change. Many of these methodological challenges still remain. Kelly and Kolstad (1999) have reviewed IAMs in the climate context, explaining their

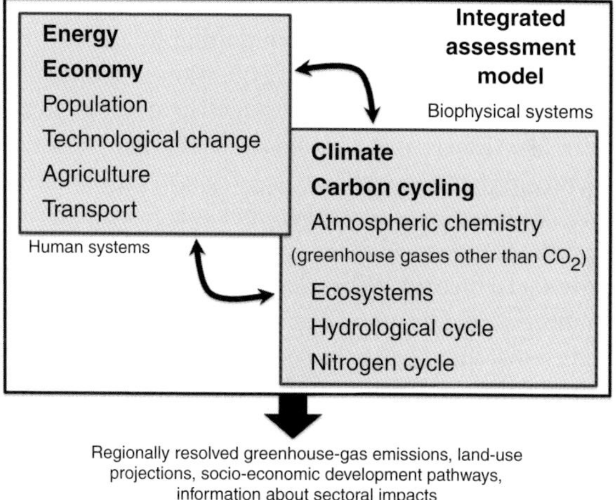

Figure 1.5 A generic representation of integrated assessment models.

assumptions and applications. Tol (2006) explains the challenges of integration in terms of issues in model coupling. Janetos *et al.* (2009) present a perspective on recent developments in integrated assessment modelling, arguing that IAMs are evolving towards the structural comprehensiveness and simulation capability of Earth system models. Much of the literature relating to IAMs focuses on issues of uncertainty (e.g. Visser *et al.*, 2000; Mastrandrea and Schneider, 2004; Tavoni and Tol, 2010). Uncertainty in integrated modelling arises from the necessarily simplified models of both the economy and the environment that are combined in these models; the assumptions made about the drivers of change, and about the nature of the relationship between the economy and the environment; the quality of the data available for input to the model; and increasingly – as models are used in more predictive ways, rather than as heuristic, exploratory tools – the variability of model output and the divergence between models. Different approaches are needed for the assessment and management of the various dimensions of uncertainty. Adding to the challenge, the linking together of very different kinds of models in integrated assessment modelling mean that multiple kinds of uncertainty are at play, and they are not always teased apart for systematic attention, despite concerns that IAMs should not be 'black boxes' (Kelly and Kolstad, 1999). Ackerman *et al.* (2009) set out some of these methodological concerns. In particular, they '*maintain that IAMs enjoy an epistemic status different from their natural science counterparts, and that economic models mix descriptive analysis and value judgments in ways that deserve close and critical scrutiny*' (p. 299).

Transdisciplinary inquiry

The critical scrutiny of methodological and theoretical aspects of research is arguably the distinguishing characteristic of the social sciences. Ironically, this may well have been a significant part of the barrier to interdisciplinary working and knowledge integration in global change research over recent decades: it marks a sharp contrast with most natural science, which stakes its claims on 'objectivity', designing its methods around an ideal of independence from context, values and world views. Gilbert and Ahrweiler (2009) explore this difference in depth, reviewing the philosophical and historical reasons why the social sciences have generally not focused on simulation modelling for their research questions. Because the ultimate theoretical interest in many of the social sciences is the individually significant, contingent and particular aspects of human experience and social reality, the starting point for integration with Earth system science, with its emphasis on systematization and computer modelling, was initially a very small potential intellectual meeting ground. The growing focus on the environment in the social sciences and humanities (ISSC, 2010) is now expanding the scope for integration across disciplinary divides.

This expansion of collaborative, transdisciplinary working is also now opening up debates about what constitutes 'good' integrated research and knowledge, and how it should be carried out. Brown (2009; p. 3) points out that these debates are '*not essential for the conduct of outstanding research, as long as the researcher is either lucky, highly instinctive, or able to slavishly follow a suitable model from previous work*'. Many people, in both the natural and social science communities, are now arguing strongly that research into a dynamically changing world, that bridges disciplines and, most importantly, informs real-world decision-making, falls into the category of research that requires close and critical scrutiny at all its stages. There are many terms for these debates, reflecting their emergence in various different fields of research. *Transdisciplinarity* is a useful overarching term; it carries within it the notions of bringing together empirical knowledge from different contributing disciplines, addressing questions that may well lie beyond the scope of any one individual discipline, and being directed to some purposive action in a real-world problem situation. There is a growing consensus (e.g. Thompson Klein *et al.*, 2001; Max-Neef,

2005; Pohl, 2010) that this kind of research should engage actively with the stakeholders involved in and affected by the problem or research question, and this in turn prompts the growing focus on values and responsibilities in research that we have already mentioned. Other related areas include *post-normal science* (e.g. Funtowicz and Ravetz, 1993), which focuses on the democratic dimensions of how to deal with situations where there is a high degree of system uncertainty and the decision stakes are high, and '*Mode 2*' *science* (e.g. Gibbons *et al.*, 1994; Nowotny *et al.*, 2003), which emphasizes the process of co-development of knowledge between academics, practitioners, and other stakeholders, particularly considering the institutional and epistemological changes that this form of societally engaged research requires. A common theme in these debates is the attention they give to the role of the citizenry in science and environmental governance, and their emphasis on transparency and accountability in science, particularly when knowledge from different specialist fields needs to be combined in order to present a more integrated picture of a complex world.

These debates may not have radically transformed the content of Earth system models, but they are increasingly influencing the way in which Earth system models are deployed in the policy context, and the way that Earth system research is being designed at a strategic level (Section 1.4). Nevertheless, it is important to recognize the extent and depth of academic concerns, particularly in the social science research community, about the process and approaches for the integration of knowledge. Debates about methodologies are taking place together with a more explicit questioning of the underlying assumptions in socio-environmental research, and there is often a tangible tension in meetings and in the transdisciplinary academic literature. In the context of global change, knowledge needs are urgent, but sensitive and supportive approaches to research integration are nevertheless needed.

'Integration of knowledge' from the social and natural sciences is in many ways an effort to square a circle, which unsurprisingly presents persistent problems. Disciplines can be regarded rather like cultures, characterized by their shared language and practices (Strathern, 2006). As with any encounter between long-established cultures, efforts at interdisciplinary engagement can be hampered by profound and sometimes bewildering differences in behaviours and motivations. Where one knowledge culture is dominant – as is currently the case for the quantitative sciences (e.g. Demeritt, 2001; Bjurström and Polk, 2011), there is even greater sensitivity around efforts to integrate or assimilate the outputs of scholarly inquiry. There is often a tacit presumption that integration for Earth system science is merely a question of quantifying the qualitative, but many social scientists argue emphatically, like Gilbert and Ahrweiler (2009), that '*the social sciences are not a "pre-science" waiting to approximate the state of the natural sciences via more and more discovery and mathematisation of the laws of the social realm*'. The early social theorist Weber described where law-finding strategies could be useful (e.g. Weber, 1988; p. 12ff.), but argued that they need to be complemented with detailed investigation and description in order to provide the requisite knowledge. Nevertheless, recurrent assumptions about social science and the role of social scientists continue to appear in the '*... sometimes colonising and patronising rhetoric of "good science"*' (using the words of Gilbert and Ahrweiler again), and they need to be recognized and confronted when they appear (Box 1.3). Steady progress is being made in bringing together the best available knowledge from all disciplines to inform responses to global change, but for the time being it is still vital to bear in mind that '*for interactions between the social and earth sciences to succeed, a certain level of tolerance and mutual understanding will be needed*' (Liverman and Roman-Cuesta, 2008).

Box 1.3 Debunking myths about human dimensions research

- Humans matter in global environmental change ...
- ... and not just because of population pressures
- Human behaviour may not be predictable ...
- ... and predicting even the predictable human behaviour may not be socially desirable
- Economists are not the only social scientists that are needed
- Social science is *science* – not politics or journalism
- Social scientists are smart enough to understand equations
- Social scientists can do large-scale research
- Social science is not always cheaper
- Social scientists can help with stakeholder engagement
- Social science may enhance the chances of environmental research being relevant, and of getting funded.

(Adapted with permission from presentations by Diana Liverman.)

Perhaps a more serious concern than the potential intellectual sensitivity of scientists in the interdisciplinary fray is the fact that the divisions of academic life into disciplines have resulted in a discouragement to asking questions that require combining modes of investigation that may be derived from more than one discipline, or even in developing fundamentally new modes and new conceptualizations. So for example, we still face huge gaps in understanding climate change adaptation from a practical, policy-relevant point of view, in part because until recently it has not been in the interest of any (discipline-bounded) kind of academic to study it.

Promising new areas for transdisciplinary integration include the conceptualization of society and the natural environment as a linked, mutually adaptive system, recognizing the complexity of the interactions between humans and the natural world. These new approaches bring prospects of entraining knowledge from very diverse fields; most scholars now agree that understanding global change over multiple scales of space and time involves historical and cultural and even philosophical perspectives in addition to the biological, physical and (geo)chemical. Wittrock (2010) emphasizes the need to take a critical, reflective view on these integrations:

> It is a great challenge for the future to maintain and strengthen intellectual sites in research and academic landscapes which are both open to cooperation across the divide between the cultural and the natural sciences and yet characterized by a measure of organized scepticism against proposals that entail that the social and human sciences should rapidly abandon core elements of their own theoretical traditions.

In the next sections, some of these 'core elements' are outlined, for the social sciences, economics and policy studies. It is not our objective, even if it were possible, to present a comprehensive review and meta-analysis of these fields of study. (Suggestions of some books that provide fuller treatments in these areas are given in Box 1.4.) In each case, a few seminal issues or areas of research have been selected to illustrate the corpus of research on human dimensions of global change. The basis for the selection is either that the dialogues for integration in Earth system science are already mature, or that there are particular areas of debate or controversy that currently impede the integration of knowledge.

Box 1.4 DPSIR – a simple systems framework in practice?

Humans causing damage to their environment also ultimately cause damage to themselves as a result of being part of that degraded environment. At the practical level, many decision-makers involved in real-world responses to socio-ecological problems use the Driver–Pressure–State–Impact–Response framework for analysing the cause–effect behaviour (DPSIR; OECD, 1993; EEA, 1999). Unlike environmental impact assessments, which attempt to identify and quantify the impacts on the natural environment of human activities (development), and risk or hazard assessments, which identify and quantify the consequences for people of a given environmental change, the DPSIR framework (see Figure 1.6) allows for the conceptual analysis of interactions in both directions, on multiple scales and indeed over a cyclic or recurrent process of changing human activities and changing environment.

The DPSIR framework highlights that effective responses may need to tackle the issues at the multiple levels. Remediation of the state-change evident in the environment is one level, but response options also include interventions to reduce the impacts felt by society, changing sector policy and the behaviour of members of society. Contexts where the framework has been used extensively include coastal-zone management, the delivery of aid programmes and, increasingly, in global-scale changes. The analysis reported in the Millennium Ecosystem Assessment (2005) was structured in this way. The outline for the forthcoming Working Group 2 contribution to the Fifth Assessment Report of the IPCC differs from previous assessment reports in being more oriented towards adaptation options in a state–impacts–response framework.

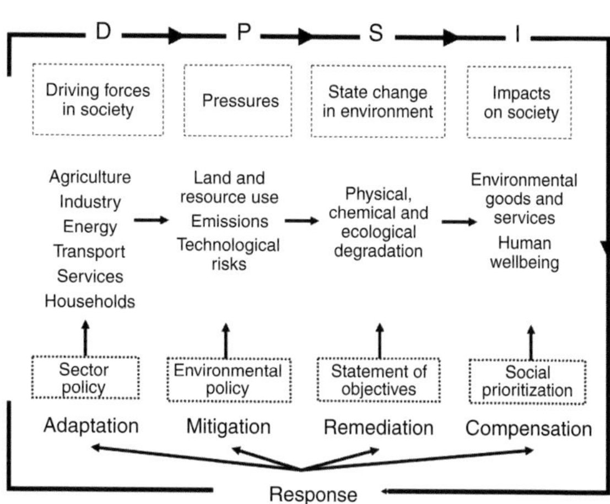

Figure 1.6 The DPSIR framework.

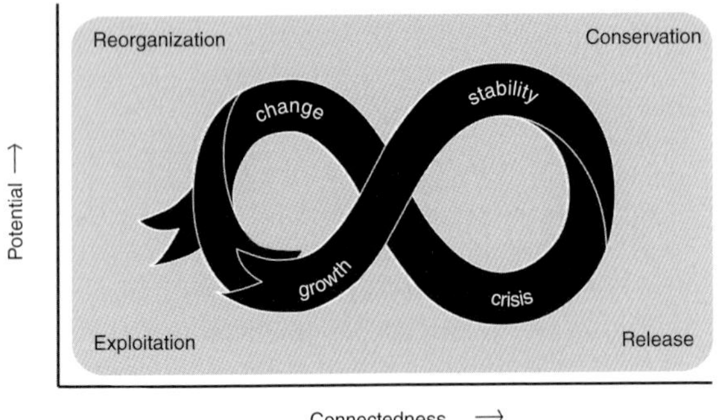

Figure 1.7 The resilience cycle in socio-ecological systems. Adapted from the panarchy model of Holling and Gunderson (2001).

1.3 Social science perspectives on the Earth system

1.3.1 Social science priorities

In order to map out the scope of contemporary social sciences and identify the key areas where research priorities are aligned with the interests of Earth system science, we have turned to two international bodies that represent social scientists' interests and shape research strategies and agendas. The first of these is the UNESCO-supported International Social Sciences Council (www.worldsocialscience.org), which highlighted several global challenges in its 2010 World Social Science Report (ISSC, 2010). The second is the IHDP, which has an explicit remit to frame worldwide social science research within the context of global environmental change (IHDP, 2007).

Both these international bodies emphasize the essential importance of the contributions social science can make in understanding societal trends and overcoming global challenges:

> Research conducted under the auspices of IHDP is predicated on the premises that global change research should address major social and economic science concerns, shape a social science of global change, and contribute knowledge to meet a number of major challenges currently facing societies (IHDP, 2007; p. 4).

> The social sciences are concerned with providing the main classificatory, descriptive and analytical tools and narratives that allow us to see, name and explain the developments that confront human societies. They allow us to decode underlying conceptions, assumptions and mental maps in the debates surrounding these developments. They may assist decision-making processes by attempting to surmount them. And they provide the instruments to gauge policies and initiatives, "and to determine what works and what does not" (ISSC, 2010; p. 9).

The lists below show the priority strategic issues that the IHDP and ISSC have identified. Several of them have an intrinsic environmental dimension; mutual gains in knowledge – and a better-founded expectation of overcoming the real-world challenges – can arise from interaction between social scientists and natural scientists. For other issues (such as social learning, institutions, equity), Earth system science may not be geared towards contributing to those bodies of empirical or theoretical knowledge, but it can still engage with them and draw from them in the effort to better understand global changes. Gudmund Hernes, President of the ISSC, called for this broadened engagement, issuing '*a plea for integrated research where the humanities and the natural and social sciences jointly address natural phenomena, social processes, institutional design, cultural interpretations, ethical norms and mindsets*' (ISSC, 2010; p. ix).

ISSC global challenges:	IHDP cross-cutting themes:
• global environmental change • poverty • equity, inequalities and the global economy • population and demographics • social unrest and violence • urbanization	• vulnerability, adaptation and resilience • governments and institutions • social learning and knowledge • thresholds and transitions

1.3.2 The shared language of systems

Understanding the social phenomena listed above requires understanding not just of social structures but also of social processes. These might include cooperation, conflict and the myriad transactions between social groups. This process orientation in the social sciences is shared with Earth system science, and offers one point for intellectual convergence between these disparate academic fields. But how can we understand the 'dynamics' of individuals, institutions, communities and society?

An immediate challenge is that there are many very different schools of thought within the social sciences, bringing sharply different theoretical perspectives to the questions of social structures, processes and capabilities for bringing about social change (often termed 'agency'). Indeed, the theme of the 2010 ISSC report (p. 3) was '*knowledge divides*', acknowledging the difficulties that this multiplicity presents, but also framing this diversity as an asset: it potentially offers a much richer picture of humans and their social interactions. Many fundamental social theories have addressed human society within the natural environment to some extent. The most important factor in a social phenomenon might be evaluated very differently by these different perspectives (e.g. Goldman and Schurman, 2000).

It is important to look back at the experiences and, in some cases, the pitfalls of these various approaches as we seek to develop new integrative approaches. Table 1.1 shows examples, in cartoon outline, of several social theories that have dealt with nature–society interactions, using the links between climatic changes and famine to illustrate the kinds of causal explanation that these different fields proffer. These examples have been selected not because they are broadly representative of all the social sciences, but because they are some of the socio-environmental discourses that have been influential beyond the academic field, and that continue to be debated now.

Within the various approaches in the social sciences, there are also very different, and often contested, ways of understanding social processes. Some theoretical framings have been oriented towards describing the dynamics of society through the identification of direct causality. Underpinning these framings is the view that a social phenomenon can be explained in terms of cause/effect relationships, which can be subjected to the same kind of scientific investigation as is used in the natural sciences (e.g. Outhwaite, 1998). The focus of these fields of study, rather like the natural sciences, is the 'mechanism', and their ideal endpoint is the abstraction from observed cause/effect relationships into generalizable laws about society and social processes. For example, both Malthusianism and environmental determinism tend to take this kind of positivist, reductive stance. Broadly speaking, these social theories were most prevalent in the past, and have largely been discredited, falling out of favour in the latter half of the twentieth century. However, causality and mechanism are again rising in the discourse of the mainstream social sciences (a sample of the debate can be seen in Hedström and Ylikoski (2010) and Abbot (2007)) and in philosophy of social science (e.g. Bhaskar, 1998). This is in part because of the growing academic attention to knowledge *for action*, where the goal of science–policy interaction is to deliver evidence for interventions in society. This goal drives a demand for predictive power in the findings of social theory, which mirrors and is reinforced by the same demands of the natural sciences. Global change research is the impetus for a major strand of integrative socio-environmental research that leans towards more 'mechanismic' naturalistic approaches (seeking 'how-possibly' explanations), if not necessarily mechanistic ones (that provide actual causal explanation). This includes work on impacts assessment and the development of quantitative socio-economic scenarios for use in climate change modelling.

For many other fields in the social sciences, giving an adequate account of a social process or phenomenon means providing as full and coherent an explanation as possible, rather than necessarily seeking to identify the causal mechanisms (that is, by eliminating alternative possible causes) and providing predictive power. What constitutes explanation is itself a long-standing debate in the social sciences. In an early but still sound contribution to the debate, Jarvie (1964) gives examples of the kinds of explanation that might be deployed for social events. To 'explain' a lynching, for instance, he notes that a social scientist might trace the origin of the habit of mobs summarily executing prisoners; investigate the intentions of the participants in the mob; note the particular dispositions in certain regions at certain times for this kind of dispensation of justice; identify the reasons for an individual's participation in this particular lynching; consider the 'social function' of such events (which is in part an explanation of the practice in society in terms of its effects, rather than its causes); assess the extent to which a description of an event fits pre-existing theory; and of course make empirical generalizations (*'all lynch mobs get out of control'*).

Jarvie's vignette shows several features of social explanation that present challenges to those seeking greater integration in global change science. First, for the vast majority of social research, explaining social events and phenomena is not a simple process of eliminating 'false' causes through experiment or observation and narrowing down to the 'true' one. The objective of the research is often precisely to obtain a richly textured and fine-grained picture. Where social events or mechanisms have a multiplicity of causal processes acting in conjunction, the outcome – and the explanation – will be contingent on all those contributory processes. This is a feature of complex

Table 1.1 Examples of different theoretical framings applied to the same socio-environmental phenomenon
This table shows why adaptation, vulnerability and impacts research follows different strands, and also why social scientists hold some deep-seated concerns about quantification of human and social phenomena. This table has been developed from a list of different social science framings first devised by Diana Liverman.

Theory: key features	Explanation of causal sequence	Current debates
Malthusianism		
Malthus (1798) argued for the need to control population growth or face a 'positive check' (warfare, disease or other catastrophe) as population returns to sustainable level.	Climatic changes *cause* population growth beyond carrying capacity *causes* famine	Does population growth need to be tackled? *'Failing to confront adequately questions of social scarcity, or confused regarding whether the scarcity in question is social or natural in character ... is a mistake.'* (Shantz, 2003); *'We ... alerted people to the importance of environmental issues and brought human numbers into the debate on the human future.'* (Ehrlich and Ehrlich, 2009)
Environmental determinism		
The idea that physical environment determines cultural development was prominent in the late nineteenth/ early twentieth centuries (e.g. Huntington, 1913). It was strongly discredited by the mid-twentieth century, for its conflation with racism and imperialism.	Climatic changes *cause* reduced crop yields *cause* famine	Where will the 'winners' and 'losers' be in climate change? *'The current global environmental crisis is now so severe and pressing that ... the deterministic role of the environment is relevant again. ... This environmental determinism distorts realities as well as results in unintended consequences, frequently with negative impacts on the least powerful.'* (Radcliffe *et al.*, 2010)
Cultural/human ecology		
Features of social organization are explained in terms of interacting ecological and social historical factors (e.g. Sauer, 1925; Steward, 1955). The field has a strong focus on process (adaptation), and also on the fusion of the ideas of ecology and the human sciences.	Climatic changes *cause* adaptive responses shaping social organization *cause* changed agricultural systems *cause* famine	How can the conceptual division of society from 'external' nature be overcome? Descola and Pálsson (1996) reviewed the field, '*emphasising the problems posed by the nature–culture dualism, some misguided attempts to respond to these problems, and potential avenues out of the current dilemmas of ecological discourse.*' Head (2007) sets the debates in contemporary context: '*It is precisely because of the pervasiveness of human activity that we need to critically re-examine the work that the metaphor of 'human impacts' is doing. Human impacts is a hard-won concept that has made a crucial contribution to our understanding of the long-term human role in earth processes. Yet ... it is neither conceptually nor empirically strong enough for the complex networks of humans and non-humans now evident.*'

Theory: key features	Explanation of causal sequence	Current debates
Mainstream (neoclassical) economics		
Key elements of theory are rational actors, basing allocation decisions under conditions of perfect information to maximize utility, with the system achieving equilibrium between supply and demand. The founding thinkers in this large and hegemonic field include Adam Smith, William Stanley Jevons, Maynard Keynes and Milton Friedman.	Market demand *determines* factors of production (in part subject to climatic changes) *determine* supply *determines* price *causes* decline in food availability *causes* famine	Are markets the best – or only – approach for managing the global environment sustainably? *'An effective, efficient and equitable collective response to climate change will require deeper international co-operation in areas including the creation of price signals and markets for carbon.'* (Stern, 2007) *'The debate over whether and how environmental economists ought to measure noninstrumental and nonanthropocentric values, such as the worth of a plant unseen by and of no apparent use to humans, will not be settled soon.'* (McAfee, 1999)
Political economy/political ecology		
Social, political and economic conditions are seen as the dominant determinants of the consequences of environmental changes. A key concern in this field is understanding and illuminating power relations that shape access to resources (e.g. Blaikie and Brookfield, 1987).	Poverty *causes* vulnerability (to climatic changes) *causes* famine	Who is responsible for the climate problem, and who should respond to it? *'What counts as nature and what works as nature politics are two arenas that are being effectively remade.'* (Goldman and Schurman, 2000) *'Starting with* a priori *judgments, theories, or biases about the importance or even primacy of certain kinds of political factors in the explanation of environmental changes, self-styled political ecologists have focused their research on environmental or natural resource politics and have missed or scanted the complex and contingent interactions of factors whereby actual environmental changes often are produced.'* (Vayda and Walters, 1999)

systems: even if the 'initial conditions' were able to be identified and well specified, determining the covering laws for social processes remains an impracticable task. Without venturing into the rich philosophical debates on this topic, Bhaskar (2008) emphasizes that causality includes both the antecedent conditions that trigger mechanisms, and the mechanisms themselves. This sets a phenomenon or event into a specific context; understanding the event entails understanding that context.

In addition, whereas functional explanations (that is, those that explain a structure or phenomenon in terms of what it *does*) predominate in natural science, many explanations of human action relate to human intentions, or human choice. In these cases, the outcome is 'contingent' upon a future goal or end at which the action is directed, not just on the past stages in the process leading up to the event. In many social scientists' view, that intentionality puts a kink in the 'arrow of time': history is not and cannot be a simple predictor of the future.

Another major challenge for integration relates to the way in which knowledge is acquired about social phenomena. The scholar's cognitive interests influence the structure of the explanations sought and made (for instance, in terms of one of the theoretical framings described in Table 1.1). The conditions under which knowledge is acquired provide a particular context to each inquiry. More importantly, the interpretation of this context by the scholar is a vital dimension of some fields of the social sciences. From this perspective, the understanding of social interactions, in the present and in history, depends on the meanings that people attribute to their actions, to their social context and, indeed, to the natural environment.

The complex causalities and the interdependence of the system in question with its context both suggest that a systems-oriented research approach is needed for an integrative understanding of global change. In the systems approach, the system – a part of reality – is conceptualized in terms of a set of components that interact with each other and with their context. The behaviour of the system as a whole depends upon the connectedness and interrelationships of its components. Fraser *et al.* (2007) provide a brief review of studies that demonstrate how the social and ecological context in which climatic problems occur is likely to be as important in determining the outcomes, if not more so, than the nature and magnitude of the climatic shock itself. They also draw attention to the methodological challenges of understanding the system dynamics, and demonstrate the use of relatively simple modelling frameworks that capture institutional, socio-economic and ecosystem components. However, systemic research is still the exception in modern science (Gallopín *et al.*, 2001).

The development of theory for systems is comparatively recent. An early approach was von Bertalanffy's (1968) development of 'general systems theory'. Von Bertalanffy, a biologist, was interested in generic patterns of system behaviour seen in wider social issues, and recognized the need to look at the subject as an interactive and adaptive system. His theories have been developed into widely used integrative methods that extend into cybernetics, information systems and complexity science, and that find application in fields as diverse as engineering, business, biomedical sciences and ecological conservation.

The concepts and terminology of both systems and complexity theory have proliferated in interdisciplinary and integrated research, but does systems theory provide the right tools for integrated global change research? Especially in the field of socio-environmental research, there is still a great deal of experimentation and debate about the rigorous application of systems theory. Systems analysis potentially provides frameworks for bridging the 'cultural divide' (e.g. Newell *et al.*, 2005; van der Leeuw *et al.*, 2011), offering scope for richer, more nuanced explanations than any single discipline can provide. It can help counter the over-specialized and over-simplified views that are widely seen as a barrier to the production of useful and usable knowledge in the context of real-world complexity and uncertainty. A strong argument for a systems approach is that it recognizes the importance of the system's context, which cannot be excluded from the scope of investigation. However, determining the boundaries of the system for investigation is far from straightforward, as both the Social Process diagram and the Bretherton diagram show. Arguably, these iconic representations of global change processes reached their limits as research-framing tools because they both incorporated their contexts as components (the human dimensions box in the Bretherton diagram and the environmental processes box in the Social Process diagram).

Despite the proliferation of systems concepts in many academic fields, there is the alarming possibility that global change research could inadvertently be divided by its common systems language (see also Park, 2011). The shared terminology often hides fundamental differences in underpinning theories or world views. Of particular relevance to global change research and Earth system modelling is the social science debate relating to methodological individualism versus 'emergentism'. For the methodological individualist, if the motivations and behaviour of one individual can be observed and explained, then society can be explained as the simple aggregate of individuals. Other thinkers argue that there are facts about the social world that are not reducible to facts about individuals. They argue for the distinct reality and integrity of the social world, which impresses itself on individual behaviour. Are social structures 'real' (and thus capable of being modelled and theorized in their own right)? If society and its dynamics cannot be explained or predicted simply as the aggregate of individual actions, this implies emergent social phenomena arising from complex interactions at the individual level – and complexity of this sort is needed to explain many aspects of social change (e.g. Elder-Vass, 2007). Adding textures to this debate is the fact that modernity has brought many different conceptualizations of 'the individual' – variously as rational, as autonomous, as social and so on (Simon, 1957). To date, a mishmash of divergent assumptions about social and individual dynamics have been embedded in global models, notably in IAMs. To climate modellers, this pragmatic approach may look like it produces useful outputs. To researchers in the other fields whose knowledge is needed for understanding the Earth system, it can result in unverifiable confusion or academic dead-ends.

Nevertheless, significant progress is being made in exploring the dynamics of social systems, in the growing sub-field of social systems theory. Social simulation is a rapidly growing field (Gilbert and Ahrweiler, 2009; Heckbert *et al.*, 2010); there are efforts to capture both the biological imperatives and the cultural dimensions that shape human life (e.g. Ziervogel *et al.*, 2005; Saqalli *et al.*, 2010). Agent-based modelling is a particularly powerful technique in this context. After all, social systems can be defined as those that involve multiple agents with varying intentions and incentives. Using models and observations together – also a core feature of Earth system science – means that concepts of emergence in social systems can be explored more systematically (e.g. Sawyer, 2005; Vogel, 2009; Read, 2010). To date, however, not much of this work has had a well-articulated environmental dimension, and the application of models of social dynamics in Earth system science is still relatively novel. Knowledge that has so far developed disparately now needs to be consolidated.

An ambitious conceptual application of socio-ecological system ideas that explicitly aims to deliver that consolidation was articulated at a Dahlem workshop held in 2005 (Costanza *et al.*, 2007), tracing the long-term historical timeline of human societies, focusing in on the long-term multi-scale dynamics of socio-environmental change. This initiative is developing into a deeply interdisciplinary international research project, linking palaeoclimatologists, archaeologists, climate scientists and historians, anthropologists, economists and ecologists and more. This project, *Integrated History and future of People on Earth* (IHOPE, www.stockholmresilience.org/ihope; see Cornell *et al.* (2010a) and van der Leeuw *et al.* (2011)) aims to recast the narratives of human history in ways that incorporate what we know of the history of the natural environment and its interactions with society.

Another area of significance for global change research is the focus that the social sciences now bring to the modelling process itself – from conceptualization through to application. Given that systems representations are necessarily simplifications of complex realities, for knowledge integration there is no substitute for deliberation and discussions – among the contributors *and* with the users of the knowledge. Rademaker (1982) expressed this as the vital need for '*pondering the imponderables*' once a model has been constructed and run. Fraser (2007) emphasizes that rather than driving towards ever more spuriously precise 'prediction' of socio-environmental systems, the modelling process should attempt both to incorporate and to expose the potential feedbacks and different assumptions brought by the scientists and stakeholders. Allison *et al.* (2009) are also explicit that large-scale analyses, like their national-level analysis of economic vulnerability to the impacts of climate change on fisheries, need to be complemented by local, site-specific assessments involving the individuals affected by the changes.

1.3.3 Resilience, adaptation and vulnerability

Since the 1990s, there has been a growing focus on coupled socio-ecological systems and their ***resilience***. The resilience framework was developed by Holling and Gunderson (2001), both biologists (like von Bertalanffy) who extended to the social world their insights from ecological processes. The framework emerged from a deliberate process to develop theories of socio-environmental change. Change is viewed in terms of oscillations between phases of growth and stability, and of crisis and reconfiguration (Figure 1.7). Holling (2001) reviews the key concepts of resilience for linked social, economic and ecological systems. Several properties shape the transformations or future states of a complex adaptive socio-ecological system. Its sparseness or richness sets the system's potential (or 'wealth'), determining the number of alternative options for the future. The connectedness of the elements in the system shapes the extent to which a system can control its destiny. The adaptive capacity of the system determines how vulnerable it is to disturbances that can exceed or break that control.

Resilience research emphasizes some key characteristics of complex systems that inform human dimensions research with an Earth systems perspective:

- *Emergence* has already been mentioned briefly. It implies that the properties of the parts of the system can be understood only within the context of the larger whole – and that the whole cannot be adequately analyzed only in terms of its parts. Irreducibility is its corollary: true novelty can emerge from the interactions between the elements of the system.
- *Multiple interacting scales*: resilience theory posits that socio-ecological systems are hierarchic, with strong coupling between the different levels. This

makes it impossible to have a unique, correct, all-encompassing (adequate) perspective on a system from any single level; plurality and uncertainty are inherent in systems behaviour. The system must therefore be analyzed or managed at more than one scale simultaneously. The concept of adaptive management (e.g. Peterson *et al.*, 1997; Folke *et al.*, 2002; Olssen *et al.*, 2004; Dearing *et al.*, 2010) is being developed in response to this challenge. The presence of global phenomena that society wants to manage raises questions about the structures and processes of global governance. This has also become a focal area for research and science–policy dialogue (e.g. Biermann, 2001; Biermann and Pattberg, 2008; Hulme, 2010 and the IHDP's project www.earthsystemgovernance.org).

- *Non-linearity*: many relations between the elements of a complex adaptive system do not show linear behaviour. The magnitude of the effects are not proportional – or even in the same direction – as the magnitude of the causes. Counterintuitive behaviour is typical of many complex systems. Accordingly, there is a strong focus in current research on thresholds, discontinuities and 'explosions' or 'collapses' of growth.
- *Multiple legitimate perspectives*: understanding an adaptive system demands a consideration of its context, and, where social systems are concerned, this means a consideration of the many social groupings with an interest in the issue. If these voices are to be considered in decision-making about global change, then this characteristic of socio-ecological systems also has implications for the ways in which global governance can operate.

Other fields of socio-environmental research in the climate context appear to be converging on the language and concepts, if not necessarily the full theoretical underpinnings, of resilience and socio-ecological systems.

Adaptation studies have conventionally been underpinned by economics and policy analysis, generally taking a sectoral perspective on society. Reflecting current lines of academic specialism and policy structures, these analyses have generally considered sectors separately. The Working Group II report of the IPCC's Fourth Assessment Report (2007) demonstrates this emphasis. Its synthesis of adaptation research is divided into chapters for agriculture, transport and infrastructure, coastal zones, and so on. It frames the challenges of responding to climate change largely in terms of the financial costs of adaptation, the limits that these costs impose and the barriers to their implementation. The adaptation plans prepared by the world's poorer countries under their commitments to the national adaptation programme of action[3] of the UN Framework Convention on Climate Change are predominantly focused on sector policy. Swart *et al.* (2009) review Europe's climate adaptation strategies, which are similarly framed in sectoral terms. They recognize that cross-sectoral conflicts present a threat to successful adaptation, with several countries identifying the need to enhance resilience through flexibility in strategic planning.

Adaptive capacity is increasingly defined as synonymous with resilience. Following from the growing acceptance of the idea of a coupled socio-ecological system, the recognition of ecological constraints on social activities and of the role of the natural environment in maintaining social resilience is becoming prominent in policy discourse. It also brings new challenges for policy integration (e.g. Swart *et al.*, 2009; Berrang-Ford *et al.*, 2011). Examples of policy documents that emphasize environmental limits and thresholds include the 2005 UK Sustainable Development strategy (Defra, 2005), the UK Department for Food and Rural Affair's reports on ecosystem services (Defra, 2007), the TEEB report (Kumar, 2010), and the UN Development Programme's Human Development Reports of 2007/2008 and 2010 (UNDP, 2008, 2010).

Vulnerability – of the individual, household, community or nation – has previously been framed very much in terms of situated social research, drawing on the fields of development and international relations. Vulnerability research often uses the language of hazard and risk, rather than of economics and sector policy. Again, the structuring of the IPCC's synthesis reports show this: vulnerability research has generally been reported in regional terms, rather than sectoral, using specific case studies or events as its evidence base. Increasingly, there is recognition that issues experienced in a particular place are likely to have complex interactions across different geographical

[3] National adaptation programmes of action (NAPAs) are the process whereby least developed countries (LDCs) identify priority activities in responding to climate change. http://unfccc.int/national_reports

scales, and that vulnerability can be exacerbated by the simultaneous action of multiple stressors.

Resilience theory has been deployed in bringing vulnerability into a global framework. Turner *et al.* (2003) explain how vulnerability can be regarded as a function of three factors: the exposure to the hazard; the sensitivity of the system, in terms of both human and environmental conditions; and the resilience of the system. This is not necessarily promoting a 'global vulnerability analysis' as such (although Allison *et al.* 2009 show how the framework can be used in a multi-country approach) but it offers new ways to build up a robust picture of human vulnerability to environmental change worldwide, and to devise governance structures to address the problems.

1.3.4 Economics and Earth system science

Economics textbooks describe the discipline simply enough as the study of resource allocation. 'Oikos', the Greek root word, means *home* or *house* or *holdings*, making economics the 'management of the holdings'. Key concepts in economics relate to stocks and capital, material throughput and transformation, value and income – in other words, the holdings themselves, the processes, and the drivers in the economic system.

Most mainstream economic theory has been predicated on the idea of equilibrium: the assumption that, unperturbed, the economic system tends towards a stable state. The transactions of exchange of goods and services in markets are shaped by people's preferences and values, and determine prices, *ceteris paribus* ('all other things being equal'). At equilibrium, the supply of goods and services efficiently meets demand. Where there is inefficiency in allocation, the price mechanism is what shifts demand or supply. Several other idealized assumptions are embedded in mainstream (classical and neoclassical) economic theory. Within resource constraints, utility maximization is the determinant of wellbeing. Decisions about resource allocation are made by a hypothetical unitary rational actor (this is one of the fields underpinned by methodological individualism). Economic demand and supply relationships depend on the assumption of perfect information flows about the goods and services available. Many fields of economics have long traditions of critique of all these assumptions, but it is this idealized body of economic theory that dominates in Earth system science.

Economics shares the language of quantification with environmental science. In that regard it has faced less of a methodological hurdle in integration in global change research than many fields of social science. Integrated assessment models that typically link modules of climate, land use, energy and economics (already mentioned briefly in Section 1.2) are widely used in global climate science and policy, as well as in many regional environmental contexts.

In the global change context, many IAMs embed macroeconomic global trade models (as discussed in Grubb *et al.*, 2002) that apply these concepts of supply and demand. They tend to draw on statistical and empirical analysis of national accounts and of the worldwide flows of raw materials. Typically, they use computable general equilibrium methods to simulate the reallocation of resources through different sectors following an imposed change, such as a policy intervention affecting land use or energy systems. Integrated assessment models have therefore become important tools for the assessment of the implications of such changes, informing decisions about policy and pricing in the emergence of markets for carbon, and potentially also for ecosystem services (the benefits that human society obtains from natural processes and functions of ecosystems). However, as Warren (2011) argues, this area of model application requires a step change in their treatment of sectors and regions to address the global interconnections. With regard to decision support about ecosystem services, Naidoo *et al.* (2008) and Carpenter *et al.* (2009) emphasize that improvement is needed in the treatment of both the economic and ecological dynamics in models. While full-complexity Earth system models do now represent the linked dynamics of ecosystems and climate well, so far they have not incorporated economic modules. However, progress in this area is being made with some intermediate complexity Earth system models and some spatially explicit multi-component IAMs (reviewed in Tol, 2006).

A discussion of the representation of economics in models is outside our scope here.[4] Our aim is to explore how economics concepts are deployed in global change science and policy. It is no surprise that a discipline concerned with the supply and demand of money meshes easily and fundamentally with politics but, mirroring

[4] Books providing an overview of the technical aspects of economic modelling include Nordhaus (1994), which explains the economic theory embedded in the DICE IAM, and Tol *et al.* (2009) which describes the theoretical and empirical foundations for modelling the economics of land use at the global scale.

the theoretical diversity in the other social sciences already discussed, a wide range of schools of economic thought exist that address the same issues through very different lenses.

A marker of the emergence of the field of environmental economics was the publication of *Blueprint for a Green Economy* (Pearce *et al.*, 1989). Perhaps the single most influential idea of this popular book was its focus on environmental externalities – costs or benefits to society that are not captured within markets. Pearce and his colleagues pointed out that a consequence of the fact that nobody owns the natural environment can be that it is valued at zero in decision-making processes, even though both environmental costs and benefits can be very significant indeed. The book described a strand of research into ways of quantifying environmental values in monetary terms and incorporating these environmental costs and benefits into decision-making processes, which already often operated on the basis of cost/benefit analysis.

In recent decades, methods have been developed for assessing the value of environmental goods and services when market prices do not exist (e.g. Pearce and Turner, 1990; Defra, 2007). These valuation and value-transfer approaches were initially applied to decisions in local contexts such as waste management and conservation-site selection, and are now accepted inputs to more complex and regional concerns like river-catchment planning and integrated coastal zone management. The thinking outlined in *Blueprint for a Green Economy* has informed a wide range of policy measures, such as the UK's landfill disposal charge, urban congestion charges and carbon emissions trading. Its neoclassical perspective has had a lasting influence, traceable in the emphasis on environmental assets, the definition and allocation of property rights, the development of markets for environmental goods and services, and a leaning towards economic incentives rather than, say, taxes or regulatory control as preferred policy instruments. The UK Government-commissioned Stern Review (Stern, 2007) that quantified the costs of responding to climate change was a product of this tradition. Thus, its key recommendations were strongly oriented towards market mechanisms: in identifying a collective global way forward, '*policy must promote sound market signals, [and] overcome market failures*' (Stern, 2007, executive summary).

Another strand that has developed in recent decades is 'ecological economics', a radical project to rewrite (macro) economics integrating natural-resource constraints. Early proponents were Daly (1974, 2009), a World Bank economist, and the ecologist Ehrlich (1989). Ecological economists base their arguments on the idea that the economy and social activity need to be seen as sub-systems of a global ecosystem. They call for an explicit recognition of ecological 'limits', and the maintenance of critical natural capital (reviewed in Ekins *et al.*, 2003). A landmark paper for this field was the assessment by Costanza *et al.* (1997) of the global total annual value of the services that nature provides to society, such as storm-surge attenuation by marshlands, nutrient cycling, the regulation of atmospheric composition and so on. Until this analysis of the 'value of nature', the magnitude of the opportunity costs of human development had never been totted up. Damage to the environment was largely incremental and – as Pearce and his colleagues had argued – it was rather invisible to the policy process, even if it was starkly evident in other ways. In their paper, this fact was pointed out in ways commensurate with other decision-making processes. In practice, of course, the notional 'total world value' of nature has little immediate relevance to policy-making. In most day-to-day practice about environmental-resource use, decision-making is done on a local level, and global-aggregate analysis misses the mark. In a rigorous environmental economic analysis prepared in response to shortcomings in the Costanza *et al.* article, Balmford *et al.* (2002) calculate a benefit:cost ratio of 100:1 for the global conservation of ecosystems. Despite the conclusive magnitude of the net economic benefit in the Balmford *et al.* analysis, development and conservation priorities have not been transformed on that basis, which suggests that decision-makers do not use esoteric global assessment numbers like these.

What caught policy-makers' attention in this debate was the articulation of the concept of ecosystem services in economic terms, which could form the basis for the creation of real markets in previously 'invisible' natural resources, and thus provide a different set of institutional levers to promote and incentivize resource conservation. Daily *et al.* (2000) described early developments in the creation of markets for ecosystem services, envisaging the trading of '*new environmental products, such as credits for clean water…*' on the stock exchange. The Millennium Ecosystem Assessment (MA, 2005) brought ideas from ecological economics into mainstream policy by emphasizing the dependence of human society on well-functioning ecosystems. The MA framed environmental processes and functions as providers of benefits to human society – the ecosystem services (Ehrlich and Mooney, 1983) that nature confers. It also framed a well-functioning nature as a scarce resource, documenting the relentless worldwide process of human-induced environmental degradation and loss of biodiversity. It stopped short of quantifying these trends in economic terms, but the policy world had taken notice. The language of ecosystem services now pervades climate, environmental and conservation policy (although few actual markets have been created so far). In 2007, the environment ministers of the G8 countries and five emerging economies launched the Potsdam Initiative for biodiversity. Among its commitments are a promise to '*initiate the process of analysing the global economic benefit of biological diversity, the costs of the loss of biodiversity and the failure to take protective measures versus the costs of effective conservation*', and a commitment to devise and support financial mechanisms for markets to trade ecosystem services. The report entitled *The Economics of Ecosystems and Biodiversity* (known as the 'TEEB report'; ten Brink *et al.*, 2009; Kumar, 2010) documents this promised analysis.

Balmford *et al.* (2011) and Fisher and Turner (2008) tease out some of the tangles that remain in the concept of ecosystem services as outlined in the Millennium Ecosystem Assessment. The MA classification (provisioning, regulating, supporting and cultural services) has drawn policy-makers' attention to the interdependencies of the many ecosystem processes and functions, but it has been conceptually problematic for ecologists and economists alike. It would be helpful – for scientists *and* policy-makers – if we made a clear distinction between ecosystem processes (what nature does), services (what nature does that we can use or depend upon) and benefits (what we actually do use and value). Any credible Earth system science response to the often-articulated demands to 'model global ecosystem services' (e.g. Perrings *et al.*, 2010) depends on understanding and accommodating this distinction.

A parallel set of critiques is focused less on the conceptual principles and methods being used in ecosystem services economics than on the policy implications and applications. Cornell (2010b) argues that the ascendance of global market mechanisms in this context is backtracking on well-established international sustainability commitments. Cornell (2011) explores the perspectives and arguments made in other critical analyses (Spash and Vatn, 2006; Norgaard, 2010) of the ecosystem services concept. In short, although the concept of ecosystem services has proved to be a powerful and useful framework highlighting the risks of human-caused environmental degradation, responses that seek to apply its economic dimensions need to recognize the complexity of the socio-ecological system, and to accommodate the issues of equity and justice that are equally essential components in the sustainable management of the global environment.

The ecosystem services concept highlights how shifts in public-policy discourse reflect and influence the value placed on the natural environment. For instance, a socio-technical agenda dominated in the 1950s, with environmental degradation a consequence of the pursuit of 'a decent standard of life'. In the later twentieth century, neoliberal public policy was strongly oriented towards contributing to economic growth, which has driven today's reliance on market mechanisms to incentivize care for the environment on a global scale. At the start of the twenty-first century, there has been an international shift towards 'redefining wellbeing', in terms of quality of life not quantity of material consumption (e.g. Stiglitz *et al.*, 2009; Dolan *et al.*, 2011). Sometimes called the happiness agenda, it implies a quest for very different indicators of societal wellbeing 'beyond GDP'.

The first issue we want to highlight in this regard is the flurry of methodological developments this has brought. Economic tools and approaches that were previously marginal interests have become prominent in this current context. These include multi-criteria analysis that includes environmental values, rather than conventional cost/benefit analysis; and participatory processes in economics, both for the deliberative optimization of economic valuations, and for a more consensual discussion of implementation of policy options. Echoing the debates in the social sciences about emergence is the idea that communities may have shared values. Incorporating these into valuation studies can only be determined through dialogue, rather than assuming fixed individual preferences (Turner, 2010). From a sustainability perspective, this focus on democratic engagement and transparency in economic valuation is a good thing. However, it shifts away from predictive global socio-environmental modelling as it has been envisaged to date.

A second point is that economic growth is increasingly becoming 'de-materialized' (Panayotou, 2003; Ausubel and Waggoner, 2008), with economies relying less on the production and consumption of 'stuff' (actual physical goods) for the generation of wealth and wellbeing. The long-run prospects for economic growth (and all the benefits it brings to society) require a stronger separation of economic goods from 'bads', such as decoupling economic growth from environmental impact, wellbeing from material consumption, and of course energy provision from fossil-fuel combustion.

The idea of a 'Green New Deal' (e.g. Barbier, 2010) or a greening of the economy (with green infrastructure built by green collar workers, in the words of political scientist Timothy Luke) has gained traction in the light of both a suppressed global economy and the growing recognition of the need to mitigate climate change. The idea does not always mean a reversion to much more local, low-impact, low-tech modes of living. The United Nations Environment Programme (UNEP, 2011) calls for '*investing two per cent of global GDP in greening ten central sectors of the economy in order to shift development and unleash public and private capital flows onto a low-carbon, resource-efficient path*'. The UK Conservatives Gummer and Goldsmith (2007) note that '*The pricing of carbon itself represents a host of exciting opportunities for reinvigorating the economic system and companies within it – for flushing out old power structures, opening space for new entrants and technologies, stimulating job creation and encouraging nimbler, lighter management.*'

Earth system science already plays a uniquely powerful role in this latter vision, whether the scientists know it or not. One fundamental way is in the pricing of carbon (other emerging ways include informing the design and monitoring of climate mitigation through biosphere management, as discussed in Chapter 7). Dietz (2007) highlights one specific issue in this regard: the social cost of carbon, which is the 'shadow price' used by policy-makers in pricing carbon, depends not on current levels but on the future trajectory of CO_2 concentrations in the atmosphere. Policy choices can determine the level of emissions, but the Earth system links between emissions and concentration are not a matter for policy guesswork. Optimal policies depend fundamentally on the scientific knowledge that only Earth system science can provide.

Although Earth system models are still a considerable way off being able to represent the interplay of climate, ecology and economics in the necessary detail, there are foreshadows of the part envisaged for projections and predictions of ecosystem services in the future 'green' economy. McAfee (1999; p. 133) wrote with concern that '*by the logic of this paradigm, nature is constructed as a world currency and ecosystems are recoded as warehouses of genetic resources for biotechnology industries.*' Perrings (2010) confirms that this is now precisely the case, although he also highlights progress in the design of governance systems that treat biodiversity as a transboundary public good.

In recent years, as climate adaptation and mitigation have become major policy issues, there has been a growing emphasis on the natural environment in development economics. Development economists have long been concerned with social questions of equity and legitimacy in resource allocation (e.g. Boyd and Juhola, 2009), along with the conventional economic attention to efficiency and effectiveness. Research and application in this field have been influenced more than mainstream economics by a rights-based ethics, rather than purely a utility function. Sen (1973) pointed out that utilitarianism does not deserve its 'egalitarian reputation' and called for economics to concern itself more with fairer treatments of inequality and welfare distribution. However, a common critique of the globalization of the economy is that social dimensions have been sidelined in favour of economic prerogatives. In 2001, Meier and Stiglitz collated a very diverse set of sociological and economic perspectives on economics and development, exploring this tension. Some of these analyses (Thomas, 2001; Yusuf and Stiglitz, 2001) also explicitly considered the natural environmental and climate change.

Many of the polarities discussed in Meier and Stiglitz's book are still evident in today's debates about environmental justice. The UNEP Green Economy report (2011) notes the persistence of inequity in the distribution of environmental costs and benefits, and emphasizes again, as Sen did 40 years ago, that equity needs to be made a parallel priority in development. Failings in integrating the social, environmental and economic dimensions can be seen in concern about the 'double jeopardy' problem where the world's poorest people are likely to be most exposed to climate hazards but also are likely to have the least adaptive capacity (O'Brien and Leichenko, 2000; LaFleur *et al.*, 2008); in the equity debates relating to the 'contraction and convergence' approach (GCI, 2005) used in the UN Framework Convention on Climate Change, which has focused attention on the emissions associated with the historic development of the world's rich countries; and of course in the discourses of payments for ecosystem services and the Kyoto and post-Kyoto mechanisms for climate mitigation (Griffiths and Martone, 2009; Ikeme, 2009; Leach, 2009; UNEP, 2011).

To close this section, we highlight a couple of contexts where social research, economics and environmental science are meeting, allowing for deeply transdisciplinary approaches to global change.

A stronger focus on human behaviour is evident in many areas of environmental change, society's responses, and environmental risk management. This can be seen in the research and policy discourse, and the expansion of environmental themes in areas such as behavioural economics and social psychology. An iconic character in this field is the Nobel Laureate economist Kahneman whose research highlighted the often-irrational decision-making of human beings, and the importance of subjective wellbeing (in other words, people can feel happy when all indicators suggest they should not be, and vice versa). Evidence is growing that climate impacts are strongly mediated by these behavioural and social dimensions. For instance, Fraser (2007) documents cases where big drought events (in terms of absolute water shortage) did not cause commensurately big effects on agriculture, as well as cases when small droughts had huge impacts, in part because of the nature of people's responses. Findings like this indicate that, in responding to climate change, people's behaviours need to change. These behavioural aspects have profound implications for the design of policy incentives for socio-economic or technological change, and for concerted societal action (Shibutani, 1961; Crosweller and Wilmhurst., in press; Cornell and Jackson, in press). This body of evidence also presents challenges to the analysis and modelling of systems; issues of human choice and agency have always been problematic in modelling, but the 'systems behaviour of human behaviour' is plagued with non-linearities and surprises.

Despite all the challenges outlined in this chapter, the research communities variously known as 'Impacts, Adaptation, Vulnerability' or 'Vulnerability, Adaptation, Resilience' are showing an entrainment of social and economic insights into Earth system science. In Chapter 5, we discuss the ways that the future scenarios being developed for the IPCC's fifth and subsequent assessment reports aim to incorporate social and economic narratives better than before. But this trend extends beyond the subset of scientists involved in the IPCC. Several major international conferences have explicitly sought to bring these academic, policy and practice communities together for more effective dialogues. These include Earth System Science 2010, Resilience 2008 and 2011, and Planet Under Pressure 2012 (Table 1.2). International strategic research planning processes have also mapped out ways to improve the transdisciplinary interaction that is needed to inform responses to global change. These include the 2010–2011 ICSU/ISSC Visioning Exercise for Earth system science, and the European Science Foundation's RESCUE Forward Look on science needs in the Anthropocene. These efforts are taking integrated Earth system science to a new level.

1.3.5 Societal responses to environmental change: the policy context for Earth system science

The changing roles of the natural and physical sciences and social and human sciences in global change research mirrors a growing policy focus on the 'human dimensions' of global environmental change. It reflects a shift from needing first to characterize the problem adequately to needing now to inform responses and solutions (Box 1.4). The recognition of the interlinked socio-ecological system increasingly pervades what previously were either environmental or social policy contexts.

For example, the notion of a coupled socio-ecological system is becoming more widely applied in the policy context. In the UK, the national sustainable development strategy, *Securing the Future* (Defra, 2005), articulates its five guiding principles within a socio-ecological framework. In its top-level principles, it makes a clear statement that society should aim to optimize its wellbeing while operating within environmental limits. This balance of nature and society is mediated and managed through good governance, good application of (scientific) knowledge, and good economics. A shift in the focus of the UNDP Human Development Reports can also be seen, from addressing poverty in narrowly socio-economic terms to a deeper consideration of its causes and consequences in terms of environmental justice (UNDP, 2008). There has also been a clear trend from the first IPCC assessment report in 1995 to the latest in 2007 towards greater integration across the socio-environmental divide. Early assessments were criticized for very piecemeal treatment of social concerns in contrast with a richly elaborated and resource-intensive theoretical treatment of the physical climate science. The current round of assessment includes a complete reconfiguration of the future scenarios used

Table 1.2 Some dialogue initiatives for integrative Earth system science
All these initiatives aim to provide a forum for people working with the complex dynamics of interconnected social–ecological systems to discuss the challenges of future societal development. They seek to highlight new directions for understanding the interactions between humans and our environment, and strengthen the much-needed interactions between the natural and social sciences, and between policy, assessment and research.

Initiative	Further information
Resilience 2008: *Resilience, Adaptation and Transformation in Turbulent Times* Stockholm, April 2008 International conference convened by the Resilience Alliance, with the Swedish Academy of Sciences and ICSU.	http://resilience2008.org WebTV coverage of discussions and key presentations are available on http://resilience.qbrick.com/
Resilience 2011: *Resilience, Innovation and Sustainability: Navigating the Complexities of Global Change* Tempe, AZ, March 2011 International conference convened by the Resilience Alliance, with Arizona State University.	www.resilience2011.org
Earth System Science 2010: *Global Change, Climate and People* Edinburgh, May 2010 The first IGBP AIMES Open Science Conference, with QUEST, the UK NERC research programme for Earth system science.	http://quest.bris.ac.uk/workshops/AIMES-OSC Conference proceedings were published in *Procedia Environmental Sciences* (2011), vol. 6 www.sciencedirect.com/science/journal/18780296
Planet under Pressure 2012: *New Knowledge Towards Solutions* London, March 2012 International conference convened by the four international global change programmes IGBP, IHDP, WCRP and Diversitas, and the ESSP.	www.planetunderpressure2012.net
Earth System Visioning: *Developing a New Vision and Strategic Framework for Earth System Research* International consultation and strategy development exercise by ICSU and the ISSC.	www.icsu-visioning.org Grand Challenges identified for Earth system science for global sustainability (Reid *et al.*, 2010).
ESF Forward Look: *Responses to Environmental and Societal Challenges for our Unstable Earth (RESCUE)* International consultation and horizon-scanning exercise by the European Science Foundation.	www.esf.org/rescue Articles based on the working group discussions are being published as a special issue of *Environmental Science and Policy*.

for the climate modelling, precisely so that a richer set of socio-economic storylines can be developed in line with the global projections of climate.

The immediate question that this shift from pure science to a solutions-oriented application should prompt for Earth system scientists is: what is policy? Who makes and implements it, and how, when and where? We can state in the broadest terms that policy is a societal decision about how to steer itself (as in Pielke, 2004), but we still need to recognize that policy is fluid and evolving.

Coordinated climate policy first developed in the 1980s. At the first World Climate Conference (Hare and White, 1979), participants were mostly invited scientists, discussing the physical science of climate change and natural hazards, and the social impacts of change. After that event, where scientific consensus coalesced about the threats of climate change, major scientific and environmental bodies including the World Meteorological Organization, UNEP and ICSU coordinated efforts rapidly to mobilize a societal response. The landmark Villach meeting in 1985 (Franz, 1997) set the foundations for the IPCC, which was established in 1988. The IPCC remains a vitally important feature now in the climate policy landscape, as the intergovernmental forum for scientific synthesis. The prime instrument of climate policy is the United Nations Framework Convention on Climate Change (UNFCCC; adopted in 1992; http://unfccc.int) with its regulatory targets and mitigation instruments, translated into regional and national policy responses (described in Schneider *et al.*, 2010). In the currently ongoing synthesis activities of IPCC's Working Group II (Impacts, Adaptation and Vulnerability), there is a stronger steer towards more directly addressing the specific policy-oriented information needs of the UNFCCC.

A major theme in this chapter and throughout this book is that global change involves much more than just the climate system. The evolution of international policy for many aspects of environmental change and development have paralleled climate policy. Third World development concerns prompted the creation of the UNDP in 1971. The 1972 Stockholm Conference on the Human Environment (ECOI) led to the creation of UNEP, devised to be the 'environmental conscience' of the UN system. The implications of land-use change and deforestation in dryland regions led to the UN Convention to Combat Desertification in 1977. The Brundtland report (1987) and the 1992 United Nations Conference on Environment and Development (the first Rio Earth Summit) marked the formalization of global policy for sustainability.

Attention is needed to the challenges of integration of all these policy strands at two levels. The first is the integration of environmental change into existing (national, sectoral) policies. The second is the integration of the many 'jigsaw-puzzle pieces' of policies targeted at different aspects of environmental change (Nilsson and Eckerberg, 2007), which could potentially have conflicting goals or instruments. Scientists aiming to engage effectively with policy need to understand this complex policy landscape. One example of a concerted effort to bridge science and its complex policy context was the 2009 International Nitrogen Initiative's expert meeting of the Convention on Long-range Transboundary Air Pollution and the Convention on Biological Diversity, which explored the scientific links that need to be addressed in developing the evidence base for these two conventions, and also made a preliminary consideration of the links to climate science for the UNFCCC (UNECE, 2009). This event showed how considerable benefits for scientific work supporting policy discussions can be reaped from increased collaboration among the various scientific communities currently tackling environmental problems as if they were separate issues.

Compared with the institutions and instruments of national-level policy, global policy faces particular challenges. For any environmental management effort to be effective, the spatial reach of governance systems must match the spatial scale of the problems they seek to address. Earth system science gives insight into the workings of global-scale problems but, in the absence of a planetary authority to steer a desirable and sustainable pathway (as suggested by Schellnhuber and Kropp, 1998), the notion of 'Earth system governance' demands attention to effective political solutions. The strategic overview that global governance institutions and instruments may implement also needs to meet local needs (Berkes, 2002), which presents dilemmas and operational obstacles. Commitments to forms of democratic participatory involvement in decision-making (for persuasion, consensus, and legitimacy) are enshrined in many global agreements and national constitutions, but are not mirrored in structures for such engagement at the global level. Adaptable governance is needed, capable of provisional and responsive decision-making in the context of uncertainty about future climate and its impacts. And a critically important question still remains open: who pays for the processes of global governance and society's responses to climate and global environmental change?

These are very real concerns and tensions in current debates, particularly in climate mitigation which is necessarily a global issue. Chapter 7 returns to some of the global policy and governance issues that shape the context of mitigation, and Chapter 8 reflects on the knowledge gaps that science can help fill. In Box 1.5, we indicate some recent books that exemplify how the social sciences and humanities are addressing global change. In our closing section, we focus on what lessons can be drawn from the transdisciplinary efforts to date that can inform twenty-first century Earth system science.

Box 1.5 Other information resources

Several recent social science and humanities books address 'the climate dimension' of human systems. Here are some that bridge the diverse fields of scholarship:

- *The Politics of Climate Change* (2009), by sociologist Anthony Giddens, starts from the premise that although the science and policy of climate change have a very high profile, the actual politics of climate change is still rudimentary. Giddens argues that the political changes needed for responses to climate change should be mapped out in the context of the political structures and institutions that currently exist. The nub of the challenge is that people will only mobilize politically for action in the face of climate change when its effects become tangible problems in their experience – by which time it is very likely to be too late to fix the problem. In Giddens' view, then, a key part of the solution lies with getting people to recognize climate risks as real and pressing.

- *The Economics of Climate Change* (2007), widely known as *'the Stern Review'* after Sir Nicholas Stern, the Treasury economist who led this major review for the UK Government, considers the economic impacts of climate change, and also of climate mitigation. Its starting point was that a clear exposition of the available economic analyses and current understanding of the global impacts was made and shared widely, then, rather than each nation trying to devise their own policy positions in isolation, achieving international agreements on action would be more straightforward. It departed from many previous economic studies by taking an explicitly risk-averse position in its modelling studies, changing the balance between present and future values, and addressing the disproportionate impacts on poor regions.
- *Interdisciplinarity and Climate Change: Transforming Knowledge and Practice for our Global Future* (Routledge, 2010), edited by philosopher Roy Bhaskar and colleagues, brings insights from the critical realist school of philosophy to the challenges of interdisciplinary and action-oriented research in many contemporary socio-environmental issues (including energy, food, consumerism and other global issues), and the problem of linking theoretical knowledge to practical action in today's society.
- Historian Clive Ponting's 2007 *New Green History of the World: The Environment and the Collapse of Great Civilizations* is a revised edition of his book published in 1991. He was one of the first scholars to take a long look back, and to consider what this meant in terms of society's prospects for the future. While he does not explicitly consider a systems perspective, he does give a careful historical exposition of the interplay of human and environmental change.

1.4 Creating usable and useful integrated research about the Earth system

In the olden days, or so goes the story (Balconi *et al.*, 2008), government shaped society, and science informed government. Now both governance and the role of science in society are seen in very different terms. States have lower importance in steering society, while international agencies and corporate actors have higher (Paavola, 2007; Biermann and Pattberg, 2008; Jordan, 2008; Moser, 2008; Paavola *et al.*, 2009). Interactions between individuals have been radically enabled through information technologies and the new media, changing patterns of power, participation, activism and learning. And alongside these changes, science has changed too. The enormous technological investment and increasing specialization in research has put some areas of scientific knowledge out of reach of non-scientists and, as von Bertalanffy noted even back in the 1960s (von Bertalanffy, 1968), it has also tended to leave scientists encapsulated in their disciplinary areas. Countering that is a significant trend in the 'democratization of science' (e.g. Funtowicz and Ravetz, 1993; Carolan, 2006), bringing a stronger focus on issues such as communication, public understanding, policy engagement as part of the scientific research process, and the handling of uncertainty and contested knowledge.

Insights from Earth system science are being channelled into a context of complex social behaviour and political positioning. The end goal of working at the science–policy interface is that society can use these insights to respond with more confidence to the current and projected environmental changes. But although the goal is clear, a great deal needs to be done to ensure that the research being done is both useful and usable.

In this chapter, we have mapped out some of the issues and recent positive developments in the expansive, varied and changing terrain of Earth system science for the Anthropocene. Here, we revisit some of the key themes, paying attention to what they mean for the application of this science.

To begin with, we emphasized the conceptual shifts that researchers in all the contributing fields have needed to make in turning their attention to the human component of the Earth system: human society is an intrinsic part of the Earth system while it is also the collective mind observing, researching and intervening in its dynamics. The balance of experimentation, observation, abstraction, theoretical coherence, and subjectivity or objectivity all determines what constitutes scientific evidence. Global environmental change has been a nexus for change in this balance, reframing our role as scientists within the system that we study.

If the science really is to influence social directions, then partial, over-simplified, or misapprehended knowledge presents tangible, potentially enormously costly risks. A precautionary approach pushes us towards more transdisciplinary integrated knowledge; useful science in today's environment *must* draw on all the available expertise and insight. The scholarly requirements for new integrative working need attention in parallel. We still have some academics (Tol, 2006; p. 2) arguing that '*Like car repair is best left to mechanics, modelling is best left to modellers*', but as Rademaker pointed out many years ago (1980; p. 503), '*Modeling is the easiest part of the job. Detecting the right questions and delivering the desired answers in the right language at the right moment is far more difficult.*'

However, we risk putting a too-utilitarian perspective on the matter. We seek understanding of our environment because it is enthralling. We want understanding of how our environment works at the global scale – but not just at that level. After all, the large-scale workings of the planets in our Solar System have long been known. The fascination lies in seeking knowledge beyond the regularities, seeing how small changes can have big and unanticipated consequences, seeing patterns in the complexity, and testing and probing our knowledge in new ways.

A systems perspective provides a common ground for this exploration. Not all scholars who 'think in systems ways' articulate it in those terms, just as not all of the theoretical literature currently produced about systems is of use to our Earth system inquiries. Nevertheless, many of the contributing fields of study now engaging in global change research have well-established foundations in systems thinking, and so also do many of the end-users of the science. Policy integration, after all, is an archetypal process of dynamic accommodation of complexity, contingency and causality.

Having the same ground underfoot is important, but traversing the shifting terrains of global change still demands attention, thought and dialogue. Like any frontier exploration, global-scale transdisciplinary research is uncomfortable at times. Absolute objectivity – in the sense of the traditional scientific ideal of single-variable experimentation under controlled conditions – is not possible in Earth system research, so other ways are needed of ensuring adequate scientific scrutiny and verification of claims. There will necessarily be debates, sometimes conflicts. Better forums are still needed for the transdisciplinary interaction and public and policy engagement that must underpin the envisioned developments in Earth system science. In terms of the science itself, this entails moving beyond pragmatic but over-blunt numeric approximations of socio-environmental processes to agreeing the rules for integration of skills and knowledge among contributing fields of study (as argued by Liverman and Roman-Cuesta, 2008). For useful outcomes, Earth system scientists also need to pay closer attention to the practices and ethics of engagement with policy and wider society (Demeritt, 2001; Hulme, 2008). In short, institutional frameworks need to continue to evolve for research itself *and* for society's responses to global socio-environmental change, better supporting these interfaces and actively encouraging the dialogues and practices that are needed.

For our science to be useful, we need much better ways of addressing the mismatch in scale of global processes and local agency (discussed well by Wilbank and Kates, 1999). Global-scale policy concerns come under the spotlight with the rising attention being given to climate mitigation interventions ranging from biofuel mega-plantations through to geoengineering. Society may decide that it has no option other than to attempt to control climate deliberately through large-scale planetary interventions, but Earth system science should have something to say about *how* this might need to be considered. If we understand climate change essentially as a glitch in the planet's radiative balance, then it is not unreasonable to seek to tweak the planet back, by modifying the physical aspects such as the greenhouse gases in atmosphere, or the reflection or absorbance of Earth, or even by optimizing the biosphere. But as scientists, we may want to issue a caution: we should be aware that a system as intricate and interlinked as Earth may not respond like clockwork.

The multiplicity of theories of biophysical change, human behaviour and the social environment are being connected to the contexts of policy and practice through the lens of global environmental change. The remainder of this book gives an overview of the science of global change as we currently understand it. We close, in optimism, with E. O. Wilson's line (1998; p. 12):

> ... If we dream, press to discover, explain and dream again, thereby plunging repeatedly into new terrain, the world will somehow become clearer and we will grasp the true strangeness of the universe. And the strangeness will all prove to be connected and make sense.

References

Abbot, A. (2007). Mechanisms and relations. *Sociologica*, **7**, 1–22.

Ackerman, F., DeCanio, S. J., Howarth, R. B. and Sheeran, K. (2009). Limitations of integrated assessment models of climate change. *Climatic Change*, **95**, 297–315.

Allison, E., Perry, A. L., Badjeck, M.-C., *et al.* (2009). Vulnerability of national economies to the impacts of climate change on fisheries. *Fish and Fisheries*, **10**, 173–196.

Ausubel, J. H. and Waggoner, P. E. (2008). Dematerialization: variety, caution, and persistence. *Proceedings of the National Academy of Sciences USA*, **105** (35), 12774–12779.

Balconi, M., Brusoni, S. and Orsenigo, L. (2008). In defence of the linear model: an essay. Centro di Ricerca sui Processi di Innovazione e Internazionalizzazione. Working Paper 216. Milan: Università Commerciale Luigi Bocconi. See: www.cespri.unibocconi.it.

Balmford, A., Bruner, A., Cooper, P. *et al.* (2002). Economic reasons for conserving wild nature. *Science*, **297**, 950–953.

Balmford, A., Fisher, B., Green, R. E. *et al.* (2011). Bringing ecosystem services into the real world: an operational framework for assessing the economic consequences of losing wild nature. *Environmental Resource Economics*, **48**, 161–175.

Balstad Miller, R. and Jacobson, H. K. (1992). Research on the human components of global change. *Global Environmental Change*, **2**(3), 170–182.

Barbier, E. B. (2010). *A Global Green New Deal: Rethinking the Economic Recovery*. Cambridge: Cambridge University Press and UNEP.

Berkes, F. (2002). Cross-scale institutional linkages: perspectives from the bottom up. In *The Drama of the Commons*, eds. E. Ostrom, T. Dietz, N. Dolšak *et al.* Washington, DC: National Academy Press, pp. 293–319.

Berrang-Ford, L., Ford, J. D. and Paterson, J. (2011). Are we adapting to climate change? *Global Environmental Change*, **21**, 25–33.

Bhaskar, R. (1998). *The Possibility of Naturalism*, 3rd edn. London: Routledge.

Bhaskar, R. (2008). *A Realist Theory of Science*, 2nd edn. London: Routledge.

Biermann, F. (2001). The emerging debate on a world environment organization: a commentary. *Global Environmental Politics*, **1**(1), 45–55.

Biermann, F. and Pattberg, P. (2008). Global environmental governance: taking stock, moving forward. *Annual Reviews in Environmental Resources*, **33**, 277–294.

Bjurström, A. and Polk, M. (2011). Physical and economic bias in climate change research: a scientometric study of IPCC Third Assessment Report. *Climatic Change*. DOI: 10.1007/s10584-011-0018-8.

Blaikie, P. and Brookfield, H. (1987). *Land Degradation and Society*. London: Methuen.

Boyd, E. and Juhola, S. (2009). Stepping up to the climate change: opportunities in re-conceptualising development futures. *Journal of International Development*, **21**, 792–804.

Brown, S. (2009). Naivety in systems engineering research: are we putting the methodological cart before the philosophical horse? 7th Annual Conference on Systems Engineering Research, Loughborough, 20–23 April 2009. See: cser.lboro.ac.uk/paper/s03-23.pdf.

Bruckmann, G. (1981). IIASA's role in global modelling. *Global Modelling: Lecture Notes in Control and Information Sciences*, **35**/1981, 192–195.

Bruinsma, J. (2009). The resource outlook to 2050: by how much do land, water and crop yields need to increase by 2050? Expert Meeting on How to Feed the World in 2050. Rome: UN Food and Agriculture Organization. See: www.fao.org/wsfs/forum2050/wsfs-background-documents/wsfs-expert-papers/en/.

Carolan, M. S. (2006). Science, expertise, and the democratization of the decision-making process. *Society and Natural Resources*, **19**, 661–668.

Carpenter, S. R., Mooney, H. A., Agard, J. *et al.* (2009). Science for managing ecosystem services: beyond the Millennium Ecosystem Assessment. *Proceedings of the National Academy of Sciences USA*, **106**(5), 1305–1312.

Chertow, M. R. (2001). The IPAT equation and its variants: changing views of technology and environmental impact. *Journal of Industrial Ecology*, **4**(4), 13–29.

Chesnais, J. C. (1992). *The Demographic Transition – Stages, Patterns and Economic Implications* (transl. E. Kreager and P. Kreager). Oxford: Oxford University Press.

Clark, W. C. (1988). Human dimensions of global environmental change. In: *Toward an Understanding of Global Change: Initial Priorities for U.S. Contributions to the International Geosphere–Biosphere Program*. Office of International Affairs (OIA), pp. 134–200.

Cohen, J. E. (1998). Should population projections consider "limiting factors" – and if so, how? *Population Development Review*, **24**, 118–138.

Commoner, B. (1972). The environmental cost of economic growth. In *Population, Resources and the Environment*, ed. R. G. Ridker. Washington, DC: Government Printing Office, pp. 339–363.

Cornell, S. E. (2010). Valuing ecosystem benefits in a changing world. *Climate Research*, **45**, 261–262.

Cornell, S. E. (2011). The rise and rise of ecosystem services: is "value" the best bridging concept between society and the natural world? *Procedia-Environmental Sciences*, **6**, 88–95.

Cornell, S. and Jackson, M. (forthcoming, 2012). Social science perspectives on natural hazards risk and uncertainty. In *Risk and Uncertainty Assessment for Natural Hazards*, eds. L. J. Hill, J. C. Rougier and R. S. J. Sparks. Cambridge: Cambridge University Press, ch. 17.

Cornell, S. E., Costanza, R., Sörlin, S. and van der Leeuw, S. (2010). Developing a systematic "science of the past" to create our future. *Global Environmental Change*, **20**, 426–427.

Costanza, R., d'Arge, R., de Groot, R. *et al.* (1997). The value of the world's ecosystem services and natural capital. *Nature*, **387**, 253–260.

Costanza, R., Graumlich, L., Steffen, W. *et al.* (2007). Sustainability or collapse: what can we learn from integrating the history of humans and the rest of nature? *Ambio*, **36**, 522–527.

Crosweller, H. S. and Wilmshurst, J. (forthcoming, 2012). The role of the risk manager. In *Risk and Uncertainty Assessment for Natural Hazards*, eds. L. J. Hill, J. C. Rougier and R. S. J. Sparks. Cambridge: Cambridge University Press, ch. 15.

Daily, G. C., Söderqvist, T., Aniyar, S. *et al.* (2000). The value of nature and the nature of value. *Science*, **289**(5478), 395–396.

Daly, H. (2009). Three anathemas on limiting economic growth. *Conservation Biology*, **23**(2), 252–253.

Daly, H. E. (1974). Steady-state economics versus growth-mania – critique of orthodox conceptions of growth, wants, scarcity, and efficiency. *Policy Sciences*, **5**(2), 149–167.

Davis, S. J. and Caldeira, K. (2010). Consumption-based accounting of CO_2 emissions. *Proceedings of the National Academy of Sciences USA*, **107**(12) 5687–5692.

Dearing, J. A., Braimoh, A. K., Reenberg, A., Turner, B. L. and van der Leeuw, S. (2010). Complex land systems: the need for long time perspectives to assess their future. *Ecology and Society*, **15**(4), 21.

Defra – Department of Environment, Food and Rural Affairs (2007). An introductory guide to valuing ecosystem services. UK Government Department of Environment, Food and Rural Affairs. See: www.defra.gov.uk/environment/natural/ecosystems-services/.

Defra (2005). *Securing the Future: the UK Sustainable Development Strategy*. London: The Stationery Office.

Demeritt, D. (2001). The construction of global warming and the politics of science. *Annals of the Association of American Geographers*, **91**(2), 307–337.

Descola, P. and Pálsson, G. (1996). *Nature and Society: Anthropological Perspectives*. London: Routledge.

Dietz, S. (2007). Review of Defra paper: 'The social cost of carbon and the shadow price of carbon: what they are, and how to use them in economic appraisal in the UK'. See: http://personal.lse.ac.uk/dietzs/.

Dietz, T. and Rosa, E. A. (1997). Effects of population and affluence on CO_2 emissions. *Proceedings of the National Academy of Sciences USA*, **94**, 175–179.

Dietz, T., Rosa, E.A. and York, R. (2007). Driving the human ecological footprint. *Frontiers in Ecology and the Environment*, **5**, 13–18.

Dolan, P., Layard, R. and Metcalfe, R. (2011). *Measuring Subjective Well-being for Public Policy*. London: UK Office for National Statistics.

EEA – European Environment Agency (1999). Environmental indicators: typology and overview. Technical Paper 25/1999, European Environment Agency. See: www.eea.europa.eu/publications/TEC25.

Ehrlich, P. R. (1989). The limits to substitution: meta-resource depletion and a new economic–ecological paradigm. *Ecological Economics*, **1**, 9–16.

Ehrlich, P. R. and Holdren, J. P. (1971). Impact of population growth. *Science*, **171**, 1212–1217.

Ehrlich, P. R. and Mooney, H. A. (1983). Extinction, substitution, and ecosystem services. *BioScience*, **33** (4), 248–254.

Ehrlich, P. R. and Ehrlich, A. H. (2009). The population bomb revisited. *Electronic Journal of Sustainable Development*, **1**(3), 63–71.

Ekins, P., Simon, S., Deutsch, L., Folke, C. and De Groot, R. (2003). A framework for the practical application of the concepts of critical natural capital and strong sustainability. *Ecological Economics*, **44**(2–3), 165–185.

Elder-Vass, D. (2007). For emergence: refining Archer's account of social structure. *Journal for the Theory of Social Behaviour*, **37**(1), 25–44.

European Research Advisory Board (2004). Interdisciplinarity in research. See: http://ec.europa.eu/research/eurab/pdf/eurab_04_009_interdisciplinarity_research_final.pdf.

FAO – Food and Agriculture Organization of the United Nations (2010). *The State of World Fisheries and Aquaculture – SOFIA 2010*. Rome: FAO.

Fisher, B. and Turner, R. K. (2008). Ecosystem services: classification for valuation. *Biological Conservation*, **141**, 1167–1169.

Folke, C., Carpenter, S., Elmqvist, T. *et al.* (2002). Resilience and sustainable development: building adaptive capacity in a world of transformations. *Ambio*, **31**(5), 437–440.

Forrester, J. W. (1971). *World Dynamics*. Cambridge: Wright-Allen Press.

Franz, W. E. (1997). The development of an international agenda for climate change: connecting science to policy. ENRP Discussion Paper E- 97–07, Kennedy School of Government, Harvard University, and International Institute for Applied Systems Analysis Interim Report IR-97–034/August 1997.

Fraser, E. D. G. (2007). Travelling in antique lands: studying past famines to understand present vulnerabilities to climate change. *Climate Change*, **83**, 495–514.

Fraser, E. D. G., Dougill, A. J., Hubacek, K. *et al.* (2011). Assessing vulnerability to climate change in dryland livelihood systems: conceptual challenges and interdisciplinary solutions. *Ecology and Society*, **16**(3), 3.

Funtowicz, S. and Ravetz, J. (1993). Science for the post-normal age. *Futures*, **25**, 739–755.

Gallopín, G. C., Funtowicz, S., O'Connor, M. and Ravetz. J. (2001). Science for the twenty-first century: from social contract to the scientific core. *International Journal of Social Science*, **168**, 219–229.

GCI – Global Commons Institute (2005). *Contraction and convergence: GCI Briefing*. London: GCI. See: www.gci.org.uk/briefings.html.

Gibbons, M., Limoges, C., Nowotny, H. *et al.* (1994). *The New Production of Knowledge: The Dynamics of Science and Research in Contemporary Societies*. London: Sage.

Giddens, A. (2009). *The Politics of Climate Change*. Cambridge: Polity Press.

Gilbert, N. and Ahrweiler, P. (2009). The epistemologies of social simulation research. In *EPOS 2006, LNAI 5466*, ed. F. Squazzoni. Berlin, Heidelberg: Springer-Verlag, pp. 12–28.

Goldman, M. and Schurman, R. A. (2000). Closing the "Great Divide": new social theory on Society and Nature. *Annual Review of Sociology*, **26**, 563–584.

Griffiths, T. and Martone, F. (2009). Seeing 'REDD'? Forests, climate change mitigation and the rights of indigenous peoples and local communities. Moreton-in-Marsh, Forest Peoples Programme. See: www.katoombagroup.org/documents/cds/redlac_2010/resources/doc_923.pdf.

Grubb, M., Köhler, J. and Anderson, D. (2002). Induced technical change in energy and environmental modeling: analytic approaches and policy implications. *Annual Reviews of Energy and Environment*, **27**, 271–308.

Gummer, J. S. and Goldsmith, Z. (2007). *Blueprint for a Green Economy: Submission to the Shadow Cabinet*. London: Quality of Life Policy Group, UK Conservative Party. See: www.qualityoflifechallenge.com.

Halpern, B. S., Walbridge, S., Selkoe, K. A. *et al.* (2008). A global map of human impact on marine ecosystems. *Science*, **319**(5865), 948–952.

Hare, F. and White, R. M. (1979). Preface. In *Proceedings of the World Climate Conference: A Conference of Experts on Climate and Mankind, World Meteorological Organization, Geneva, 12–23 Feb 1979*. Report Number WMO-No-537. Geneva: WMO.

Head, L. (2007). Cultural ecology: the problematic human and the terms of engagement. *Progress in Human Geography*, **31**, 837–846.

Heckbert, S., Baynes, T. and Reeson, A. (2010). Agent-based modeling in ecological economics. *Annals of the New York Academy of Sciences: Ecological Economics Reviews*, **1185**, 39–53.

Hedström, P. and Ylikoski, P. (2010). Causal mechanisms in the social sciences. *Annual Review of Sociology*, **36**, 49–67.

Holling, C. S. (2001). Understanding the complexity of economic, ecological, and social systems. *Ecosystems*, **4**, 390–405.

Holling, C. S. and Gunderson, L. H. (2001). Resilience and adaptive cycles. In *Panarchy: Understanding Transformations in Human and Natural Systems*, eds. L. Gunderson and C. S. Holling. Washington, DC: Island Press.

Hulme, M. (2008). Geographical work at the boundaries of climate change. *Transactions of the Institute of British Geographers, New Series* **33**, 5–11.

Hulme, M. (2010). Problems with making and governing global kinds of knowledge. *Global Environmental Change – Human and Policy Dimensions*, **20**(4SI), 558–564.

Huntington, E. (1913). Changes of climate and history. *The American Historical Review*, **18**(2), 213–232.

IHDP (2007). *IHDP Strategic Plan 2007–2015*. Bonn: IHDP. See: www.ihdp.unu.edu/article/read/strategic-plan.

Ikeme, J. (2009). Equity, environmental justice and sustainability: incomplete approaches in climate change politics. *Global Environmental Change*, **13**(3), 195–206.

IPCC (2000). *Special Report on Emissions Scenarios (SRES), a special report of Working Group III of the Intergovernmental Panel on Climate Change*. Cambridge: Cambridge University Press. Technical Summary, ch. 7. See: www.grida.no/publications/other/ipcc_sr/?src=/climate/ipcc/emission/.

IPCC (2001). *Special Report on Emissions Scenarios: a Special Report of Working Group III of the Intergovernmental Panel on Climate Change*. Cambridge: Cambridge University Press.

IPCC (2007). *Climate Change 2007: Synthesis Report*, eds. R. K. Pachauri and A. Reisinger. Geneva: IPCC.

ISSC – International Social Science Council (2010). *The World Social Science Report: Knowledge Divides*. Paris: United Nations Educational, Scientific and Cultural Organization.

Janetos, A. C., Clarke, L., Collins, B. *et al.* (2009). Science challenges and future directions: climate change integrated assessment research. Report PNNL-18417, US Department of Energy, Office of Science. See: http://science.energy.gov/~/media/ber/pdf/ia_workshop_low_res_06_25_09.pdf.

Jarvie, I. C. (1964). Explanation in social science. *British Journal for the Philosophy of Science*, **15**(57), 62–72.

Jordan, A. J. (2008). The governance of sustainable development: taking stock and looking forwards. *Environment and Planning C*, **26**, 17–33.

Kaya, Y. and Yokobori., K., eds. (1993). *Environment, Energy, and Economy: Strategies for Sustainability*. United Nations University Press: Tokyo.

Kelly, D. L. and Kolstad, C. D. (1999). Integrated assessment models for climate change control. In *International Yearbook of Environmental and Resource Economics 1999/2000: A Survey of Current Issues*, eds. H. Folmer and T. Tietenberg. Cheltenham: Edward Elgar, pp. 171–196.

Kuhn, W., Wiegandt, E. and Luterbacher, U. (1992). *Pathways of Understanding, the Interactions of Humanity and Global Environmental Change*. Saginaw: CIESIN.

Kumar, P., ed. (2010). *The Economics of Ecosystems and Biodiversity: the Ecological and Economic Foundations*. London: Earthscan.

LaFleur, V., Purvis, N. and Jones, A. (2008). Double jeopardy: what the climate crisis means for the poor. Brookings Blum Roundtable on Development, 5. The Brookings Institution. See: www.brookings.edu/reports/2009/02_climate_change_poverty.aspx.

Leach, A. J. (2009). The welfare implications of climate change policy. *Journal of Environmental Economics and Management*, **57**(2), 151–165.

Liverman, D. M. and Roman-Cuesta, R. M. (2008). Human interactions with the Earth system: people and pixels revisited. *Earth Surface Processes and Landforms*, **33**, 1458–1471.

Lutz, W., Sanderson, W. and Scherbov, S. (2008). The coming acceleration of global population ageing. *Nature*, **451**, 716–719.

Mastrandrea, M. D. and Schneider, S. H. (2004). Probabilistic integrated assessment of 'dangerous' climate change. *Science*, **304**(5670), 571–575.

Max-Neef, M. A. (2005). Foundations of transdisciplinarity. *Ecological Economics*, **53**, 5–16.

McAfee, K. (1999). Selling nature to save it? Biodiversity and green developmentalism. *Environment and Planning A*, **17**, 133–154.

Meadows, D., Richardson, J. and Bruckmann, G. (1982). *Groping in the Dark: the First Decade of Global Modeling*. Chichester: John Wiley and Sons.

Meadows, D. H., Meadows, D. L., Randers, J. and Behrens III, W. W. (1972). *The Limits to Growth*. New York, NY: Universe Books.

Meier, G. M. and Stiglitz, J. E. (2001). *Frontiers of Development Economics: the Future in Perspective*. Oxford/New York, NY: Oxford University Press and The World Bank.

Millennium Ecosystem Assessment – MA (2005). *Ecosystems and Human Well-Being: Current State and Trends*. Washington, DC: Island Press.

Moser, S. C. (2008). Resilience in the face of global environmental change. CARRI Research Paper No.2, Oak Ridge National Laboratory, Oak Ridge, TN. See: www.resilientus.org/publications/reports.html.

Moss, R. H., Edmonds, J. A., Hibbard, K. A. *et al.* (2010). The next generation of scenarios for climate change research and assessment. *Nature*, **463**, 747–756.

Naidoo, R., Balmford, A., Costanza, R. *et al.* (2008). Global mapping of ecosystem services and conservation priorities. *Proceedings of the National Academy of Sciences USA*, **105**(28), 9495–9500.

NASA Advisory Council (1988). *Earth System Sciences: A Closer View*. Report of the NASA Earth System Sciences Committee (F. Bretherton, Chair). Washington DC: National Aeronautics and Space Agency.

National Research Council – NRC (1999). *Human Dimensions of Global Change: Research Pathways for the Next Decade*. Washington DC: National Academies Press.

Newell, B., Crumley, C.L., Hassan, N., *et al.* (2005). A conceptual template for human–environment research. *Global Environmental Change*, **15**, 299–307.

Nilsson, M. and Eckerberg, K., eds. (2007). *Environmental Policy Integration in Practice: Shaping Institutions for Learning*. London: Earthscan.

Nordhaus, W. D. (1994). *Managing the Global Commons: the Economics of Climate Change*. Boston, MA: MIT Press.

Norgaard, R. (2010). Ecosystem services: from eye-opening metaphor to complexity blinder. *Ecological Economics*, **69**, 1219–1227.

Nowotny, H., Scott, P. and Gibbons, M. (2003). Introduction – 'Mode 2' revisited: the new production of knowledge. *Minerva*, **41**, 179–194.

O'Brien, K. and Leichenko, R. (2000). Double exposure: assessing the impacts of climate change within the context of globalization. *Global Environmental Change*, **10**, 221–232.

OECD – Organization for Economic Cooperation and Development (1993). *OECD Core Set of Indicators for Environmental Performance Reviews*. OECD Environment Monograph 83. Paris: OECD. See: www.oecd.org.

Olsson, P., Folke, C. and Berkes, F. (2004). Adaptive co-management for building resilience in social-ecological systems. *Environmental Management*, **34**(1), 75–90.

Outhwaite, W. (1998). Realism and social science. In *Critical Realism: Essential Readings*, eds. M. Archer, R. Bhaskar, A. Collier, T. Lawson and A. Norrie. London: Routledge, pp. 282–296.

Paavola, J. (2007). Institutions and environmental governance: a reconceptualization. *Ecological Economics*, **63**, 93–103.

Paavola, J., Gouldson, A. and Kluvankova-Oravska, T. (2009). Interplay of actors, scales, frameworks and regimes in the governance of biodiversity. *Environmental Policy and Governance*, **19**(3), 148–158.

Panayotou, T. (2003). Economic growth and the environment. In *Economic Survey of Europe*, No. 2, ch. 2. United Nations Economic Commission for Europe. See: www.unece.org/ead/pub/surv_032.htm.

Park, A. (2011). Beware paradigm creep and buzzword mutation. *Forestry Chronicle*, **87**(3), 337–344.

Pearce, D. W. and Turner, R. K. (1990). *Economics of Natural Resources and the Environment*. Baltimore, MD: Johns Hopkins University Press.

Pearce, D. W., Markandya, A. and Barbier, E. B. (1989). *Blueprint for a Green Economy*. London: Earthscan.

Perrings, C. (2010). The economics of biodiversity: the evolving agenda. *Environment and Development Economics*, **15**, 721–746.

Perrings, C., Naeem, S., Ahrestani, F. *et al.* (2010). Ecosystem services for 2020. *Science*, **330**, 323–324.

Peterson, G., De Leo, G. A., Hellmann, J. J. *et al.* (1997). Uncertainty, climate change, and adaptive management. *Conservation Ecology*, **1**(2), 4.

Pielke, R., Jr (2004). *The Honest Broker: Making Sense of Science in Policy and Politics*. Cambridge: Cambridge University Press.

Pitcher, H. M. (2009). Measuring income and projecting energy use. *Climatic Change*, **97**(1–2), 49–58.

Pohl, C. (2010). From transdisciplinarity to transdisciplinary research. *Transdisciplinary Journal of Engineering and Science*, **1**(1), 74–83.

Ponting, C. (2007). *New Green History of the World: The Environment and the Collapse of Great Civilizations*. London: Vintage.

Radcliffe, S. A., Watson, E. E., Simmons, I., Fernandez-Armesto, F. and Sluyter, A. (2010). Environmentalist thinking and/in geography. *Progress in Human Geography*, **34**(1), 98–116.

Rademaker, O. (1980). On the methodology of global modeling. In *Input–Output approaches in global modeling. Proceedings of the Fifth IIASA Symposium on Global Modeling, September 26–29, 1977*, ed. G. Bruckmann. Oxford: Pergamon Press, pp. 499–508.

Rademaker, O. (1982). On modelling ill-known social phenomena. In *Groping in the Dark: the First decade of Global Modeling*, eds. D. Meadows, J. Richardson and G. Bruckmann. Chichester: John Wiley and Sons, pp. 206–208.

Rainbird, P. (2002). A message for our future? The Rapa Nui (Easter Island) ecodisaster and Pacific island environments. *World Archaeology*, **33**(3), 436–451.

Ramankutty, N. and Foley, J. A. (1999). Estimating historical changes in global land cover: croplands from 1700 to 1992. *Global Biogeochemical Cycles*, **13**, 997–1027.

Read, D. (2010). Agent-based and multi-agent simulations: coming of age or in search of an identity. *Computational and Mathematical Organizational Theory*, **16**, 329–347.

Reid, W. V., Chen, D., Goldfarb, L. *et al.* (2010). Earth system science for global sustainability: Grand Challenges. *Science*, **330**, 916–917.

Saqalli, M., Bielders, C. L., Gerard, B. and Defourney, P. (2010). Simulating rural environmentally and socio-economically constrained multi-activity and multi-decision societies in a low-data context: a challenge through empirical agent-based modelling. *Journal of Artificial Societies and Social Simulation*, **13**(2), art. 1.

Sauer, C. O. (1925). The morphology of landscape. *University of California Publications in Geography*, **2**, 19–53.

Sawyer, R. (2005). *Social Emergence: Societies as Complex Systems*. Cambridge: Cambridge University Press.

Schellnhuber, H.-J. and Kropp, J. P. (1998). Geocybernetics: controlling a complex dynamical system under uncertainty. *Naturwissenschaften*, **85**(9), 411–425.

Schneider, S. H., Rosencranz, A., Mastrandrea, M. D. and Kuntz-Duriseti, K. (2010). *Climate Change Science and Policy*. Washington, DC: Island Press.

Sen, A. (1973). Welfare economics, utilitarianism and equity. In *On Economic Inequality*. Oxford University Press, Oxford. (1997 expanded edn.), ch. 1.

Shantz, J. (2003). Scarcity and the emergence of fundamentalist ecology. *Critique of Anthropology*, **23**(2), 144–154.

Shibutani, T. (1988). Social psychology and social control. In *Society and Personality: An Interactionist Approach to Social Psychology*. Piscataway, NJ: Transaction Publishers, ch. 18.

Simon, H. A. (1957). *Models of Man: Social and Rational*. New York: John Wiley and Sons, Inc.

Smil, V. (2001). *Feeding the World*. Cambridge, MA: MIT Press.

Spash, C. L. and Vatn, A. (2006). Transferring environmental value estimates: issues and alternatives. *Ecological Economics*, **60**, 379–388.

Steffen, W., Sanderson, R. A., Tyson, P. D. *et al.* (2004). *Global Change and the Earth System: A Planet Under Pressure*. Berlin: Springer-Verlag.

Stern, N. (2007). *The Economics of Climate Change: The Stern Review*. London: UK Government Cabinet Office/ HM Treasury.

Strathern, M. (2006). A community of critics? Thoughts on new knowledge. *Journal of the Royal Anthropological Institute (New Series)*, **12**, 191–209.

Steward, J. H. (1955). *Theory of Culture Change: The Methodology of Multilinear Evolution*. Urbana, IL: University of Illinois Press.

Stiglitz, J. E., Sen, A. and Fitoussi, J.-P. (2009). Report by the Commission on the Measurement of Economic Performance and Social Progress. See: www.stiglitz-sen-fitoussi.fr.

Swart, R., Biesbroek, R., Binnerup, S. *et al.* (2009). *Europe Adapts to Climate Change: Comparing National Adaptation Strategies*. PEER Report No. 1. Helsinki: Partnership for European Environmental Research.

Tavoni, M. and Tol, R. S. J. (2010). Counting only the hits? The risk of underestimating the costs of a stringent climate policy. *Climatic Change*, **100**(3–4), 769–778.

ten Brink, P., Berghöfer, A., Schröter-Schlaack, C. *et al.* (2009). The Economics of Ecosystems and Biodiversity in National and International Policy Making: Summary. See: www.teebweb.org/.

Thomas, V. (2001). Revisiting the challenge of development. In *Frontiers of Development Economics: The Future in Perspective*, eds. G. M. Meier and J. E. Stiglitz. Oxford/ New York, NY: Oxford University Press and The World Bank, pp. 149–182.

Thompson Klein, J., Grossenbacher-Mansuy, W., Häberli, R. *et al.* (2001). *Transdisciplinarity: Joint Problem Solving Among Science, Technology, and Society. An Effective Way for Managing Complexity*. Basel: Birkhäuser.

Toffler, A. (1984). Foreword. In *Order out of Chaos: Man's New Dialogue with Nature*, eds. I. Prigogine and I. Stengers. London: Flamingo, pp. xi–xxvi.

Tol, R. S. J. (2006). *Integrated Assessment Modelling*. Working Paper FNU-102. Research Unit Sustainability and Global Change, Hamburg University, Hamburg. See: www.fnu.zmaw.de/fileadmin/fnu-files/publication/ working-papers/efieaiamwp.pdf.

Tol, R. S. J., Hertel, T. W. and Rose, S. K. (2009). *Economic Analysis of Land Use in Global Climate Change Policy*. London: Routledge.

Turner, B. L. II, Kasperson, R. E., Matson, P. A. *et al.* (2003). A framework for vulnerability analysis in sustainability science. *Proceedings of the National Academy of Sciences USA*, **100**(14), 8074–8079.

Turner, K. (2010). *A Pluralistic Approach to Ecosystem Services Evaluation*. CSERGE Working Paper EDM 10-07. Norwich: The Centre for Social and Economic Research on the Global Environment.

UNDP (2008). *The Human Development Report 2007/8. Fighting Climate Change: Human Solidarity in a Divided World*. New York: UNDP.

UNDP (2010). *The Human Development Report 2010. The Real Wealth of Nations: Pathways to Human Development*. New York: UNDP.

UN Economic Commission for Europe (2009). Links between air pollution and biodiversity. Informal Document No. 21, CLRTAP Executive Body, 27th session, 14–18 December 2009. See: http://live.unece.org/env/lrtap/executivebody/welcome.27.html.

UNEP (2011). Towards a Green Economy: Pathways to Sustainable Development and Poverty Eradication. See: www.unep.org/greeneconomy.

United Nations (2011). World Population Prospects – The 2010 Revision. See: http://esa.un.org/unpd/wpp/index.htm.

UN World Commission on Environment and Development (1987). *Our Common Future*. (The Brundtland Report). Oxford: Oxford University Press.

US Census Bureau, Population Division (2011). International Data Base – World population growth rates: 1950–2050. See: www.census.gov/ipc/www/idb/ worldgrgraph.php.

van der Leeuw, S., Costanza, R., Aulenbach, S. *et al.* (2011). Toward an integrated history to guide the future. *Ecology and Society*, **16**(4), 2.

van Ierland, E., Brink, C., Hordijk, L. and Kroeze, C. (2002). Environmental economics for environmental protection. Short Communication, The International Conference on Environmental Concerns and Emerging Abatement Technologies 2001: Collection of Short Communications. *The Scientific World*, **2**, 1254–1266.

Vayda, A. P. and Walters, B. B. (1999). Against political ecology. *Human Ecology*, **21**(1), 167–179.

Visser, H., Folkert, R. J. M., Hoekstra, J. and De Wolff, J. J. (2000). Identifying key sources of uncertainty in climate change projections. *Climatic Change*, **45**(3–4), 421–457.

Vogel, M. P. (2009). Understanding emergent social phenomena comparatively: the need for computational simulation. *European Journal of Social Sciences*, **7**(4), 84–92.

von Bertalanffy, K. L. (1968). *General System Theory: Foundations, Development, Applications*. New York, NY: George Braziller.

Warren, R. (2011). The role of interactions in a world implementing adaptation and mitigation solutions to climate change. *Philosophical Transactions of the Royal Society A*, **369**(1934), 217–241.

Weber, M. (1988) *Gesammelte Aufsätze zur Wissenschaftslehre* (Collected essays on the methodology of the social sciences). Tübingen: Mohr Siebeck/UTB für Wissenschaft.

Wilbanks, T. J. and Kates, R. W. (1999). Global change in local places: how scale matters. *Climatic Change*, **43**(3), 601–628.

Wilson, E. O. (1998). *Consilience: The Unity of Knowledge*. London: Abacus.

Wittrock, B. (2010). Shifting involvements: rethinking the social, the human and the natural. Background Paper for 2010 World Social Science Report, ISSC/2010/WS/8. International Social Sciences Council.

World Commission on Dams (2000). *Dams and Development: A New Framework for Decision-Making*. London: Earthscan.

Young, O., Berkhout, F., Gallopin, G. *et al.* (2006). The globalization of socio-ecological systems: an agenda for scientific research. *Global Environmental Change*, **16**(3), 304–316.

Yusuf, S. and Stiglitz, J. E. (2001). Development issues: settled and open. In *Frontiers of Development Economics: The Future in Perspective*, eds. G. M. Meier and J. E. Stiglitz. Oxford/New York, NY: Oxford University Press and The World Bank, pp. 227–268.

Zalasiewicz, J., Williams, M., Steffen, W. and Crutzen, P. (2010). The new world of the Anthropocene. *Environmental Science and Technology*, **44**(7), 2228–2231.

Ziervogel, G., Bithell, M., Washington, R. and Downing, T. E. (2005). Agent-based social simulation: a method for assessing the impact of seasonal climate forecast applications among smallholder farmers. *Agricultural Systems*, **81**, 1–26.

Chapter

2

Fundamentals of climate change science

I. Colin Prentice, Peter G. Baines, Marko Scholze and Martin J. Wooster

This chapter provides a high-level summary of the state of knowledge regarding observations, processes and models of climate, terrestrial ecosystems and the global carbon cycle. We focus strongly on observations (at various timescales, including palaeo timescales as appropriate), and what can be learned from their interpretation in the light of the established principles of climate science and terrestrial ecosystem science. The field is very broad and therefore we have had to be highly selective. We discuss aspects pertinent to understanding recent and contemporary changes in climate and the global carbon cycle, with emphasis on the terrestrial component.

2.1 Observing and studying climate

2.1.1 Background and history of climate science

Like the weather, everyone has an interest in climate and knows something about it. Climate is generally understood as 'average weather'. By definition, climate cannot change from year to year; but it can (and does) change over decades and centuries.

Until the 1970s, the study of climate was largely descriptive. The data were concentrated in certain regions, and often anecdotal. Nonetheless, as Lamb (1982) and others described, these data already showed the existence of a great deal of variability in climate on many timescales, and that this variability has had a pervasive impact on human societies.

Climate also has a dominant effect on ecosystems. The patterns of terrestrial biomes, from dense tropical forests to high-latitude and mountain tundra and deserts, reflect spatial patterns of average temperature and rainfall and show that climate has had a profound role in shaping the ecology and evolution of land plants. Relationships between vegetation and climate formed the basis for Köppen's (1918) classification of world climates, which allowed climate to be inferred from vegetation at a time when direct climate observations were sparse.

Long-term changes in climate have been recognized since the beginnings of modern geology. The alternation of ice ages and the intervening warmer periods (interglacials) was first established by nineteenth-century fieldwork in the Alps (Krüger, 2008). Much more detailed and temporally resolved knowledge about recent climate changes, and their consequences for terrestrial ecosystems, began to accrue from the early twentieth century thanks to the application of pollen analysis (von Post, 1918) to continuous sediments. The invention of radiocarbon dating (Libby, 1952) provided an absolute timescale for these changes. Quaternary scientists have developed an extensive toolkit allowing the analysis of changes in many different aspects of the terrestrial, marine and atmospheric environments (Bradley, 1999). The temporal resolution achieved is typically decadal to centennial, but can be annual in some natural archives such as corals, annually laminated lake sediments and speleothems.

Climate, ecosystem and palaeoenvironmental science all have a long history as empirical subjects. But two developments of the 1970s radically changed the study of climate, and began to convert it into a quantitative and predictive science. The first was the advent of global satellite observations; the second was the development of numerical models that aimed to describe global climate and its variability. Parallel breakthroughs transformed

Understanding the Earth System: Global Change Science for Application, eds. Sarah E. Cornell, I. Colin Prentice, Joanna I. House and Catherine J. Downy. Published by Cambridge University Press

the science of terrestrial ecosystems. Remotely sensed observations of land-surface properties, and global-scale models of ecosystem dynamics, appeared first in the 1970s and 1990s, respectively, and now form established research fields. Palaeoclimatology and palaeoecology progressed from description to analysis with the application of climate models to past climates, providing a stimulus for the global synthesis of terrestrial and marine palaeodata (CLIMAP, 1981; Kutzbach and Guetter, 1986; COHMAP Project Members, 1988; Wright *et al.*, 1993). Advances in ice-drilling technology and mass spectrometry allowed the fundamental advance represented by the 400,000-year long Vostok core from Antarctica (Petit *et al.*, 1997). Much research since then has centred on the unique palaeoenvironmental records preserved in ice.

Climate models evolved from weather forecasting, a technology developed initially for military applications, which soon found a wider use. Through the 1960s and 1970s, weather forecasting evolved to become reliant on numerical modelling. The Global Atmospheric Research Programme (GARP, 1969) was originally conceived with weather forecasting as the main beneficiary. It had become apparent that poor coverage and quality of observations (used as the initial conditions for forecasting) severely limited the accuracy of forecasts. Global data coverage is needed to forecast the weather as long as a week ahead. Although this was computationally impossible at the time, the idea encouraged the development of models that could run for more than a few days – a non-trivial objective, as it required the imposition of constraints to prevent the model from drifting into unrealistic states. At first, climate models only included the atmospheric circulation; sea-surface conditions were prescribed. The first climate model to include ocean circulation was developed by Manabe and Bryan (1969). Increasingly realistic model results began to appear during the 1970s, and the development of both climate models and global climate observations has continued apace since then.

The available climate data covering the past 30 years are vastly more detailed than those for earlier times. The global meteorological network now provides information on the state and movement of the atmosphere (pressure, sea-surface and air temperature, velocity), and this together with advanced weather-forecasting models enables a 'reanalysis' of the atmosphere, providing three-dimensional fields of meteorological variables on a sub-daily basis (Kalnay *et al.*, 1996, Uppala *et al.*, 2005). In recent years, the deployment of Argo robotic floats has also provided an ongoing description of the motion and properties of the upper 2 km of the global ocean (www.argo.ucsd.edu), allowing a major improvement in the quantitative description of the ocean circulation (Wunsch *et al.*, 2009). The combination of advanced modelling with the longer historical time series of available observations for sea-surface temperature (SST), sea-ice and sea-level pressure has made it possible to develop a reanalysis of monthly climates from the late nineteenth century through to the present (Compo *et al.*, 2011).

2.1.2 Earth system observations in the modern era

Interest in climate and its interactions with biogeochemical cycles and human activities has been further stimulated since the 1980s by the emergence of anthropogenic climate change as an issue of global concern. The idea that fossil-fuel burning would increase atmospheric concentrations of carbon dioxide (CO_2), and thereby raise temperatures at the Earth's surface, was already proposed in the late nineteenth century by Arrhenius (1896). But an instrument precise enough to detect such an increase was not built until the 1950s. The new measurements (Keeling, 1958) confirmed that CO_2 concentration was indeed rising, and revealed a seasonal cycle, which was identified as the 'breathing' of the continental biosphere (Keeling *et al.*, 1989). Over several decades, observations of atmospheric constituents – CO_2 and other greenhouse gases, including methane (CH_4) and nitrous oxide (N_2O), and tracers of the carbon cycle such as the stable isotope $^{13}CO_2$ – have been made at a small but increasing number of sites worldwide. Measurements made continuously *in situ* at some locations are supplemented by measurements on flasks of air, periodically filled at a number of remote locations and sent to laboratories for analysis. Modelling of the large-scale exchanges of CO_2 between terrestrial ecosystems and the atmosphere has been encouraged by the desire to establish the role of the terrestrial biosphere in taking up CO_2 in the fossil-fuel era. Models can benefit from testing against highly precise measurements of CO_2 concentration at different locations (Prentice *et al.*, 2000), but such opportunities have been underexploited and, as a result, the uncertainties in terrestrial carbon-cycle models are still large (Friedlingstein *et al.*, 2006; Denman *et al.*, 2007; Sitch *et al.*, 2008).

Direct measurement of the concentrations of trace species near the Earth's surface remains the most precise way to quantify changes in the composition of the atmosphere. But, increasingly, satellite observations have provided complementary information on total atmospheric column concentrations of a range of atmospheric constituents including CH_4, ozone (O_3), nitrogen dioxide (NO_2), formaldehyde (HCHO) and other compounds. Satellite observations have also been crucial in observing the biosphere. A key breakthrough was due to Tucker (1979), who realized that spectral reflectances from the Advanced Very High Resolution Radiometer (AVHRR) onboard a weather satellite could be used to measure the 'greenness' of the land surface. The normalized difference vegetation index (NDVI), based on the ratio of reflectances in the red and near-infrared bands, provides a record of variations in greenness that extends back to the early 1980s.

The full range of variables now measured from space includes sea-surface temperature and salinity, soil moisture content, snow cover, vertical temperature profiles in the atmosphere, lightning strikes, fires, sea-surface chlorophyll content, and a great deal more (Chuvieco, 2008). It is remarkable how much can be inferred entirely from observations of the electromagnetic spectrum.

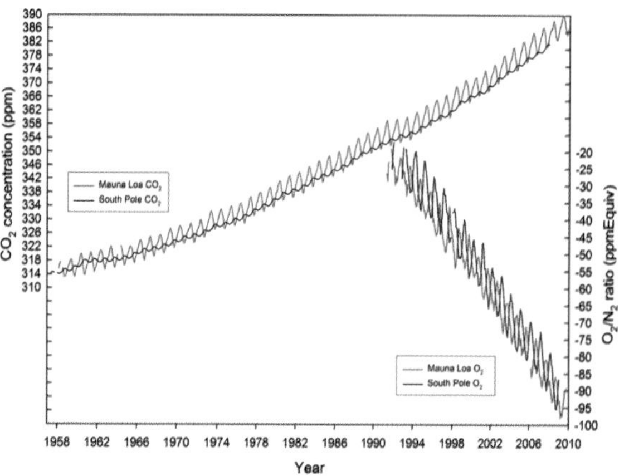

Figure 2.1 Time series of measured CO_2 and O_2 in the atmosphere at Mauna Loa, Hawaii and the South Pole. Figure kindly provided by A.C. Manning.

2.1.3 Observations of past atmospheric composition

The observational record of many atmospheric constituents has been extended backwards in time by analyzing tiny air bubbles trapped in polar firn (consolidating snow) and ice. Antarctica's remoteness from natural and man-made sources of potential contamination has made it especially suitable for this purpose, and Antarctica has yielded by far the longest ice-core record, the EPICA Dome C core, extending back to 820,000 years ago. Ice cores from Greenland have also been extremely valuable in providing higher-resolution records of some atmospheric constituents back to 120,000 years ago.

Ice-core records have established that the contemporary CO_2 rise – and the continuing rise in CH_4 and N_2O concentrations – began in the late nineteenth century, at the time of the Industrial Revolution. The CO_2 rise has accelerated, albeit with some ups and downs in the rate of increase, towards the present (MacFarling Meure *et al.*, 2006). Changes in the atmospheric O_2:N_2 ratio (Figure 2.1) have been measurable to significant precision only since the 1990s (Keeling *et al.*, 1996). These confirm that the CO_2 increase has involved combustion and thus the consumption of O_2. Measured changes in the carbon isotopes of atmospheric CO_2 show a progressive shift towards the stable-isotope (^{13}C) signature of terrestrial biospheric carbon (including fossil fuel, which was originally formed from biomass). Analyses of tree rings, formed prior to the atmospheric nuclear-weapons tests that confound the signal, have revealed a progressive depletion of the content of the radioactive isotope (^{14}C). This isotope is generated in the atmosphere by cosmic rays and taken up into the biosphere; but it is absent from fossil fuel due to radioactive decay (Suess, 1955; Stuiver and Quay, 1981; Keeling *et al.*, 1989; Levin and Kromer, 1997). These various findings unequivocally link the CO_2 rise during the industrial era to human activities.

The exquisite sensitivity obtainable from mass spectrometers makes it possible to measure arcane quantities, such as the amounts of the two stable isotopes of carbon monoxide (CO), which have been used to provide a biomass burning record of the past 650 years (Wang *et al.*, 2010). Ice-core records have also provided a much longer extension in time, with concentrations of CO_2, CH_4 and N_2O now recorded over the full length of the EPICA ice-core record spanning eight glacial–interglacial cycles (Lüthi *et al.*, 2008; Loulergue *et al.*, 2008; Schilt *et al.*, 2010).

2.1.4 Reconstructions of past climate

An extensive body of palaeoclimatic information is based on indirect records, i.e. records of physical, chemical or biological quantities that respond to changes in climate. For the past millennium, tree-ring series can be matched up to allow dating with annual precision and combined with other annual recorders such as corals, annually laminated sediments, and speleothems to yield time series of past climates (with temperature foremost among the controls in temperate regions, and precipitation often foremost in subtropical and tropical regions). Complex statistical manipulations can be involved in order to remove artefacts, such as growth-related trends in tree-ring widths. The lack of a standard data-processing methodology has been a significant limitation of this field, potentially leading to unproductive controversy, as in some discussions of the 'hockey-stick' curve for global temperatures first published by Mann *et al.* (1999). However, many independent reconstructions of the climate of the past millennium have now been made, using different combinations of data and statistical methods. As a result, Jansen *et al.* (2007) were able to publish a robust consensus picture of northern-hemisphere average temperature variations during the past 850 years. Most of the more recent reconstructions show greater variability on a century timescale than the original 'hockey stick', but all agree on the exceptional degree of warming in recent decades compared with the past millennium.

Extension of climate records further back in time, albeit with reduced time resolution and dating precision, has been possible thanks to the refinement and synthesis of long-standing methods involving the analysis of biotic assemblages and geochemical properties of terrestrial and marine sediments. These include pollen and macroscopic charcoal analysis, and foraminiferal assemblages and temperature-dependent biomarkers from marine sediments. These types of data yield information directly related to ecosystems of the past, and can also be used to derive quantitative reconstructions of past climates (CLIMAP, 1981; MARGO project members, 2009; Bartlein *et al.*, 2011). Water isotope signals in ice cores (variations in the ^{18}O and ^{2}H content of ice) provide signals of temperature variation in the polar regions.

2.2 Fundamentals of climatology

2.2.1 Energy balance and radiative forcing

Virtually all life and motion on Earth depends on radiation from the Sun. The peak wavelength is near 0.5 μm, in the visible range, and the distribution of wavelengths (spectrum) of this radiation is approximately that of a 'black body' with a temperature of 5800 K, the mean surface temperature of the Sun. Total incoming solar radiation (insolation, or shortwave radiation) has a mean strength of ~ 1366 W m^{-2}. This quantity, known as the 'solar constant', actually varies by about ± 0.5 W m^{-2} on a somewhat irregular cycle with a period near 11 years (Gray *et al.*, 2010). Solar output peaks with sunspot numbers, due to internal dynamical processes within the Sun. This fractional change in energy flux is very small, but it is not uniform across the spectrum and is concentrated in the ultraviolet (high-energy) range. As a result, it has a larger influence on the Earth's climate than one might expect. Solar variations on longer timescales (decades to centuries) have been inferred from historical records of the numbers of sunspots per cycle. Solar variations are now thought to have been one of the principal factors responsible for natural centennial-to-millennial-scale variability in temperature. In particular, the small difference of about 0.6 K between the Mediaeval Warm Period (*c.* 950–1250) and the Little Ice Age (*c.* 1600–1850) has been explained mainly by variations in solar output (Crowley, 2000; Haigh, 2003).

The radiation balance of the atmosphere, which determines the temperature near the Earth's surface, is summarized in Figure 2.2. The flux of radiant energy from the Sun is partly reflected and absorbed by the atmosphere, and the remainder is partly reflected and absorbed at the surface. The reflected component reduces the effective solar radiation and plays no part in the dynamics of climate. For the absorbed component, an approximately equal flux of energy is re-radiated to space from the combined system of the Earth's surface and atmosphere, but at a correspondingly lower temperature. The radiation balance at the top of the atmosphere is

$$\pi r^2 F_{SW}\downarrow(1-\alpha) \approx 4\pi r^2 F_{LW}\uparrow = 4\pi r^2 \sigma T_E^4 \qquad (2.1)$$

where r is the radius of the Earth, $F_{SW}\downarrow$ is the shortwave radiation (W m^{-2}) and α is the mean albedo of the Earth (~ 0.3): i.e. the fraction of the shortwave radiation that is reflected back to space. Over a long enough period this shortwave radiation flux must be equalled by the mean outgoing longwave radiation flux $F_{LW}\uparrow$ from the Earth. The Stefan–Boltzmann law tells us that this flux must be equal to σT_E^4 where T_E is the radiative equilibrium temperature (the temperature of the Earth as seen from space) and σ is the Stefan–Boltzmann constant

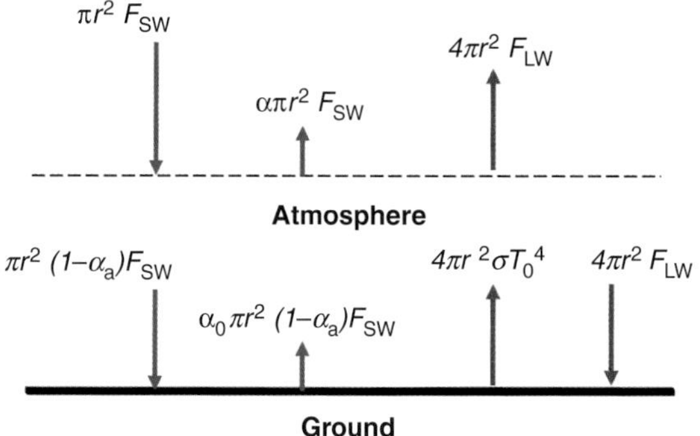

Figure 2.2 Radiation balance of the atmosphere. Blue arrows denote shortwave fluxes, red arrows longwave fluxes. α denotes the planetary albedo, α_0 and α_a the ground and atmosphere albedos, respectively. T_0 is surface temperature.

(5.67×10^{-8} W m^{-2} K^{-4}). The peak wavelength of the longwave radiation is at 11.3 μm, in the infrared range, and its spectrum is approximately that of a black body with $T_E \approx 255$ K.

Inside the atmosphere, and at the Earth's surface, the mean temperature may differ from T_E. Solar radiation is reflected in the atmosphere by liquid water drops (i.e. clouds or rain), and by aerosols, which are suspensions of liquid droplets or solid particles in the air. Two kinds of process produce aerosols: *direct injection* e.g. of sea-salt particles from evaporating sea-spray, and dust by deflation of unvegetated soils; and *chemical reactions* of gaseous materials within the atmosphere, such as the transformation of volcanically or industrially emitted sulfur dioxide into sulfate. The albedo in Equation (2.1) is composed of reflection in the atmosphere (water drops and reflective aerosols), plus reflection from the ground, which is largest from surfaces covered by ice or snow. Aerosols are generally more prominent over land than the ocean, and like water drops they may reflect longwave radiation downward as well as shortwave radiation upward. Some aerosols, such as soot (black carbon) from fires, also absorb shortwave radiation. However, whereas the shortwave solar radiation can pass through the atmosphere with relatively little attenuation, the longwave radiation is rapidly attenuated by the tri-atomic gases H_2O (water vapour), CO_2, O_3 (ozone) and nitrous oxide (N_2O), and also by CH_4 (methane). These molecules possess rotational and vibrational states that allow absorption and emission of infrared radiation. The upward longwave radiation from the ground is to a large extent absorbed by the atmosphere, and partly re-radiated back down again. At any particular level the net downward radiative flux may be expressed in terms of upward and downward short- and longwave fluxes, so at the surface:

$$F_0 = F_{SW}\downarrow (1-\alpha_0) - \sigma T_0^4 + F_{LW}\downarrow \quad (2.2)$$

where the subscript ($_0$) denotes quantities at the surface. Compared with equation (2.1), this equation has the extra term $F_{LW}\downarrow$ for the downward longwave flux from the atmosphere. This term increases the surface temperature T_0 to a value greater than T_E. This process is known as the greenhouse effect and, as a result of it, the climate near the ground is about 30 K warmer than it otherwise would be. In other words, the greenhouse effect makes the world habitable. Without the greenhouse effect the global mean temperature would be about −18 °C, and life as we know it would be impossible.

The magnitude of the greenhouse effect depends on the composition of the atmosphere, and can be quantified in terms of radiative forcing. Radiative forcing is an increase in the downward longwave flux that results from a given change in atmospheric composition, relative to a defined baseline. In discussions of contemporary climate change, the baseline is the pre-industrial atmosphere. Each greenhouse gas has a particular radiative efficiency (in W m^{-2} ppm^{-1}) so the radiative forcing due to changes in its concentration can be calculated (e.g. Myhre *et al.*, 1998). The radiative efficiencies of CO_2, CH_4 and N_2O (Forster *et al.*, 2007) are 0.014, 0.37 and 3.03 W m^{-2} ppm^{-1}. Since 1870, the atmospheric concentrations of these gases have risen by 29%, 150% and 21%, respectively (MacFarling Meure *et al.*, 2006). Taking into account additional positive radiative forcing from an estimated increase in tropospheric O_3 concentration, and negative radiative forcing due to an increase in reflective aerosols, and a number of additional, minor terms, the best estimate of current total radiative forcing relative to pre-industrial time is +1.6 W m^{-2} (Forster *et al.*, 2007). This radiative forcing inevitably produces an *enhanced greenhouse effect*, commonly known as global warming.

2.2.2 Orbital variations

The tilt of the Earth's axis (currently 23.45°), the eccentricity of the Earth's orbit around the Sun, and the time of year when the Earth is closest to the Sun (currently

3 January) vary predictably on timescales of millennia – with profound implications for climate (Berger, 1988). Milankovitch (1941) first set out the 'astronomical theory' for the alternation of ice ages and interglacials that has characterized the past few million years. He proposed that the July insolation at 65° N drives the extent of ice sheets in high latitudes of the northern hemisphere. The involvement of orbital variations in climate change is now universally accepted. The detailed mechanisms that govern the waxing and waning of large continental ice sheets are still debated, however. These mechanisms, discussed further in Chapters 3 and 4, certainly involve natural changes in the greenhouse effect due to changes in the atmospheric concentration of greenhouse gases (most importantly CO_2, but also CH_4 and N_2O). These changes arise as a feedback from climate changes set in train by the variations of the Earth's orbit.

Many other consequences of the orbital variations are well understood. For example, the end of the last glacial period, 11,700 years ago, occurred at a time of high tilt (favouring high summertime and total annual insolation in high latitudes) when the Earth was closest to the Sun in July (favouring high summer insolation in the northern hemisphere). This anomaly remained strong during several millennia, and was responsible for warmer-than-present conditions in northern high latitudes during the first millennia of the present (Holocene) interglacial (Bartlein *et al.*, 2011). During the first part of the last interglacial (about 130,000–125,000 years ago) the insolation anomaly was similar but even stronger, due to higher orbital eccentricity. As a result, high northern latitudes were even warmer than in the Holocene; Arctic sea ice disappeared, and part of the Greenland ice sheet melted away (Jansen *et al.*, 2007). Insolation anomalies also had indirect consequences for climate, through their effects on the atmospheric circulation, which can now be analyzed with the help of climate models.

2.2.3 Drivers of the atmospheric circulation

The atmospheric convection resulting from radiative heating of the ground can cause the air to rise to heights of up to 16 km in the tropics, where insolation is most intense. Air that can be in contact with the ground as a result of convective motion of various sorts is known as the troposphere, which reaches up to a level where the atmospheric pressure is about a fifth of what it is at sea level. Air above this level is the stratosphere, which has its own circulation and in general does not mix with the troposphere (apart from occasional mid-latitude features known as tropopause folds).

Here we confine attention to the atmospheric circulation in the troposphere, which is ultimately driven by the spatial distribution of insolation and thus by solar geometry. Over the course of a day, at any time of year, the tropical regions receive much more solar radiation than regions at high latitudes. However, because of the tilt of the Earth's axis, the time of year makes a big difference to the relative amount of radiation received by the mid and high latitudes of both hemispheres. The zonal means of the incident and absorbed radiation for the two extreme seasons DJF (December–February) and JJA (June–August) are shown in Figure 2.3. If the Earth had no atmosphere and ocean, radiation would determine the local temperature at each point, and the winter hemisphere would be very much colder than the summer. Instead, the atmospheric and oceanic circulations bodily transport heat from the summer hemisphere to the winter hemisphere. A more uniform

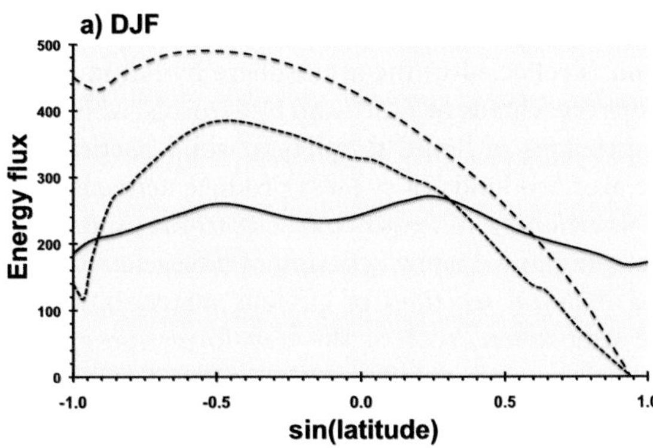

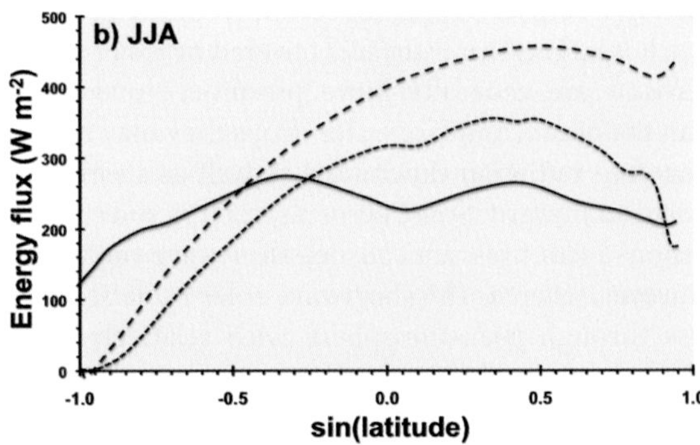

Figure 2.3 Earth's radiation budget (W m^{-2}) as a function of latitude. Dashed curves, incident solar energy flux; dotted curves, absorbed solar energy flux; solid curves, emitted longwave energy flux. Means over (a) DJF and (b) JJA. Note that insolation in DJF exceeds that in JJA, reflecting the fact that perihelion occurs in January. Adapted from James (1994).

global temperature results, as indicated by the (approximately horizontal) solid curves in Figure 2.3. Note that the insolation at high latitudes is quite high in the summer hemisphere, but that the net absorbed radiation remains small because of the high albedo caused by an ice and snow-covered environment.

2.2.4 Water vapour, clouds and the hydrological cycle

Water is a crucial component of the climate system. Two-thirds of the surface of the Earth is ocean, and much of this is in the tropics. A large proportion of the global radiation reaching the surface is absorbed by the tropical oceans. Evaporation from the tropical ocean surface pumps water vapour into the atmosphere. When this water vapour is carried aloft, it condenses and releases energy in the form of latent heat. This energy release heats the air, and drives much of the resulting atmospheric motion that transports warm air poleward and cold air equatorward. More than half of the energy driving this circulation is in the form of latent heat.

Quite independently, water vapour is important to climate because it is the dominant greenhouse gas, responsible for about half of the total greenhouse effect and two to three times more than the direct greenhouse effect of CO_2 (Schmidt *et al.*, 2010). Water vapour is concentrated at low levels, and in the tropics. In general, the atmosphere contains more water vapour than CO_2. In the low-level tropics, the extreme case, the atmosphere contains about 100 times more water vapour than CO_2. *When the CO_2 content is increased and consequently raises the global mean temperature, it enables the atmosphere to hold yet more water vapour.* This effect is known as water-vapour feedback. Relative humidity (water-vapour concentration as a percentage of its saturation value) tends to remain constant as the temperature varies. Because it is a greenhouse gas, the resulting extra water vapour causes the global temperature to increase further. Water-vapour processes are an integral part of all climate models. Furthermore, climate models are in good agreement with one another about the magnitude of the water-vapour feedback (Soden *et al.*, 2002).

Clouds complicate the physical picture enormously and are, in contrast with water vapour, extremely difficult to model. In particular, they contain their own feedback properties, where an increase of stratus cloud cover (for example) blocks both the incoming solar and outgoing infrared radiation. Cloud height is a factor here, and if the blockage of the incoming solar radiation is more important, the net result may be cooling, with generation of more cloud. There are substantial differences between different models' estimates of the feedback due to clouds (Soden and Vecchi, 2011).

Water evaporated from the ocean is eventually returned to the surface as precipitation. Some of this precipitation occurs over land, where the water may be stored temporarily in the form of snow, surface water or groundwater but eventually either is evaporated again (about two-thirds), or returns to the ocean as runoff (about one-third). Evaporation from the land includes a large fraction in the form of transpiration, i.e. upward transport of water through plants with the evaporation occurring at the leaf surfaces. The sum of transpiration and other forms of evaporation (e.g. from bare ground, and wet leaf surfaces) is called evapotranspiration. The whole evaporation-driven process is known as the global hydrological cycle, illustrated in Figure 2.4.

2.2.5 Climate sensitivity

Climate sensitivity is the global warming expected for a doubling of atmospheric CO_2 concentration, after the atmosphere and ocean have come into equilibrium with the change. The doubling of CO_2 can also be expressed as a radiative forcing (ΔF) leading to a change in the equilibrium temperature (ΔT); the ratio $S = \Delta T/\Delta F$ is known as the climate sensitivity parameter. Without feedbacks, Equation (2.1) would yield a temperature change of $\Delta T_0 \approx 1$ K only for a doubling of CO_2. Climate sensitivity is increased by positive feedbacks such as the water-vapour feedback (see Knutti and Hegerl (2008) for a review).

The climate sensitivity shown by a given climate model can be calculated, and there are differences between the values obtained by different models. Typical values lie in the range of 2 to 4.5 K. The differences between models are mainly due to different representations of cloud processes. It is also possible to create versions of climate models with much higher climate sensitivities, and it has been argued that the upper bound of climate sensitivity will be very difficult to constrain by improving the representation of processes in models (Roe and Baker, 2007). However, a variety of independent observational data-based constraints show that very high sensitivities are unlikely (Knutti *et al.*, 2006; Edwards *et al.*, 2007). Hegerl *et al.* (2007) gave a best estimate of 3 K, taking a variety of data-based constraints into account. Recent syntheses

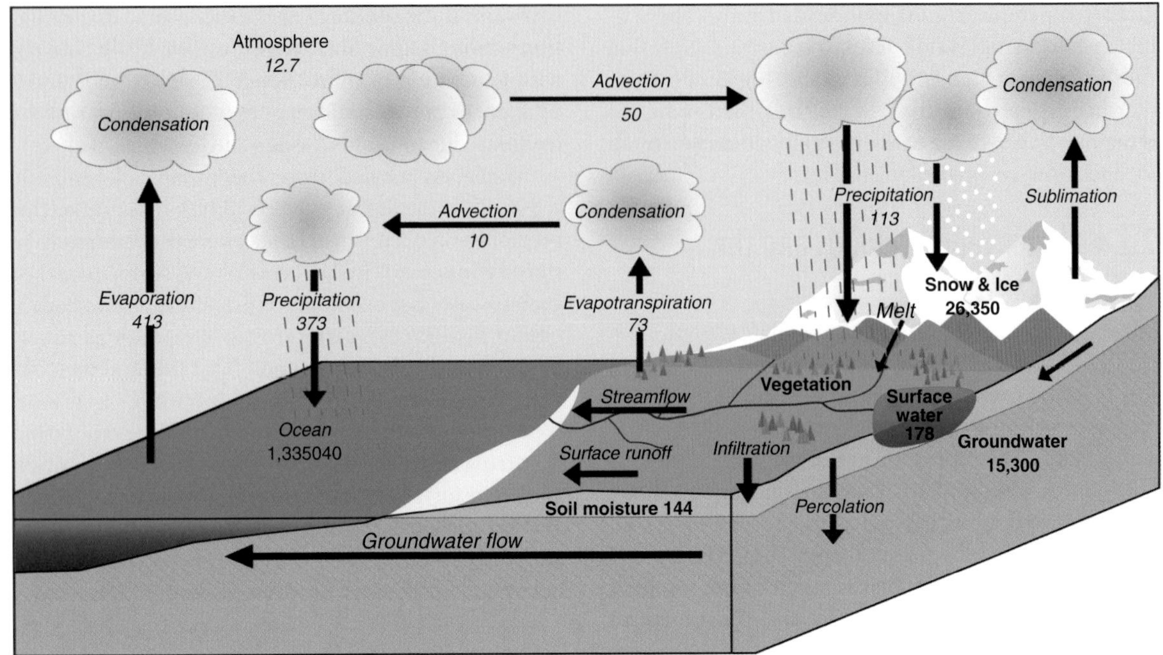

Figure 2.4 Mean fluxes and storages of water in the global hydrological cycle (Chahine, 1992; Trenberth *et al.*, 2007b). Fluxes denoted by arrows and numbers, in units of 10^{15} kg a^{-1}; storages by numbers, in units of 10^{15} kg. Runoff over land includes snowfall. Runoff includes streamflow, overland flow and groundwater flow.

of palaeoclimate data from the Last Glacial Maximum (MARGO Project Members, 2009; Bartlein *et al.*, 2011) yield a median estimate of 2.3 K, and completely eliminate values greater than 6 K (Schmittner, 2011). The use of palaeodata to constrain the climate sensitivity is discussed in more depth in Chapter 3.

2.2.6 The atmospheric circulation

The atmospheric circulation has a complex three-dimensional flow pattern, which varies with the seasons. Descriptions (e.g. Peixoto and Oort, 1992) usually begin with a zonally averaged mean picture of the transport in the troposphere. However, as the seasonal cycle in the atmosphere is so large, it is more meaningful to describe the extreme seasons of December–January–February and June–July–August rather than an annual mean. In general, the interim season of March to May resembles a weaker form of DJF, and September to November a weaker form of JJA.

The heat transported polewards by the atmosphere varies through the seasons. The atmosphere carries a much larger fraction of this heat flux than the oceans. Figure 2.5 (a) shows the zonally averaged atmospheric component of this circulation for the season DJF, inferred from analysis by Pauluis *et al.* (2008, 2010). The contours denote the meridional transport as a function of latitude and height. The circulation forms a single cell in the southern (summer) hemisphere from the tropics to the pole, but with a sinking region over a broad range of latitudes. In the northern (winter) hemisphere the pattern is more complex, with a sinking region corresponding to the Hadley circulation in the tropics. In mid latitudes there is rising motion driven by latent heat through the large-scale synoptic eddies that dominate the flow to carry heat (and latent heat) polewards, and cold air equatorwards (Palmén and Newton, 1969). There is considerable variation with longitude in this flow, as seen in Figures 2.5(b) and (c).

The mean distribution of rainfall for DJF is shown in Figure 2.5(b). The regions of largest rainfall are in the tropics and are generally where the water is warmest at the downstream end of long fetches across the oceans and, for DJF, centred south of the equator. These rainfall regions are the dominant locations where air rises from near ground level to the upper troposphere, forming the rising branch of the Hadley circulation. Airflow may proceed poleward as depicted in Figure 2.5(a), or

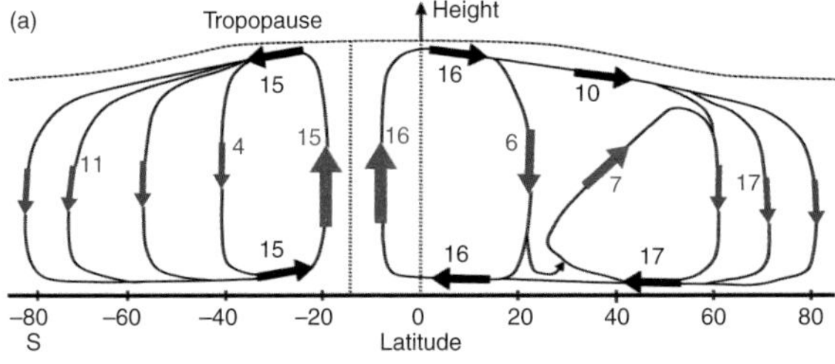

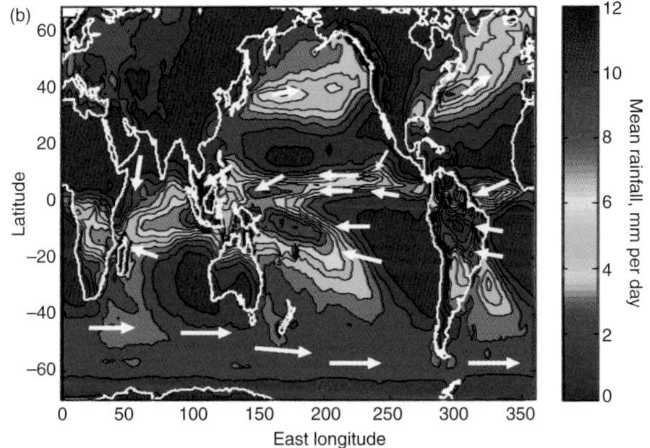

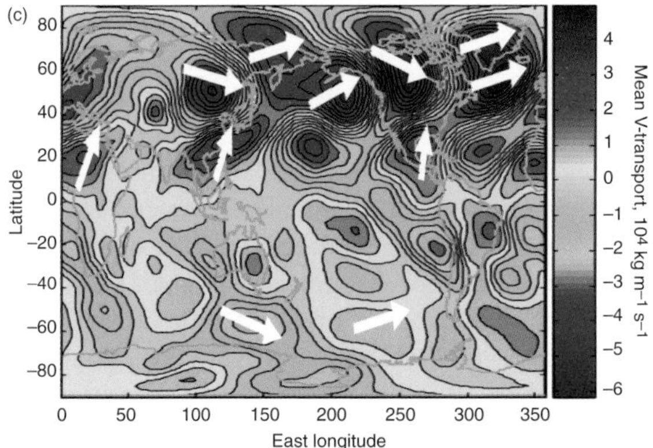

Figure 2.5 Mean circulation for DJF. (a) Schematic of the zonal mean circulation as a function of latitude and height. Red arrows denote rising motion, blue arrows descending. Units are 10^{10} kg s^{-1}. (b) Mean rainfall (GPCPv2.1 data; Adler *et al.*, 2003) in mm d^{-1} with white arrows denoting surface wind direction. The main rainfall regions are at the downstream end of long fetches over warm tropical oceans. (c) Spatial pattern of northward transport ('v-transport', following the convention that u is the eastward and v the northward component of flow) in the upper troposphere (pressure levels 550–80 hPa), in units of 10^4 kg m^{-1} s^{-1}. White arrows denote the approximate actual direction of flow. The three near-northward-pointing arrows in the tropics denote branches of the Hadley circulation, emanating from regions of large rainfall. In the lower troposphere the pattern of flow is similar to this in mid to high latitudes, but very different in the tropics. Note that northward motion is concentrated in the storm tracks in the Pacific and Atlantic sectors in mid latitudes in the northern hemisphere, which is the location of the increased transport seen in (a). Data from NCEP/NCAR re-analysis.

eastward as part of the Walker circulation. The north–south components that comprise the Hadley circulation are shown in Figure 2.5(c), where red denotes mean northward flow in the upper troposphere (above 550 hPa), and blue southward flow (the white arrows denote the actual direction of the flow, including the zonal component). As expected from Figure 2.5(a), the flow in the northern hemisphere is much stronger than that in the southern. The three red regions centred near 20° N denote the three main branches of the Hadley circulation for DJF. Between 30° N and 40° N these poleward flows weaken, but they become stronger in the main storm tracks over the Atlantic and Pacific oceans, where they constitute the main pathways in which warmer air moves poleward.

Figures 2.6(a)–(c) show the corresponding flow patterns for JJA. Again, the circulation in the winter hemisphere is much stronger than the summer. The flow pattern in the northern (summer) hemisphere now resembles that in the southern, with the rising path evident in mid latitudes. The main tropical precipitation regions are centred north of the equator, and the four blue regions (centred near 20° S) with southward pointing white arrows in Figure 2.6(c) denote the upper branches of the Hadley circulation from these rainfall regions.

Much of this seasonal difference is manifested in the monsoons (Johnson, 1989), which involve a complete reversal in the low-level flow over the Indian Ocean with corresponding changes aloft as shown in Figures 2.5(c) and 2.6(c). In JJA, the Indian and East Asian monsoons are fed by northward flow across the Equator along the African coast, which turns eastward and provides a warm moist airstream over India. A similar but smaller-scale phenomenon occurs in the Atlantic and West Africa. In DJF the lower tropospheric flow over the Indian Ocean reverses, driving the Australian monsoon that feeds a warm moist airstream over Indonesia and northern Australia.

Although the qualitative features of the atmospheric circulation are 'hard-wired', there is strong evidence for variations in the strength and location of particular features associated with climate changes. One well-understood consequence of orbital variations is that whenever precession amplifies the seasonal cycle in a given hemisphere, the monsoon circulations (and associated intense summer rainfall) in that hemisphere are amplified in response to the resulting increased land–sea contrast (Kutzbach and Guetter, 1986). This mechanism was responsible for the much wetter and more

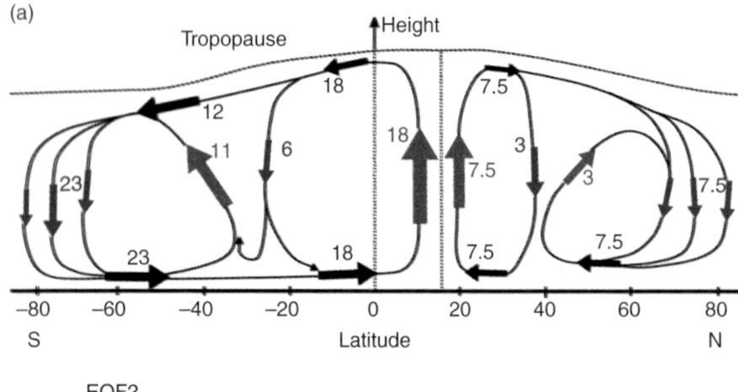

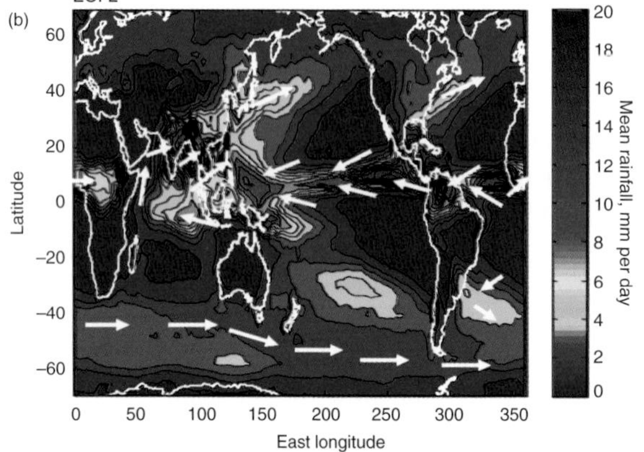

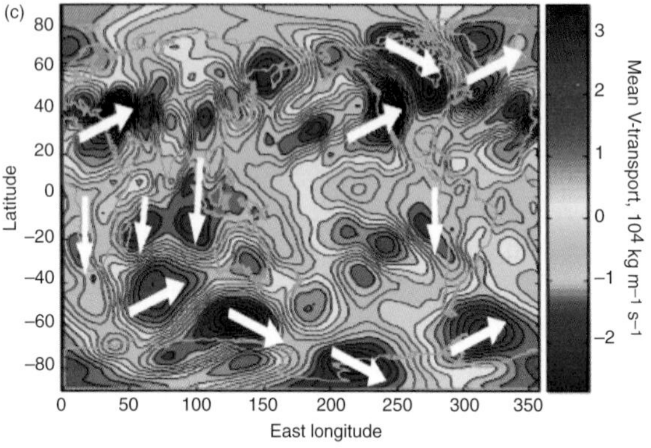

Figure 2.6 Mean circulation, as for Figure 2.5, but for JJA. In (a), slantwise convection driven by mid-latitude latent-heat release occurs in both hemispheres. In (c), the four arrows in the tropics denote upper level flow of the Hadley circulation. As in Figure 2.5, the corresponding flow in the lower troposphere is similar to this in mid to high latitudes, but very different in the tropics.

vegetated conditions that prevailed in the early to mid Holocene in what is now the Sahara Desert (Kutzbach and Street-Perrott, 1985; Jolly *et al.*, 1998).

2.2.7 The ocean circulation

The ocean circulation is shown schematically in Figure 2.7. There is no large annual cycle, so this figure shows the annual mean. The horizontal pattern is shown in (a) and the vertical structure in (b), with the depth colour-coded in the same way. The currents in the surface and upper levels, coloured red, are mostly (80%) driven by the wind. Strong poleward currents occur on the western boundaries of the ocean basins, including the Gulf Stream in the Atlantic, the Kuroshio in the Pacific, the East Australian Current and the Agulhas Current in the southern Indian Ocean. In the northwestern Indian Ocean the Somali Current reverses direction following the seasonal cycle of the overlying winds, which in JJA feed the Indian monsoon.

Convection-driven movements occur in locations where the surface waters are so cold that they become denser than those at depths below, causing water to sink from the surface to levels near the ocean bottom. Such regions may be seen in Figure 2.7 as places where the streamlines change from red (near the surface) through yellow and green to dark blue (near the bottom). In the northern hemisphere, two such regions are depicted, in the Greenland–Iceland–Norwegian Sea and the Labrador Sea. In the southern hemisphere, the two main regions are the Weddell and Ross seas, plus smaller ones such as the Mertz Glacier Polynya near 145° E. The resulting three-dimensional circulation is seen in Figure 2.7(b), which gives estimates of mean transports in the deep flows in sverdrup (1 Sv = $10^6 m^3 s^{-1}$). Sinking in the northern North Atlantic forms the water mass known as North Atlantic Deep Water (NADW). This contains the largest deep-water transport (17.6 ± 3.1 Sv), which moves southward at depth and occupies much of the volume of the Atlantic Ocean. On a timescale of ~ 1,000 years NADW reaches the Southern Ocean where it joins the Antarctic Circumpolar Current, circling the globe, and slowly returns to the surface in a process known as the Atlantic Meridional Overturning Circulation (AMOC). The sinking water at near-Antarctic locations forms Antarctic Bottom Water (AABW), and flows northward in all three oceans. In the Atlantic Ocean, AABW lies beneath the NADW but eventually merges with it and returns south. In the Pacific and Indian oceans it slowly rises and eventually returns to the Southern Ocean at intermediate depths (yellow tracks in Figure 2.7).

2.2.8 Abrupt climate change and the AMOC

The release of heat during NADW formation warms the North Atlantic and surrounding lands. It has long been understood theoretically that the AMOC can exist in

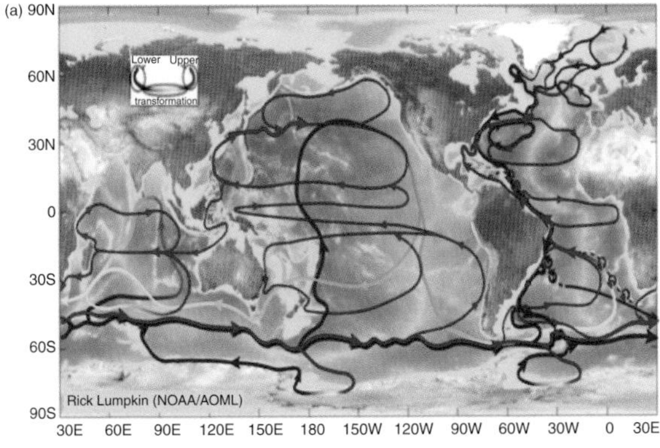

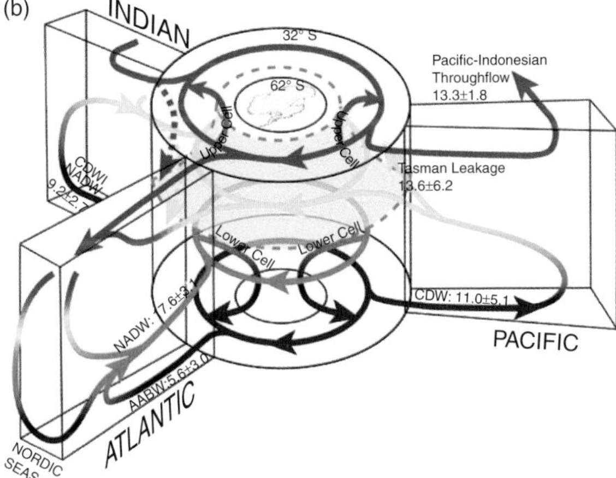

Figure 2.7 Schematic of the annual mean global overturning circulation of the ocean. Colours denote depth of flow, with red near surface to dark blue near the bottom. (a) Plan view (C. F. Lumpkin, personal communication, 2010); (b) three-dimensional picture with transports in sverdrup (from Lumpkin and Speer, 2007, following Schmitz, 1996). Water masses: AABW = Antarctic Bottom Water, CDW = Circumpolar Deep Water, NADW = North Atlantic Deep Water.

more than one stable state, implying that under certain conditions it could 'flip' between states. It is now clear that abrupt changes in the AMOC have actually happened, and that they caused abrupt, sometimes large regional changes in climate (Alley, 2007; Broecker, 2010).

The key control of the AMOC is the flux of freshwater to the North Atlantic. A sufficiently large freshwater flux can inhibit NADW formation and thereby push the AMOC into a 'weak' or 'shut-down' state; removal of this flux causes the AMOC to 'turn' on again. During the last ice age, when the North Atlantic was fringed by large ice sheets, the AMOC was generally in a weak state. But abrupt warming events occurred at intervals of about 1,500 years, reflecting recoveries of the AMOC. These warming events are referred to as *Dansgaard–Oeschger events*, or D–O events. They are very well documented in ice cores from Greenland and sediment cores from the Atlantic Ocean, but their effects can be traced globally (Harrison and Sanchez-Goñi, 2010). The warming, which amounted to 5 to 15 K in the North Atlantic region, typically took place during 20–50 years; the warm phases lasted 200–2,000 years. It is presumed that they were caused by variations in the flux of melting ice, but the detailed mechanisms are not yet clear (see Chapter 3). Among the intervening cold periods, the coldest are known as *Heinrich events* and were marked by the passage of icebergs (which left their mark in the form of ice-rafted debris in the sediments) far south in the North Atlantic.

The most prominent abrupt climate change of recent geological times was the Younger Dryas, an abrupt cooling that occurred during the transition from the last ice age to the current (Holocene) interglacial (Watts, 1980; Johnsen *et al.*, 1992; Lehman and Keigwin, 1992). This event was named for the Arctic tundra plant *Dryas*, which suddenly reappears in sediments of southern Scandinavia after the first development of forests, giving evidence of a major ecosystem change in response to a changed climate. In modern terminology the Younger Dryas has the properties of a Heinrich event, including a final abrupt warming that took place in less than 30 years and marks the beginning of the Holocene interglacial. The accepted hypothesis is that the Younger Dryas was caused by a massive influx of low-salinity, low-density water from the melting of the North American continental ice sheet, leading to a temporary weakening of the AMOC and much-reduced northward heat transport (Broecker *et al.*, 1985).

2.2.9 Observed patterns of climate variability and change during the past 50–100 years

Observations of SST can be used to demonstrate the nature of recent climate variability and change. Because heat storage in the upper layers of the ocean is much larger than on land, SST is a powerful influence on the climate of the land. Historical time series of SST data are presented here after treatment with the multivariate technique known as empirical orthogonal function (EOF) analysis or principal component analysis (PCA). This breaks down the variance of the quantity of interest (here SST) into a sum of mutually uncorrelated spatial patterns, each multiplied by an associated time series. The patterns are ordered by the amount of variance each accounts for, so that the dominant patterns emerge as the first few EOFs. (Note that the signs and units of EOFs are arbitrary – what matters are the percentage of variance explained by each EOF, and the spatial and temporal patterns of values.) Two sets of SST observations have been compiled to produce a complete description of global SST back to the nineteenth century: HadISST1 (Rayner *et al.*, 2003) from the Hadley Centre, and erSSTv3 (Smith *et al.*, 2008) from the National Oceanic and Atmospheric Administration (NOAA). The two give results with only minor differences. Interannual variability is considered first, using HadISST1 over the recent period for which the data are most accurate and abundant; then decadal variability, using erSSTv3 for the whole period.

Interannual variability in global climate is dominated by the El Niño Southern Oscillation (ENSO). Philander (1990) provided an introduction to ENSO, which already had a vast literature. During El Niño, heat storage in the equatorial Pacific Ocean is shifted eastwards. The 'warm pool' in the western equatorial Pacific, with waters warmer than 27 °C (seen in Figure 2.8), extends toward South America, causing changes in rainfall and wind patterns over much of the globe. This causes a decrease in the Walker circulation in the Pacific, in which the trade winds, rising motion over Indonesia and return flow in the upper troposphere are all reduced. El Niño usually begins in June and ends by the following May. It is often followed by La Niña, which shows the opposite pattern. The EOF analysis was carried out on the anomalies (differences from the long-term mean) of SST for 'e-years' (June to May) between June 1979 and May 2007 (Figure 2.9). The first EOF represents ENSO, with warm anomalies in the eastern equatorial Pacific and cold anomalies in the western equatorial Pacific and mid latitudes. El Niño events have positive values in the time series.

The second EOF represents a nearly ubiquitous global warming and has a nearly consistent upward trend, particularly since the early 1990s. The third EOF has a time series resembling that of ENSO with most of the variance centred in the Pacific. This pattern is the 'El Niño Modoki' (Ashok *et al.*, 2007) or 'Central Pacific' pattern. It may be understood best as a modification of the structure of the larger ENSO events.

Analysis of five-year running means of SST anomalies back to 1900 (Figure 2.10) filters out the ENSO signal and turns the spotlight on longer-term signals. The dominant pattern – accounting for more than half the total variation in the data set – is now the global-warming pattern, with a time series closely resembling the curve of global mean temperature as presented by the IPCC (Trenberth *et al.*, 2007a). The spatial pattern of global warming is approximately uniform except for the region close to Antarctica, and a small region in the north of the North Atlantic. Model studies of global warming have indicated that it would be expected to cause a decrease in the transport of the AMOC, accompanied by a cooling near the regions where water sinks. This is precisely what is seen in Figure 2.10 (a). Recent work has indicated that some features of the curve prior to 1950 may be attributed to variations in methods used to observe SST (Thompson *et al.*, 2008), and future versions of this data set will probably show some differences. But the data over the most recent 50 years are more reliable, and the overall pattern is robust.

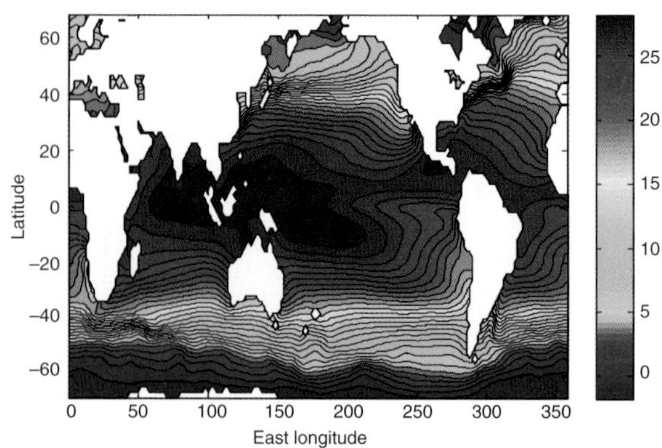

Figure 2.8 Mean SST (°C) over 108 e-years, June 1900 to May 2009. Data from erSSTv3.

The second pattern is recognizable as the Pacific Decadal Oscillation, or Interdecadal Pacific Oscillation, and resembles a smoothed-out version of ENSO. The origin of this pattern is unknown. It may be just a low-frequency average of the ENSO phenomenon, with the same dynamics; it may also be influenced by the 11-year solar cycle (van Loon *et al.*, 2004; Meehl *et al.*, 2009). The third pattern is the Atlantic Meridional Oscillation, associated with variations in the AMOC. The signal is largest in the Atlantic, where the two hemispheres have opposite sign, and this is mirrored to a lesser extent in the other oceans.

In conclusion, the pattern of climate variation from year to year is still dominated by ENSO. Other patterns of regional extent are also recognized to be important

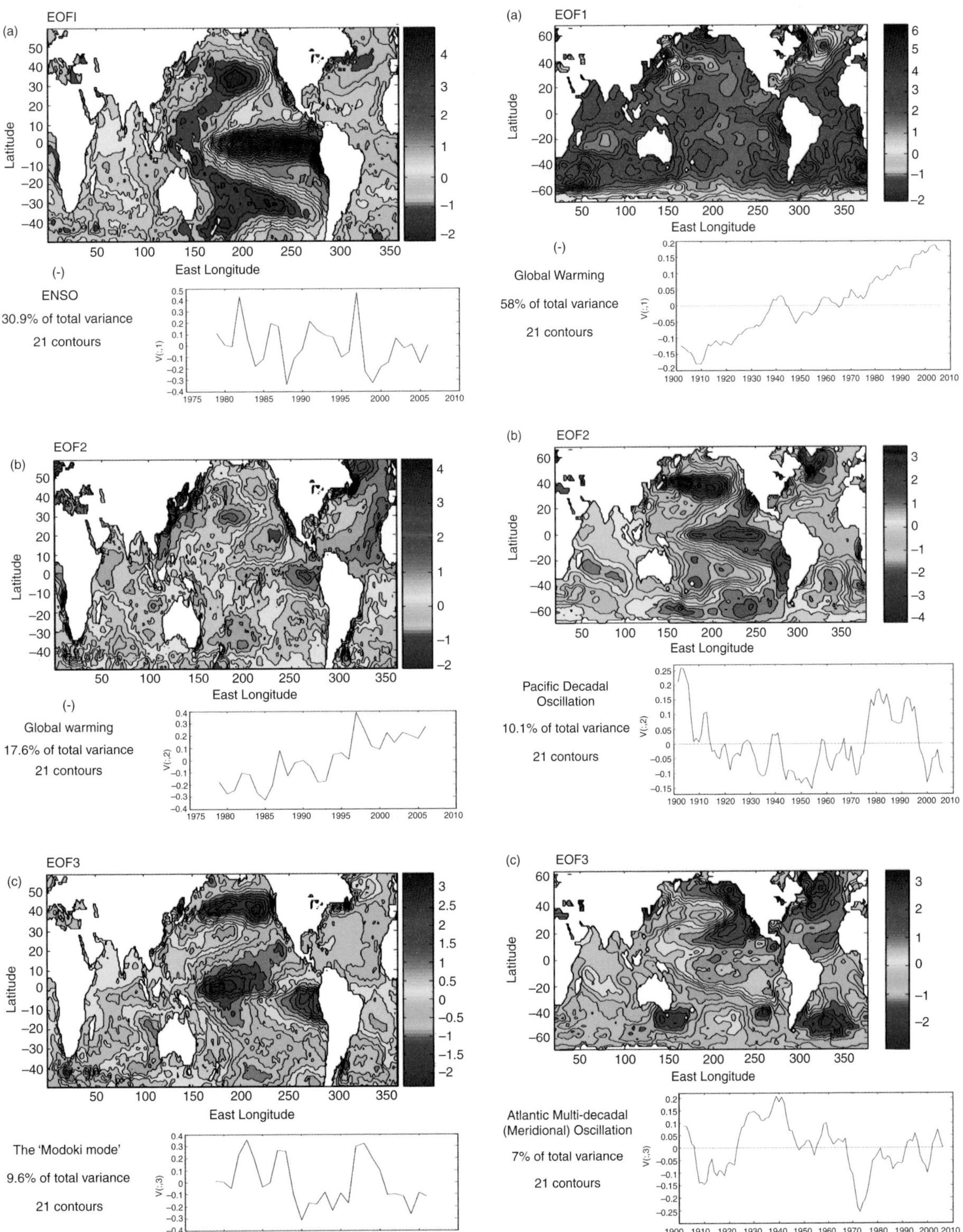

Figure 2.9 (a)–(c) The first three EOFs (EOF1–EOF3) of annual mean SST over 28 'e-years' from June 1979 to May 2007 (data from HadISST1). The domain is 60° N to 50° S.

Figure 2.10 (a)–(c) The first three EOFs (EOF1–EOF3) of 5-year running means of SST over June 1900 to May 2009 (data from erSSTv3).

in the variability of observed weather patterns, such as the influence of the North Atlantic Oscillation on winter temperatures in Europe and North America. Thus, it is not surprising that climate change is not a matter of everyday experience. When decades of climate records are analyzed, the signal of global warming emerges nonetheless.

2.2.10 The nature of global warming

The evidence that the observed global warming in the present day is a consequence of radiative forcing does not depend on there being no other causes of climate change (there are many), the present rate of climate change being unique (it isn't), or the pre-industrial climate being static (it wasn't). The key evidence is presented in Hegerl *et al.* (2007). Climate models that include the known processes governing the Earth's radiation balance and climate, and when driven by the known changes in radiative forcing during the industrial period (greenhouse gases, sulfate aerosols, solar variations, volcanic events), *can* reproduce the magnitude of the observed recent warming over the oceans and on each of the continents. They *can* reproduce many aspects of the climate changes that have taken place during the past 150,000 years, including the exceptional high-latitude warmth of the last interglacial, the response of the AMOC and regional temperature variations to meltwater pulses, the nature and pattern of ice-age climates and the shifts in monsoon strength during the Holocene. They *can* reproduce the 'ups and downs' of the climate of the last millennium, including the timing of the Mediaeval Warm Period and the Little Ice Age (Ammann *et al.*, 2007) and the transient cooling during the 1950s. And they *can* reproduce spatial features of the observed global warming since the 1970s, including the absence of warming in part of the North Atlantic and the drying trend in already-dry regions such as the Mediterranean basin. What they *cannot* do is to reproduce this observed warming if the recent greenhouse gas increases are left out.

The observed drying trend in some regions deserves further consideration. It is not intuitively obvious that global warming should cause rainfall reductions. Globally, the expected trend is for rainfall to increase as temperature increases. However, the phenomenon of wet areas getting wetter while dry areas get drier is both observed and modelled (Kundzewicz *et al.*, 2007). A simple explanation is as follows (Held and Soden, 2006; John *et al.*, 2009). An increase in the temperature of the atmosphere implies that the air can hold more water vapour. For an increase in temperature δT, from the Clausius–Clapeyron equation, the saturation vapour pressure e_s of water vapour increases by $\delta e_s/e_s \approx \gamma\delta T$ where $\gamma \approx 0.07\ \mathrm{K}^{-1}$. In other words, the amount of water vapour that the air can hold increases by 7% per degree. As relative humidity tends to remain constant, this equation also gives a good measure of the actual increase of atmospheric water vapour content with temperature (Wentz *et al.*, 2007). Moisture (water vapour plus liquid water) in the atmosphere is produced by evaporation E from the surface, and is removed by precipitation P. The divergence of the (vertically integrated) flux of moisture in the atmosphere, $\mathbf{F}$, is div $\mathbf{F} = E - P$ where E and P denote mean spatial patterns. The pattern of the flow, and hence of the moisture flux, does not change greatly in response to warming. So the change in moisture flux $\mathbf{F}$ is approximated by $\delta\mathbf{F} \approx (\delta e_s/e_s)\mathbf{F} \approx \gamma\delta T\mathbf{F}$, and the change due to global warming of $P - E$ is then, approximately, $\delta(P - E) \approx \gamma\delta T(P - E)$. In other words, the change simply enhances the existing pattern of $P - E$. The fields of E and P are very different: evaporation varies slowly over long distances whereas precipitation varies over much shorter distances, and this distinction must also apply to the changes in P and E. Hence the changes in rainfall patterns should mirror the existing rainfall distribution. This is shown from observations (1979–2009) in Figure 2.11, which can be compared with the global rainfall pattern (Figures 2.5(b) and 2.6(b)).

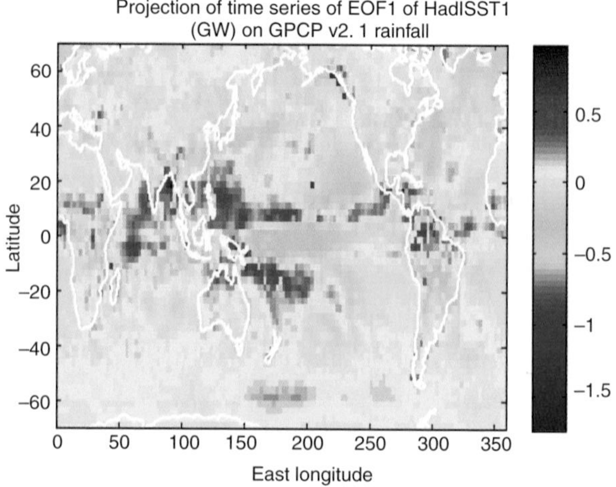

Figure 2.11 Spatial pattern of the trend in annual mean rainfall (GPCPv2.1 data) associated with the trend in EOF1 from Figure 2.10 (a). Red denotes more rainfall, green/blue less.

2.3 Fundamentals of terrestrial ecosystem science

2.3.1 Primary production: basics

Ecosystems are powered by green plants, which use solar energy to synthesize carbohydrates from CO_2 derived from the atmosphere and water derived from the soil (the process of photosynthesis). Ecosystems also include herbivores, which derive energy from consuming plants; predators, which derive energy from eating other animals; and decomposers – primarily bacteria and fungi that break down the detritus formed from dead plant and animal tissues and derive energy by converting this material back to CO_2. The overwhelmingly dominant processes in this cycle on land are photosynthesis, plant respiration and decomposition. Photosynthesis by the whole ecosystem is called gross primary production (GPP). About half of GPP in terrestrial ecosystems is converted back to CO_2 by autotrophic respiration (R_A) by the plants, which is necessary for growing and maintaining plant tissues. The rest is net primary production (NPP), which builds plant tissues. Eventually, NPP becomes detritus, and CO_2 is returned to the atmosphere principally by heterotrophic respiration (R_H) carried out by decomposers (microbes) in the soil. Less than 5% of NPP is combusted in fires and returned directly to the atmosphere as CO_2.

Photosynthesis depends on the absorption of photosynthetically active radiation (PAR), which constitutes about half of the incoming shortwave radiation, by chlorophyll in plant leaves. Photosynthesis requires a sufficient concentration of CO_2 in the atmosphere, and conditions warm enough for all of the reactions to proceed. Gross primary production by vascular plants – plants with water-conducting systems, which dominate in all but the least productive terrestrial environments – also requires a continuous supply of water to be evaporated through the tiny pores on the leaf surface (stomata). When the stomata are open they let CO_2 in, but they also let water out. The water comes from the soil water store; it is taken up by the roots and transported upward through conducting pipes (xylem elements) in the stems and leaves. The driving force for water transport (transpiration) is the water potential difference between the soil and the leaves. This in turn is set up by the loss of water through the stomata. The opening and closing of stomata is tightly regulated and their behaviour is consistent with optimization of the balance between carbon gain and water loss (Medlyn *et al.*, 2011). Evapotranspiration is ultimately controlled either by energy availability or water availability. In dry seasons, losses of water by evapotranspiration can exceed precipitation; then the store of water in the soil becomes depleted, and stomata tend to become more and more closed.

Conversion of the products of photosynthesis (sugars) into plant material depends on conditions light and warm enough for growth, wet enough to allow transpiration, and an adequate supply of nutrients. Land plants require many different nutrients but most are needed in tiny quantities and are not usually in short supply. The nutrients most commonly limiting to NPP are nitrogen (N), in one of the reactive forms including nitrate and ammonia that can be taken up by plants (especially important in cold climates); phosphorus (P), in a chemical form such that it can be taken up by roots and their fungal associates (especially important on highly weathered soils in the tropics and subtropics); and several base cations, especially potassium (K). Reactive N arrives in ecosystems mainly through the action of bacteria and archaea with the ability to derive reactive N from N_2 in the air (biological N_2 fixation). Additional reactive N arrives by atmospheric deposition. Some of this is natural, due to reactions associated with lightning, and also the release of nitrogen oxides from soil and in fires and their subsequent transport in the atmosphere. These natural sources are now supplemented in the most industrialized regions by the atmospheric deposition of N released from burning fossil fuels, so some regions now have a massive oversupply of reactive N (Galloway *et al.*, 2008). Phosphorus and base cations are obtained mainly from soil minerals derived from the underlying parent materials. Some ecosystems acquire significant extra inputs of P and base cations from atmospheric dust derived from remote deserts and other unvegetated or partially vegetated surfaces (Harrison *et al.*, 2001).

2.3.2 Biomes and plant functional types

Satellite observations of the fraction of absorbed photosynthetically active radiation (fAPAR; the proportion of incident PAR that is absorbed by plants) rely on the reflectance of the land surface in different spectral bands. Chlorophyll absorbs most strongly in the red, and not in the near infrared, and this discrimination allows satellite-borne spectrometers to measure the greenness of the land surface. Observations of fAPAR provide evidence for major climatic controls on

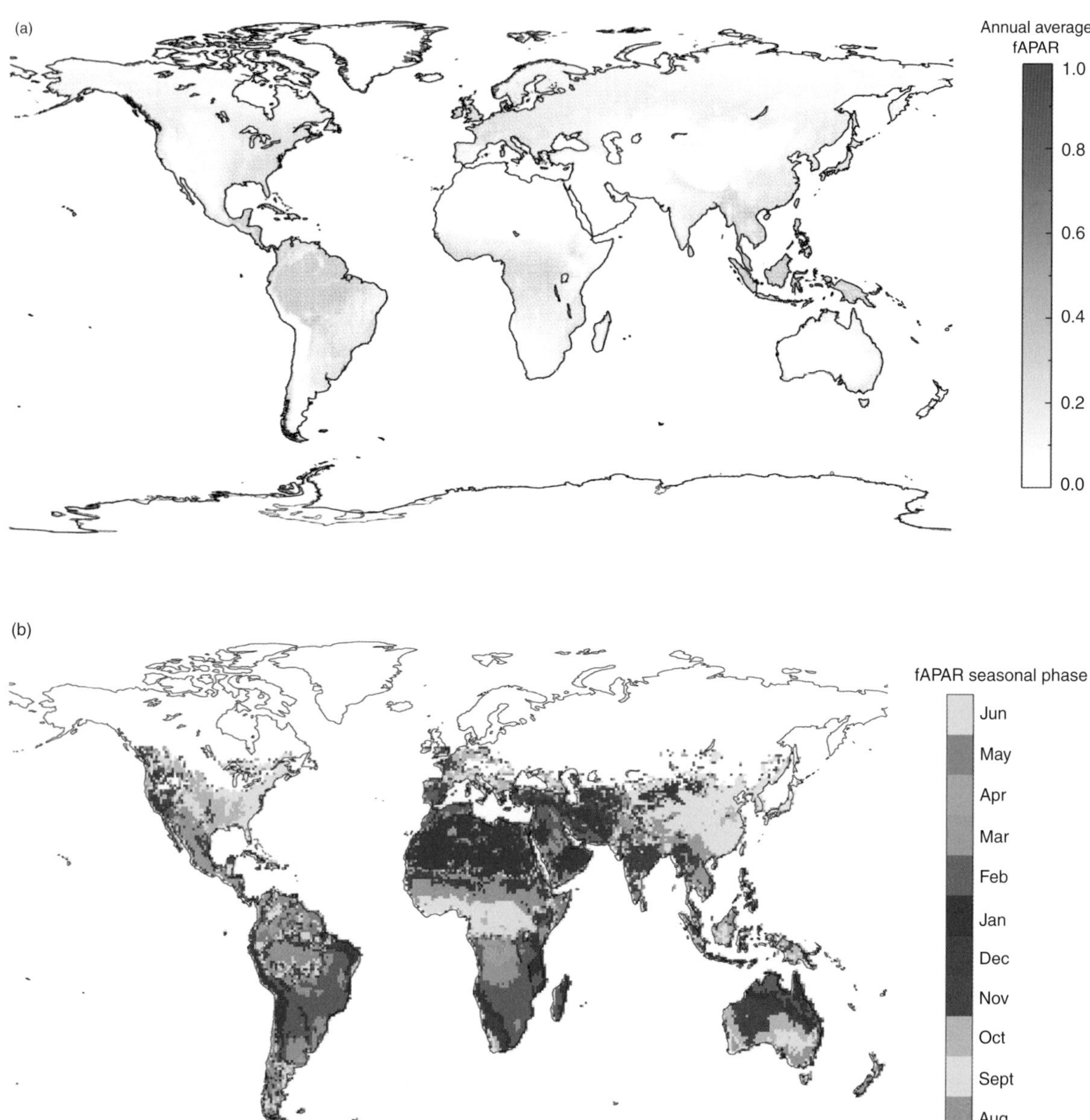

Figure 2.12 Global patterns in the annual average and seasonal cycle of fAPAR, an index of green vegetation cover: (a) annual average; (b) timing (centre of the seasonal fAPAR distribution); (c) a measure of the seasonal concentration. Data for 1998–2005 from http://oceancolor.gsfc.nasa.gov/SeaWiFS/. Annual average fAPAR is the weighted average of monthly fAPAR values, with weighting by monthly incident PAR. Timing and seasonal concentration are calculated as in Harrison *et al.* (2003). Timing and seasonal concentration are not shown for high latitudes or regions with significant snow cover, which interferes with the estimation of fAPAR from space.

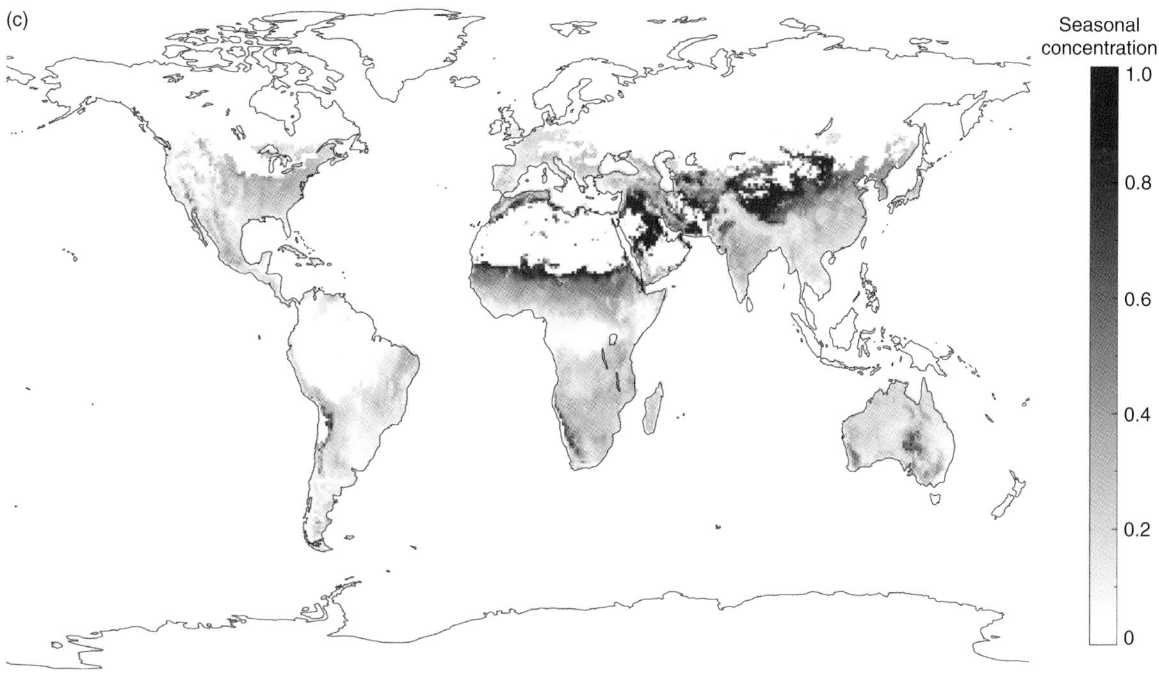

Figure 2.12 *(cont.)*

the structure and productivity of terrestrial ecosystems (Figure 2.12). The fAPAR is greatest in climates with abundant year-round rainfall, and lowest in deserts and in extremely cold climates (high latitudes and elevations). The timing of maximum fAPAR in seasonal climates varies, notably between Mediterranean-type climates with winter rainfall maxima and wet/dry subtropical and tropical climates with summer rainfall maxima. These two types of climate support biomes with different adaptations to drought. Mediterranean-type ecosystems are characterized by the dominance of deep-rooted trees and shrubs with hard leaves; wet/dry subtropical and tropical ecosystems are tree–grass mixtures (savannas) or grasslands.

The plant traits that distinguish biomes confer tolerance of different stresses: extreme cold temperatures in winter (with different mechanisms allowing tolerance to different degrees of cold); lack of warmth in summer (plants from very cold climates hug the ground, or adopt a cushion form that allows them to heat up in the sun); excessive heating (plants from hot, dry environments, where cooling by evaporation is not an option, have small leaves that avoid heat damage); and drought, under which plants show a huge variety of special forms and responses. Harrison *et al.* (2010b) reviewed the plant traits that are diagnostic of biomes, their physical and physiological basis, and 'bioclimate variables' that express aspects of climate that are important in determining the distribution of different plant types. Figure 2.13 summarizes the global pattern of variation in three variables that relate to three main axes of variation in vegetation composition.

Another key distinction among vascular plants is between two variants of photosynthesis. The main carbon-fixing step in photosynthesis is carried out by just one enzyme (rubisco) that is common to all green plants. But some terrestrial plants (including most tropical grasses, but almost no trees) have a particular anatomical and biochemical mechanism called C_4 photosynthesis that allows them to concentrate CO_2 where it is most needed. They thereby avoid a problem experienced by ordinary plants with C_3 photosynthesis, namely the wastage of CO_2 in photorespiration, whereby rubisco reacts with the very abundant O_2 instead of CO_2. The C_4 photosynthesis mechanism is advantageous at high temperatures, especially under dry conditions, and under low CO_2 concentrations such as were found during the peaks of the ice ages that have punctuated the past million years.

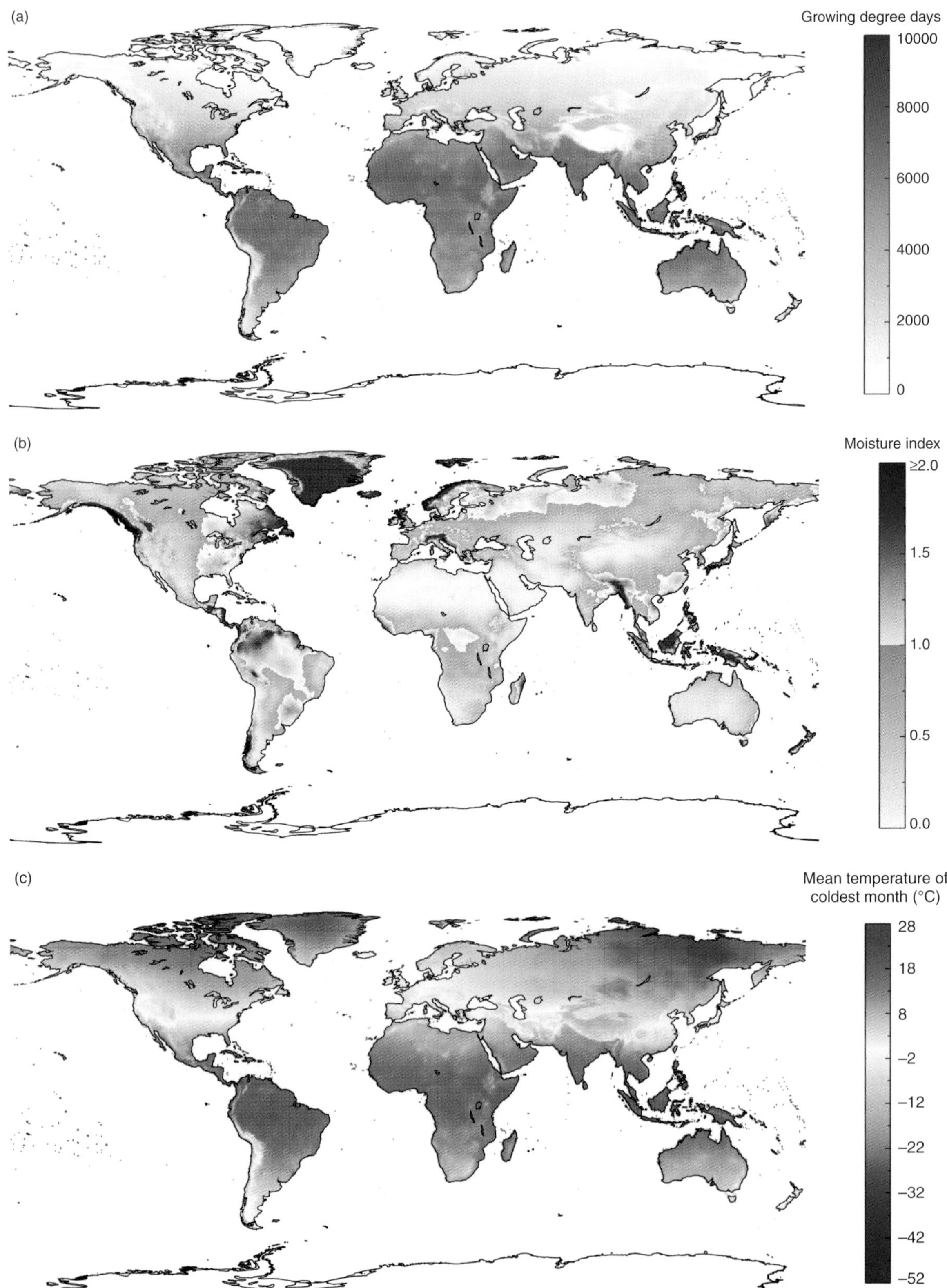

Figure 2.13 Global pattern of bioclimate variables: (a) annual sum of daily temperatures above zero, an index of growing-season length and warmth; (b) moisture index (here defined as the ratio of annual precipitation to annual equilibrium evapotranspiration), an index of plant-available water supply; (c) mean temperature of the coldest month, an index of winter cold. Data sources and calculations as in Gallego-Sala *et al.* (2010).

2.3.3 Environmental controls of primary production

Two simplified expressions for GPP, the water-use efficiency (WUE) and light-use efficiency (LUE) formulae, can be derived on the assumption that plant functional types are distributed so that potential GPP is achieved everywhere. Both can be derived from the fundamental equations of C_3 photosynthesis, which were developed by Farquhar *et al.* (1980) and form the basis for all subsequent modelling. To derive the WUE formula, we start by writing a pair of diffusion equations for CO_2 and water transport across the stomata:

$$A = g_s (c_a - c_i) \quad (2.3)$$

$$E = 1.6 g_s D \quad (2.4)$$

where A is assimilation of CO_2 (photosynthesis), g_s is the stomatal conductance to CO_2, c_a and c_i are the mole fractions of CO_2 outside and inside the leaf, E now represents just transpiration, and D is the vapour pressure deficit: $D = e_s - e$ where e is the actual water vapour content of the air surrounding the leaf. The factor 1.6 arises because water diffuses through stomata faster than CO_2. Applying these equations to a whole ecosystem, their ratio A/E is proportional to $(c_a - c_i)/D$. The ratio c_i/c_a ranges only between about 0.5 and 0.9, so A/E is approximately proportional to c_a/D. This ratio declines towards drier environments but so does the coverage of green plants – inevitably, because a dense vegetation cover in a dry environment would soon run out of water – and so does the absolute value of GPP. Thus, a first-order prediction of GPP can be made on the assumption that it is proportional to the total water used by the ecosystem, which can either be measured as the difference between precipitation and runoff, or computed as a function of precipitation and atmospheric evaporative demand (Zhang *et al.*, 2008).

Alternatively, GPP can be calculated from absorbed PAR (LUE). Over periods of days to weeks, it is assumed that the amount of rubisco in the leaves adjusts to prevailing conditions so that as much as possible of the incident PAR is used in photosynthesis (Dewar, 1996; Haxeltine and Prentice, 1996). This allows the use of the following formula for photosynthesis, known as the electron-transport limited formula:

$$A = \phi_o I \text{fAPAR} \, (c_i - \Gamma^*)/(c_i + 2\Gamma^*) \quad (2.5)$$

where ϕ_o is the intrinsic quantum efficiency of photosynthesis, I is incident PAR, and Γ^* is the CO_2 compensation point, i.e. the value of c_i below which no photosynthesis occurs. This is 43 ppm at 25 °C and standard atmospheric pressure (Bernacchi *et al.*, 2003). Given the relatively minor variation of the term involving c_i under natural conditions, we can estimate GPP from the product of fAPAR and I.

The two approaches described above can both fit observational data quite well (H. Wang, I. C. Prentice and J. Ni, unpublished results), but they make different predictions of the effect of changing CO_2 concentration (c_a). In WUE, a doubling of CO_2 leads to a doubling of GPP. In LUE, a doubling of CO_2 leads to a more modest increase of GPP (34%, if we start at c_a = 280 ppm and assume a typical value of $c_i/c_a = 0.7$). This difference is explicable. Stomatal conductance declines with increasing c_a, as is theoretically predicted and generally observed (Medlyn *et al.*, 2011). Applying WUE, we assumed that vegetation cover would increase everywhere, fully compensating for the reduction in g_s. This is not possible for dense vegetation. Applying LUE, we assumed that vegetation cover (fAPAR) would not change at all. This is unlikely for sparse vegetation. So these calculations give upper and lower bounds for the effect of CO_2 on GPP.

If stomatal conductance declines and vegetation cover does not increase then runoff (the part of precipitation that is not evaporated, which provides freshwater supplies) must increase (Gedney *et al.*, 2006). There is thus a trade-off between two 'beneficial' effects of rising CO_2, on plant growth and on the proportion of precipitation that is routed to streams and available for use. But it is still not clear to what extent rising CO_2 and other environmental factors have contributed to large-scale increases and decreases in runoff (Peel, 2009).

The direct observational basis for CO_2 effects at the ecosystem level is also rather sparse. We can infer long-term consequences of climate change by studying spatial gradients of climate (substituting space for time), but CO_2 concentration is similar globally, precluding this approach. Free Air Carbon Dioxide Enrichment (FACE) experiments in temperate forests have shown that a 200 ppm increase in CO_2 concentration leads to a ~ 23% increase in NPP (Norby *et al.*, 2005). This is similar to the prediction of the LUE model. It can also be predicted that boreal forests should respond less, and tropical forests more, but these predictions have not been tested (Hickler *et al.*, 2008). There is experimental evidence that nutrient availability can constrain the response of NPP to increasing CO_2, but the magnitude and timescales of this effect are still not established, leading to a significant uncertainty in predictions of the CO_2 effect on NPP at a global scale.

Effects of CO_2 are observable indirectly in the palaeovegetation record. Carbon dioxide fertilization of a magnitude similar to that predicted by current models is required in order to explain changes in terrestrial carbon storage between the Last Glacial Maximum and the Holocene (Prentice and Harrison, 2009). Low CO_2 during the last ice age was at least partly responsible for the great expansion of C_4-dominated grasslands in the tropics and the retreat of forests to the wettest areas (Harrison and Prentice, 2003). Conversely, the rising CO_2 concentration over recent decades is probably contributing to the widely observed phenomenon of 'woody thickening' whereby trees and shrubs are gaining ground at the expense of grasses in tropical and subtropical savannas (Bond and Midgley, 2000).

2.3.4 Climate, people and fire

Vegetation fires burned 3.3 to 4.3 million square kilometres of the Earth surface each year between 1997 and 2008 (Giglio *et al.*, 2010). Such wildland fires are often termed a 'disturbance' to ecosystems but the term is rather misleading, because fire occurrence is natural in most ecosystems and has occurred throughout the geological history of land plants (Bowman *et al.*, 2009). Much of the carbon released by burning is reabsorbed during the following growing season(s), and fires also release nutrients from plants that are then available to support vegetation regrowth. Many plants have features adapted to their region's fire regime, including trees and shrubs living under frequent-fire regimes that resprout rapidly after burning, savanna trees with thick bark that protects them from damage by ground fires, and many species that require fire to germinate their seeds. The global distribution of savannas is strongly dependent on high fire frequency, which maintains the balance between trees and grasses (Bond, 2008).

Fire has long been used by humans for many purposes, including land clearance, and many fires today are started (either purposefully or not) by people. This has led to the incorrect belief that fire is mainly anthropogenic. But, until recently, there were no comprehensive data on fire that would allow this belief to be challenged. The sporadic, patchy nature of fires, and their large-scale and frequent occurrence in regions with few systematic ground observations, makes data from Earth-observing satellites especially important for quantifying and understanding fire.

Recent developments in remote sensing have dramatically increased our knowledge of fire's extent and variability, and transformed our view of its controls. Globally, the pattern of burned area is overwhelmingly determined by climate. There is a unimodal relationship between annual average burned area and precipitation: in dry climates, fire is limited by low NPP, resulting in insufficient fuel, while in wet climates fire is limited by the fuel being too wet to burn (van der Werf *et al.*, 2008, Prentice *et al.*, 2011). A detailed spatial analysis of remotely sensed burned area over southern Africa has confirmed a strong control by climate (Archibald *et al.*, 2009) and a *negative* relationship between burned area and human population density. It was found that although human activities are responsible for very many ignition events, the area burned depends above all on whether the condition of the vegetation is suitable to support a rapidly spreading fire; while the predominant effect of human settlement is to reduce the availability and continuity of fuel, thus inhibiting fire spread (Harrison *et al.*, 2010a). This contemporary analysis is consistent with palaeodata from ice cores (CO isotopes: Wang *et al.*, 2010) and sedimentary charcoal records (Marlon *et al.*, 2008). Both of these independent sources of palaeofire data show that the present prevalence of biomass burning is at a historic low, having declined dramatically starting in the 1870s, i.e. around the time when European-style agricultural practices became adopted over large regions.

Fires show strong temporal variability on all timescales, including strong seasonality (Roberts *et al.*, 2009). Summer fires typically dominate at high latitudes and Mediterranean-type climates, with winter (dry-season) fires dominating in more subtropical and tropical wet/dry climates. In many areas, fire also shows a strong interannual variability, with sometimes complex geographic patterns (van der Werf *et al.*, 2006). Such areas include tropical rain forests, where fire has been used as an agent of deforestation. However, the 'windows of opportunity' for extensive fires in this biome are infrequent and/or short (e.g. Lewis *et al.*, 2011), and generally confined to the strongest climate anomalies such as El Niño events (Fuller and Murphy, 2006). Paradoxically, therefore, human-caused 'deforestation fires' are under especially strong climatic control (van der Werf *et al.*, 2008; Wooster *et al.*, 2011).

Figure 2.14 combines two data sources to illustrate the spatial variation of the Earth's fire regimes: the Global Fire Emissions Database burned-area product, derived from a combination of MODIS (Moderate Resolution Imaging Spectroradiometer) active-fire and optical remote sensing, and a MODIS product that shows the radiative power generated by individual fires (fire radiative power). Fire radiative power (FRP) relates to the amount of energy released by burning, and when integrated over time is greater in areas where total fuel consumption is higher. Combining both data sets allows the spatial distribution of fire frequency and also fuel consumption per unit area to be derived. This new map can be considered as a fire-regime map, akin to an observationally based version of the map published by Lavorel *et al.* (2007). It shows large, climatically related differences among fire regimes, which are important in controlling the dynamics of vegetation in different biomes.

The data in Figure 2.14 are based on remote-sensing records that have become available only over the last decade. Sedimentary-charcoal data provide a longer-term perspective. Data of the past two millennia show that there is a close association between biomass burning, as measured by relative charcoal deposition, and large-scale temperature anomalies (Marlon *et al.*, 2008). The Roman and Mediaeval warm periods showed biomass-burning maxima and the post-Roman cold period and the Little Ice Age showed minima, even though the magnitude of temperature differences among these periods is only a few tenths of a degree when averaged across the northern hemisphere. The Mediaeval maximum and Little Ice Age minimum of burning are also shown in the ice-core record of Wang *et al.* (2010). Over the past 20,000 years, sedimentary charcoal records a large increase in global biomass burning from the LGM to the Holocene, and other changes in zonal averages of biomass burning that can be predicted by a climate model (A. L. Daniau, P. J. Bartlein, S. P. Harrison *et al.*, unpublished results).

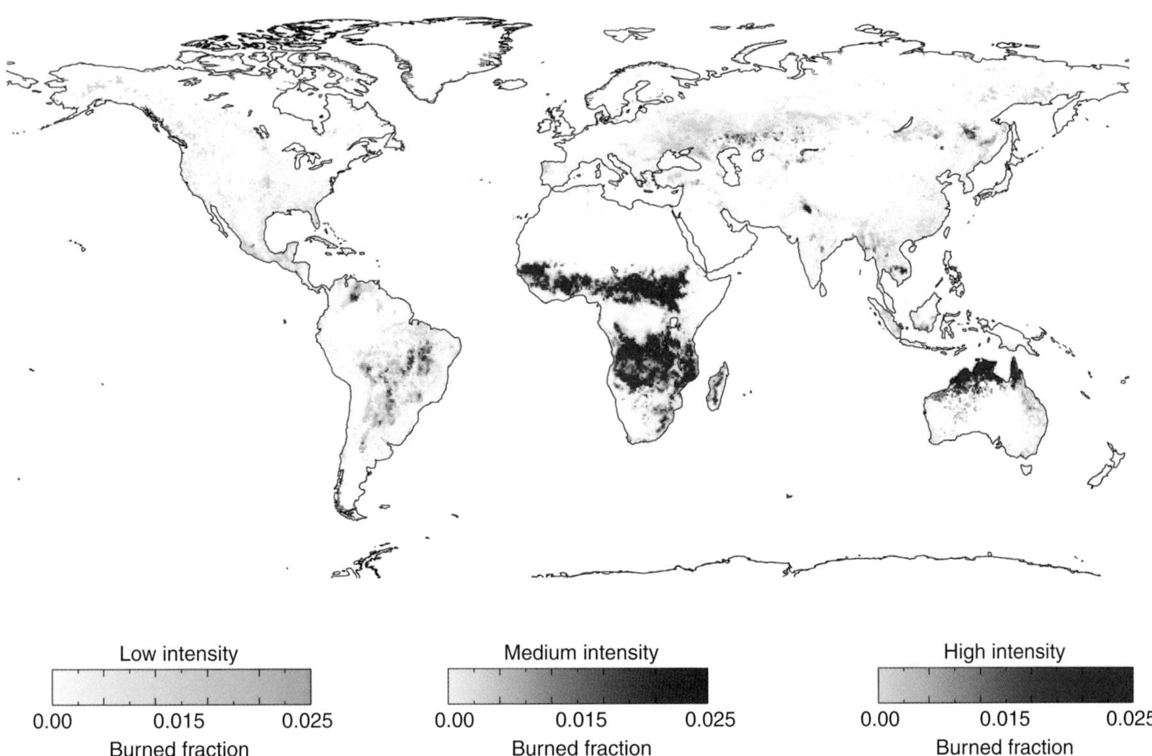

Figure 2.14 Global distribution of fire regimes. The two remotely sensed variables combined here are the burned-area data from the Global Fire Emissions Database (GFED), and fire radiative power (FRP) data – otherwise known as fire intensity – from MODIS, categorized into three intensity classes. Areas having high annual burned areas are those subject to frequent fires, whereas intensity is higher in fires consuming larger amounts of fuel.

2.4 The global carbon cycle

2.4.1 Principles and observations

By far the largest annual fluxes in the global carbon cycle (Figure 2.15) are the exchanges of CO_2 between the atmosphere and land due to GPP and autotrophic and heterotrophic respiration, and the exchanges between the atmosphere and the ocean due to CO_2 uptake in cold regions and release in warm regions. In both cases, the large 'downward' and 'upward' fluxes are approximately in balance. For simplicity, Figure 2.15 omits small (≤ 0.2 Pg C a^{-1}) CO_2 fluxes, including the small source from volcanic activity and small sinks due to the accumulation of peat and deep-sea sediments.

Because the global mixing of CO_2 by the atmospheric circulation takes 1–2 years, there is significant information in the different seasonal cycles of measured atmospheric CO_2 concentrations at different stations (Figure 2.16), especially between different latitudes. The seasonal cycle of atmospheric CO_2 is almost completely dominated by terrestrial ecosystems because the seasonality of fossil-fuel use is slight, and seasonal changes in CO_2 metabolism by the marine biosphere are strongly buffered so they do not greatly affect the atmosphere–ocean exchange of CO_2. In high-northern latitudes the amplitude of the CO_2 seasonal cycle can be as large as 15 ppm. The seasonal cycle arises because of the offset in timing between PAR (maximum at midsummer) and temperature, especially soil temperature (for decomposition – maximum in late summer). Gross primary production is strongly driven by PAR whereas respiration, especially R_H, is driven mainly by temperature. The signal is large in the north because this offset is greatest in boreal climates, and because of the large area of land in the north. In mid-northern latitudes the period of CO_2 drawdown is longer and the amplitude smaller. The signal is weaker in the tropics and weakest in the southern extratropics, where it is in antiphase with the north.

By comparison with these large two-way fluxes, human emissions of CO_2 are quite small. They are important because the release of CO_2 from fossil-fuel burning constitutes a one-way transfer of fossil carbon (which has been isolated for millions of years from the faster-cycling system of the atmosphere, ocean and land) back into the system, and therefore adds to the total amount of carbon in circulation. Emission of CO_2 due to deforestation is also significant for the carbon cycle if the land-use change is permanent, i.e. the forest is not allowed to grow back.

Carbon dioxide emissions from fossil-fuel burning are known with reasonable accuracy from energy statistics and fuel combustion characteristics. They were 7.7 ± 0.4 Pg C a^{-1} on average during 2000–2008 (Le Quéré *et al.*, 2009) and 8.4 ± 0.5 Pg C a^{-1} in 2009 (Friedlingstein *et al.*, 2010). Land-use change was responsible for a further 1.1 ± 0.7 Pg C a^{-1} during the 2000s (Friedlingstein *et al.*, 2010). This is a net amount, composed of a larger gross flux from deforestation that has been partly compensated by uptake in regrowing forests elsewhere (Pan *et al.*, 2011).

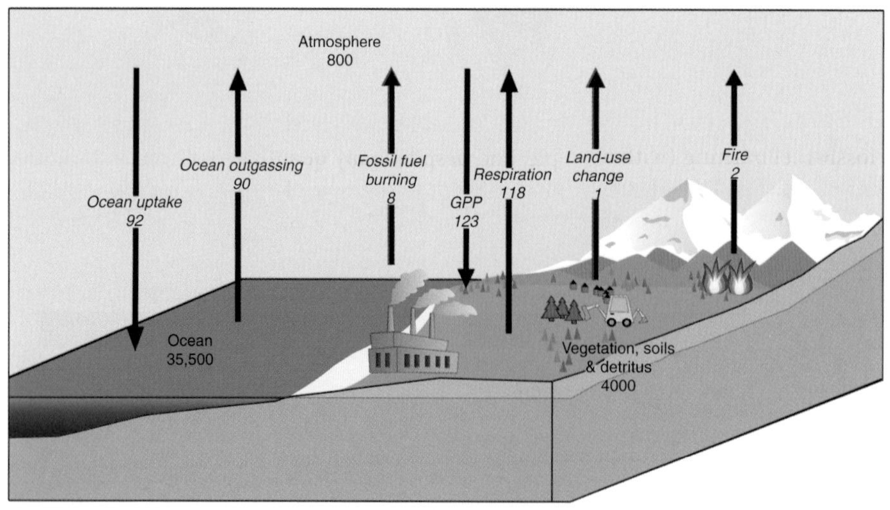

Figure 2.15 Schematic of the global carbon cycle in the 2000s. Stocks (to nearest 100 Pg C): atmosphere, mean of the 2000s based on Mauna Loa data; vegetation, soils and detritus, mean of IPCC Third Assessment Report values (Prentice *et al.*, 2001) plus permafrost soils (Tarnecai *et al.*, 2009); ocean, GLODAP data (Key *et al.*, 2004). Fluxes (to nearest 1 Pg C a^{-1}): GPP, Beer *et al.* (2010); fire, van der Werf *et al.* (2010); fossil-fuel burning, land-use change and ocean uptake, Friedlingstein *et al.* (2010).

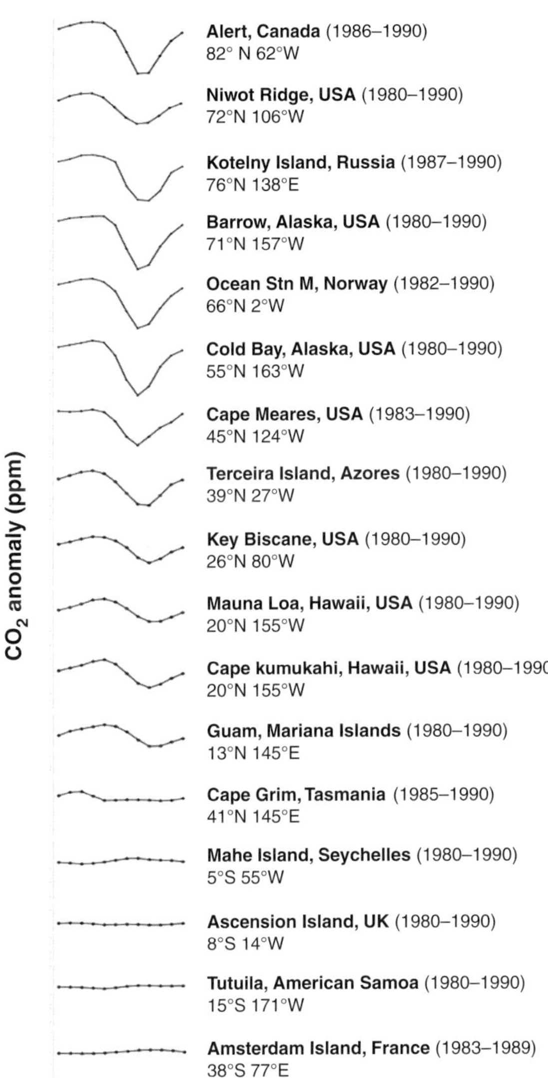

Figure 2.16 Mean seasonal cycles of CO_2 concentration at different locations (as deviations from the annual mean).

Emissions of CO_2 from fossil-fuel burning (with a small addition due to cement production) and land-use change cause CO_2 to accumulate in the atmosphere. But not all of the CO_2 emitted remains there. On average, over each decade, about a quarter is being taken up by the land and another quarter by the oceans. Uptake of CO_2 by land ecosystems involves a release of O_2 whereas CO_2 dissolving in the ocean is O_2-neutral, so precise measurements of O_2:N_2 ratios allow the uptake of CO_2 to be partitioned into land and ocean components (Keeling and Shertz, 1992; Keeling *et al.*, 1996; Battle *et al.*, 2000; Prentice *et al.*, 2001; Manning and Keeling, 2006). Carbon-13 measurements on atmospheric CO_2 can also be used for this purpose, as land carbon has a lower ^{13}C content than ocean carbon. This approach gives similar results. Ocean models also show good agreement, with one another and with the independent observations (Denman *et al.*, 2007; Gruber *et al.*, 2009). Thus, there are several different ways to construct a decadal-average 'carbon budget', all of which now give closely similar numbers. Table 2.1 is based on Friedlingstein *et al.* (2010).

2.4.2 Mechanisms and timescales of CO_2 uptake

What causes the uptake of anthropogenic CO_2 by the oceans and land? For the ocean, it is a well understood physical–chemical process. So long as CO_2 in the atmosphere is increasing, a CO_2 concentration gradient is set up at the ocean surface. This gradient drives immediate uptake into the surface waters, followed (more slowly, over 200–2,000 years) by mixing into the depths (Archer *et al.*, 2009). The expected uptake can be calculated by ocean models and corresponds to that observed (Denman *et al.*, 2007). Further measurements in the ocean confirm the presence of excess CO_2 in an amount consistent with expected total uptake of anthropogenic CO_2 over the industrial era (Sabine and Tanhua, 2010). For the land, the leading hypothesis attributes net CO_2 uptake to the effect of rising CO_2 concentration on photosynthesis (Friedlingstein *et al.*, 1995). When CO_2 is increasing, GPP and thus NPP are expected to be increasing as well – up to a saturation level that has not been reached. Increasing NPP must drive an increase in the soil-carbon pool, but with a time lag. So R_H must increase as well, but the increase lags behind NPP. It is the imbalance of NPP and R_H that causes net CO_2 uptake, according to this hypothesis. Current terrestrial models incorporate this principle, and despite many quantitative uncertainties, they do indicate CO_2 uptake of similar magnitude to that inferred from atmospheric observations (McGuire *et al.*, 2001; Le Quéré *et al.*, 2009).

An important (and often misunderstood) point about ocean and land uptake by each of these mechanisms is that they are *caused by the continuing increase* in atmospheric CO_2 concentration. If atmospheric CO_2 concentration were stable, these 'sinks' would not exist. So, for example, it would be wrong to assume that CO_2 concentration could be stabilized by reducing emissions to the level of current sinks, because the

Table 2.1 Global carbon budgets for 1990–1999 and 2000–2009

All numbers are in Pg C a^{-1}. Positive numbers denote annual fluxes into the atmosphere, negative numbers uptake from the atmosphere. These numbers are averages of annual values calculated by Friedlingstein *et al.* (2010). Ocean uptake is based on the average of four ocean biogeochemistry models; land uptake is calculated as a residual. Closely similar decadal budgets can be obtained by model-independent methods, including the use of observed O_2: N_2 ratios to quantify ocean versus land uptake (Denman *et al.*, 2007).

	1990s	2000s
Fossil-fuel burning and cement	6.4	7.7
Land-use change (net)	1.5	1.1
Land uptake	−2.5	−2.4
Ocean uptake	−2.2	−2.3
Atmospheric increase	3.1	4.1

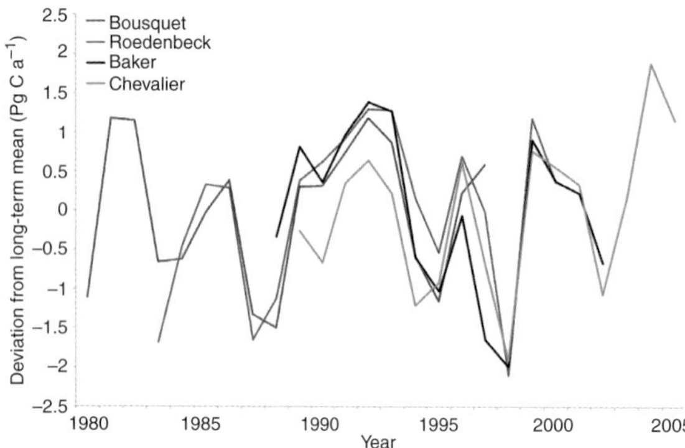

Figure 2.17 Interannual variability in the carbon balance of the land, inferred from atmospheric CO_2 measurements (as deviations from the long-term mean) by Bousquet *et al.* (2000), Rödenbeck *et al.* (2003), Baker *et al.* (2006) and Chevallier *et al.* (2010).

sinks depend on the CO_2 concentration continuing to increase.

The rate of increase of CO_2 in the atmosphere varies considerably from year to year. These variations are related to ENSO, with high growth rates in El Niño years (Bacastow, 1976). The main cause of these variations is the response of terrestrial ecosystems – reduced NPP and/or increased R_H under the hot and dry conditions that accompany El Niño over much of the tropics. Atmospheric inversions exploit the spatial pattern of interannual variability in CO_2 uptake to calculate a land and an ocean contribution (Denman *et al.*, 2007; Figure 2.17), the former being dominant. The ^{13}C content of atmospheric CO_2 also varies, in the way expected, i.e. 'light' carbon increases in the atmosphere when the growth rate increases (Fung *et al.*, 1997). Interannual variability of land CO_2 uptake, similar to that observed, is produced by terrestrial biosphere models (Sitch *et al.*, 2008; Le Quéré *et al.*, 2009; Prentice *et al.*, 2011).

The average flux of CO_2 to the atmosphere through biomass burning (bypassing respiration) averages about 2 Pg C a^{-1} (van der Werf *et al.*, 2010). Deforestation fires, especially those associated with the strongest El Niño events (van der Werf *et al.*, 2008), create variability in this flux. As a result, fire is estimated to contribute roughly one-third of the total interannual variability of land carbon uptake (van der Werf *et al.*, 2006; Prentice *et al.*, 2011).

Dissolution of CO_2 in the ocean, and physiologically stimulated CO_2 uptake by the land, are relatively rapid processes. Other processes can take up further CO_2 but are far slower (Archer *et al.*, 2009). On timescales of ~ 5,000 years, ocean acidification dissolves carbonate ions from sediments, increasing the ocean's buffering capacity so that more CO_2 can be dissolved. Over 80% of emitted CO_2 can be taken up on this timescale. On timescales of ~ 100,000 years, the effect of increased CO_2 on rock weathering is expected to remove the rest.

These processes can be observed 'in action' in the record of past changes in CO_2. Over the past 8,000 years a small (20 ppm) increase in atmospheric CO_2 was driven at least in part by ocean *outgassing* of CO_2, in response to the *removal* of CO_2 by terrestrial biosphere growth after the Last Glacial Maximum (Joos *et al.*, 2004). The Palaeocene–Eocene Thermal Maximum (PETM), an abrupt event that began 56.3 million years ago, provides a case study showing that a period as long as 100,000 to 200,000 years was required to eliminate an initial large (~ 5,000 Pg C) pulse of atmospheric CO_2, as shown by the time taken for the observed 'light' ^{13}C anomaly to disappear (Archer *et al.*, 2009). See McInerney and Wing (2011) and Chapter 3 for more information about the PETM.

The processes that lead to the uptake of the 'last part' of the CO_2 emitted are too slow to be of practical interest. To first order, the amount of CO_2 that remains in the atmosphere for millennia is simply proportional to the amount that is added (Allen *et al.*, 2009; Meinshausen *et al.*, 2009). The other biogenic greenhouse gases, CH_4 and N_2O, behave quite differently. Their concentrations decay in a simple way, with atmospheric lifetimes of 8.7 and 114 years, respectively. In contrast, the concentration of CO_2 does not have a single atmospheric lifetime (Archer *et al.*, 2009). Although the concept of global-warming potential (GWP) is widely used to establish an 'exchange rate' between different greenhouse gases, using CO_2 as a reference, it needs to be borne in mind that there the equivalence is strongly dependent on timescale. The GWP most often used applies to the 100-year timescale, but this is arbitrary, and some emitted CO_2 is expected to remain in the atmosphere for very much longer than 100 years.

2.4.3 Recent trends and implications

Atmospheric CO_2 concentration has increased by ~90 ppm and global mean annual temperature by ~0.7 K since the beginning of the twentieth century. This is a relatively modest warming by the standards of the past million years, and therefore would be expected to have led to relatively minor effects on land and ocean surface processes. Nevertheless, some effects are detectable in observations, including a remarkably large number of cases where plant and animal species distributions have begun to shift in a direction consistent with the observed climate change (Rosenzweig *et al.*, 2007).

There is an approximate linear relationship between NDVI and fAPAR (Knorr and Heimann, 1995). Northern high latitudes experienced a 'greening' during the 1980s and 1990s as shown by NDVI records (Myneni *et al.*, 1997). This greening can be predicted as a consequence of warming, leading to a longer growing season (Lucht *et al.*, 2002). The trend was interrupted by a short-lived cooling due to the eruption of Mount Pinatubo. Other regions have shown opposite trends, including 'browning' in regions such as southern Africa where precipitation has been declining (Knorr *et al.*, 2005). It has been claimed that NPP generally declined over the past decade, from a model based on remotely sensed fAPAR (Zhao and Running, 2010). However the results are dominated by interannual variability, the model ignores the expected gradual increase in NPP due to increasing CO_2 concentration, and the reported trend is not statistically significant.

The observational record of fire is not long enough to determine the existence or absence of a global trend, but an increase in the incidence of large wildfires has been reported from several regions (e.g. Running, 2006). It is plausible that biomass burning might have increased as a result of the warming that has taken place over the past 30 years. But in terms of global totals, human effects may have been more important on this timescale – land-management changes either tending to promote fire (e.g. agricultural abandonment, deforestation fires) or to suppress it (grazing, cropping, land fragmentation). Projections of future trends in fire suggest a continuing increase due to warming (Pechony and Shindell, 2010) although some areas may see reductions due to increasing rainfall (Krawchuk *et al.*, 2009).

The trend in atmospheric CO_2 concentration is dominated by the emissions term and has shown a remarkably constant airborne fraction through time (Knorr, 2009). One calculation has indicated that the airborne fraction is increasing (Le Quéré *et al.*, 2009) although this result is on the margin of significance (Gloor *et al.*, 2010). An increase in the airborne fraction of CO_2 would be in line with the theoretical expectation that the effectiveness of ocean uptake of emissions should gradually decline as the pH of the ocean declines, while the effectiveness of land uptake should also gradually decline as the CO_2 effect on NPP tends towards saturation. Scenario projections of future carbon uptake by the land have suggested a decline in this uptake under higher temperatures (Cramer *et al.*, 2001), and scenario runs with coupled climate–carbon-cycle models have suggested a future increase in the airborne fraction (Friedlingstein *et al.*, 2006). But there are still large differences among the predictions of different models (Sitch *et al.*, 2008).

Alternative approaches to quantifying the 'climate–carbon-cycle feedback' using palaeodata are discussed by Friedlingstein and Prentice (2010) and in Chapters 3 and 4. Work is under way (discussed further in Chapter 5) to set up international benchmarking standards for terrestrial carbon-cycle models, which should help to narrow the large uncertainty in the modelled response of NPP, especially, to rising CO_2 and warming (Denman *et al.*, 2007). At the least, models should be required to reproduce trends in global atmospheric CO_2 concentration (given emissions), interannual variability in the CO_2 growth rate, and the long-term increasing trend in the amplitude of the seasonal cycle of CO_2 (McGuire *et al.*, 2001), as well as the key results of FACE experiments. Narrowing this uncertainty is important because it influences the relationship between future emissions and concentrations of CO_2. Nevertheless, it should be noted that *the major control on future CO_2 concentration is the rate of abatement of emissions* (House *et al.*, 2008), even when the wide range of model responses reported by Friedlingstein *et al.* (2006) is taken into account.

2.5 Prognosis

2.5.1 Features of climate change

Scenarios of twenty-first-century climate change have been developed for modelling purposes, based on different assumptions about the trajectory of future greenhouse-gas (especially CO_2) emissions and concentrations. These are some of the most discussed outputs of the IPCC. All scenarios implicitly accept that (a) CO_2 concentration is rising due to anthropogenic CO_2 emissions; (b) the rise in CO_2 concentrations is producing an increasing radiative forcing that is the main cause of recent global warming; and (c) the rise in CO_2 concentrations cannot be halted immediately. Scenarios differ in the extent to which they include processes or policies that would bring down emissions and eventually stabilize radiative forcing, at some lower or higher level.

Thus, all scenarios involve continued global warming, with some tending to stabilize the warming at lower levels than others. Here we examine additional aspects of climate change that are independent of particular scenarios, and predictable from observations and theory.

We have already noted the observed (and predicted) tendency for wet areas to become wetter and dry areas to become drier as warming proceeds. In addition, it is often stated that global warming will be accompanied by an increase of 'extreme events'. Increasing atmospheric water vapour (Wentz *et al.*, 2007) must certainly increase the number and intensity of rainfall events (Allan and Soden, 2008) because a warmer atmosphere has more latent energy, particularly in the tropics. Warming also implies an inevitable statistical tendency towards more frequent periods with excessively high temperatures. The European heatwave of 2003 was a case in point. The warmer the mean temperature over a given period, the more likely that a given temperature threshold will be exceeded. This simple arithmetic effect, rather than any change in variability, remains the dominant cause for the increasing exceedance of climate norms (Simolo *et al.*, 2011). There is evidence for hurricanes increasing in intensity, but not in frequency (Emanuel, 2005; Hoyos *et al.*, 2008). No trend for strengthening or weakening of ENSO has been observed; there is no accepted theory that would predict the future of ENSO (apart from the prediction that it will continue to exist). Climate models disagree as to whether ENSO is expected to be stronger or weaker in future.

Climate change nevertheless is to some extent predictable, as are its consequences for ecosystems. The greater the radiative forcing that is applied, the larger the climate change inevitably will be. Climate change also unavoidably causes feedbacks as discussed in Chapter 4, and potentially a wide range of impacts as discussed in Chapters 5 and 6; even though many of these impacts and feedbacks are still much less well quantified, or confidently predicted, compared with climate change itself.

2.5.2 Non-linearities and surprises

It is known from observational and modelling studies of the past that abrupt changes of climate can occur in response to smooth forcing of climate. The D–O events are an example. Most concrete examples refer to one particular instability in the climate system, namely the propensity of the AMOC to exist in alternative stable states (Alley, 2007). There is observational evidence for weakening of the AMOC (Bryden *et al.*, 2005) although it is not clear whether this really is a trend. Early climate models predicted that rapid warming could induce a shutdown of the AMOC (with the paradoxical effect of strongly cooling the region around the North Atlantic), but today's models do not show this. Many do show weakening of the AMOC, and therefore a slower rate of warming around the North Atlantic.

Several other potential 'tipping elements' have been identified (Lenton *et al.*, 2008), including the much-discussed (but still poorly quantified) potential for greatly enhanced sea-level rise in the long run due to melting of the West Antarctic and Greenland ice sheets (Notz, 2009; Overpeck *et al.*, 2006). Other tipping elements have regional consequences, but the prognoses for all of them are very uncertain. Ecosystems show a variety of threshold behaviours at the local scale (especially, spatially and temporally sharp transitions between high- and low-fire regimes) but these do not translate into wholesale shifts over large regions. In particular, there is no scientific basis for the widely assumed existence of a 'safe limit' of 2 degrees, or any other specific level, of warming. Rather, given present knowledge, there seems to be an approximate proportionality between climate changes and their effects on ecosystem processes, so that even small changes do have effects, while larger changes lead to larger effects (Scholze *et al.*, 2006). This perspective does not, of course, conflict with the evidence – discussed in Chapter 6 – that the societally relevant impacts of climate change are likely to become steeply more challenging with each degree of warming.

References

Adler, R. F., Huffman, G. J., Chang, A. *et al.* (2003). The Version 2.1 global precipitation climatology project (GPCP) monthly precipitation analysis (1979-present). *Journal of Hydrometeorology*, **4**, 1147–1167.

Allan, R. P. and Soden, B. (2008). Atmospheric warming and the amplification of precipitation extremes. *Science*, **321**, 1481–1484.

Allen, M. R., Frame, D. J., Huntingford, C. *et al.* (2009). Warming caused by cumulative carbon emissions towards the trillionth tonne. *Nature*, **458**, 1163–1166.

Alley, R. B. (2007). Wally was right: predictive ability of the North Atlantic "conveyor belt" hypothesis for abrupt climate change. *Annual Review of Earth and Planetary Science*, **35**, 241–272.

Ammann, C. M., Joos, F., Schimel, D. S., Otto-Bliesner, B. L. and Tomas, R. A. (2007). Solar influence on climate during the past millennium: results from transient simulations with the NCAR Climate System Model. *Proceedings of the National Academy of Science*, **104**, 3713–3718.

Archer, D., Eby, M., Brovkin, V., *et al.* (2009). Atmospheric lifetime of fossil fuel carbon dioxide. *Annual Review of Earth and Planetary Science*, **37**, 117–134.

Archibald, S., Roy, D. P., van Wilgen, B. W. and Scholes, R. J. (2009). What limits fire? An examination of drivers of burnt area in southern Africa. *Global Change Biology*, **15**, 613–630.

Arrhenius, S. (1896). On the influence of carbonic acid in the air upon the temperature of the ground. *Philosophical Magazine and Journal of Science, Series 5*, **41**, 237–276.

Ashok, K., Behera, S. K., Rao, S. A., Weng, H. and Yamagata, T. (2007). El Niño Modoki and its possible teleconnection. *Journal of Geophysical Research*, **112**, C11007.

Bacastow, R. B. (1976). Modulation of atmospheric carbon cycle by the Southern Oscillation. *Nature*, **261**, 116–118.

Baker, D. F., Doney, S. C. and Schimel, D. S. (2006). Variational data assimilation for atmospheric CO_2. *Tellus B*, **58**, 359–365.

Bartlein, P. J., Harrison, S. P., Brewer, S. *et al.* (2011). Pollen-based continental climate reconstructions at 6 and 21 ka: a global synthesis. *Climate Dynamics*, **37**, 775–802.

Battle, M., Bender, M. L., Tans, P. P. *et al.* (2000). Global carbon sinks and their variability inferred from atmospheric O_2 and $\delta^{13}C$. *Science*, **287**, 2467–2470.

Beer, C., Reichstein, M., Tomelleri, E., *et al.* (2010). Terrestrial gross carbon dioxide uptake: global distribution and covariation with climate. *Science*, **329**, 834–838.

Berger, A. (1988). Milankovitch theory and climate. *Reviews of Geophysics*, **6**, 624–657.

Bernacchi, C. J., Pimentel, C. and Long, S. P. (2003). *In vivo* temperature response functions of parameters required to model RuBP-limited photosynthesis. *Plant, Cell and Environment*, **26**, 1419–1430.

Bond, W. J. (2008). What limits trees in C_4 grasslands and savannas? *Annual Review of Ecology, Evolution and Systematics*, **39**, 641–659.

Bond, W. J. and Midgley, G. (2000). A proposed CO_2-controlled mechanism of woody plant invasion in grasslands and savannas. *Global Change Biology*, **6**, 865–869.

Bousquet, P., Peylin, P., Ciais, P. *et al.* (2000). Regional changes in carbon dioxide fluxes of land and oceans since 1980. *Science*, **290**, 1342–1346.

Bowman, D. M. J. S., Balch, J. K., Artaxo, P. *et al.* (2009). Fire in the Earth system. *Science*, **324**, 481–484.

Bradley, R. S. (1999). *Quaternary Paleoclimatology: Methods of Paleoclimatic Reconstruction*, 2nd edn. San Diego, CA: Academic Press.

Broecker, W. S. (2010). *The Great Ocean Conveyor*. Princeton, NJ: Princeton University Press.

Broecker, W. S., Peteet, D. M. and Rind, D. (1985). Does the ocean–atmosphere system have more than one stable mode of operation? *Nature*, **315**, 21–26.

Bryden, H. L., Longworth, H. R. and Cunningham, S. A. (2005). Slowing of the Atlantic meridional overturning circulation at 25° N. *Nature*, **438**, 655–657.

Chahine, M. (1992). The hydrological cycle and its influence on climate. *Nature*, **359**, 373–380.

Chevallier, F., Ciais, P., Conway, T. J. *et al.* (2010). CO_2 surface fluxes at grid point scale estimated from a global 21-year reanalysis of atmospheric measurements. *Journal of Geophysical Research*, **115**, D21307.

Chuvieco, E. (2008). *Earth Observation of Global Change*. Berlin: Springer-Verlag.

CLIMAP (1981). Seasonal reconstructions of the Earth's surface at the last glacial maximum. *Map Series*,

Technical Report MC-36. Boulder, CO: Geological Society of America.

COHMAP Members (1988). Climatic changes of the last 18,000 years: observations and model simulations. *Science*, **241**, 1043–1052.

Compo, G. P., Whitaker, J. S., Sardeshmukh, P. D. *et al.* (2011). The twentieth century reanalysis project. *Quarterly Journal of the Royal Meteorological Society*, **137**, 1–28.

Cramer, W., Bondeau, A., Woodward, F. I. *et al.* (2001). Global responses of terrestrial ecosystems to changes in CO_2 and climate. *Global Change Biology*, **7**, 357–373.

Crowley, T. J. (2000). Causes of climate change over the past 1000 years. *Science*, **289**, 270–277.

Denman, K. L., Brasseur, G., Chidthaisong, A. *et al.* (2007). Couplings between changes in the climate system and biogeochemistry. In *Climate Change 2007: The Physical Science Basis. Contribution of Working Group I to the Fourth Assessment Report of the Intergovernmental Panel on Climate Change*, eds. S. Solomon, D. Qin, M. Manning *et al.* Cambridge: Cambridge University Press, pp. 499–587.

Dewar, R. C. (1996). The correlation between plant growth and intercepted radiation: an interpretation in terms of optimal plant nitrogen content. *Annals of Botany*, **78**, 125–136.

Edwards, T. L., Crucifix, M. and Harrison, S. P. (2007). Using the past to constrain the future: how the palaeorecord can improve estimates of global warming. *Progress in Physical Geography*, **31**, 481–500.

Emanuel, K. (2005). Increasing destructiveness of tropical cyclones over the past 30 years. *Nature*, **436**, 686–688.

Farquhar, G. D., Caemmerer, S. and Berry, J. A. (1980). A biochemical model of photosynthetic CO_2 assimilation in leaves of C_3 species. *Planta*, **149**, 78–90.

Forster, P. M., Ramaswamy, V., Artaxo, P. *et al.* (2007). Changes in atmospheric composition and radiative forcing. In *Climate Change 2007: The Physical Science Basis. Contribution of Working Group I to the Fourth Assessment Report of the Intergovernmental Panel on Climate Change*, eds. S. Solomon, D. Qin, M. Manning *et al.* Cambridge: Cambridge University Press.

Friedlingstein, P. and Prentice, I. C. (2010). Carbon–climate feedbacks: a review of model and observation based estimates. *Current Opinion in Environmental Sustainability*, **2**, 251–257.

Friedlingstein, P., Fung, I., Holland, E. *et al.* (1995). On the contribution of CO_2 fertilization to the missing biospheric sink. *Global Biogeochemical Cycles*, **9**, 541–556.

Friedlingstein, P., Cox, P., Betts, R. *et al.* (2006). Climate–carbon cycle feedback analysis: results from the C4MIP model intercomparison. *Journal of Climate*, **19**, 3337–3353.

Friedlingstein, P., Houghton, R. A., Marland, G. *et al.* (2010). Update on CO_2 emissions. *Nature Geoscience*, **3**, 811–812.

Fuller, D. and Murphy, K. (2006). The ENSO–fire dynamic in insular southeast Asia. *Climatic Change*, **74**, 435–455.

Fung, I., Field, C. B., Berry, J. A. *et al.* (1997). Carbon-13 exchanges between the atmosphere and biosphere. *Global Biogeochemical Cycles*, **11**, 507–533.

Gallego-Sala, A.V., Clark, J. M., House, J. I. *et al.* (2010). Bioclimatic envelope model of climate change impacts on blanket peatland distribution in Great Britain. *Climate Research*, **45**, 151–162.

Galloway, J. N., Townsend, A. R., Erisman, J. W. *et al.* (2008). Transformation of the nitrogen cycle: recent trends, questions, and potential solutions. *Science*, **320**(5878), 889–892.

GARP (1969). An introduction to GARP, *GARP Publication Series*, **1**, WMO/ICSU.

Gedney, N., Cox, P. M., Betts, R. A. *et al.* (2006). Detection of a direct carbon dioxide effect in continental river runoff records. *Nature*, **439**, 835–838.

Giglio, L., Randerson, J. T., van der Werf, G. R. *et al.* (2010). Assessing variability and long-term trends in burned area by merging multiple satellite fire products. *Biogeosciences*, **7**, 1171–1186.

Gloor, M., Sarmiento, J. L. and Gruber, N. (2010). What can be learned about carbon cycle climate feedbacks from the CO_2 airborne fraction? *Atmospheric Chemistry and Physics*, **10**, 7739–7751.

Gray, L. J., Beer, J., Geller, M. *et al.* (2010). Solar influences on climate. *Reviews of Geophysics*, **48**, RG4001.

Gruber, N., Gloor, M., Mikalof Fletcher, S. E. *et al.* (2009). Oceanic sources, sinks, and transport of atmospheric CO_2. *Global Biogeochemical Cycles*, **23**, GB1005.

Haigh, J. D. (2003). The effects of solar variability on the Earth's climate. *Philosophical Transactions of the Royal Society of London, Series A*, **361**, 95–111.

Harrison, S. P. and Prentice, I. C. (2003). Climate and CO_2 controls on global vegetation distribution at the last glacial maximum: analysis based on palaeovegetation data, biome modeling and palaeoclimate simulations. *Global Change Biology*, **9**, 983–1004.

Harrison, S. P. and Sanchez-Goñi, M. F. (2010). Global patterns of vegetation response to millennial-scale variability and rapid climate change during the last glacial period. *Quaternary Science Reviews*, **29**, 2957–2980.

Harrison, S. P., Kohfeld, K. E., Roelandt, C. and Claquin, T. (2001). The role of dust in climate changes today, at the last glacial maximum and in the future. *Earth-Science Reviews*, **54**, 43–80.

Harrison, S. P., Kutzbach, J. E., Liu, Z. *et al.* (2003). Mid-Holocene climates of the Americas: a dynamical response to changed seasonality. *Climate Dynamics*, **20**, 663–688.

Harrison, S. P., Marlon, J. L. and Bartlein, P. J. (2010a). Fire in the Earth system. In *Changing Climates, Earth Systems and Society*, ed. J. Dodson. Dordrecht: Springer-Verlag, pp. 21–48.

Harrison, S. P., Prentice, I. C., Barboni, D. *et al.* (2010b). Ecophysiological and bioclimatic foundations for a global plant functional classification. *Journal of Vegetation Science*, **21**, 300–317.

Haxeltine, A. and Prentice, I. C. (1996). A general model for the light use efficiency of primary production. *Functional Ecology*, **10**, 551–561.

Hegerl, G. C., Zwiers, F. W., Braconnot, P. *et al.* (2007). Understanding and attributing climate change. *Climate Change 2007: The Physical Science Basis. Contribution of Working Group I to the Fourth Assessment Report of the Intergovernmental Panel on Climate Change*, eds. S. Solomon, D. Qin, M. Manning *et al.* Cambridge: Cambridge University Press.

Held, I. M. and Soden, B. J. (2006). Robust responses of the hydrological cycle due to global warming. *Journal of Climate*, **19**, 5686–5699.

Hickler, T., Smith, B., Prentice, I. C. *et al.* (2008). CO_2 fertilization in temperate FACE experiments not representative of boreal and tropical forests. *Global Change Biology*, **14**, 1531–1542.

House, J. I., Huntingford, C., Knorr, W. *et al.* (2008). What do recent advances in quantifying climate and carbon cycle uncertainties mean for climate policy? *Environmental Research Letters*, **3**, 044002.

Hoyos, C. D., Agudelo, P. A., Webster, P. and Curry, J. A. (2008). Deconvolution of the factors contributing to the increase in hurricane intensity. *Science*, **312**, 94–97.

James, I. N. (1994). *Introduction to Circulating Atmospheres.* Cambridge: Cambridge University Press.

Jansen, E., Overpeck, J., Briffa, K. R. *et al.* (2007). Palaeoclimate. *Climate Change 2007: The Physical Science Basis. Contribution of Working Group I to the Fourth Assessment Report of the Intergovernmental Panel on Climate Change*, eds. S. Solomon, D. Qin, M. Manning *et al.* Cambridge: Cambridge University Press.

John, V. O., Allan, R. P. and Soden, B. J. (2009). How robust are observed and simulated precipitation responses to tropical ocean warming? *Geophysical Research Letters*, **36**, L14702.

Johnsen, S. J., Clausen, H. B., Dansgaard, W. *et al.* (1992). Irregular interstadials recorded in a new Greenland ice core. *Nature*, **359**, 407–12.

Johnson, D. R. (1989). The forcing and maintenance of global monsoonal circulations: an isentropic analysis. *Advances in Geophysics*, **31**, 43–316.

Jolly, D., Prentice, I. C., Bonnefille, R. *et al.* (1998). Biome reconstruction from pollen and plant macrofossil data for Africa and the Arabian peninsula at 0 and 6000 years. *Journal of Biogeography*, **25**, 1007–1027.

Joos, F., Gerber, S., Prentice, I. C., Otto-Bliesner, B. and Valdes, P. J. (2004). Transient simulations of Holocene atmospheric carbon dioxide and terrestrial carbon since the Last Glacial Maximum. *Global Biogeochemical Cycles*, **18**, GB2002.

Kalnay, E., Kanamitsu, M., Kistler, R. *et al.* (1996). The NCEP/NCAR 40-year reanalysis project. *Bulletin of the American Meteorological Society*, **77**, 437–471.

Keeling, C. D. (1958). The concentration and isotopic abundances of atmospheric carbon dioxide in rural areas. *Geochimica et Cosmochimica Acta*, **13**, 322–334.

Keeling, R. F. and Shertz, S. R. (1992). Seasonal and interannual variations in atmospheric oxygen and implications for the global carbon cycle. *Nature*, **358**, 723–727.

Keeling, R. F., Piper, S. C. and Heimann, M. (1996). Global and hemispheric CO_2 sinks deduced from changes in atmospheric O_2 concentration. *Nature*, **381**, 218–221.

Keeling, C. D., Bacastow, R. B., Carter, A. F. *et al.* (1989). A three dimensional model of atmospheric CO_2 transport based on observed winds: 1. Analysis of observational data. In *Aspects of Climate Variability in the Pacific and the Western Americas*, ed. D. H. Peterson. Washington, DC: American Geophysical Union, pp. 165–236.

Key, R. M., Kozyr, A., Sabine, C. L., *et al.* (2004). A global ocean carbon climatology: Results from GLODAP. *Global Biogeochemical Cycles*, **18**, GB4031.

Knorr, W. (2009). Is the airborne fraction of anthropogenic CO_2 emissions increasing? *Geophysical Research Letters*, **36**, L21710.

Knorr, W. and Heimann, M. (1995). Impact of drought stress and other factors on seasonal land biosphere CO_2 exchange studied through an atmospheric tracer transport model. *Tellus B*, **47**, 471–489.

Knorr, W., Scholze, M., Gobron, N., Pinty, B. and Kaminski, T. (2005). Global-scale drought caused atmospheric CO_2 increase. *EOS Transactions*, **86**, 178–181.

Knutti, R. and Hegerl, G. C. (2008). The equilibrium sensitivity of the Earth's temperature to radiation changes. *Nature Geoscience*, **1**, 735–743.

Knutti, R., Meehl, G. A., Allen, M. R. and Stainforth, D. A. (2006). Constraining climate sensitivity from the seasonal cycle in surface temperature. *Journal of Climate*, **19**, 4224–4233.

Köppen, W. (1918). Klassifikation der Klimate nach Temperatur, Niederschlag und Jahresablauf (Classification of climates according to temperature, precipitation and seasonal cycle). *Petermanns Geographische Mitteilungen*, **64**, 193–203, 243–248.

Krawchuk, M., Moritz, M., Parisien, M.-A., Van Dorn, J. and Hayhoe, K. (2009). Global pyrogeography: the current and future distribution of wildfire. *PLoS ONE*, **4**, e5102.

Krüger, T. (2008). *Die Entdeckung der Eiszeiten*. Basel: Schwalbe Verlag.

Kundzewicz, Z. W., José Mata, L., Arnell, N. *et al.* (2007). Freshwater resources and their management. In *Climate Change 2007: Impacts, Adaptation and Vulnerability. Contribution of Working Group II to the Fourth Assessment Report of the Intergovernmental Panel on Climate Change*, eds. M. L. Parry, O. F. Canziani, J. P. Palutikof, P. J. van der Linden and C. E. Hanson. Cambridge: Cambridge University Press.

Kutzbach, J. E. and Street-Perrott, F. A. (1985). Milankovitch forcing of fluctuations in the level of tropical lakes from 18 to 0 kyr BP. *Nature*, **317**, 130–134.

Kutzbach, J. E. and Guetter, P. J. (1986) The influence of changing orbital parameters and surface boundary conditions on climate simulations for the past 18,000 years. *Journal of the Atmospheric Sciences*, **43**, 1726–1759.

Lamb, H. H. (1982). *Climate, History and the Modern World*. London: Methuen.

Lavorel, S., Flannigan, M. D., Lambin, E. F. and Scholes, M. C. (2007). Vulnerability of land systems to fire: interactions between humans, climate, the atmosphere and ecosystems. *Mitigation and Adaptation Strategies for Global Change*, **12**, 33–53.

Le Quéré, C., Raupach, M. R., Canadell, J. G. *et al.* (2009). Trends in the sources and sinks of carbon dioxide. *Nature Geoscience*, **2**, 831–836.

Lehman, S. J. and Keigwin, L. D. (1992). Sudden changes in North Atlantic circulation during the last deglaciation. *Nature*, **356**, 757–762.

Lenton, T. M., Held, H., Kriegler, E. *et al.* (2008). Tipping elements in the Earth's climate system. *Proceedings of the National Academy of Science*, **105**, 1786–1793.

Levin, I. and Kromer, B. (1997). Twenty years of high precision atmospheric $^{14}CO_2$ observations at Schauinsland station, Germany. *Radiocarbon*, **39**, 205–218.

Lewis, S. L., Brado, P. M., Phillips, O. L., van der Heijden, G. M. F. and Nepstad, D. (2011). The 2010 Amazon drought. *Science*, **331**, 554.

Libby, W. J. (1952). *Radiocarbon Dating*. Chicago, IL: University of Chicago Press.

Loulergue, L., Schilt, A., Spahni, R. *et al.* (2008). Orbital and millennial-scale features of atmospheric CH_4 over the past 800,000 years. *Nature*, **453**, 383–386.

Lucht, W., Prentice, I. C., Myneni, R. B. *et al.* (2002). Climatic control of the high-latitude vegetation greening trend and Pinatubo effect. *Science*, **296**, 1687–1689.

Lumpkin, R. and Speer, K. (2007). Global ocean meridional overturning. *Journal of Physical Oceanography*, **37**, 2550–2562.

Lüthi, D., Le Floch, M., Bereiter, B. *et al.* (2008). High-resolution carbon dioxide concentration record 650,000–800,000 years before present. *Nature*, **453**, 379–382.

MacFarling Meure, C., Etheridge, D., Trudinger, C. *et al.* (2006). Law Dome CO_2, CH_4 and N_2O ice core records extended to 2000 years BP. *Geophysical Research Letters*, **33**, L14810.

Manabe, S. and Bryan, K. (1969). Climate calculations with a combined ocean–atmosphere model. *Journal of the Atmospheric Sciences*, **26**, 786–789.

Mann, M. E., Bradley, R. S. and Hughes, M. K. (1999). Northern hemisphere temperatures during the past millennium: inferences, uncertainties, and limitations. *Geophysical Research Letters*, **26**, 759–762.

Manning, A. C. and Keeling, R. F. (2006). Global oceanic and land biotic carbon sinks from the Scripp. atmospheric oxygen flask sampling network. *Tellus B*, **58**, 95–116.

MARGO Project Members (2009). Constraints on the magnitude and patterns of ocean cooling at the last glacial maximum. *Nature Geoscience*, **2**, 127–132.

Marlon, J. R., Bartlein, P. J., Carcaillet, C. *et al.* (2008). Climate and human influences on biomass burning over the past two millennia. *Nature Geoscience*, **1**, 697–702.

McGuire, A. D., Sitch, S., Clein, J. S. *et al.* (2001). Carbon balance of the terrestrial biosphere in the twentieth century: analyses of CO_2, climate and land-use effects with four process-based ecosystem models. *Global Biogeochemical Cycles*, **15**, 183–206.

McInerney, F. A. and Wing, S. L. (2011). The Paleocene–Eocene Thermal Maximum: a perturbation of carbon cycle, climate, and biosphere with implications for the future. *Annual Review of Earth and Planetary Science*, **39**, 489–516.

Medlyn, B. E., Duursma, R. A., Eamus, D. *et al.* (2011). Reconciling the optimal and empirical approaches to modelling stomatal conductance. *Global Change Biology*, **17**, 2134–2144.

Meehl, G. A., Arblaster, J. M., Mathes, K., Sassi, F. and van Loon, H. (2009). Amplifying the Pacific climate system response to a small 11 year solar cycle forcing. *Science*, **325**, 1114–1118.

Meinshausen, M., Meinshausen, N., Hare, W., *et al.* (2009). Greenhouse-gas emission targets for limiting global warming to 2 °C. *Nature*, **458**, 1158–1162.

Milankovitch, M. (1941). Kanon der Erdebestrahlung und seine Anwendung auf das Eiszeitenproblem. *Edition Spéciale, Académie Royale Serbe*, **133**.

Myhre, G., Highwood, E. J., Shine, K. P. and Stordal, F. (1998). New estimates of radiative forcing due to well mixed greenhouse gases. *Geophysical Research Letters*, **25**, 2715–2718.

Myneni, R. C., Keeling, C. D., Tucker, C. J., Asrar, G. and Nemani, R. R. (1997). Increased plant growth in the northern high latitudes from 1981 to 1991. *Nature*, **386**: 698–702.

Norby, R. J., DeLucia, E. H., Gielen, B. *et al.* (2005). Forest response to elevated CO_2 is conserved across a broad range of productivity. *Proceedings of the National Academy of Science* **102**, 18052–18056.

Notz, D. (2009). The future of ice sheets and sea ice: between reversible retreat and unstoppable loss. *Proceedings of the National Academy of Science*, **106**, 20590–20595.

Overpeck, J. T., Otto-Bliesner, B. L., Miller, G.H. *et al.* (2006). Ice sheet instability and rapid sea-level rise. *Science*, **311**, 1747–1750.

Palmén, E. and Newton, C. W. (1969). *Atmospheric Circulation Systems*. New York, NY: Academic Press,.

Pan, Y., Birdsey, R. A., Fang, J. *et al.* (2011). A large and persistent carbon sink in the world's forests. *Science*, **333**, 988–993.

Pauluis, O., Czaja, C. and Korty, R. (2008). The global atmospheric circulation on moist isentropes. *Nature*, **321**, 1075–1078.

Pauluis, O., Czaja, A. and Korty, R. (2010). The global atmospheric circulation in moist isentropic coordinates. *Journal of Climate*, **23**, 3077–3093.

Pechony, O. and Shindell, D. T. (2010). Driving forces of global wildfires over the past millennium and the forthcoming century, *Proceedings of the National Academy of Sciences*, **107**, 3382–3387.

Peel, M. C. (2009). Hydrology: catchment vegetation and runoff. *Progress in Physical Geography*, **33**, 837–844.

Peixoto, J. P. and Oort, A. H. (1992). *Physics of Climate*. Melville, NY: AIP Press.

Petit, J. R., Basile, I., Leruyuet, A. *et al.* (1997). Four climate cycles in Vostok ice core. *Nature*, **387**, 359–360.

Philander, S. G. (1990). *El Niño, La Niña and the Southern Oscillation*. San Diego, CA: Academic Press.

Prentice, I. C. and Harrison, S. P. (2009). Ecosystem effects of CO_2 concentration: evidence from past climates. *Climate of the Past*, **5**, 297–307.

Prentice, I. C., Heimann, M. and Sitch, S. (2000). The carbon balance of the terrestrial biosphere: ecosystem models and atmospheric observations. *Ecological Applications*, **10**, 1553–1573.

Prentice, I. C., Farquhar, G. D., Fasham, M. J. R. *et al.* (2001). The carbon cycle and atmospheric carbon dioxide. In *Climate Change 2001: The Scientific Basis*, eds. J.T. Houghton, Y. Ding, D. J. Griggs *et al.* Cambridge: Cambridge University Press.

Prentice, I. C., Kelley, D. I., Harrison, S. P. *et al.* (2011). Modeling fire and the terrestrial carbon balance. *Global Biogeochemical Cycles*, GB3005.

Rayner, N. A., Parker, D. E., Horton, E. B. *et al.* (2003). Global analyses of sea surface temperature, sea ice, and night marine air temperature since the late nineteenth century. *Journal of Geophysical Research*, **108**, 4407.

Roberts, G., Wooster, M. J. and Lagoudakis, E. (2009). Annual and diurnal biomass burning temporal dynamics. *Biogeosciences*, **6**, 849–866.

Rödenbeck, C., Houweling, S., Gloor, M. and Heimann, M. (2003). CO_2 flux history 1982–2001 inferred from atmospheric data using a global inversion of atmospheric transport. *Atmospheric Chemistry and Physics*, **3**, 1919–1964.

Roe, G. H. and Baker, M. B. (2007). Why is climate sensitivity so unpredictable? *Science*, **318**, 629–632.

Rosenzweig, C., Casassa, G., Karoly, D. *et al.* (2007). Assessment of observed changes and responses in natural and managed systems. In *Climate Change 2007: Impacts, Adaptation and Vulnerability. Contribution of Working Group II to the Fourth Assessment Report of the Intergovernmental Panel on Climate Change*, eds. M. L. Parry, O. F. Canziani, J. P. Palutikof, P. J. van der Linden and C. E. Hanson. Cambridge: Cambridge University Press.

Running, S. W. (2006). Is global warming causing more, larger wildfires? *Science*, **313**, 927–928.

Sabine, C. J. and Tanhua, T. (2010). Estimation of anthropogenic CO_2 inventories in the ocean. *Annual Review of Marine Sciences*, **2**, 175–198.

Schilt, A., Baumgartner, M., Blunier, T. *et al.* (2010). Glacial–interglacial and millennial scale variations in the atmospheric nitrous oxide concentration during the last 800,000 years. *Quaternary Science Reviews*, **29**, 182–192.

Schmidt, G. A., Ruedy, R. A., Miller, R. L. and Lacis, A. A. (2010). Attribution of the present-day total greenhouse effect. *Journal of Geophysical Research*, **115**, D20106.

Schmittner, A., Urban, N. M., Shakun, J. D. *et al.* (2011). Climate sensitivity estimated from temperature reconstructions of the last glacial maximum. *Science*, **334**, 1385–1388.

Schmitz, W.J., Jr (1996). On the world ocean circulation. Vol. II: The Pacific and Indian oceans–a global update. Technical Report WHOI-96-O8, Woods Hole Oceanographic Institution.

Scholze, M., Knorr, W., Arnell, N. W. and Prentice, I. C. (2006). A climate change risk analysis for world ecosystems. *Proceedings of the National Academy of Sciences*, **103**, 13 116–13 120.

Simolo, C., Brunetti, M., Maugeri, M. and Nanni, T. (2011). Evolution of extreme temperatures in a warming climate. *Geophysical Research Letters*, **38**, L16701.

Sitch, S., Huntingford, C., Gedney, N. *et al.* (2008). Evaluation of the terrestrial carbon cycle, future plant geography and climate–carbon cycle feedbacks using 5 Dynamic Global Vegetation Models (DGVMs). *Global Change Biology*, **14**, 1–25.

Smith, T. M., Reynolds, R. W., Peterson, T. C. and Lawrimore, J. (2008). Improvements to NOAA's historical merged land-ocean surface temperature analysis (1880–2006). *Journal of Climate*, **21**, 2283–2296.

Soden, B. J. and Vecchi, G. A. (2011). The vertical distribution of cloud feedback in coupled ocean–atmosphere models. *Geophysical Research Letters*, **38**, L12704.

Soden, B. J., Wetherald, R. T., Stenchikov, G. L. and Robock, A. (2002). Global cooling after the eruption of Mount Pinatubo: a test of climate feedback by water vapor. *Science*, **296**, 727–730.

Stuiver, M. and Quay, P. D. (1981). Atmospheric ^{14}C changes resulting from fossil fuel CO_2 release and cosmic ray flux variability. *Earth and Planetary Science Letters*, **53**, 349–362.

Suess, H. E. (1955). Radiocarbon content in modern wood. *Science*, **122**, 415–417.

Tarnocai, C., Canadell, J. G., Schuur, E. A. G., *et al.* (2009). Soil organic carbon pools in the northern circumpolar permafrost regon. *Global Biogeochemical Cycles*, **23**, GB2023.

Thompson, D. W. J., Kennedy, J. J., Wallace, J. M., and Jones, P. D. (2008). A large discontinuity in the mid-twentieth century in observed global-mean surface temperature. *Nature*, **453**, 646–649.

Trenberth, K., Jones, P. D., Ambenje, P. *et al.* (2007a). Observations: surface and atmospheric climate change. In *Climate Change 2007: The Physical Science Basis. Contribution of Working Group I to the Fourth Assessment Report of the Intergovernmental Panel on Climate Change*, eds. S. Solomon, D. Qin, M. Manning, *et al.* Cambridge: Cambridge University Press.

Trenberth, K., Smith, L., Qian, L., Dai, A. and Fasullo, J. (2007b). Estimates of the global water budget and its annual cycle using observational and model data. *Journal of Hydrometeorology*, **8**, 758–769.

Tucker, C. J. (1979). Red and photographic infrared linear combinations for monitoring vegetation. *Remote Sensing of Environment*, **8**, 127–150.

Uppala, S. M., Kållberg, P. W., Simmons, A. J. *et al.* (2005). The ERA-40 re-analysis. *Quarterly Journal of the Royal Meteorological Society*, **131**, 2961–3012.

van der Werf, G. R., Randerson, J. T., Giglio, L. *et al.* (2006). Interannual variability in global biomass burning emissions from 1997 to 2004. *Atmospheric Chemistry and Physics*, **6**, 3423–2441.

van der Werf, G. R., Randerson, J. T., Giglio, L., Gobron, N. and Dolman, A. J. (2008). Climate controls on the variability of fires in the tropics and subtropics. *Global Biogeochemical Cycles*, **22**, GB3028.

van der Werf, G. R., Randerson, J. T., Giglio, L. *et al.* (2010). Global fire emissions and the contribution of deforestation, savanna, forest, agricultural, and peat fires (1997–2009). *Atmospheric Chemistry and Physics*, **10**, 11707–11735.

van Loon, H., Meehl, G. A. and Arblaster, J. M. (2004). A decadal solar effect in the tropics in July–August. *Journal of Atmospheric and Solar-Terrestrial Physics*, **66**, 1767–1778.

von Post, L. (1918). Skogsträdspollen i sydsvenska torvmosselagerföljder. *Forhandlinger ved de skandinaviske naturforskeres 16. møte i Kristiania den 10.-15. juli 1916* (Scandinavian Scientist Conference Proceedings), pp. 432–465.

Wang, J., Chappellaz, J., Park, K. and Mak, J. E. (2010). Large variations in southern hemisphere biomass burning during the last 650 years. *Science*, **330**, 1663–1666.

Watts, W. A. (1980). Regional variations in the response of vegetation of late-glacial climate events in Europe. In *The Late-Glacial of Northwest Europe*, eds, J. J. Lowe, J. M. Gray and J. E. Robinson. New York, NY: Pergamon, pp. 1–22.

Wentz, F. J., Ricciardulli, L., Hilburn, K. and Mears, C. (2007). How much more rain will global warming bring? *Science*, **317**, 233–235.

Wooster, M. J., Perry, G. L. W. and Zoumas, A. (2011). Fire, drought and El Niño relationships on Borneo during the pre-MODIS era (1980–2000). *Biogeosciences Discussions*, **8**, 975–1013.

Wright, H. E., Jr, Kutzbach, J. E., Webb III, T. *et al.*, eds. (1993). *Global Climates Since the Last Glacial Maximum*. Minneapolis, MN: University of Minnesota.

Wunsch, C., Heimbach, P., Ponte, R. and Fukumori, I. (2009). The global general circulation of the ocean estimated by the ECCO-consortium. *Oceanography*, **22**, 88–103.

Zhang, L., Potter, N., Hickel, K., Zhang, Y. Q. and Shao, Q. X. (2008). Water balance modelling over variable timescales based on the Budyko framework: I. Model development and testing. *Journal of Hydrology*, **360**, 117–131.

Zhao, M. and Running, S. (2010). Drought-induced reduction in global terrestrial net primary production from 2000 through 2009. *Science*, **329**: 940–943.

Chapter 3

How has climate responded to natural perturbations?

Eric W. Wolff, Sandy P. Harrison, Reto Knutti, Maria Fernanda Sanchez-Goñi, Oliver Wild, Anne-Laure Daniau, Valérie Masson-Delmotte, I. Colin Prentice and Renato Spahni

In this chapter, we describe and explain some of the patterns observed in the behaviour of Earth's climate system. We explain some of the causes of the climate's natural variability, setting contemporary climate change in its longer-term context. We describe the various lines of evidence about climate forcing and the feedbacks that determine the responses to perturbations, and the way in which reconstructions of past climates can be used in combination with models and contemporary observations of change.

3.1 Introduction

Human activity is creating a major perturbation to the Earth, directly affecting the composition of the atmosphere, and the nature of the land surface. These direct effects are expected in turn to cause impacts on numerous aspects of the Earth: regional climates, the distribution of ice and vegetation types, and perhaps the circulation of the oceans. Numerous interactions within the Earth system must be understood to enable prediction of the effects of the imposed changes. Models used for prediction are underpinned by a physical understanding of the climate. Aspects of these models are generally tuned to the Earth we experience today, but it is their representation of Earth's response to change that really interests us.

By observing the Earth, both directly in the present and indirectly in the past, we learn about processes and feedbacks that models need to represent; and we can test whether the real Earth has responded to perturbations with the speed and magnitude that our models display. The ultimate goal is to use such observations to test models quantitatively, and to calibrate some of their less-constrained parameters. This goal cannot be fully realized unless we have knowledge of both the perturbation and the spatial pattern and magnitude of the response. This chapter concentrates on observations of how the Earth's climate has responded to perturbations in the past. To set the scene, we first discuss what constitutes a perturbation, and how we can observe its consequences. After a summary of some of the major features of climate evolution, we discuss how observations of the past have been used to constrain (quantitatively or qualitatively) some large-scale features of climate. We then discuss a few examples of particular perturbations in the past for which observations have informed, or in some cases transformed, our view of the Earth.

3.2 Climate perturbations

Earth's climate is the result of the physical requirement to maintain a balance between energy reaching and leaving the atmosphere. The Sun supplies the incoming energy. Some of this energy is reflected by light-coloured surfaces in the atmosphere (e.g. clouds) or at the surface (e.g. ice). The remainder is radiated away. Heat is transported around the Earth in both the atmosphere and the ocean, and the patterns of this transport modulate climate at any given location. The controls on the Earth's surface temperature, including the extremely important role of greenhouse gases, are discussed in more detail in Chapter 2. However, from this simple analysis, it is possible to define the kinds of perturbation that could have altered our climate over time (Figure 3.1).

The total amount of incoming energy depends on the strength of the Sun, and on the distance of the Earth

Understanding the Earth System: Global Change Science for Application, eds. Sarah E. Cornell, I. Colin Prentice, Joanna I. House and Catherine J. Downy. Published by Cambridge University Press

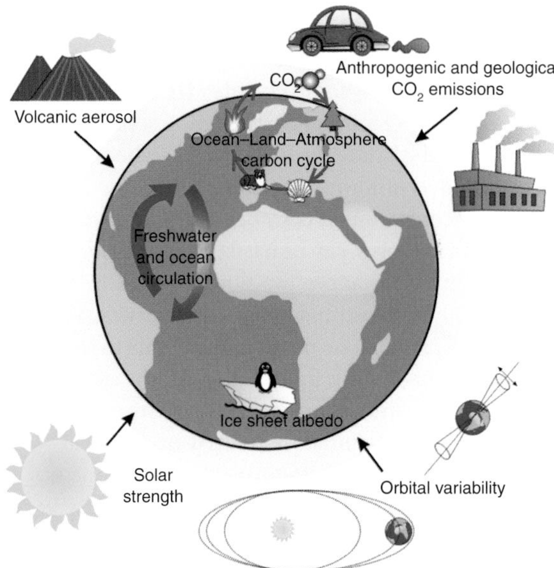

Figure 3.1 A summary of external and internal perturbations to which Earth's climate system may respond.

from the Sun. The energy output of the Sun has slowly increased over its 4.5-billion-year lifetime, but there are also small perturbations to its output over different timescales, of which the best known is the 11-year sunspot cycle. The Earth's orbit around the Sun varies over thousands to millions of years in ways that slightly affect the total energy received on Earth, and that more strongly affect the way the incoming energy is distributed in latitude and season. Perturbations to the amount of energy arriving at the top of the atmosphere have generally been smooth and quasi-cyclical on the timescales of orbital and solar-output variations.

The Earth's albedo (the proportion of radiation that is reflected back into space) can be increased by turning dark surfaces such as ocean into highly reflective ice, by changing the nature of the land surface (e.g. from forest to desert) or by placing more reflecting particles (including dust and cloud droplets) in the atmosphere. Mostly, such changes occur as reactions or feedbacks to other perturbations. Volcanic eruptions directly perturb the climate by emitting gases and particles into the atmosphere; one of the effects is to reflect sunlight before it reaches the Earth's surface.

Water vapour is the most important of the greenhouse gases, reacting on fast timescales, principally to temperature. The concentrations of longer-lived greenhouse gases – carbon dioxide (CO_2), methane (CH_4), nitrous oxide (N_2O) – can be perturbed on longer timescales, either by inputs from geological reservoirs, or by changes in partitioning between the atmosphere and other sectors of the Earth system.

In considering perturbations that affect climate, it is useful to think about the nature of forcings and feedbacks. A range of 'primary' forcings can be considered external to the climate system: these include changes in the amount of incoming solar radiation (insolation) caused by changes in the Earth's orbit, or changes in solar activity. Volcanic activity, which releases trace gases and particulates into the atmosphere, is also generally thought of as an external forcing on the climate system. Anthropogenic and geologic emissions of greenhouse gases are also generally considered as primary forcings.

Changes in external forcing cause responses in various components of the Earth system (Figure 3.1), with response times that range between days (in the case of the atmosphere), months to centuries (in the case of vegetation), years to millennia (in the case of the oceans), and many millennia (in the case of large ice sheets). Each of these changes in itself may cause a further change (a feedback) to the climate system. The fast-responding components (such as sea-ice change and water-vapour change) are treated as feedbacks, diagnosed in models and observations as part of the response to the primary forcing. The more slowly responding parts of the Earth system can be considered as forcings or feedbacks depending on the circumstances. Thus, the ice sheets take many millennia to build up and decay in response to orbitally induced changes in insolation. On this multi-millennial timescale, changes in ice-sheet extent and height are a response to climate changes. On shorter timescales (hundreds to thousands of years), they themselves cool regional climates and alter atmospheric circulation, so can be considered as a forcing. Ice sheets also affect the input of freshwater to the oceans, which in turn can affect ocean circulation and heat transport. This freshwater input can also be treated as an observed forcing, or as a modelled feedback.

Much of the climate change literature, including assessments such as those of the Intergovernmental Panel on Climate Change (IPCC, 2007), is centred around trying to predict the response of the Earth system to a rather well-characterized perturbation: the addition of a known amount of various greenhouse gases, most importantly CO_2, to the atmosphere. In this chapter we assess how study of past perturbations can inform and calibrate such assessments.

In order to understand the response to a perturbation in the past we need:

(a) a good quantification of the perturbation itself;
(b) observations of the spatial and temporal extent of the response, and how long after the perturbation the response is expressed;
(c) modelling tools, based on our understanding of physical, chemical and biological processes, that allow us to estimate the expected response to such a perturbation; and
(d) quantitative comparisons between the modelled estimates and the observations, to determine the adequacy of our current understanding of the processes leading to climate change.

3.3 Methods for observing and understanding the past

3.3.1 The recent past

For the last few decades, remote sensing has provided information on numerous climate-related variables with excellent spatial coverage. It offers information on the atmosphere (including, in many cases, variations in composition and properties with altitude), and the land, ocean and ice surface. The information from remote sensing may be supplemented or validated by ground-based networks of observations: they give a much poorer spatial view than satellite data, but can sometimes provide more precise data than remote sensing can achieve, as well as providing a longer historical perspective. The subsurface ocean is impervious to remote sensing but moorings give good temporal data at a spot, and cruises give spatial patterns for a snapshot in time. The Argo ocean observing system, put in place over the last decade, consists of a few thousand small drifting probes that circulate around the world with the ocean currents and dive up to 2,000 metres in depth measuring temperature and salinity.

Before the satellite remote-sensing era, there are a few decades over which some direct observations are adequate: in particular, it has been possible to extend the instrumental temperature record for the globe back to the end of the nineteenth century (Hansen *et al.*, 2010). However, beyond about 150 years ago (a little longer for parts of Europe and Asia), palaeo-observations are the only source of information about both perturbations and responses.

3.3.2 Quantifying past forcings

Palaeo-observations provide key information about the external forcings on climate: (i) Earth's orbital parameters (determining the incoming solar radiation by season and latitude through time) can be calculated accurately over millions of years (Laskar *et al.*, 2004); (ii) the concentration of cosmogenic isotopes, particularly ^{10}Be in ice cores, can be used (Usoskin *et al.*, 2009), after making allowances for slow changes in Earth's magnetic field, to estimate past solar activity. Various assumptions have to be made to relate this to the solar parameters that matter for climate; (iii) the strength of individual volcanic eruptions can be estimated from the fallout flux of sulfate measured in ice cores (e.g. Gao *et al.*, 2008), which in turn can be converted into a stratospheric loading, although care has to be taken to distinguish local eruptions from those with a global impact.

Two other factors – greenhouse gases and ice sheets – are commonly treated as forcings, although (before the present, anthropogenic, perturbation of greenhouse gases) they really are feedbacks. The concentrations of CO_2 (Lüthi *et al.*, 2008), CH_4 (Loulergue *et al.*, 2008) and N_2O (Schilt *et al.*, 2010a) are directly available from ice cores for the last 800,000 years. Beyond this time, CO_2 is estimated indirectly from marine and terrestrial records such as the δ^{13}C of alkenones (Pagani *et al.*, 2009) or δ^{10}B of foraminifera (Pearson *et al.*, 2009) in marine sediments.

Palaeo-ice-sheet extent and volume are reconstructed from a combination of different lines of evidence. The oxygen isotope content of benthic foraminifera in marine sediment cores (Lisiecki and Raymo, 2005) records a combination of global ice volume and deep-water temperature, while individual recorders (such as corals) and particular marine records (e.g. Rohling *et al.*, 2009) can be used to reconstruct sea level. Changes in the margins of each palaeo-ice sheet can be estimated from geomorphic and stratigraphic evidence (e.g. Dyke *et al.*, 2002; Svendsen *et al.*, 2004), although the evidence from earlier glaciations has often been obliterated by the most recent one. Models of mantle viscosity are employed in order to make globally consistent estimates of past ice-sheet extent and orography (e.g. Peltier, 1994); however, there may be several ice-sheet configurations consistent with known margins and relative sea-level constraints.

3.3.3 Palaeo-observations

Most palaeo-observations consist of a time series at a single site of a measurement such as vegetation type in a pollen sequence, isotopic content of water in an ice core, or elemental ratios in the shells of creatures in marine sediments. These measures show variability and change in response to perturbations, with a response that is governed by a range of climatic and other factors. Some of these measures have been shown to be controlled mainly by a particular climate variable, such as summer or mean annual temperature, sea-surface temperature (SST), sea-ice presence or precipitation amount.

Terrestrial archives include tree rings, lake sediments, loess and peat sequences, and speleothems. Marine sediments and corals provide information on water properties and sea level. Ice cores contain direct evidence for atmospheric-gas composition and snow accumulation rate, and indirect evidence for local temperature and other properties (Figure 3.2). Because water isotopes in Greenland and Antarctic ice cores give a physically based measure of temperature, they are often used as reference temperature records, but it should be borne in mind that they represent the polar regions, not the entire hemisphere.

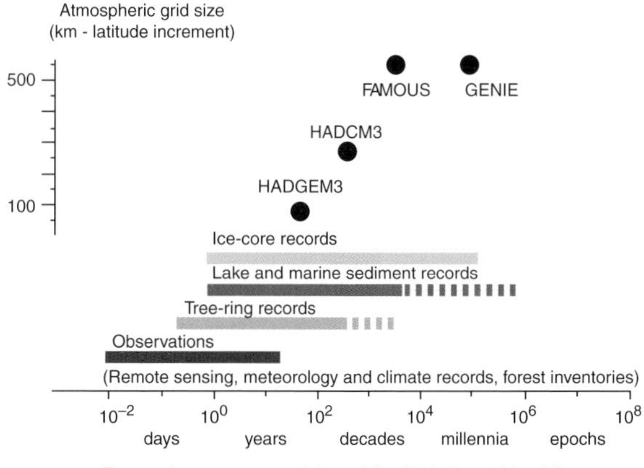

Figure 3.2 The temporal range of different observational methods, and the typical speeds and spatial scales of models. For the observations, the coloured bars represent the range of timescales typically studied. The black dots show the typical grid box size (*y*-axis) and computational speed (*x*-axis) for the major UK Earth system models. Most of the models can be run at a variety of resolutions and on platforms with varying speeds: the values illustrated here are typical ones reported in the recent literature. The temporal range for the models is the number of model years that can be run in 100 days of elapsed time in typical configuration.

In some cases, the palaeorecord lends itself to qualitative inferences about changes in climate: the presence of desiccation layers in peat bogs, for example, indicating intervals of drier climate (Borgmark, 2005). Quantitative reconstructions of climate variables commonly rely on calibration of the recent record using modern climate observations. For example, interpretation of changes in tree-ring widths is based on correlating recent tree-ring series (detrended to remove changes due to increase in tree size) against observations from a local meteorological station and then applying the derived relationship back through time (e.g. Briffa *et al.*, 1983; Cook *et al.*, 1990). In the case of terrestrial pollen or marine-plankton assemblages, empirical relationships between climate variables and the abundances of taxa are established using a spatially distributed modern 'training' data set and these relationships are then applied to temporal variations (Bartlein *et al.*, 2011). One problem with this approach is that non-climatic factors can affect taxon distribution – for example, atmospheric CO_2 concentration influences the competition between C_3 and C_4 plants and this directly affects terrestrial plant distribution, e.g. in glacial intervals when atmospheric CO_2 was very different from present (Harrison and Prentice, 2003; Prentice and Harrison, 2009). This problem can be overcome by inverting a process-based model, which can take account of the direct effect of CO_2 levels on plant growth and competition, to estimate the climate that is consistent with observed plant distributions at a specified level of CO_2 (Wu *et al.*, 2007).

While a single record can provide a temporal pattern, it requires considerable effort, and the synthesis of numerous palaeorecords, to determine a spatial pattern. The pattern of a particular climate variable for a single time is referred to as a 'time slice', and is of particular use for comparison with model outputs, which are often visualized as a sequence of time slices. Pollen and plant macrofossil data have been summarized in the form of maps of vegetation types (Harrison and Sanchez-Goñi, 2010; Prentice *et al.*, 2011; Prentice *et al.* 2000) (Figure 3.3). There are still very few global data sets showing spatial patterns of climate. Two recent compilations (Bartlein *et al.*, 2011; MARGO Project Members, 2009) provide spatial reconstructions of temperature at the Last Glacial Maximum (LGM, 21,000 years ago). The Bartlein *et al.* (2011) data set also provides temperature reconstructions over land for the mid Holocene (6,000 years ago) and additional climate variables for the LGM and mid Holocene. The LGM reconstructions over land show mean annual temperature lower than present (Figure 3.3), with maximum temperature lowering adjacent to the greatly expanded northern-hemisphere ice sheets and smaller changes in the tropics. This pattern is echoed in the reconstructed sea-surface temperatures (also shown in Figure 3.3).

The changing extent of sea ice can be reconstructed from changes in marine-plankton assemblages, particularly in the abundance of diatoms that live on sea ice (Gersonde *et al.*, 2005) and dinoflagellates that live close to the sea-ice margin (de Vernal *et al.*, 2005). It is more difficult to reconstruct ocean-circulation patterns, although broad-scale changes have been inferred indirectly using nutrient (such as $\delta^{13}C$ or

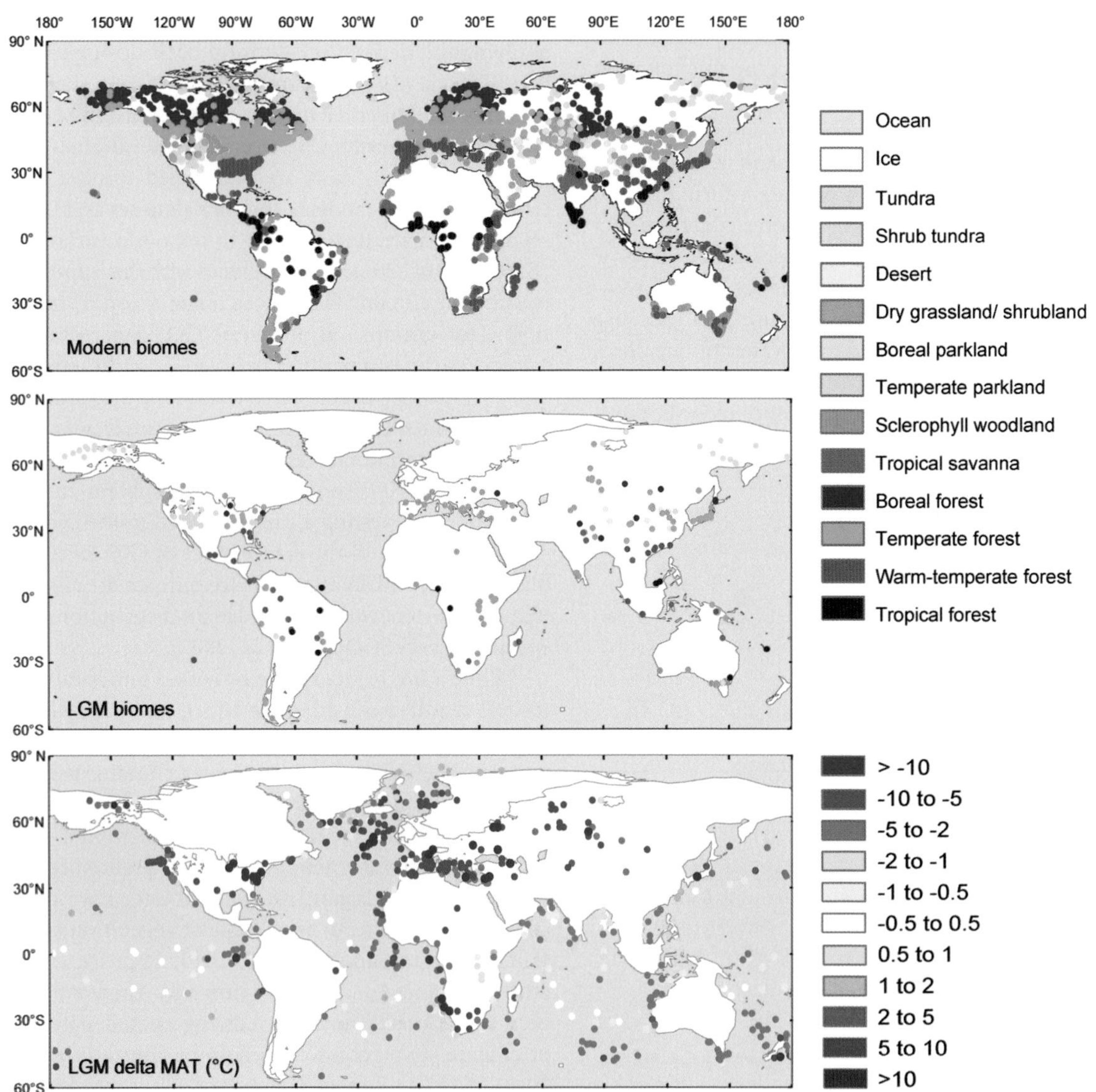

Figure 3.3 Reconstructed changes in vegetation and temperature at the Last Glacial Maximum (LGM, 21,000 years ago) compared to present. Reconstructed vegetation patterns today (top panel: modern biomes) and at the LGM (middle panel: LGM biomes), based on pollen assemblages from individual sites, reclassified using the biome scheme from Prentice *et al.*, (2011). Changes in mean annual temperature (LGM delta MAT) on land points were reconstructed from pollen and plant macrofossil data (Bartlein *et al.*, 2011) and changes in mean annual SST were reconstructed from biological and geochemical climate proxies (MARGO Project Members, 2009). Both MAT and SST values are for the difference between LGM and the modern day (LGM – modern).

Ca/Cd ratios in plankton) or transport (e.g. ^{14}C, $^{231}Pa/^{230}Th$ ratios) tracers.

3.3.4 Palaeoclimate modelling

Models provide the tools for understanding the response to perturbations. They range from simple conceptual models through to full numerical Earth system models of the type that are also used for climate prediction. Chapter 5 summarizes the structure and development of such models. Here we highlight a few issues which affect our ability to compare palaeo-observations with model outputs.

A first limitation for using models to understand past climate change is the high computing demand of current, high-resolution models. The time and computing resources required to make simulations over thousands of years, using these state-of-the-art models, are too great to be practical (Figure 3.2). Two approaches have been used to deal with this limitation.

One approach is to make 'time-slice' simulations for times where there is a clear contrast with present climate. The mid Holocene and the LGM have been widely used as targets: the mid Holocene because it is a time when the latitudinal and seasonal pattern of insolation was different from today, and the LGM because of the very strong contrast in greenhouse-gas and ice-sheet forcing (Braconnot *et al.*, 2007). Achieving an equilibrium climate in a coupled model requires typically several hundred years of model spin-up, so even these simulations are computationally demanding. The advantage of this approach is that one can use highly specified models in which an equilibrium climate, as well as short-timescale variability, can be investigated. The disadvantage is that it treats each climate state as an equilibrium, and fails to use the rich evidence about rates and patterns of change that is contained in the palaeorecord. Nevertheless, time-slice simulations of the LGM and mid Holocene are included as modelling targets in the current phase of the Coupled Model Intercomparison Project (Braconnot *et al.*, 2011) and considerable effort has gone into synthesizing data for these periods.

The other approach is either to reduce the resolution of the model, or to devise a model with less complexity and resolution. The former tactic has been used, for example, in the design of FAMOUS (Smith *et al.*, 2008), a variant of the Hadley Centre's HADCM3 model, which runs at half the resolution of HADCM3 but 10 times faster. The latter tactic has led to the development of Earth system models of intermediate complexity (EMICs) (Claussen *et al.*, 2002). EMICs may also include dynamic representations of components (such as continental ice sheets) that are important for palaeoclimate evolution, but have to be prescribed as constant in more complex climate models. These models have been used for transient runs covering multiple glacial cycles (Ganopolski *et al.*, 2010; Holden *et al.*, 2010). However, EMICs sacrifice resolution and complexity in the ocean and atmosphere to achieve computational efficiency. Many processes are parameterized in EMICs, making it hard to capture non-linear responses correctly. The GENIE model (Lenton *et al.*, 2006) is the EMIC developed in the UK and used widely in QUEST.

A second limitation for using models to understand past climate changes is that models do not simulate the behaviour of the environmental sensors directly. While palaeoenvironmental records can be translated into estimates of traditional climate variables using statistical models, these estimates may be invalid when there is no modern analogue for a past climate state, or subject to large uncertainties when the climatic controls on the sensor are imperfectly understood. An alternative approach is to forward-model the environmental sensor explicitly, either in a separate model (e.g. for tree-ring widths (Evans *et al.*, 2006)) or within the climate model itself. Several climate models now explicitly simulate vegetation dynamics (e.g. Krinner *et al.*, 2005), carbon cycling (Friedlingstein *et al.*, 2006) or water isotopes (e.g. Tindall *et al.*, 2009).

A third limitation is that of discrepancies in scale. Palaeodata are usually point measurements representing catchment-scale or regional climate (*c.* 10 to 100 km distance). Models, on the other hand, describe climate at often much coarser resolution (see Figure 3.2). Scale issues also affect model topography: climate models do not represent the true elevation of mountain areas and thus their predictions of climate and environment will not necessarily be consistent with reconstructions from such areas. Averaging or gridding palaeoclimate reconstructions (Bartlein *et al.*, 2011) and statistical down-scaling of climate-model outputs (e.g. Vrac *et al.*, 2007) have been used to bridge the scale gap between observations and simulations. Climate simulations are best used to address broad-scale features of the palaeorecord, rather than individual time series.

3.4 How climate has altered in the past

The past has given us a set of natural experiments which allow us to examine the workings of the Earth system and climate. Here we summarize briefly what has happened to the climate. We have restricted ourselves to the relatively recent geological past: most emphasis is on the last million years. There are two reasons for this restriction: it is easier to draw lessons from a time when the Earth was similar to today – with about the same solar strength, similar continental topography and landmass distributions and therefore ocean-transport pathways. More pragmatically, for earlier periods our knowledge of events and their timing is too uncertain to allow us to clearly relate perturbations to effects.

The most prominent feature of the Cenozoic Era (the last 65.5 million years) has been a general cooling trend from a maximum global temperature about 50 million years ago (Zachos *et al.*, 2008) (Figure 3.4a). This cooling led to the initiation of an ephemeral and eventually, about 34 million years ago, a permanent ice sheet over Antarctica. The cooling coincided with a long decrease in atmospheric CO_2 concentration and this probably played a significant role in the cooling (DeConto and Pollard, 2003), although the measures used to estimate CO_2 concentrations before the ice-core era are difficult to interpret and sometimes contradictory. Changes in ocean circulation as continents moved were likely also a factor. We will not use the Cenozoic cooling trend as a case study, because of the uncertain quantification of the CO_2 change in this period and the different palaeogeography. However, one particular short-lived warming feature, known as the Palaeocene–Eocene Thermal Maximum (PETM, ~56 million years ago) does stand out in the record, and will be discussed later as an example of the effects and timescales associated with a large release of carbon into the ocean and atmosphere.

It was only at the start of the Quaternary Period and the Pleistocene Epoch, 2.6 million years ago, that large ice sheets started to appear and disappear periodically in the northern hemisphere. Oxygen isotopes in marine sediments (Lisiecki and Raymo, 2005) show the global ice volume growing and shrinking initially in 40,000 year cycles, and finally, in the last few hundred thousand years, at periods of roughly 100,000 years (Figure 3.4b). It is only in the last 800,000 years that we have marine, terrestrial and ice-core data, and we can clearly observe the linkages between cycles in global ice volume, climate and greenhouse gases. These linkages will be discussed later in order to inform our knowledge of biogeochemical feedbacks in the Earth system, a theme also taken up in Chapter 4.

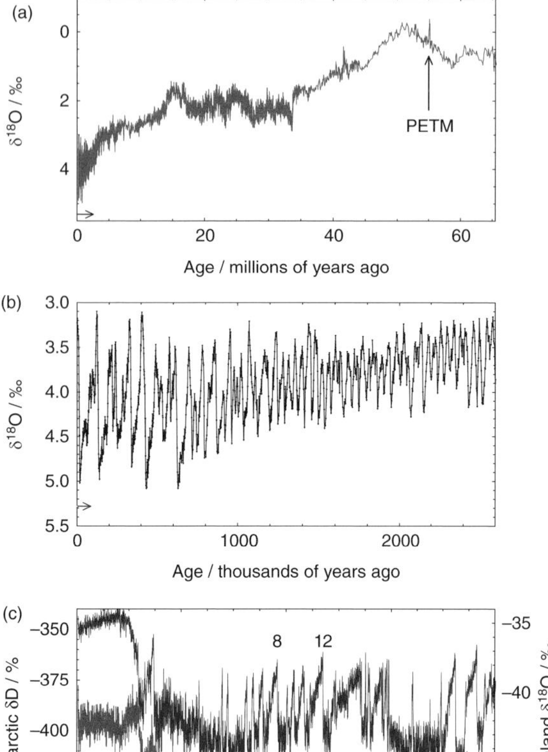

Figure 3.4 The history of Cenozoic climate. In all cases, warm/low-ice climate is up and time runs from right (the deeper past) to left (the present). (a) The last 65 million years; stacked benthic foraminiferal oxygen isotope data, five-point smoothed (Zachos *et al.*, 2008), representing temperature between 65 and about 35 million years ago, and a combination of temperature and global ice volume after 35 million years ago; the Palaeocene–Eocene Thermal Maximum (PETM) is marked. (b) Zoom on the last 2.6 million years (Pleistocene and Holocene); stacked benthic foraminiferal oxygen isotope data (Lisiecki and Raymo, 2005) showing the succession of 40,000-year and then 100,000-year cycles. (c) Zoom on the last 80,000 years from polar ice core water isotopic data (a measure of temperature); δD from the Antarctic EPICA Dome C core (Jouzel *et al.*, 2007) (red, left axis) and $\delta^{18}O$ from the Greenland NorthGRIP core (North Greenland Ice Core Project Members, 2004) (blue, right axis); the Last Glacial Maximum (LGM) is marked, as are two of the numbered rapid millennial-scale Greenland interstadials (8, 12); the range between the coldest and warmest points in each record represents about 10 K for Dome C and 20 K for NorthGRIP.

The most recent glacial–interglacial cycle started with the last interglacial (LIG; *c.* 130,000 to 115,000 years ago), which was a period of warmth in both polar regions, and acts as a case study of how polar warmth affects sea level. The last glacial period (Figure 3.4c) was characterized by a series of millennial-length events, most prominently observed as a series of abrupt warmings (followed by somewhat less abrupt coolings) in the Greenland ice core and other North Atlantic records, but with subtle counterpart climate changes in Antarctica (e.g. Wolff *et al.*, 2010). These events, probably resulting from reorganization of ocean heat transport, are examples of rapid climate changes and allow us to consider the stability of ocean circulation patterns.

The LGM was the period of the glacial cycle with the greatest ice volume and lowest sea level (~120 m below present); ice sheets extended far south over large areas of North America and northern Europe. It was followed by the termination into the present interglacial, the Holocene, beginning about 11,700 years ago. The present interglacial includes significant regional climate trends and millennial-scale variability (Wanner *et al.*, 2008), but without the major reorganizations of the glacial period. Variability in the Holocene is presumably due to changes in natural forcings: slow trends in orbital forcing, variations in solar irradiance and volcanic eruptions, as well as internal changes that were less severe than those of the glacial period. There are far more data available for the Holocene and the LGM than for earlier times, allowing us to use these periods to discuss aspects of climate such as climate sensitivity, and the relative response of the poles compared to the rest of the world.

The aim of the following sections is to use the observational record to determine whether our models (both conceptual and numerical) are capable of capturing the kinds of behaviour seen in the past, and to discuss how these observations can be used to improve quantification in models. However, it is worth remembering that, at the very least, the past is a guide to what is possible: assuming the boundary conditions are similar, then anything that has happened in the past can happen again. This role of palaeo-observations – showing for example that abrupt changes in ocean heat transport can occur and lead to major climate and ecological impacts; that climate–carbon-cycle feedbacks have been strong in the past; that large carbon releases can occur in nature – should not be underestimated. Having said this, it is also true that changes in boundary conditions could change the likelihood of events recurring, while they may also allow events that have not previously been observed. The past provides no exact analogue for the future.

3.5 Response of climate change to forcing

In this section, we use palaeoclimate data to assess quantitatively the effect of forcings on climate, starting at the global scale, then narrowing down to aspects of the climate system which produce effects at regional to local scales.

3.5.1 The global scale: climate sensitivity

The concept of radiative forcing can be used to quantify both natural and anthropogenic climate drivers. Radiative forcing values are expressed as a change in the radiative balance at the top of the atmosphere in W m^{-2} relative to pre-industrial conditions (defined as AD 1750). Climate sensitivity is the equilibrium change in annual mean global surface temperature in response to a doubling of atmospheric CO_2 concentration, after physical feedbacks (on timescales of years to a few centuries) have been taken into account. Because changes in many local climate variables (e.g. tropical and Antarctic temperatures) approximately scale with those of global temperature, climate sensitivity provides a useful first-order estimate of how strongly the climate system responds to an external perturbation.

The climate sensitivity is a property of the real climate system, but can also be considered a property of a climate model, whose value depends on the particular model, and on the equations and parameter values used in the model. The concept of the climate sensitivity is rather simple, but one difficulty in using it is that the notion of 'global average temperature' is a model construct: we cannot measure any such value in the real world except by averaging large numbers of local measurements. Measurements become increasingly sparse further back in time, so constructing this value becomes very difficult. Another difficulty is the concept of equilibrium: the current climate, for example, is not in equilibrium with the imposed forcing; the climate sensitivity cannot be measured directly from a transient state without taking into account the energy imbalance.

The climate sensitivity (S) can be considered as the initial (black-body radiative) response (ΔT_0, estimated as ~ 1 K) to the forcing from a doubling of CO_2 (estimated as about 3.7 W m^{-2}), modulated by feedbacks (represented by a feedback factor, f, or a feedback gain, g). As defined here, g can be considered as the sum of gains (g_i) from each different feedback:

$$S = f\Delta T_0 = \Delta T_0/(1 - g) = \Delta T_0/(1 - \Sigma g_i) \quad (3.1)$$

The importance of this equation is that it illustrates that the climate sensitivity is all about the strength of feedbacks. If any g_i is negative, a negative feedback is in operation. If g is large then the climate system is more sensitive. It is also important to be aware of which feedbacks are included: in the usual definition, the four most important feedbacks – water vapour (a greenhouse gas that increases with temperature), lapse rate (the atmospheric vertical temperature profile), clouds and albedo (changes in sea ice and snow cover) are included. This choice is motivated by the main feedbacks that climate models include and that are relevant on timescales of years to a few centuries. Feedbacks acting on longer timescales, such as changes in the carbon cycle and changes in ice-sheet area, are not included in the definition. They are not ignored, but rather are treated as prescribed forcings rather than as feedbacks.

The climate sensitivity S embodies some complex physics in a single number. This has some consequences for its use. First, it makes the simplifying assumption that all forcings are equal – that the response to a forcing from greenhouse-gas increases is the same as the response to a forcing of the same magnitude from volcanic aerosol; and that the response is independent of the initial climate state to which the forcing is applied. Neither of these assumptions is likely to be entirely correct. Secondly, it is possible to get the 'right' climate sensitivity for the wrong reasons: thus, models that overemphasize one feedback and underestimate another will have the same climate sensitivity, but a different pattern of response.

A further complication, particularly relevant for palaeoclimate, is the issue of response timescales and the associated ambiguity in separating forcings and feedbacks. The concepts of radiative forcing, feedbacks and the global energy balance assume that all feedbacks can be summed up and scale linearly with the global temperature increase (Knutti and Hegerl, 2008), which is approximately correct for the four feedbacks listed above. But there are other feedbacks that have their own intrinsic, mostly longer, timescale, such as changes in ice sheets or vegetation. For example, if we keep the radiative forcing fixed at several W m^{-2}, it may take 10,000 years for the Greenland ice sheet to melt, i.e. the ice-sheet feedback will change largely independently of the temperature response of the rest of the system. For model projections of future climate, those long-term feedbacks have often been neglected, or have been externally prescribed. For palaeoclimate modelling, however, the difficulty is whether an ice-sheet forcing, for example, should be prescribed in much the same way as other radiative forcings, or whether the ice-sheet change should be modelled and treated as an internal feedback.

In yet another category are changes in the carbon cycle that are not considered in the definition of climate sensitivity, which can only describe the response of climate to a *fixed* CO_2 concentration change. In other words, climate sensitivity includes physical but not biogeochemical feedbacks. The separation into forcing and feedback is useful but is not unique. Classical climate sensitivity can be estimated from the LGM by prescribing changes in ice sheets, dust, vegetation and so on as forcings, but the only real external change at the LGM was the orbital forcing, and the ice sheets, the carbon cycle, dust, vegetation were all feedbacks resulting from it. Both views can be defended, but the estimated numbers will differ because they describe a different set of feedbacks (see sections 3.6.3 and 3.6.4).

Despite these limitations, climate sensitivity is a useful way to assess how well we can estimate the global response of surface temperature to a forcing. The IPCC Fourth Assessment Report (IPCC, 2007, page 65) stated that climate sensitivity was likely (> 66% probability) to be in the range 2 to 4.5 K, with a best estimate of 3 K, and very likely (>90%) above 1.5 K. Values significantly larger than 4.5 K could not be ruled out. This statement reflects a fundamental difficulty, that the quantities that we measure are related in a non-linear way to climate sensitivity, making it difficult to rule out high values for climate sensitivities (e.g. Knutti and Hegerl, 2008; Roe and Baker, 2007). The above equation shows that climate sensitivity gets very large if the gain g approaches unity. The ranges and distributions obtained from different lines of evidence are therefore typically skewed, with long tails towards high values for climate sensitivity.

Figure 3.5 summarizes the ranges and probability density functions for climate sensitivity as obtained by many studies. Climate sensitivity can be estimated by relating the observed warming trends in surface and ocean temperature to estimated changes in radiative forcing, from simulated feedbacks in global models, from the temperature response to volcanic eruptions (see Section 3.6.1) and from palaeoclimate evidence, either based on data, or models, or a combination of the two (see Section 3.5.2). The different lines of evidence agree reasonably well (Knutti and Hegerl, 2008).

The most obvious period to look at is the observed warming over the past century. Because the general state of the climate system – ice sheets, ocean circulation pathways, and so on – has not changed on this timescale, and the response timescales involved are similar to those over the next century, the separation of forcing and feedback is quite clear. Many of the problems listed above are minimized. However, the current climate is not in equilibrium with the forcing, the forced signal is small (< 1 K) and the magnitude of the forcing is uncertain, mostly due to uncertainties in the aerosol component (Anderson *et al.*, 2003; Knutti *et al.*, 2002). Volcanic eruptions are times in the recent past when a single forcing may have an overwhelming effect, and there are several studies of the cooling that resulted from the eruption of Mount Pinatubo (see Section 3.6.1). However, the response to a volcanic eruption is transient and therefore not well suited to deduce an equilibrium response.

The underlying climate sensitivity in models arises from the physics and parameterizations that allow them to reproduce modern climatology (both its spatial structure, and its temporal variability). This approach alone gives insufficient confidence that the system will respond as expected to a perturbation: therefore, there are sound reasons to use the palaeoclimate record to validate the existing range and hopefully to narrow it. The uncertainty at the first order of global mean temperature, as well as the way that value translates into regional and hydrological change, are at the heart of the difficulty that scientists face in defining what concentration of CO_2 *would* (as opposed to *might*) cause 'dangerous climate change'.

3.5.2 Climate sensitivity and palaeo-observations

The different sources of information about climate sensitivity were reviewed in two papers (Edwards *et al.*, 2007; Knutti and Hegerl, 2008). We summarize the findings of those reviews, supplementing them with recent work.

In theory, one could use observations of a particular past palaeoclimate state to determine the climate sensitivity directly. Increasingly, more sophisticated studies are taking advantage of the availability of greater computing

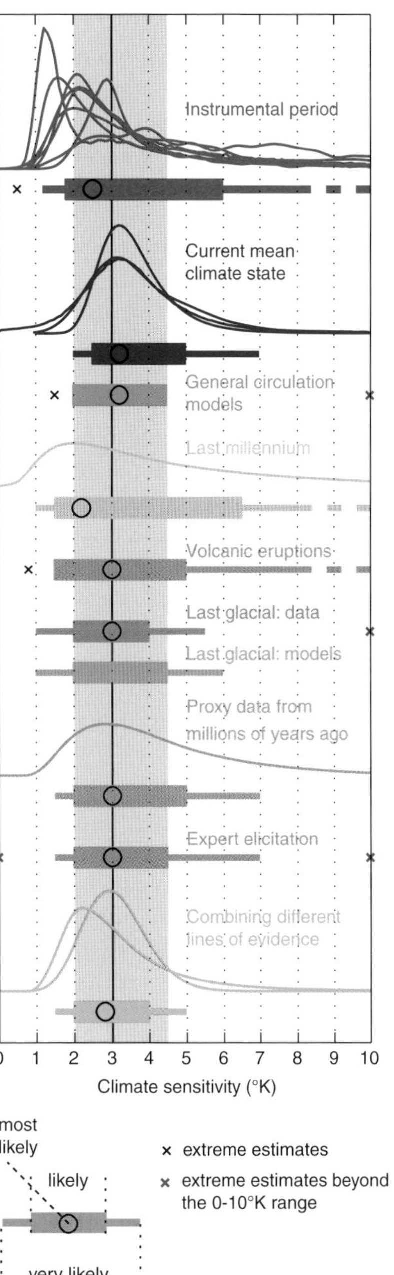

Figure 3.5 Distributions and ranges for climate sensitivity from different lines of evidence. (Figure first published in *Nature Geoscience*, *1* (11), 735–743 (Knutti and Hegerl, 2008).) The most likely values (circles), likely (bars, > 66% probability) and very likely (lines, > 90% probability) ranges are subjective estimates based on the available distributions and uncertainty estimates from individual studies, taking into account the model structure, observations and statistical methods used. Single extreme estimates or outliers (some not credible) are marked with crosses. The IPCC likely range and most likely value are indicated by the vertical grey bar and black line, respectively (Knutti and Hegerl, 2008).

capacity to explore 'perturbed physics ensembles'. This is where the parameters describing different Earth system processes within a model are varied within reasonable constraints; the models that show good agreement with data for the palaeoclimatic time period are retained, and an assessment of the climate sensitivity can be made.

Whichever approach is used, each climate period poses specific difficulties. The mid Holocene has been the subject of data syntheses and modelling, but is problematic because of the tiny net global forcing (Hargreaves and Annan, 2009). The problem of small variations also applies to studies of the last millennium (Hegerl *et al.*, 2006). The LGM is an attractive target because the forcings and response were large. However, some of the forcings (notably the change in albedo due to enhanced ice sheets over northern continents) are not directly relevant to the future.

A first-order estimate of climate sensitivity can nevertheless be obtained by directly relating the forcing to the temperature response of the LGM. Recent calculations with improved estimates of radiative forcing suggested a most likely value for climate sensitivity of between 2 and 3 K (with 5–95% confidence range of 1.4–5.2 K) (Köhler *et al.*, 2010). Studies based on climate models provide results that are fairly consistent with the above. One difficulty is that some models indicate that even the feedbacks that are not related to ice sheets and vegetation could differ significantly between the present and the LGM (Hargreaves *et al.*, 2007). A recent study that compared the century-timescale response of a model ensemble with the now well-characterized temperature anomalies at the LGM gave a similar median and range to previous work, and effectively rules out climate sensitivities greater than 6 K (Schmittner *et al.*, 2011).

Further back in time, we run into the difficulty that both the forcing and the response are poorly known. It has therefore been difficult to constrain climate sensitivity significantly based on pre-Quaternary climate data (Knutti and Hegerl, 2008). A recent modelling study of the mid-Pliocene warm period (about 3 million years ago) assessed some of the factors involved, and pointed out that the net response should take account of factors, in particular changes in ice sheets and vegetation, that do not appear in the traditional formulation of climate sensitivity. Doing so could lead to a response 30–50% higher than the traditional calculation would imply (Lunt *et al.*, 2010). Similar arguments for much larger climate sensitivity values have been made for other time periods and in other studies. While suggesting the importance of assessing future change in large ice sheets and vegetation, these arguments do not help to narrow the range of climate sensitivity as used in the climate literature.

Most importantly, palaeodata provide confidence that the real value of climate sensitivity is very unlikely to be sitting in the long tail that such estimates generally possess. More extensive data syntheses and quantitative climate reconstructions should offer further possibilities for refining estimates of climate sensitivity and also characterizing the long-term feedbacks involving ice sheets, vegetation and the carbon cycle. For example, it has recently become possible to estimate (Shakun and Carlson, 2010) mean global temperature and the latitudinal distribution of change over the entire last glacial termination (19,000 to 11,000 years ago); and the entire period from the LGM to the pre-industrial has been successfully simulated with an EMIC (Timm and Timmermann, 2007). It is even now possible to carry out simulations with EMICs that cover complete glacial cycles (Ganopolski *et al.*, 2010; Holden *et al.*, 2010), and to model an entire cycle using a series of snapshots from a full global climate model (Singarayer and Valdes, 2010), allowing quantitative comparisons with data syntheses for this longer period (e.g. Lang and Wolff, 2011).

3.5.3 Large-scale features of the climate: polar amplification

Some large-scale features of climate change are predictable both from conceptual understanding and by numerical models. Such large-scale features are generally related to the existence and strength of one or more feedbacks, so their presence in palaeodata can serve as a guide to improvements in the representation of such feedbacks in models.

One candidate is the amplification of climate change in the polar regions. Polar amplification is seen in model simulations of the effect of doubled CO_2 (Winton, 2006). For the Arctic, maximum amplification is predicted at the highest latitudes over ocean areas, reaching a factor 2–4 in the range of models tested (Holland and Bitz, 2003). A smaller amplification is also predicted under doubled CO_2 over Greenland, with only a small amplification over the Antarctic continent (Masson-Delmotte *et al.*, 2006), partly because there is already only a small area around Antarctica with sea ice in summer.

Observations show that the surface amplification of warming over the Arctic is already occurring, with a spatial and seasonal pattern comparable to that predicted (Serreze *et al.*, 2009). The Antarctic is a different case: although there has been a clear warming over the Antarctic Peninsula, this is not the case over the rest of the continent. Moreover, the amount of autumn sea ice has increased in the last 30 years around about two-thirds of the continent. It is believed that this increase is due to changed atmospheric circulation, probably a result of stratospheric ozone depletion (Turner *et al.*, 2009). This second anthropogenic environmental effect may so far have masked any amplification over the ocean surrounding Antarctica.

Palaeorecords can be used to test whether polar amplification is characteristic of past climates. Ice cores provide climate data over the landmasses within the sea-ice zone, data on sea-surface temperatures and sea-ice extent are available from marine sediments, and vegetation records provide information on temperature changes over the adjacent high-latitude land areas in the northern hemisphere.

The polar climate of the LGM has been extensively studied in climate models, most notably in the Palaeoclimate Model Intercomparison Project Phase 2 (PMIP2) project. After accounting for the change in surface elevation of the Greenland and Antarctic ice caps in the boundary conditions specified for PMIP2, an amplification of around 2 was predicted for Greenland LGM temperature relative to pre-industrial control, compared to the global mean (Masson-Delmotte *et al.*, 2006). Water isotope data in Greenland ice cores (Johnsen *et al.*, 1992), calibrated through borehole temperature profiles (Cuffey *et al.*, 1995; Johnsen *et al.*, 1995), suggest that the LGM temperature anomaly in Greenland was as large as –19 to –22 K; the actual elevation change also contributing to this change remains uncertain. A typical estimate (from multiple data sources) of the global average LGM anomaly is around –5 K (Shakun and Carlson, 2010). Thus, the data confirm a very large (larger than predicted) polar amplification over Greenland. No explanation stands out in the literature, but we speculate that the LGM climate anomaly in Greenland also includes a millennial-scale climate shift (Section 3.6.6) related to changes in ocean heat transport that are not explicitly modelled.

For central East Antarctica, as in the doubled CO_2 runs, little polar amplification is seen in the LGM PMIP2 model runs (after correction for the altitude of the ice sheet used in the boundary conditions) (Masson-Delmotte *et al.*, 2006). This presumably reflects the fact that central Antarctica is far removed, in both distance and altitude, from the sea ice surrounding the continent, or from changes in land ice around the continental margin and shelf. Ice-core data (e.g. Jouzel *et al.*, 2007) suggest a rather larger temperature change (about 9 ± 2 °C at Dome C) than is seen in the models, or in the global average.

A recent study (Miller *et al.*, 2010) has considered other time periods to assess whether Arctic amplification is apparent in the data. The early Holocene (~8,000 years ago) shows enhanced warming at high northern latitudes, but this is expected from the latitudinal pattern of the insolation anomaly; the global temperature was little different from the pre-industrial. In the LIG, the Arctic experienced greatly enhanced warmth compared to the global average (CAPE-Last Interglacial Project Members, 2006; Turney and Jones, 2010), but again this is an expected consequence of the insolation pattern. Estimates for the mid Pliocene (~3.5 million years ago) suggest Arctic amplification by a factor 3 (global temperature anomaly estimated at 4 K, Arctic at 12 K) (Miller *et al.*, 2010).

Thus, where it can be diagnosed, polar amplification is observed in both cooler and warmer climates. Until recently, it was assumed that the underlying physical basis for amplification was related to ice-albedo feedback resulting from reduction in sea-ice extent (Serreze *et al.*, 2009). As climate warms, there is less sea ice at the end of summer, and the ocean has absorbed more heat during the summer. The result is that ice formation in the autumn is delayed, removing part of the insulating layer between ocean and atmosphere that would otherwise be present, and there is more upward heat transfer in winter from the warmer ocean into the cold lower atmosphere. Changes in land-surface albedo, associated either with changes in snow cover directly or with vegetation changes (trees replacing low-tundra vegetation) that effectively shielded the snow from the atmosphere, were thought to produce a similar but smaller feedback. However, recent modelling work (Graversen and Wang, 2009) shows that significant amplification occurs even when surface albedo is held constant. This appears to be due to a number of factors, including an enhanced impact of changes in water vapour and cloud cover at high latitude, and the stability of the Arctic atmosphere, concentrating effects in surface layers.

In summary, it is now becoming clear that there are a number of mechanisms that contribute to polar amplification, and that albedo feedback is not the most important of these (Serreze and Barry, 2011). The palaeorecord confirms that the overall effect of such mechanisms is being captured by models, but improved observations of the actual sea-ice extent in the past (since sea ice is central to many of the processes involved) and better understanding of the seasonality of the climate signal, would allow more confidence in our understanding of the relative importance of each mechanism.

3.5.4 Large-scale features of the climate: land–sea contrast

Models show temperatures over land increase more (by a factor of 1.36 to 1.84) than temperatures over the sea in response to a warming perturbation (Crook *et al.*, 2011; Sutton *et al.*, 2007). Although the pattern of land–sea contrast differs for forcings other than CO_2, it is always present (Joshi and Gregory, 2008). This is an equilibrium response (not just a transient one), explained by the fact that over wet surfaces (including the ocean) some of the extra heat is used for evaporation rather than surface warming; possibly with additional effects from land-surface feedbacks.

Observations over the last few decades confirm that air temperature has increased more over land than over the sea (Sutton *et al.*, 2007). Comparison of different coupled model simulations of the LGM show more cooling over land than over the ocean (averaged over 30 °N to 30 °S (Braconnot *et al.*, 2007)). A multi-measure assessment suggests that this asymmetry is also present in the data (Braconnot *et al.*, 2012). The range in the land–sea ratio of temperature change found in palaeoclimate simulations is similar to that found for future climate projections (Laine *et al.*, 2009), although comparisons with the LGM temperature reconstructions show that some of the higher simulated values for this ratio are unrealistic.

3.5.5 Large-scale features of climate: hydrological cycle

Changes in the hydrological cycle are important both in terms of regional climates, and as a feedback on other components of the Earth system such as the strength of the thermohaline circulation. The global hydrological cycle is largely controlled by energy availability (Allen and Ingram, 2002) and it is expected that global warming will be accompanied by an intensification of the global hydrological cycle. However, the patterns of the precipitation response to greenhouse warming vary considerably in space and between models (IPCC, 2007). There is a general tendency for precipitation to increase in high latitudes and some parts of the tropics, and to decrease in the subtropics. Observations of twentieth-century changes in the hydrological cycle vary regionally and by variable, and statements about the global trend have tended to be contradictory (e.g. Huntington, 2006). However, the hope is that analyses of changes in the hydrological cycle in response to the much larger changes experienced in the past will help in assessing the reliability of future projections.

Palaeodata, including lake, river flow and vegetation records, as well as measurements of oxygen isotopes in speleothems, indicate large changes in the regional expression of the monsoons over the last glacial–interglacial cycle (e.g. Gasse, 2000; Prentice *et al.* 2000; Wang *et al.*, 2008). Models reproduce some aspects of these changes as a response to orbital forcing and ocean feedback (Braconnot *et al.*, 2007), but they underestimate the magnitude of the observed regional changes in Africa and Asia (Braconnot *et al.*, 2007, 2012). Simulated changes in precipitation are highly sensitive to the location of oceanic warming (Zhao *et al.*, 2005; Zhao and Harrison, 2011), as well as the partitioning between latent and sensible heating (Ohgaito and Abe-Ouchi, 2007). Improved data as well as data–model comparisons could help to resolve such issues, as well as uncertainties in the role of land-surface feedbacks in amplifying monsoons (e.g. Otto *et al.*, 2009). Chapter 4 discusses issues surrounding the modelling of biogeophysical feedbacks from the land surface.

3.5.6 Atmospheric and oceanic circulation

Atmospheric circulation and ocean overturning are critical aspects of the climate system, well diagnosed in models and crucial for the study of feedbacks. However, they are hard to infer directly from palaeodata.

Attempts have been made to use palaeodata to assess changes in the strength and latitude of the southern-hemisphere westerlies. Such changes are important not only for understanding past climates, but also because the location of the westerlies features in one hypothesized mechanism for lowered atmospheric CO_2 content during glacial periods (Toggweiler *et al.*, 2006). Lake, pollen and marine records from the LGM have been interpreted both as showing an equatorward shift (e.g. Moreno *et al.*, 1999) and a poleward shift (e.g. Lamy *et al.*, 1999) in the southern-hemisphere westerlies. Recent work has questioned whether the records involved, typically assessing the latitudinal changes in precipitation, are suitable for assessing the location of the westerlies (Sime *et al.*, 2010). Changes in the amount of wind-blown dust found in offshore marine cores, which provide more direct evidence about circulation, show strengthened westerly flow at the LGM (e.g. Carter *et al.*, 2000; Shulmeister *et al.*, 2004) but do not indicate a major shift in the position of the westerlies. Model results are also contradictory: an analysis of the PMIP2 model simulations suggest very modest changes overall (Rojas *et al.*, 2009). Improved interpretation and data syntheses could resolve this very important area of discrepancy among models.

The Atlantic Meridional Overturning Circulation (AMOC) is important climatically because it is accompanied by a net northward transport of sensible heat in the ocean and is intimately coupled to the transport of latent heat in the atmosphere and thus to the hydrological cycle. During the LGM, the factors controlling the structure of the overturning circulation (freshwater budgets and atmospheric circulation in the northern and southern high latitudes) were very different from today. The broad-scale changes in LGM circulation (shallowing of North Atlantic Deep Water (NADW), northward extension of Antarctic Bottom Water into the Atlantic Ocean, weakening of the North Atlantic overturning cell (Lynch-Stieglitz *et al.*, 2007)) are reasonably well known. The rates of the shallow and deep overturning circulations are less well constrained, although the balance of evidence suggests an ocean circulation slightly slower than at present. State-of-the-art models produce widely different Atlantic circulation scenarios when forced with LGM atmospheric CO_2 concentration and ice sheets. Some show a stronger overturning cell associated with NADW formation either extending slightly deeper or unchanged in vertical extent; others show a weaker and shallower overturning cell associated with NADW production (Otto-Bliesner *et al.*, 2007). This is another area where an advance in data collection and interpretation could resolve an important model uncertainty.

3.6 Case studies of climate perturbations and responses

In this section, we discuss specific examples of perturbations that have special resonance for particular aspects of Earth system science. For each we will discuss their character and cause, and then describe the observed responses.

3.6.1 Response to volcanic perturbation – Mount Pinatubo

The eruption of Mount Pinatubo in the Philippines in June 1991 provided a notable opportunity to study the response of the climate system to a natural, short-term forcing. This eruption lasted more than a week and injected a large volume of ash and gases into the upper troposphere and lower stratosphere. The thermal stability of the lower stratosphere and the absence of cloud-related removal processes can lead to relatively long lifetimes for ash that reaches these altitudes. The ash cloud was observed by satellite instruments to circle the Earth within a month in a relatively coherent plume between 20° S and 30° N (McCormick and Veiga, 1992). Over the following months, the ash cloud dispersed over a wider area and concentrations decreased, but the ash signature was detectable from surface radiation measurements for the next two to three years.

The presence of this sulfate-rich ash in the lower stratosphere led to an increase in atmospheric albedo, with increased reflection, scattering and absorption of incoming solar radiation. The peak radiative forcing is estimated at -3 W m^{-2}, of similar magnitude to the total forcing to date from greenhouse gases, but much shorter-lived. Effects included reducing the shortwave radiation reaching the Earth's surface, increasing the diffuse fraction of this radiation, and local warming of the lower stratosphere itself. The reduction in shortwave radiation reduced surface heating: radiative-flux anomalies from the Earth Radiation Budget Experiment (ERBE) showed that the net effect of these changes was a strong cooling (Minnis *et al.*, 1993). Global mean temperatures in the lower troposphere were 0.3 to 0.4 K lower over the following two to three years (Soden *et al.*, 2002). These changes also influenced the hydrological cycle, affecting tropospheric water vapour, precipitation and soil moisture. The cooling led to reductions in leaf-area index and the length of the growing season at high latitudes (Lucht *et al.*, 2002). The effect on the global carbon cycle was a marked temporary slowing of the atmospheric CO_2 increase (Keeling *et al.*, 1995). This slowdown is now explained by a combination of two effects: reduced soil decomposition due to the cooling (Jones *et al.*, 2003) and increased primary production due to the greater effectiveness of diffuse radiation for photosynthesis (Farquhar and Roderick, 2003; Gu *et al.*, 2002; Mercado *et al.*, 2009). The radiative effects of the aerosol led to a net warming of the lower stratosphere, although subsequent depletion of ozone due to heterogeneous chemical reactions on the aerosol surface contributed a cooling effect which partly offset this warming. These changes in lower stratospheric heating rates altered stratospheric dynamics, and the greater latitudinal temperature gradient in the wintertime led to changes in atmospheric circulation (Stenchikov *et al.*, 2002), and in particular to a poleward shift in the jet streams in 1991–1992 associated with a strengthened positive phase of the Arctic Oscillation. These dynamical changes probably account for the increase in surface temperatures observed at high latitudes in the winter following the eruption (Robock and Mao, 1995).

Less direct impacts on climate have also been suggested through changes in tropospheric composition resulting from changes in shortwave radiation due to the aerosol layer and reduced stratospheric ozone. The rise in concentration of CH_4 slowed markedly in 1991, consistent with both decreased wetland sources due to surface-temperature changes and changes in atmospheric-oxidation capacity; however, the contribution made by the eruption remains unclear, due to other climate influences and to a large uncertainty in the anthropogenic contributions to CH_4 variability during recent decades (e.g. Telford *et al.*, 2010).

Unambiguous attribution of the observed changes to the eruption is difficult, especially in light of the strong warm phase of the El Niño Southern Oscillation (ENSO) that occurred in 1992–1993. However, the modelling studies exploring the climate response suggest that the eruption was responsible for many of the observed changes. Careful observation of the effects of future eruptions will allow confirmation and further cement our understanding of the climate response to such a perturbation.

3.6.2 Glacial–interglacial cycles and biogeochemical cycles

The strongest climate signal observed in the last million years is the succession of glacial–interglacial cycles. These are a persistent feature of the Quaternary Period (Figure 3.4(b)), although both their period and amplitude are greater in the later part of the Quaternary. Here we will only discuss these stronger cycles, in particular the last 800,000 years for which we have ice-core as well as extensive data from terrestrial and marine sediments.

The definition of glacials and interglacials referred originally to the waxing and waning of continental ice sheets, particularly over North America (the Laurentide), the Alps and northern Europe (Fennoscandian). Glacial–interglacial cycles are recorded in geomorphological evidence of ice-sheet extent, and globally in records of sea level (e.g. Rohling *et al.*, 2009). The same cyclicity can be observed in many other records, including palaeotemperature records worldwide (Lang and Wolff, 2011). The range of sea level between interglacials and glacials has been around 120 m for the most recent cycles, while the span of global average temperature has been ~5 K (Shakun and Carlson, 2010).

In many records the cycles, recurring roughly every 100,000 years, have a sawtooth shape, with relatively rapid warmings (termination) initiating short interglacials (typically 10,000 to 30,000 years long) superimposed on a generally colder and more icy background (Figure 3.6). In the next sections we will discuss two major features of glacial–interglacial cycles. The first issue concerns the perturbation that led to the very strong response (Section 3.6.3). The second is the biogeochemical feedbacks that are implicit in the records of trace gases in Antarctic ice cores through the last 800,000 years (Section 3.6.4).

3.6.3 Glacial cycles: an astronomical perturbation?

The prevailing paradigm in palaeoclimate science is that the glacial cycles are paced by astronomical forcing: the changes in solar input at different latitudes and seasons that are driven by changes in Earth's orbit around the Sun and by the tilt of Earth's axis. Indeed, making the assumption that there have not been major changes on 100,000 year periods in solar strength or volcanic activity, and that there have not been exotic perturbations such as large meteorite impacts, this astronomical forcing is the only candidate available.

Three main astronomical perturbations can be identified. (1) Earth's orbit around the Sun is not quite circular, and the extent to which it is an ellipse varies (with periods around 100,000 and 400,000 years) such that the eccentricity ranges from almost zero to 0.06; this span causes a global-scale change in annual receipt of solar energy of about 0.2%. (2) The tilt angle (obliquity) of Earth's rotational axis (compared to the plane of its orbit) varies, with a period of 41,000 years, between about 22° and 25°, effectively changing the latitude of the tropics and the polar circles. The strongest effect of high obliquities is to amplify the seasonal cycle at high latitudes. (3) The third major effect, precession of the equinoxes, depends on the eccentricity of the orbit. The effect of precession is that the season when Earth is closest to the Sun slowly varies. When

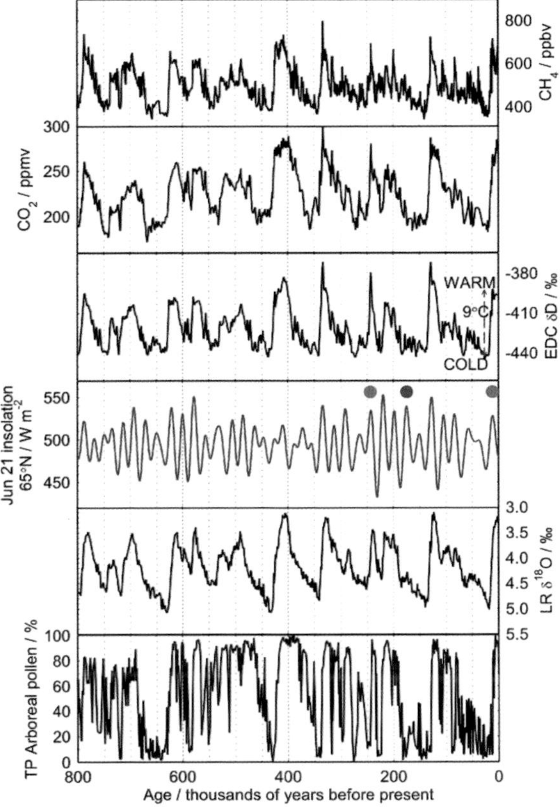

Figure 3.6 Climate and trace gases across glacial–interglacial cycles of the last 800,000 years. Stacked benthic foraminiferal oxygen isotope data (labelled LR) (Lisiecki and Raymo, 2005) represents a combination of global ice volume and deep-ocean temperature; arboreal pollen from Tenaghi Philippon (TP), Greece, represents a local climate signal for southern Europe; δD from EPICA Dome C (EDC) is an indirect measure of Antarctic temperature (Jouzel *et al.*, 2007); Dome C and Vostok CO_2 data were compiled in Lüthi *et al.* (2008); Dome C CH_4 data were compiled in Loulergue *et al.* (2008). The increases of the last 200 years are not shown; the 2009 South Pole concentration of CO_2 was 384 ppm, and of CH_4 1744 ppb. The final panel shows (in blue) insolation in mid-summer at 65° N (Laskar *et al.*, 2004). The green dots show two insolation rises where a glacial termination occurs (as seen in the LR and EDC curves above and below it) as predicted by Milanković theory; the red dot shows an equally large insolation rise with no accompanying termination, illustrating that the theory does not have predictive power.

the Earth–Sun distance is at a minimum in northern-hemisphere summer, that hemisphere will receive particularly high insolation in summer (and low insolation in winter), i.e. a greater seasonal contrast, while the southern hemisphere has the opposite response, and therefore a smaller-than-average seasonal contrast. Precession varies with two superimposed periods of 19,000 and 23,000 years.

These astronomical cycles combine to provide a time-varying seasonal and latitudinal distribution of insolation (Berger and Loutre, 1991; Laskar *et al.*, 2004). Milanković (Milanković, 1998; translated from the 1941 original) first made accurate calculations of insolation, and promoted the idea that the control of glaciations depended on the summer insolation at the latitude (65° N) at which an ice sheet would be expected to nucleate (low summer insolation implies that winter snow can persist through summer). It was only with the production of marine isotope records with good age control that these ideas gained ground, because such records proved to be dominated by just those frequencies that astronomical theory predicts (Hays *et al.*, 1976).

A comprehensive understanding of the glacial cycles still eludes us. Outstanding issues include the following:

(a) Although all the astronomical frequencies are seen in marine isotope and ice-core records, the strongest, and the one that corresponds most obviously to large climate change, is at 100,000 years. Yet the 100,000-year eccentricity cycle gives only a tiny change in insolation at the latitude and season Milanković theory highlights.

(b) Before about 1 million years ago, the prime frequency in the data is close to 40,000 years, but there is no astronomical reason for a change in the dominant periodicity.

(c) While the 100,000-year period emerges from the records, the spacing between maxima in benthic isotope or Antarctic temperature records varies from 40,000 to 120,000 years (but is never exactly 100,000 years). The result of this is that, while glacial terminations generally occur when northern-hemisphere insolation is rising (Cheng *et al.*, 2009), no insolation time series appears to have predictive power about which precessional cycle will be associated with an interglacial (Figure 3.6).

The original paper that identified the astronomical frequencies in marine isotope records (Hays *et al.*, 1976) referred to astronomical variations as 'pacemakers' of the ice ages. Since then the idea has developed that the Earth system may have a natural timescale of order 100,000 years (perhaps embedded in the timescales for the growth of large ice sheets (Paillard, 1998) or for the carbon cycle (Toggweiler, 2008), both major amplifiers, as we shall discuss later), and that this may allow the system to respond to its maximum only at multiples of the cycles with greater forcing. In extreme versions of this idea, it is suggested that the system would cycle at roughly 100,000 year in the absence of any orbital forcing, and the role of the astronomical changes is only to tune the timing of major shifts (Tziperman *et al.*, 2006).

The discussion above implies that the initial perturbation leading to glacial–interglacial cycling is some combination of orbital changes, acting mainly at seasonal and hemispheric (but not global) scales, and internal variability. It illustrates that feedbacks can turn regional forcings into global responses, and highlights the difficulty of trying to extend definitions of Earth system sensitivity.

A final issue with the causes of glacial cycles is that it has become apparent that millennial-scale changes (Section 3.6.6), most probably caused by reorganizations of ocean heat transport, play an integral role in terminations, and may even act as the trigger for them (Wolff *et al.*, 2009): albeit one that works only when the astronomical conditions are favourable.

3.6.4 Glacial cycles: biogeochemical feedbacks

The forcing induced by astronomical variations is much too small *in global average* to induce the very large changes seen in the climate and Earth system. Rather, regional and seasonal changes in external forcing, or even redistribution of heat, set off a chain of amplifiers that lead to major change. We know the identity of the main amplifiers: they are increased albedo due to expanded ice sheets, and reductions in greenhouse gases.

There have been several attempts to estimate the relative contributions of these amplifiers to the LGM–Holocene climate change. A typical example (Hewitt and Mitchell, 1997), using an atmosphere–mixed-layer ocean model, estimated that one-third of the forcing was due to changes in greenhouse gases, and almost two-thirds to changes in albedo associated with the changes in ice-sheet extent. More recent estimates, based on analyses of the PMIP2 coupled ocean–atmosphere simulations, indicate that the effect of lowered greenhouse gases and changes in albedo were comparable (Braconnot *et al.*, 2012). Further work has used the entire 800,000 years of the Antarctic ice-core records. Fortunately, we know very well from ice cores the concentrations of the major greenhouse gases (CO_2, CH_4 and N_2O) over the last 800 ka (Figure 3.6). We can also make estimates of ice-sheet extent, although these have generally been based on interpolating, using the marine isotope record, from spatial reconstructions of the LGM ice sheets. Modelling studies (Holden *et al.*, 2010; Masson-Delmotte *et al.*, 2010; Singarayer and Valdes, 2010), driven by these observations and estimates, also suggest that the changes in greenhouse gases were responsible for about 50% of the glacial–interglacial variability.

Now we look in more detail at the relationship between climate and greenhouse-gas records. The Antarctic temperature record based on ice-core deuterium measurements (Jouzel *et al.*, 2007), and the ice-core record of atmospheric CO_2 (Lüthi *et al.*, 2008) (Figure 3.6 and 3.7), are very similar ($r^2 = 0.82$). Carbon dioxide is low during glacial maxima, and high during interglacials (although the natural variations have always been within an envelope of 170–300 ppm). During the last deglaciation, the similarity runs to the details of the 'stutter' during the Antarctic Cold Reversal. Antarctic temperature appears to begin its rise slightly before CO_2, suggesting that any trigger acts first on the climate rather than the carbon cycle, but for most of the ~5000 years of rising values, the two variables change together. These observations are absolutely consistent with a feedback between temperature (specifically southern high-latitude temperature) and CO_2.

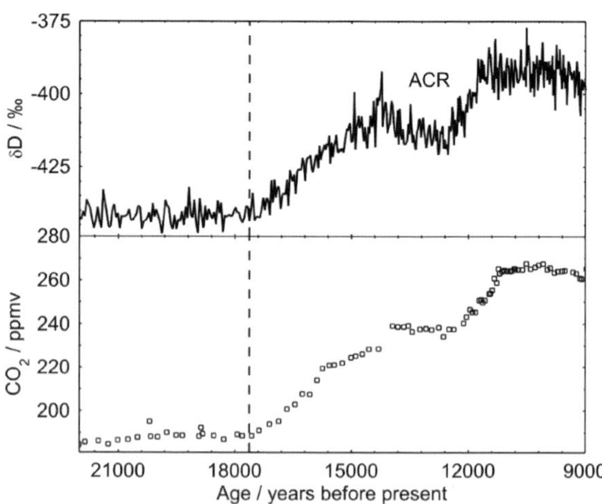

Figure 3.7 Climate–carbon-cycle feedback in action over the last glacial termination. Antarctic EPICA Dome C climate (represented by δD) and CO_2 are shown. These are based on the data in Monnin *et al.* (2001) but the data are on the EDC3 (Parrenin *et al.*, 2007) scale for the age of the ice, and the corresponding EDC3_gas_a (Loulergue *et al.*, 2007), for the age of the gas in the air bubbles within the ice. ACR is the Antarctic Cold Reversal. Note that the uncertainty on the relative age of the two plots is several centuries.

This very close relationship, in which the millennial variations of CO_2 resemble those of Antarctic temperature records rather than the abrupt jumps of many northern records, suggests that Southern Ocean processes might be major players in the temperature–CO_2 feedback on glacial–interglacial timescales. However, despite several decades of research, there are still many candidates for the processes involved, and there is no consensus about which plays the leading role (Kohfeld and Ridgwell, 2009). The terrestrial biosphere grew during deglaciations implying a withdrawal of CO_2 from the atmosphere (Prentice and Harrison, 2009). The biosphere growth leaves a net change of as much as 120 ppm to explain through ocean processes. Measurements of changes in $^{13}CO_2$ (Elsig *et al.*, 2009; Lourantou *et al.*, 2010a; Lourantou *et al.*, 2010b) can assist with understanding the causes of change but have not yet given a conclusive answer. Most of the proposed mechanisms involve either change in the ventilation that brings CO_2 from the deep ocean to the surface or biogeochemical changes that alter the export of carbon from the surface to the deep ocean.

Some data or model constraints can be placed on each mechanism. One of the biogeochemical processes invoked is iron fertilization. In the Southern Ocean (in particular), phytoplankton productivity is currently limited by a lack of iron (compared to an abundance of major nutrients). Iron supply at the LGM was increased, due to the much higher flux of terrestrial dust that is shown by both marine and ice-core data (Kohfeld and Harrison, 2001; Wolff *et al.*, 2006) and consistent with strong winds and reduced continental vegetation cover. However, both data (Röthlisberger *et al.*, 2004) and models (e.g. Bopp *et al.*, 2003; Oka *et al.*, 2011) show that the effect of the additional iron supply on Southern Ocean productivity can produce a CO_2 drawdown of 25 ppm at most. Furthermore, analysis of Southern Ocean palaeo-productivity data showed that much of the CO_2 lowering after the LIG took place before any sign of the increase in export production (the biologically mediated removal of carbon from near the sea surface) that would be required to drive this mechanism (Kohfeld *et al.*, 2005). Another hypothesis proposes that a feedback between climate and the latitude of the southern-hemisphere westerly winds could be responsible for a large part of the CO_2 change (Toggweiler *et al.*, 2006). However, a study using a physical and biogeochemically enabled ocean model suggested that such a mechanism could only contribute a few ppm of CO_2 lowering (Tschumi *et al.*, 2008), while the uncertainties in interpretations of the paleo-observations and differences between the results of coupled climate models (Rojas *et al.*, 2009) question whether such a latitudinal shift in winds actually occurred. It is generally considered probable that several mechanisms each played a small contributory role (Kohfeld and Ridgwell, 2009).

During one more recent period, centred on the seventeenth century, a small drop in CO_2 concentration was associated with a climate cooling (the Little Ice Age). If the CO_2 drop resulted from the cooling, then it should be possible to estimate the strength of the CO_2 feedback (Cox and Jones, 2008). However, the uncertainty in this calculation is large (Frank *et al.*, 2010; Friedlingstein and Prentice, 2010). A median estimate of 7.7 ppm K^{-1} has been proposed (Frank *et al.*, 2010). The LGM represents a much stronger signal: if the CO_2 change of about 100 ppm is the net result of a global mean temperature change of 5 K, then the calculated sensitivity is 20 ppm K^{-1}, but this number is relevant for the future only if all the mechanisms remain active under future warming. For example, the amount of dust contributed to the Southern Ocean from the atmosphere today is negligible in comparison to the LGM. It is unlikely that any further reduction in dust flux would significantly affect the productivity of the ocean; this mechanism must have saturated. Similarly there may be limits to the net release of carbon from the deep ocean that changes in circulation can cause, and we do not know if we have already reached such limits.

Two other greenhouse gases are measured in ice cores. Nitrous oxide (N_2O) poses some measurement difficulties, but a good outline of the millennial- and orbital-scale variations is now available (Schilt *et al.*, 2010a). Like other biogenic greenhouse gases, its natural emissions increase when it is warm, so that it gives a net positive feedback. The natural source of N_2O is from microbial transformations in soils (nitrification and especially denitrification), with a smaller contribution from the same processes in the ocean (Chapter 4). Nitrous oxide shows lower concentrations in glacials than interglacials (200–280 ppb) and strong millennial-scale variability linked to Dansgaard–Oeschger (D–O) cycles (Flückiger *et al.*, 2004; Schilt *et al.*, 2010a).

The methane (CH_4) record for the last 800,000 years (Loulergue *et al.*, 2008) also shows high values during warm periods and low values during cold periods (350–750 ppb). Methane shows an extremely strong millennial-scale variability that parallels the millennial-scale D–O events seen in Greenland ice-core climate records (see Section 3.6.6). Changes in CH_4 concentration result from changes in the source, or changes in the atmospheric sink (which is dominated by removal by the OH radical). Major natural sources include wet soils, tropical and boreal wetlands and biomass burning. Methane isotope measurements (Sowers, 2006; Fischer *et al.*, 2008; Petrenko *et al.*, 2009) have ruled out methane clathrates stored in ocean sediments or permafrost as a major contributor to the glacial–interglacial changes. Some modelling studies have suggested that changes in the strength of the atmospheric sink for CH_4 may be important. Recent papers emphasize this less (Singarayer *et al.*, 2011), but it has proved difficult to find any independent method to test the relative importance of source and sink changes (Levine *et al.*, 2011).

Taking all the concentration, isotopic and modelling evidence into account, it appears likely that the major player in the glacial–interglacial changes in CH_4 is wetland emissions (Fischer *et al.*, 2008; Singarayer *et al.*, 2011). The net emissions of CH_4 from wetlands are a subtle balance between emissions from different regions, and both tropical changes resulting from variations in monsoon strength, and boreal changes resulting partly from the retreat of the LGM ice sheets, are likely involved. Modelling studies can reproduce the observed orbital-scale changes (Singarayer *et al.*, 2011), although they have not yet addressed the large and rapid millennial changes in CH_4. We do not know how robust such results are to different models of the climate, terrestrial biosphere and wetlands. Further data – both isotopic, and comparative Greenland and Antarctic concentration data that give information about the latitude of sources – are likely to become available soon, and these will place much tighter constraints on possible explanations.

Some of the changes involved over glacial timescales are of limited relevance to the future (for example, the reaction in areas uncovered on retreat of the Laurentide ice sheet). However, a quantitative constraint of the reaction of tropical wetlands to monsoon changes will be useful for attempts to predict future climate–CH_4 feedback. We can also supply a limited constraint to concerns about large potential releases of CH_4 from Arctic permafrost under warming scenarios. During part of the LIG, Arctic summer (Otto-Bliesner *et al.*, 2006) and Greenland mean annual (North Greenland Ice Core Project Members, 2004) temperatures were raised by around 5 K compared to the present. Methane concentrations (as measured in Antarctic ice cores) remained below pre-industrial levels at this time, suggesting that any threshold for major release is beyond this range of warming.

3.6.5 Warm periods and the cryosphere

Climate models project greater warming over land and at high latitudes in the present century under a range of scenarios for CO_2 emissions (IPCC, 2007). Thus, warming of several degrees is likely over Greenland and the Arctic Ocean, with somewhat smaller values over Antarctica. Increased temperatures are expected to persist for centuries to millennia (Solomon *et al.*, 2009).

Such warming in polar regions is expected to have a strong effect on sea-ice extent, but the effect on ice sheets is much less certain because ice dynamics is incompletely understood. The IPCC (IPCC, 2007) declined to make a definitive statement about the ice-sheet contribution to future sea-level rise, '*because a basis... [for doing so] ... in the published literature is lacking*'. However, they did note the possibility of a '*virtually complete elimination*' of the Greenland ice sheet if warm temperatures and negative surface-mass balance were sustained for millennia (Gregory *et al.*, 2004).

Given the current status of ice-sheet modelling, and the long timescale of the processes leading to ice-sheet wastage, it seems obvious to look for possible analogues in the palaeoclimate record. There is no well-documented analogue for the expected change in greenhouse-gas concentrations, nor for the kind of global-scale warming expected in the next century and beyond. However, we can make a simplifying assumption that the Greenland and West Antarctic ice sheets depend mainly on their local polar climate. It then becomes particularly interesting to examine the LIG.

Arctic summer temperatures were 3 to 5 K warmer than present during the LIG (Otto-Bliesner *et al.*, 2006), and annual mean temperatures over Greenland were increased by a similar amount in the latter part of the LIG (North Greenland Ice Core Project Members, 2004). This conclusion might be modulated by improved understanding of changes in the calibration of water isotopes against temperature, or of changes in ice-sheet elevation (Masson-Delmotte *et al.*, 2011). Antarctic temperature was also elevated in the early part of the LIG, again probably by 4 K and perhaps as much as 6 K (Sime *et al.*, 2009). Maximum warmth in the two polar regions was probably not simultaneous, but we can be sure that it lasted for at least centuries to millennia in each case, long enough for the ice sheets to react. Estimates of sea level during the LIG have shown that it was between 4 and 9 metres higher than today (combining the conclusions of earlier studies with those of the most recent synthesis (Kopp *et al.*, 2009)). Thus, observations of the LIG show that an increase in sea level of several metres is expected under a sustained polar warming of order 3–6 K, which is well within the range of projections for the next century. The LIG data do not yet give us a timescale for such a change, nor is it clear what was the balance of contribution from Greenland and Antarctica.

The Pliocene, 5.3 to 2.6 million years ago, was a warm epoch when CO_2 concentrations are estimated to have been 365–415 ppm (i.e. close to the concentration today) (Pagani *et al.*, 2009). Only a reduction towards pre-industrial concentrations, and a reduction in global temperature from a peak around 3–4 K warmer than present, allowed the Greenland ice sheet to form (Lunt *et al.*, 2008). This has implications for the kind of reductions in CO_2 concentration that would be necessary to reverse Greenland ice-sheet loss, should it occur.

3.6.6 Abrupt climate changes during glacial periods

Apart from the glacial–interglacial changes, the most prominent signal in records of the last few glacial cycles is millennial and abrupt. Abrupt changes between cold stadials and relatively warm interstadials (D–O cycles) are most prominently seen in records of the last glacial period in Greenland ice cores (Johnsen *et al.*, 1992) (Figure 3.4c). The same sharp events can be identified in marine (e.g. McManus *et al.*, 1999) and terrestrial (e.g. Fleitmann *et al.*, 2009) records. They were present during at least the last 800,000 years (Jouzel *et al.*, 2007; Loulergue *et al.*, 2008), although here, we focus on the abrupt climate changes of the last glacial period and deglaciation, 73,500 to 11,700 years ago, for which the most data are available. The final D–O cycle is identified with the Bølling–Allerød warm period and Younger Dryas cold period, which occurred during the last glacial termination.

The shape of all the D–O cycles is basically the same although their durations vary (Figure 3.8). Every D–O cycle was characterized by an abrupt warming in the atmosphere of Greenland, the D–O event, taking place typically in 40 years, with a range of amplitudes from 6 to 16 K (Capron *et al.*, 2010; Wolff *et al.*, 2010), followed by a gradual cooling. The D–O event and the gradual cooling together form a Greenland interstadial (GI). The final precipitous cooling and cold period form the Greenland stadial (GS). The GI durations ranged from < 500 to > 2,000 years. The signals seen in marine and terrestrial records show that all D–O events were accompanied by sea-surface temperature reductions in the North Atlantic (Figure 3.8), and by substantial changes in atmospheric circulation. All D–O events have a subdued counterpart in Antarctica (EPICA Community Members, 2006), and the cold phases in Greenland coincide with warming in Antarctica, while the warm phases correspond to cooling in Antarctica (Blunier and Brook, 2001). This pattern of events is consistent with expected impacts from changes in the AMOC. Reductions in ocean heat transport during GSs allow the Southern Ocean to warm gradually; the resumption of full heat transport during GIs leads to a loss of heat from the Southern Ocean (Stocker and Johnsen, 2003).

Among the 25 identified and numbered GSs, six of them coincide with periods of massive iceberg discharges in the North Atlantic, called Heinrich events (HEs), occurring every 7 to 10,000 years and presumably reflecting partial collapse and fragmentation of the Fennoscandian and Laurentide ice sheets. Heinrich events are recognized by the presence in marine sediments of layers of ice-rafted debris, known as Heinrich layers (Bond *et al.*, 1992). They occur during particularly long-duration GSs, and this suggests that D–O cycles and HEs are related. There is strong evidence that during HE1, 18,000 years ago, there was an almost complete shutdown of the AMOC (Gherardi *et al.*, 2005; McManus *et al.*, 2004).

All the D–O cycles (including those that relate to HEs) are associated with rapid shifts in terrestrial ecosystems (Harrison and Sanchez-Goñi, 2010), with large changes in CH_4 concentration in the atmosphere (e.g. Flückiger *et al.*, 2004) and changes in Asian monsoon intensity (Wang *et al.*, 2008). The response of the vegetation is fast, ~50 to 100 years, and tightly parallels SST changes (Figure 3.8) indicating a dynamic equilibrium between vegetation and climate over these timescales (Sanchez-Goñi *et al.*, 2008). The strongest response is seen in the vegetation changes and, by extension, in the climate of the northern extratropics and particularly in Europe (Harrison and Sanchez-Goñi, 2010).

A recent review highlighted three candidate mechanisms for the generation of D–O cycles: tropical

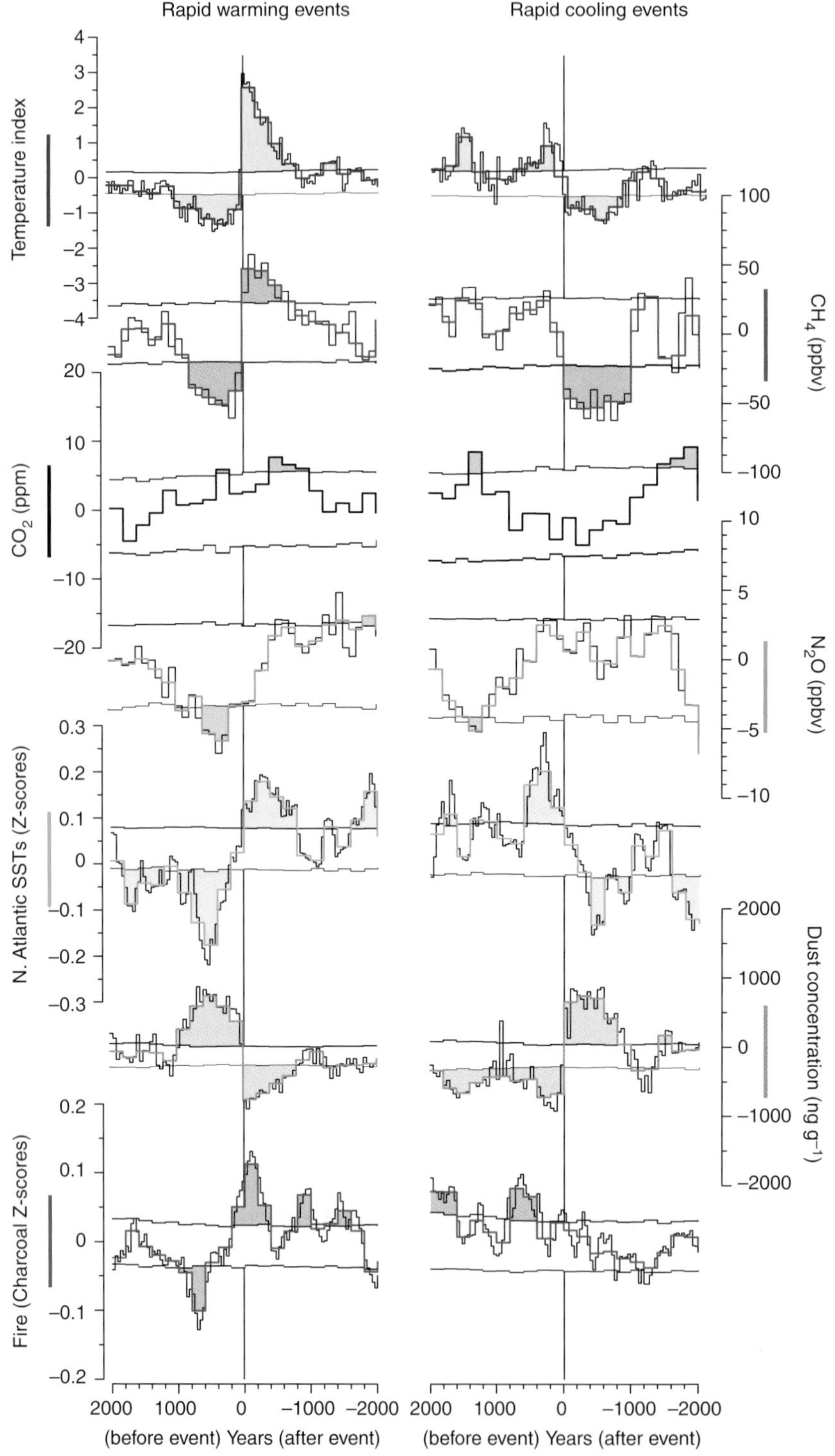

Figure 3.8 Superimposed epoch analysis of records of Dansgaard–Oeschger cycles, showing the consistent response of each record to the multiple rapid warmings and coolings during the interval 80,000 to 10,000 years ago. Shading indicates significant patterns in the response of the time series to the events of abrupt warming and rapid cooling. The Greenland oxygen isotope record (North Greenland Ice Core Project Members, 2004), an index of regional temperature, shows the characteristic saw-tooth pattern of an individual D–O cycle. Warming of the surface North Atlantic begins somewhat earlier than, and peaks after, the rapid warming seen in Greenland. Changes in CH_4 concentration (Loulergue *et al.*, 2008) and Antarctic dust (Lambert *et al.*, 2008) – which are sensitive to changes in surface hydrology and vegetation cover – are synchronous with the rapid changes in climate. Changes in CO_2 (Ahn and Brook, 2008) and N_2O (Schilt *et al.*, 2010b) – atmospheric constituents with longer lifetimes – show more gradual responses. Global biomass burning (Daniau *et al.*, 2010) shows a rapid response to warming but a more complex response to cooling, characterized by an initial decline and a gradual recovery.

processes, sea-ice processes and changes in ocean circulation (Clement and Peterson, 2008). While sea ice is certainly involved as a major feedback, most studies ascribe the major driving role in all D–O cycles to changes in the AMOC (e.g. Kageyama *et al.*, 2010). These are in turn often related to changes in freshwater discharge from the northern-hemisphere ice sheets, either by icebergs (particularly for HEs), or by routing of lake and ice-sheet drainage. However, quantitative estimates of any such change in the AMOC have proved elusive.

It is likely that the D–O cycles are *not* the result of an external perturbation. Earlier suggestions that D–O events were periodic (at about a 1,500-year period) are not supported by studies using the most recent age scale (Ditlevsen *et al.*, 2007). Dansgaard–Oeschger events are too close together for there to be a consistent orbital trigger, although the possibility that there was some weak solar or volcanic impetus that pushed the system into a new state cannot be entirely ruled out. It seems more likely that D–O events are the result of internal reorganizations in the ice-sheet–ocean–atmosphere system. The only D–O cycle for which there is an estimate of the mean *global* temperature change is the Younger Dryas GS relative to the preceding Bølling–Allerød GI. The change is only ~0.6 K (Shakun and Carlson, 2010). Thus, the regionally strong signals shown during this interval presumably resulted from regional feedbacks, including that from Arctic sea ice, producing only a small net global signal.

Recognizing that the ultimate causes of D–O events and HEs are still uncertain, we can move along the causal chain and examine the events as a response to a perturbation in the AMOC. Numerous modelling experiments have now been carried out in which freshwater is introduced into the ocean ('freshwater hosing'). Many of these (Kageyama *et al.*, 2010) are capable of producing a change in the AMOC, and a pattern of climate response similar to that during D–O cycles; although the observed high speed of the warming response at Greenland and other locations on recovery of the AMOC is generally not reproduced in models (Otto-Bliesner and Brady, 2010). The simulated response is dependent on the exact boundary conditions (Kageyama *et al.*, 2010), i.e. the ice-sheet configuration, greenhouse-gas and dust concentrations, and the orbital parameters.

The climate events of the last glacial cycle indicate that rapid (decades) and very large (up to 16 K in Greenland) regional changes in climate are possible. Furthermore, the effects of the changes were global (Blunier and Brook, 2001; Harrison and Sanchez-Goñi, 2010), even if the strongest manifestation was in the North Atlantic and surrounding lands (e.g. Fletcher *et al.*, 2010; Martrat *et al.*, 2004). But we have evidence for such changes only at times when ice sheets existed that could supply large anomalies in freshwater fluxes to the North Atlantic. The events tell us that major reorganizations of the AMOC can occur. Although this finding is not directly applicable to possible future changes with different causes, a correct description of the past events is likely only if we have good model representation of the key processes, leading to a correct sensitivity of the system. In this respect, it is a concern that many models *can* produce the pattern of change during the last deglaciation (Liu *et al.*, 2009; Menviel *et al.*, 2011) and HEs (Kageyama *et al.*, 2010), but require much larger inputs of freshwater than appear realistic (Valdes, 2011). This discrepancy suggests that the models are too stable to change in ocean circulation. Further progress in this area will require a breakthrough in quantifying past changes in the AMOC (Kageyama *et al.*, 2010).

3.6.7 PETM and the carbon cycle

The Palaeocene–Eocene Thermal Maximum (PETM) was an abrupt warming event, superimposed on a longer-term warming trend (Figure 3.4a), that occurred about 56 million years ago. Recent reviews (McInerney and Wing, 2011; Zachos *et al.*, 2008) agree that global temperatures increased by *c.* 5 to 8 K, and that elevated temperatures persisted for between 100,000 and 200,000 years. The abrupt warming is associated with a large (>3 ‰) negative anomaly in $\delta^{13}C$, indicating that there was a large release of light carbon (with low ^{13}C) into the atmosphere. The total release was of a magnitude comparable to that possible from burning fossil-fuel reserves (Haywood *et al.*, 2011), although estimates are only loosely constrained to between 3,000 and 7,000 Pg C. There are varying estimates of the duration of the carbon release, in the range of 10,000 to 20,000 years. Recent model estimates, constrained by observations, suggest that the rates of carbon injection during the PETM (0.3–1.7 Pg C a^{-1}) (Cui *et al.*, 2011) were lower than the present rate of fossil-fuel emissions of CO_2. Dissolution of sea-floor carbonates (Zachos *et al.*, 2005) shows that ocean acidification occurred, and this is presumed to have contributed to the extinction of many benthic foraminifera; major changes occurred in both marine and terrestrial biota (McInerney and Wing, 2011).

The PETM demonstrates that large and rapid climate changes could be triggered by perturbations of the carbon cycle even during times when the climate was warm and in an ice-free world. However, the trigger for the CO_2 release is still not clear. An early and popular idea was that this was the result of destabilization of methane clathrates as ocean temperatures warmed from the late Paleocene to a maximum in the early tocene (Dickens *et al.*, 1995). There have been a number of objections to this as the main source, recently summarized (McInerney and Wing, 2011) and argued against (Dickens, 2011). The nature of the carbon source is not relevant to the discussion in this chapter, which relates to its potential as an analogue for anthropogenic carbon release. However, because the nature of the source determines the carbon isotope content of the release, and therefore by mass-balance arguments, controls the inferred magnitude of the release, resolution of this issue is required in order to scale the perturbation and its effects.

Although many important details remain unresolved, the PETM shows that injection of large amounts of CO_2 into the atmosphere (whatever the source and trigger) results in global warming, ocean acidification and widespread benthic extinctions. The recovery from this perturbation took between 100,000 and 200,000 years. This is consistent with the time estimated for the anthropogenically increased CO_2 concentration to return to pre-industrial values (Archer *et al.*, 2009).

3.7 Natural perturbations as a guide to the future behaviour of the Earth system

The Earth's history offers us a range of natural experiments in which perturbations, either externally forced or internally generated, have led to significant changes in climate. The examples we have discussed range from the one-off events like a volcanic eruption that allow us to probe climate responses over periods of years to decades, to major events in Earth's history lasting millennia or longer. The challenge is to find situations in which we have a good understanding of the magnitude of both the forcing and the response. Palaeoclimate research until now has provided mainly qualitative insights: for example, the observations of rapid climate change most probably resulting from changes in AMOC give us the confidence to know that such changes are possible, but do not yet provide quantitative constraints on the magnitude of freshwater inputs required to affect the AMOC. On the other hand, the use of extensive spatially distributed quantitative data from the land, oceans and ice sheets has made it possible to place new quantitative constraints on climate sensitivity.

We have identified a number of climate scenarios that appear to have great potential in providing constraints, both at global and regional level, and for particular aspects of the system. Particular new measurements, data syntheses and modelling advances have been identified that would further enhance their potential. None of the scenarios is a perfect analogue for the current situation: they all have either a forcing of a different nature, very different boundary conditions from those of the present century, or very uncertain magnitude of forcing or response. Nonetheless, together they provide a rich seam of information on change, rates of change, and on the way that different parts of the system react together.

References

Ahn, J. and Brook, E. J. (2008). Atmospheric CO_2 and climate on millennial time scales during the last glacial period. *Science*, **322**, 83–85.

Allen, M. R. and Ingram, W. J. (2002). Constraints on future changes in climate and the hydrologic cycle. *Nature*, **419**, 224–232.

Anderson, T. L., Charlson, R. J., Schwartz, S. E. *et al.* (2003). Climate forcing by aerosols – a hazy picture. *Science*, **300**, 1103–1104.

Archer, D., Eby, M., Brovkin, V. *et al.* (2009). Atmospheric lifetime of fossil fuel carbon dioxide. *Annual Review of Earth and Planetary Sciences*, **37**, 117–134.

Bartlein, P. J., Harrison, S. P., Brewer, S. *et al.* (2011). Pollen-based continental climate reconstructions at 6 and 21 ka: a global synthesis. *Climate Dynamics*, **37**, 775–802.

Berger, A. and Loutre, M. F. (1991). Insolation values for the climate of the last 10 million of years. *Quaternary Science Reviews*, **10**, 297–317.

Blunier, T. and Brook, E. J. (2001). Timing of millennial-scale climate change in Antarctica and Greenland during the last glacial period. *Science*, **291**, 109–112.

Bond, G., Heinrich, H., Broecker, W. *et al.* (1992). Evidence for massive discharges of icebergs into the North Atlantic ocean during the last glacial period. *Nature*, **360**, 245–249.

Bopp. L., Kohfeld, K. E., Le Quéré, C. and Aumont, O. (2003). Dust impact on marine biota and atmospheric CO_2 during glacial periods. *Paleoceanography*, **18**, 1046, doi: 10.1029/2002PA000810.

Borgmark, A. (2005). Holocene climate variability and periodicities in south–central Sweden, as interpreted from peat humification analysis. *Holocene*, **15**(3), 387–395.

Braconnot, P., Otto-Bliesner, B., Harrison, S. *et al.* (2007). Results of PMIP2 coupled simulations of the mid-Holocene and last glacial maximum – Part 1: experiments and large-scale features. *Climate of the Past*, **3**, 261–277.

Braconnot, P., Harrison, S. P., Otto-Bliesner, B. *et al.* (2011). The Paleoclimate Modeling Intercomparison Project contribution to CMIP5. *CLIVAR Exchanges*, **16**, 15–19.

Braconnot, P., Harrison, S. P., Kageyama, M. *et al.* (2012). Evaluation of climate models using the palaeoclimate data. *Nature Climate Change*. doi 10.1038/NCLIMATE 1456.

Briffa, K. R., Jones, P. D., Wigley, T. M. L., Pilcher, J. R. and Baillie, M. G. L. (1983). Climate reconstruction from tree rings. 1. Basic methodology and preliminary results for England. *Journal of Climatology*, **3**(3), 233–242.

CAPE-Last Interglacial Project Members (2006). Last interglacial Arctic warmth confirms polar amplification of climate change. *Quaternary Science Reviews*, **25**, 1383–1400.

Capron, E. Landais, A., Chappellaz, J. *et al.* (2010). Millennial and sub-millennial scale climatic variations recorded in polar ice cores over the last glacial period. *Climate of the Past*, **6**, 345–365.

Carter, L., Neil, H. L. and McCave, I. N. (2000). Glacial to interglacial changes in non-carbonate and carbonate accumulation in the SW Pacific Ocean, New Zealand. *Palaeogeography Palaeoclimatology Palaeoecology*, **162**, 333–356.

Cheng, H., Edwards, R. L., Broecker, W. S. *et al.* (2009). Ice age terminations. *Nature*, **326**, 248–252.

Claussen, M. Mysak, L.A., Weaver, A. J. *et al.* (2002). Earth system models of intermediate complexity: closing the gap in the spectrum of climate system models. *Climate Dynamics*, **18**, 579–586.

Clement, A. C. and Peterson, L. C. (2008). Mechanisms of abrupt climate change of the last glacial period. *Reviews of Geophysics*, **46**, RG4002.

Cook, E. R., Briffa, K., Shiyatov, S. and Mazepa, V. (1990). Tree-ring standardization and growth-trend estimation, in *Methods of Dendrochronology: Applications in the Environmental Sciences*, eds. Cook, E. R. and Kairiukstis, L. A. New York, NY: Springer-Verlag, pp. 104–123.

Cox, P. and Jones, C. (2008). Climate change – illuminating the modern dance of climate and CO_2. *Science*, **321**, 1642–1644.

Crook, J. A., Forster, P. M. and Stuber, N. (2011). Spatial patterns of modeled climate feedback and contributions to temperature response and polar amplification. *Journal of Climate*, **24**(14), 3575–3592.

Cuffey, K. M., Clow, G. D., Alley, R. B. *et al.* (1995). Large Arctic temperature change at the Wisconsin–Holocene glacial transition. *Science*, **270**, 455–458.

Cui, Y., Kump, L. R., Ridgwell, A. J. *et al.* (2011). Slow release of fossil carbon during the Palaeocene–Eocene Thermal Maximum. *Nature Geoscience*, **4**, 481–485.

Daniau, A. L., Harrison, S. P. and Bartlein, P. J. (2010). Fire regimes during the Last Glacial. *Quaternary Science Reviews*, **29**, 2918–2930.

de Vernal, A., Eynaud, F., Henry, M. *et al.* (2005). Reconstruction of sea-surface conditions at middle to high latitudes of the northern hemisphere during the last glacial maximum (LGM) based on dinoflagellate cyst assemblages. *Quaternary Science Reviews*, **24**, 897–924.

DeConto, R. M. and Pollard, D. (2003). Rapid Cenozoic glaciation of Antarctica induced by declining atmospheric CO_2. *Nature*, **421**, 245–249.

Dickens, G. R. (2011). Down the Rabbit Hole: toward appropriate discussion of methane release from gas hydrate systems during the Paleocene–Eocene thermal maximum and other past hyperthermal events. *Climate of the Past*, **7**, 831–846.

Dickens, G. R., Oneil, J. R., Rea, D. K. and Owen, R. M. (1995). Dissociation of oceanic methane hydrate as a cause of the carbon-isotope excursion at the end of the Paleocene. *Paleoceanography*, **10**, 965–971.

Ditlevsen, P. D., Andersen, K. K. and Svensson, A. (2007). The DO-climate events are probably noise induced: statistical investigation of the claimed 1470 years cycle. *Climate of the Past*, **3**, 129–134.

Dyke, A. S., Andrews, J. T., Clark, P. U. *et al.* (2002). The Laurentide and Innuitian ice sheets during the last glacial maximum. *Quaternary Science Reviews*, **21**, 9–31.

Edwards, T. L., Crucifix, M. and Harrison, S. P. (2007). Using the past to constrain the future: how the palaeorecord can improve estimates of global warming. *Progress in Physical Geography*, **31**, 481–500.

Elsig, J., Schmitt, J., Leuenberger, D. *et al.* (2009). Stable isotope constraints on Holocene carbon cycle changes from an Antarctic ice core. *Nature*, **461**, 507–510.

EPICA Community Members (2006). One-to-one hemispheric coupling of millennial polar climate variability during the last glacial. *Nature*, **444**, 195–198.

Evans, M. N., Reichert, B. K., Kaplan, A. *et al.* (2006). A forward modeling approach to paleoclimatic interpretation of tree-ring data. *Journal of Geophysical Research*, **111**, G03008.

Farquhar, G. D. and Roderick, M. L. (2003). Atmospheric science: Pinatubo, diffuse light, and the carbon cycle. *Science*, **299**, 1997–1998.

Fischer, H., Behrens, M., Bock, M. *et al.* (2008). Changing boreal methane sources and constant biomass burning during the last termination. *Nature*, **452**, 864–867.

Fleitmann, D., Cheng, H., Badertscher, S. *et al.* (2009). Timing and climatic impact of Greenland interstadials recorded in stalagmites from northern Turkey. *Geophysical Research Letters*, **36**, L19707.

Fletcher, W. J., Sanchez Goñi, M. F., Allen, J. R. M. *et al.* (2010). Millennial-scale variability during the last glacial in vegetation records from Europe. *Quaternary Science Reviews*, **29**, 2839–2864.

Flückiger, J., Blunier, T., Stauffer, B. *et al.* (2004). N_2O and CH_4 variations during the last glacial epoch: insight into global processes. *Global Biogeochemical Cycles*, **18**, GB1020.

Frank, D. C., Esper, J., Raible, C. C. *et al.* (2010). Ensemble reconstruction constraints on the global carbon cycle sensitivity to climate. *Nature*, **463**, 527–530.

Friedlingstein, P. and Prentice, I. C. (2010). Carbon–climate feedbacks: a review of model and observation based estimates. *Current Opinion in Environmental Sustainability*, **2**, 251–257.

Friedlingstein, P. P., Cox, P., Betts, R. *et al.* (2006). Climate–carbon cycle feedback analysis: results from the C4MIP model intercomparison. *Journal of Climate*, **19**, 3337–3353.

Ganopolski, A., Calov, R. and Claussen, M. (2010). Simulation of the last glacial cycle with a coupled climate ice-sheet model of intermediate complexity. *Climate of the Past*, **6**, 229–244.

Gao, C., Robock, A. and Ammann, C. (2008). Volcanic forcing of climate over the past 1500 years: an improved ice core-based index for climate models. *Journal of Geophysical Research*, **113**, D23111.

Gasse, F. (2000). Hydrological changes in the African tropics since the last glacial maximum. *Quaternary Science Reviews*, **19**, 189–211.

Gersonde, R., Crosta, X., Abelmann, A. and Armand, L. (2005). Sea-surface temperature and sea ice distribution of the Southern Ocean at the EPILOG last glacial maximum – a circum-Antarctic view based on siliceous microfossil records. *Quaternary Science Reviews*, **24**, 869–896.

Gherardi, J. M., Labeyrie, L., McManus, J. F. *et al.* (2005). Evidence from the northeastern Atlantic basin for variability in the rate of the meridional overturning circulation through the last deglaciation. *Earth and Planetary Science Letters*, **240**, 710–723.

Graversen, R. G. and Wang, M. (2009). Polar amplification in a coupled climate model with locked albedo. *Climate Dynamics*, **33**, 629–643.

Gregory, J. M., Huybrechts, P. and Raper, S. C. B. (2004). Climatology – threatened loss of the Greenland ice-sheet. *Nature*, **428**, 616.

Gu, L. H., Baldocchi, D., Verma, S. B., *et al.* (2002). Advantages of diffuse radiation for terrestrial ecosystem productivity. *Journal of Geophysical Research*, **107**, 4050.

Hansen, J., Ruedy, R., Sato, M. and Lo, K. (2010). Global surface temperature change. *Reviews of Geophysics*, **48**, RG4004.

Hargreaves, J. C., Abe-Ouchi, A. and Annan, J. D. (2007). Linking glacial and future climates through an ensemble of GCM simulations. *Climate of the Past*, **3**, 77–87.

Hargreaves, J. C. and Annan, J. D. (2009). On the importance of paleoclimate modelling for improving predictions of future climate change. *Climate of the Past*, **5**, 803–814.

Harrison, S. P. and Prentice, I. C. (2003). Climate and CO_2 controls on global vegetation distribution at the last glacial maximum: analysis based on palaeovegetation data, biome modelling and palaeoclimate simulations. *Global Change Biology*, **9**, 983–1004.

Harrison, S. P. and Sanchez-Goñi, M. F. (2010). Global patterns of vegetation response to millennial-scale variability and rapid climate change during the last glacial period. *Quaternary Science Reviews*, **29**, 2957–2980.

Hays, J. D., Imbrie, J. and Shackleton, N. J. (1976). Variations in the Earth's orbit: pacemaker of the ice ages. *Science*, **194**, 1121–1132.

Haywood, A. M., Ridgwell, A., Lunt, D. J. *et al.* (2011). Are there pre-Quaternary geological analogues for a future greenhouse warming? *Philosophical Transactions of the Royal Society A-Mathematical Physical and Engineering Sciences*, **369**, 933–956.

Hegerl, G. C., Crowley, T. J., Hyde, W. T. and Frame, D. J. (2006). Climate sensitivity constrained by temperature reconstructions over the past seven centuries. *Nature*, **440**, 1029–1032.

Hewitt, C. D. and Mitchell, J. F. B. (1997). Radiative forcing and response of a GCM to ice age boundary conditions: cloud feedback and climate sensitivity. *Climate Dynamics*, **13**, 821–834.

Holden, P. B., Edwards, N. R., Wolff, E. W. *et al.* (2010). Interhemispheric coupling and warm Antarctic interglacials. *Climate of the Past*, **6**, 431–443.

Holland, M. M. and Bitz, C. M. (2003). Polar amplification of climate change in coupled models. *Climate Dynamics*, **21**, 221–232.

Huntington, T. G. (2006). Evidence for intensification of the global water cycle: review and synthesis. *Journal of Hydrology*, **319**, 83–95.

IPCC (2007). *Climate Change 2007: The Physical Science Basis. Contribution of Working Group I to the Fourth Assessment Report of the Intergovernmental Panel on Climate Change.* Cambridge: Cambridge University Press.

Johnsen, S. J., Clausen, H. B., Dansgaard, W. *et al.* (1992). Irregular glacial interstadials recorded in a new Greenland ice core. *Nature*, **359**, 311–313.

Johnsen, S. J., Dahl Jensen, D., Dansgaard, W. and Gundestrup, N. (1995). Greenland palaeotemperatures derived from GRIP bore hole temperature and ice core isotope profiles. *Tellus B*, **47**, 624–629.

Jones, C. D., Cox, P. and Huntingford, C. (2003). Uncertainty in climate–carbon-cycle projections associated with the sensitivity of soil respiration to temperature. *Tellus B*, **55**, 642–648.

Joshi, M. and Gregory, J. (2008). Dependence of the land–sea contrast in surface climate response on the nature of the forcing. *Geophysical Research Letters*, **35**, L24802.

Jouzel, J. Masson-Delmotte V., Cattani, O. *et al.* (2007). Orbital and millennial Antarctic climate variability over the last 800 000 years. *Science*, **317**, 793–796.

Kageyama, M., Paul, A., Roche, D. M. and Van Meerbeeck, C. J. (2010). Modelling glacial climatic millennial-scale variability related to changes in the Atlantic meridional overturning circulation: a review. *Quaternary Science Reviews*, **29**, 2931–2956.

Keeling, C. D., Whorf, T. P., Wahlen, M. and Vanderplicht, J. (1995). Interannual extremes in the rate of rise of atmospheric carbon dioxide since 1980. *Nature*, **375**, 666–670.

Knutti, R. and Hegerl, G. C. (2008). The equilibrium sensitivity of the Earth's temperature to radiation changes. *Nature Geoscience*, **1**, 735–743.

Knutti, R., Stocker, T. F., Joos, F. and Plattner, G. K. (2002). Constraints on radiative forcing and future climate change from observations and climate model ensembles. *Nature*, **416**, 719–723.

Kohfeld, K. E. and Harrison, S. P. (2001). DIRTMAP: the geological record of dust. *Earth-Science Reviews*, **54**, 81–114.

Kohfeld, K. E. and Ridgwell, A. (2009). Glacial–interglacial variability in atmospheric CO_2. In *Surface Ocean-Lower Atmosphere Processes*, ed. C. Le Quéré and E. S. Saltzman. Washington, DC: AGU, pp. 251–286.

Kohfeld, K. E., Le Quéré, C., Harrison, S. P. and Anderson, R. F. (2005). Role of marine biology in glacial–interglacial CO_2 cycles. *Science*, **308**, 74–78.

Köhler, P., Bintanja, R., Fischer, H. *et al.* (2010). What caused Earth's temperature variations during the last 800,000 years? Data-based evidence on radiative forcing and constraints on climate sensitivity. *Quaternary Science Reviews*, **29**, 129–145.

Kopp, R. E., Simons, F. J., Mitrovica, J. X., Maloof, A. C. and Oppenheimer, M. (2009). Probabilistic assessment of sea level during the last interglacial stage. *Nature*, **462**, 863–867.

Krinner, G., Viovy, N., de Noblet-Ducoudre, N. *et al.* (2005). A dynamic global vegetation model for studies of the coupled atmosphere–biosphere system. *Global Biogeochemical Cycles*, **19**, GB1015.

Lainé, A., Kageyama, M., Braconnot, P. and Alkama, R. (2009). Impact of greenhouse gas concentration changes on surface energetics in IPSL-CM4: regional warming patterns, land–sea warming ratios, and glacial–interglacial differences. *Journal of Climate*, **22**, 4621–4635.

Lambert, F., Delmonte, B., Petit, J. R. *et al.* (2008). Dust–climate couplings over the past 800,000 years from the EPICA Dome C ice core. *Nature*, **452**, 616–619.

Lamy, F., Hebbeln, D. and Wefer, G. (1999). High-resolution marine record of climatic change in mid-latitude Chile during the last 28,000 years based on terrigenous sediment parameters. *Quaternary Research*, **51**, 83–93.

Lang, N. and Wolff, E. W. (2011). Interglacial and glacial variability from the last 800 ka in marine, ice and terrestrial archives. *Climate of the Past*, **7**, 361–380.

Laskar, J., Robutel, P., Joutel, F. *et al.* (2004). A long-term numerical solution for the insolation quantities of the Earth. *Astronomy & Astrophysics*, **428**, 261–285.

Lenton, T. M., Williamson, M. S., Edwards, N. R. *et al.* (2006). Millennial timescale carbon cycle and climate change in an efficient Earth system model. *Climate Dynamics*, **26**, 687–711.

Levine, J. G., Wolff, E. W., Jones, A. E. *et al.* (2011). In search of an ice-core signal to differentiate between source-driven and sink-driven changes in atmospheric methane. *Journal of Geophysical Research*, **116**, D05305.

Lisiecki, L. E. and Raymo, M. E. (2005). A Pliocene–Pleistocene stack of 57 globally distributed benthic delta O-18 records. *Paleoceanography*, **20**, PA1003; doi:10.1029/2004PA001071.

Liu, Z., Otto-Bliesner, B. L., He, F. *et al.* (2009). Transient simulation of last deglaciation with a new mechanism for Bølling–Allerød warming. *Science*, **325**, 310–314.

Loulergue, L., Parrenin, F., Blunier, T. *et al.* (2007). New constraints on the gas age–ice age difference along the EPICA ice cores, 0–50 kyr. *Climate of the Past*, **3**, 527–540.

Loulergue, L., Schilt, A., Spahni, R. *et al.* (2008). Orbital and millennial-scale features of atmospheric CH_4 over the last 800,000 years. *Nature*, **453**, 383–386.

Lourantou, A., Chappellaz, J., Barnola, J. M., Masson-Delmotte, V. and Raynaud, D. (2010a). Changes in atmospheric CO_2 and its carbon isotopic ratio during the penultimate deglaciation. *Quaternary Science Reviews*, **29**, 1983–1992.

Lourantou, A., Lavric, J. V., Köhler, P. *et al.* (2010b). Constraint of the CO_2 rise by new atmospheric carbon isotopic measurements during the last deglaciation. *Global Biogeochemical Cycles*, **24**, GB2015.

Lucht, W., Prentice, I. C., Myneni, R. B. *et al.* (2002). Climatic control of the high-latitude vegetation greening trend and Pinatubo effect. *Science*, **296**, 1687–1689.

Lunt, D. J., Foster, G. L., Haywood, A. M. and Stone, E. J. (2008). Late Pliocene Greenland glaciation controlled by a decline in atmospheric CO_2 levels. *Nature*, **454**, 1102–1105.

Lunt, D. J., Haywood, A. M., Schmidt, G. A. *et al.* (2010). Earth system sensitivity inferred from Pliocene modelling and data. *Nature Geoscience*, **3**, 60–64.

Lüthi, D., Le Floch, M., Stocker, T. F. *et al.* (2008). High-resolution carbon dioxide concentration record 650,000–800,000 years before present. *Nature*, **453**, 379–382.

Lynch-Stieglitz, J., Adkins, J. F., Curry, W. B. *et al.* (2007). Atlantic meridional overturning circulation during the last glacial maximum. *Science*, **316**, 66–69.

MARGO Project Members (2009). Constraints on the magnitude and patterns of ocean cooling at the last glacial maximum. *Nature Geoscience*, **2**, 127–132.

Martrat, B., Grimalt, J. O., Lopez-Martinez, C. *et al.* (2004). Abrupt temperature changes in the Western Mediterranean over the past 250,000 years. *Science*, **306**, 1762–1765.

Masson-Delmotte, V., Kageyama, M., Braconnot, P. *et al.* (2006). Past and future polar amplification of climate change: climate model intercomparisons and ice-core constraints. *Climate Dynamics*, **26**, 513–529.

Masson-Delmotte, V., Stenni, B., Pol, K. *et al.* (2010). EPICA Dome C record of glacial and interglacial intensities. *Quaternary Science Reviews*, **29**, 113–128.

Masson-Delmotte, V., Braconnot, P., Hoffmann, G. *et al.* (2011). Sensitivity of interglacial Greenland temperature and delta-^{18}O to orbital and CO_2 forcing: climate simulations and ice core data. *Climate of the Past*, **7**, 1041–1059.

McCormick, M. P. and Veiga, R. E. (1992). SAGE-II measurements of early Pinatubo aerosols. *Geophysical Research Letters*, **19**, 155–158.

McInerney, F. A. and Wing, S. L. (2011). The Paleocene–Eocene Thermal Maximum: a perturbation of carbon cycle, climate, and biosphere with implications for the future. *Annual Review of Earth and Planetary Sciences*, **39**, 489–516.

McManus, J. F., Oppo, D. W. and Cullen, J. L. (1999). A 0.5 million-year record of millennial scale climate variability in the North Atlantic. *Science*, **283**, 971–975.

McManus, J. F., Francois, R., Gherardi, J. M., Keigwin, L. D. and Brown-Leger, S. (2004). Collapse and rapid resumption of Atlantic meridional circulation linked to deglacial climate changes. *Nature*, **428**, 834–837.

Menviel, L., Timmermann, A., Timm, O. E. and Mouchet, A. (2011). Deconstructing the last glacial termination: the role of millennial and orbital-scale forcings. *Quaternary Science Reviews*, **30**, 1155–1172.

Mercado, L. M., Bellouin, N., Sitch, S. *et al.* (2009). Impact of changes in diffuse radiation on the global land carbon sink. *Nature*, **458**, 1014–1017.

Milanković, M. (1998). *Canon of Insolation and the Ice-Age Problem (translated from German edition of 1941).* Belgrade: Agency for Textbooks.

Miller, G. H., Alley, R. B., Brigham-Grette, J. *et al.* (2010). Arctic amplification: can the past constrain the future? *Quaternary Science Reviews*, **29**, 1779–1790.

Minnis, P., Harrison, E. F., Stowe, L. L. *et al.* (1993). Radiative climate forcing by the Mount Pinatubo eruption. *Science*, **259**, 1411–1415.

Monnin, E., Indermuhle, A., Dallenbach, A. *et al.* (2001). Atmospheric CO_2 concentrations over the last glacial termination. *Science*, **291**, 112–114.

Moreno, P. I., Lowell, T. V., Jacobson, G. L. and Denton, G. H. (1999). Abrupt vegetation and climate changes during the last glacial maximum and last termination in the Chilean Lake District: a case study from Canal de la Puntilla (41 degrees S). *Geografiska Annaler Series A-Physical Geography A*, **81**, 285–311.

North Greenland Ice Core Project Members (2004). High-resolution record of northern hemisphere climate extending into the last interglacial period. *Nature*, **431**, 147–151.

Ohgaito, R. and Abe-Ouchi, A. (2007). The role of ocean thermodynamics and dynamics in Asian summer monsoon changes during the mid-Holocene. *Climate Dynamics*, **29**, 39–50.

Oka, A., Abe-Ouchi, A., Chikamoto, M. O. and Ide, T. (2011). Mechanisms controlling export production at the LGM: effects of changes in oceanic physical fields and atmospheric dust deposition. *Global Biogeochemical Cycles*, **25**, GB2009.

Otto-Bliesner, B. L. and Brady, E. C. (2010). The sensitivity of the climate response to the magnitude and location of freshwater forcing: last glacial maximum experiments. *Quaternary Science Reviews*, **29**, 56–73.

Otto-Bliesner, B. L., Marsha, S. J., Overpeck, J. T., Miller, G. H. and Hu, A. X. (2006). Simulating Arctic climate warmth and icefield retreat in the last interglaciation. *Science*, **311**, 1751–1753.

Otto-Bliesner, B. L., Hewitt, C. D., Marchitto, T. M. *et al.* (2007). Last glacial maximum ocean thermohaline circulation: PMIP2 model intercomparisons and data constraints. *Geophysical Research Letters*, **34**, 1–6.

Otto, J., Raddatz, T., Claussen, M., Brovkin, V. and Gayler, V. (2009). Separation of atmosphere–ocean–vegetation feedbacks and synergies for mid-Holocene climate. *Geophysical Research Letters*, **36**, L09701.

Pagani, M., Liu, Z. H., LaRiviere, J. and Ravelo, A. C. (2009). High Earth-system climate sensitivity determined from Pliocene carbon dioxide concentrations. *Nature Geoscience*, **3**, 27–30.

Paillard, D. (1998). The timing of Pleistocene glaciations from a simple multiple-state climate model. *Nature*, **391**, 378–381.

Parrenin, F., Barnola, J.M., Beer, J. *et al.* (2007). The EDC3 chronology for the EPICA Dome C ice core. *Climate of the Past*, **3**, 485–497.

Pearson, P. N., Foster, G. L. and Wade, B. S. (2009). Atmospheric carbon dioxide through the Eocene–Oligocene climate transition. *Nature*, **461**, 1110–1113.

Peltier, W. R. (1994). Ice age paleotopography. *Science*, **265**, 195–201.

Petrenko, V. V., Smith, A. M., Brook, E. J. *et al.* (2009). $^{14}CH_4$ measurements in Greenland ice: investigating last glacial termination CH_4 sources. *Science*, **324**, 506–508.

Prentice, I. C., Jolly, D. BIOME6000 participants (2000). Mid-Holocene and glacial-maximum vegetation geography of the northern continents and Africa. *Journal of Biogeography*, **27**, 507–519.

Prentice, I. C. and Harrison, S. P. (2009). Ecosystem effects of CO_2 concentration: evidence from past climates. *Climate of the Past*, **5**, 297–307.

Prentice, I. C., Harrison, S. P. and Bartlein, P. J. (2011). Global vegetation and terrestrial carbon cycle changes after the last ice age. *New Phytologist*, **189**, 988–998.

Robock, A. and Mao, J. P. (1995). The volcanic signal in surface temperature observations. *Journal of Climate*, **8**, 1086–1103.

Roe, G. H. and Baker, M. B. (2007). Why is climate sensitivity so unpredictable? *Science*, **318**, 629–632.

Rohling, E. J., Grant, K., Bolshaw, M. *et al.* (2009). Antarctic temperature and global sea level closely coupled over the past five glacial cycles. *Nature Geoscience*, **2**, 500–504.

Rojas, M., Moreno, P. I., Kageyama, M. *et al.* (2009). The Southern Westerlies during the last glacial maximum in PMIP2 simulations. *Climate Dynamics*, **32**, 525–548.

Röthlisberger, R., Bigler, M., Wolff, E. W. *et al.* (2004). Ice core evidence for the extent of past atmospheric CO_2 change due to iron fertilisation. *Geophysical Research Letters*, **31**, L16207.

Sanchez-Goñi, M. F., Landais, A., Fletcher, W. J. *et al.* (2008). Contrasting impacts of Dansgaard–Oeschger events over a western European latitudinal transect modulated by orbital parameters. *Quaternary Science Reviews*, **27**, 1136–1151.

Schilt, A., Baumgartner, M., Blunier, T., *et al.* (2010a). Glacial–interglacial and millennial-scale variations in the atmospheric nitrous oxide concentration during the last 800,000 years. *Quaternary Science Reviews*, **29**, 182–192.

Schilt, A., Baumgartner, M., Schwander, J., *et al.* (2010b). Atmospheric nitrous oxide during the last 140,000 years. *Earth and Planetary Science Letters*, **300**, 33–43.

Schmittner, A., Urban, N. M., Shakun, J. D. *et al.* (2011). Climate sensitivity estimated from temperature reconstructions of the last glacial maximum. *Science*, **334**, 1385–1388.

Serreze, M. C. and Barry, R. G. (2011). Processes and impacts of Arctic amplification: a research synthesis. *Global and Planetary Change*, **77**(1–2), 85–96.

Serreze, M. C., Barrett, A. P., Stroeve, J. C., Kindig, D. N. and Holland, M. M. (2009). The emergence of surface-based Arctic amplification. *Cryosphere*, **3**, 11–19.

Shakun, J. D. and Carlson, A. E. (2010). A global perspective on last glacial maximum to Holocene climate change. *Quaternary Science Reviews*, **29**, 1801–1816.

Shulmeister, J., Goodwin, I., Renwick, J. *et al.* (2004). The southern hemisphere westerlies in the Australasian sector over the last glacial cycle: a synthesis. *Quaternary International*, **118**, 23–53.

Sime, L. C., Wolff, E. W., Oliver, K. I. C. and Tindall, J. C. (2009). Evidence for warmer interglacials in East Antarctic ice cores. *Nature*, **462**, 342–345.

Sime, L. C., Kohfeld, K., Le Quéré, C. *et al.* (2010). Strengthening of the Southern Ocean westerlies during glaciations. *IPY Open Science Conference Abstracts*. See: http://ipy-osc.no/abstract/382303.

Singarayer, J. S. and Valdes, P. J. (2010). High-latitude climate sensitivity to ice-sheet forcing over the last 120 kyr. *Quaternary Science Reviews*, **29**, 43–55.

Singarayer, J. S., Valdes, P. J., Friedlingstein, P., Nelson, S. and Beerling, D. J. (2011). Late-Holocene methane rise caused by orbitally-controlled increase in tropical sources. *Nature*, **470**, 82–85.

Smith, R. S., Gregory, J. M. and Osprey, A. (2008). A description of the FAMOUS (version XDBUA) climate model and control run. *Geoscientific Model Development*, **1**, 53–68.

Soden, B. J., Wetherald, R. T., Stenchikov, G. L. and Robock, A. (2002). Global cooling after the eruption of Mount Pinatubo: a test of climate feedback by water vapor. *Science*, **296**, 727–730.

Solomon, S., Plattner, G. K., Knutti, R. and Friedlingstein, P. (2009). Irreversible climate change due to carbon dioxide emissions. *Proceedings of the National Academy of Sciences of the United States of America*, **106**, 1704–1709.

Sowers, T. (2006). Late Quaternary atmospheric CH_4 isotope record suggests marine clathrates are stable. *Science*, **311**, 838–840.

Stenchikov, G., Robock, A., Ramaswamy, V. *et al.* (2002). Arctic Oscillation response to the 1991 Mount Pinatubo eruption: effects of volcanic aerosols and ozone depletion. *Journal of Geophysical Research*, **107**, 4803.

Stocker, T. F. and Johnsen, S. J. (2003). A minimum thermodynamic model for the bipolar seesaw. *Paleoceanography*, **18**, 1087.

Sutton, R. T., Dong, B. W. and Gregory, J. M. (2007). Land/sea warming ratio in response to climate change: IPCC AR4 model results and comparison with observations. *Geophysical Research Letters*, **34**, L02701.

Svendsen, J. I., Alexanderson, H., Astakhov, V. I. *et al.* (2004). Late Quaternary ice sheet history of northern Eurasia. *Quaternary Science Reviews*, **23**, 1229–1271.

Telford, P. J., Lathiere, J., Abraham, N. L. *et al.* (2010). Effects of climate-induced changes in isoprene emissions after the eruption of Mount Pinatubo. *Atmospheric Chemistry and Physics*, **10**, 7117–7125.

Timm, O. and Timmermann, A. (2007). Simulation of the last 21000 years using accelerated transient boundary conditions. *Journal of Climate*, **20**, 4377–4401.

Tindall, J. C., Valdes, P. J. and Sime, L. C. (2009). Stable water isotopes in HADCM3: the isotopic signature of ENSO and the tropical amount effect. *Journal of Geophysical Research*, **114**, D04111.

Toggweiler, J. R. (2008). Origin of the 100,000-year timescale in Antarctic temperatures and atmospheric CO_2. *Paleoceanography*, **23**, PA2211.

Toggweiler, J. R., Russell, J. L. and Carson, S. R. (2006). Midlatitude westerlies, atmospheric CO_2, and climate change during the ice ages. *Paleoceanography*, **21**, PA2005

Tschumi, T., Joos, F. and Parekh, P. (2008). How important are southern hemisphere wind changes for low glacial carbon dioxide? A model study. *Paleoceanography*, **23**, PA4208.

Turner, J., Comiso, J. C., Marshall, G. J. *et al.* (2009). Non-annular atmospheric circulation change induced by stratospheric ozone depletion and its role in the recent increase of Antarctic sea ice extent. *Geophysical Research Letters*, **36**, L08502.

Turney, C. S. M. and Jones, R. T. (2010). Does the Agulhas Current amplify global temperatures during super-interglacials? *Journal of Quaternary Science*, **25**, 839–843.

Tziperman, E., Raymo, M. E., Huybers, P. and Wunsch, C. (2006). Consequences of pacing the Pleistocene 100 kyr ice ages by nonlinear phase locking to Milankovitch forcing. *Paleoceanography*, **21**, PA4206.

Usoskin, I. G., Horiuchi, K., Solanki, S., Kovaltsov, G. A. and Bard, E. (2009). On the common solar signal in different cosmogenic isotope data sets. *Journal of Geophysical Research*, **114**, A03112.

Valdes, P. (2011). Built for stability. *Nature Geoscience*, **4**, 414–416.

Vrac, M., Marbaix, P., Paillard, D. and Naveau, P. (2007). Non-linear statistical downscaling of present and LGM precipitation and temperatures over Europe. *Climate of the Past*, **3**, 669–682.

Wang, Y. J., Cheng, H., Edwards, R. L. *et al.* (2008). Millennial- and orbital-scale changes in the East Asian monsoon over the past 224,000 years. *Nature*, **451**, 1090–1093.

Wanner, H., Beer, J., Bütikoter, J. *et al.* (2008). Mid- to Late Holocene climate change: an overview. *Quaternary Science Reviews*, **27**, 1791–1828.

Winton, M. (2006). Amplified Arctic climate change: what does surface albedo feedback have to do with it? *Geophysical Research Letters*, **33**, L03701.

Wolff, E. W., Fischer, H., Fundel, F. *et al.* (2006). Southern Ocean sea-ice extent, productivity and iron flux over the past eight glacial cycles. *Nature*, **440**, 491–496.

Wolff, E. W., Fischer, H. and Röthlisberger, R. (2009). Glacial terminations as southern warmings without northern control. *Nature Geoscience*, **2**, 206–209.

Wolff, E. W., Chappellaz, J., Blunier, T., Rasmussen, S. O. and Svensson, A. (2010). Millennial-scale variability during the last glacial: the ice core record. *Quaternary Science Reviews*, **29**, 2828–2838.

Wu, H. B., Guiot, J. L., Brewer, S. and Guo, Z. T. (2007). Climatic changes in Eurasia and Africa at the Last Glacial Maximum and mid-Holocene: reconstruction from pollen data using inverse vegetation modelling. *Climate Dynamics*, **29**, 211–229.

Zachos, J. C., Rohl, U., Schellenberg, S. A. *et al.* (2005). Rapid acidification of the ocean during the Paleocene–Eocene thermal maximum. *Science*, **308**, 1611–1615.

Zachos, J. C., Dickens, G. R. and Zeebe, R. E. (2008). An early Cenozoic perspective on greenhouse warming and carbon-cycle dynamics. *Nature*, **451**, 279–283.

Zhao, Y., Braconnot, P., Marti, O. *et al.* (2005). A multi-model analysis of the role of the ocean on the African and Indian monsoon during the mid-Holocene. *Climate Dynamics*, **25**, 777–800.

Zhao, Y. and Harrison, S. P. (2011). Mid-Holocene monsoons: a multi-model analysis of the interhemispheric differences in the responses to orbital forcing and ocean feedbacks. *Climate Dynamics*, doi: 10.1007/s00382-011-1193-z.

Chapter

4 The Earth system feedbacks that matter for contemporary climate

Pierre Friedlingstein, Angela V. Gallego-Sala, Eleanor M. Blyth, Fiona E. Hewer, Sonia I. Seneviratne, Allan Spessa, Parvadha Suntharalingam and Marko Scholze

Here, we discuss the feedback processes that determine the nature and rates of climatic changes in response to climate forcing. We explain the differences in the characteristic behaviour of biophysical and biogeochemical feedbacks, and describe the means by which feedbacks can be identified using models and observational data, and their strength quantified. Improved understanding of these kinds of processes is essential for better predictions of future climate.

4.1 Introduction

Chapter 2 dealt with physical feedbacks that contribute to climate sensitivity, and with the general principles of terrestrial ecosystem function, fire regimes and the carbon cycle. Chapter 3 discussed feedbacks of many kinds, including some operating on timescales of thousands of years, such as those involving ice-sheet dynamics. This chapter focuses more closely on feedbacks due to biogeophysical and biogeochemical processes involving the land and ocean, operating on timescales that are relevant to climate change in the twenty-first century, with a brief glance at additional relatively rapid feedbacks involving human behaviour. These feedbacks are not included in the climate sensitivity, yet they will contribute to determining the effect of any change in greenhouse-gas emissions on climate. Quantifying these feedbacks is an active research area and it is not yet possible to provide a complete picture of their magnitudes. But recent research has narrowed some of the largest uncertainties and ruled out some of the more extreme possibilities, while drawing attention to the general tendency of feedbacks that involve microbial processes to be positive, that is, to amplify climate change.

4.1.1 Background

The importance of feedbacks in the Earth system has been recognized for more than a century. In a seminal paper published in 1896, Svante Arrhenius calculated the '*variation of temperature that would ensue in consequence of a given variation of the carbonic acid in the air*'. He already considered in his calculations the amplification of the carbon dioxide (CO_2)-induced warming due to the resulting changes in the atmospheric water-vapour content, and in the surface albedo due to changes in snow cover (Arrhenius, 1896). Without explicitly naming them, Arrhenius thereby identified two major physical feedbacks in the climate system. These feedbacks, along with a few others (clouds, the atmospheric lapse rate), were further quantified more than 50 years later by modelling studies (e.g. Manabe and Wetherald, 1967; Sellers, 1969; Cess, 1975).

Climate sensitivity is defined as the equilibrium temperature change $\Delta T_{2\times CO_2}$ for a change in radiative forcing $\Delta F_{2\times CO_2}$ equivalent to a doubling of the CO_2 concentration:

$$\Delta T_{2\times CO_2} = \frac{\Delta F_{2\times CO_2}}{\lambda} \tag{4.1}$$

In this equation, λ is called the feedback parameter. Without physical feedbacks, λ is approximately equal to 4 W m^{-2} K^{-1}: this is λ_{BB}, the 'black body' value of λ. The parameter λ_{BB} is simply the derivative of the Stefan–Boltzmann law, assuming the Earth to be an idealized black body radiating energy to space. With this value of λ, doubling atmospheric CO_2 would translate into a

Understanding the Earth System: Global Change Science for Application, eds. Sarah E. Cornell, I. Colin Prentice, Joanna I. House and Catherine J. Downy. Published by Cambridge University Press

warming of only ~ 1 K. Results from both observations and climate models indicate that the actual value of λ is much smaller, leading to a larger warming.

One of the earliest attempts to quantify the climate sensitivity was in the report of the US National Academy of Sciences, chaired by Jule Charney (NRC, 1979). Based on the GFDL and GISS general circulation models available at that time, they estimated a range of 1.5 to 4 K, with the most likely value near 3 K. The most recent estimates are not much different. The Fourth Assessment Report of the IPCC gave a likely range of 2 to 4.5 K with a best estimate of 3 K (IPCC, 2007). New observational constraints from palaeoclimate data yield a narrower range (1.47–2.6 K likely range, with a median estimate of 2.3 K; Schmittner *et al.*, 2011). Chapter 3 provides further background on the history and methods of estimating climate sensitivity, explaining the use of palaeoclimate constraints.

To understand how the climate sensitivity arises as a consequence of different physical feedbacks, it is necessary to analyse results of climate models that include these feedbacks. The strength of a given physical feedback can be estimated by comparison of the modelled climate response obtained with and without accounting for that feedback. Using a three-dimensional general circulation model of the atmosphere, Hansen *et al.* (1984) proposed a framework for the estimation and comparison of different feedbacks. They borrowed a terminology from electronics and defined the net feedback factor f by

$$\Delta T_{eq} = f \Delta T_0 \tag{4.2}$$

where ΔT_{eq} is the equilibrium change in global mean surface temperature and ΔT_0 is the change in temperature in the absence of feedback (for example, if atmospheric water-vapour content is held constant). The gain of this feedback, g, is the relative change in temperature:

$$g = \frac{\Delta T_{eq} - \Delta T_0}{\Delta T_{eq}} \tag{4.3}$$

which can be rearranged in terms of the equilibrium change in temperature to give:

$$\Delta T_{eq} = \frac{\Delta T_0}{1 - g} \tag{4.4}$$

It follows that feedback and gain are related, with $f = 1/(1 - g)$.

A positive value of g means a positive feedback. In such a case the equilibrium temperature, ΔT_{eq}, will be higher than the one in the absence of feedback, ΔT_0. Negative feedbacks have negative values of g, resulting in a lower temperature change relative to the no-feedback situation. It can be shown that the gain for each physical feedback is the ratio of its individual climate feedback parameter λ_i to λ_{BB}:

$$\Delta T_{eq} = \frac{\Delta T_0}{1 + \Sigma \frac{\lambda_i}{\lambda_{BB}}} \tag{4.5}$$

This framework makes it possible to compare the different physical feedbacks and to assess their respective contributions to the climate sensitivity and its uncertainty. This gain theory continues to be applied in present-day studies, for example in Bony *et al.* (2006), an analysis that highlights the particularly large uncertainty arising from cloud feedbacks.

During recent decades, additional feedbacks involving biological and chemical components of the Earth system have been identified and quantified, using a variety of models ranging from 'back-of-the-envelope' calculations to full Earth system model simulations as discussed in Chapter 5. The climate–carbon-cycle feedback is of principal importance. Climate change alters the efficiency of land and ocean in taking up the CO_2 that is added to the atmosphere. Evidence from both observations and modelling attests that overall this is a positive feedback: CO_2 uptake is reduced by warming, leading to a faster growth rate in atmospheric CO_2 concentration and amplified warming.

Feedbacks of this type are by definition not included in the climate sensitivity, which is framed in terms of a fixed relationship between temperature change and radiative forcing change. The climate–carbon-cycle feedback was quantified by Friedlingstein *et al.* (2003, 2006) in terms of a number of sensitivity coefficients. The gain of the climate–carbon-cycle feedback is not directly comparable to the gain of physical feedbacks, because the climate–carbon-cycle feedback directly affects the radiative forcing (ΔF) and not just the response (ΔT) to a given ΔF. Gregory *et al.* (2009) reconciled the two concepts in a unified framework, allowing the comparison of physical and non-physical feedback strengths and highlighting the importance of the various contributions to the carbon-cycle feedback (Figure 4.1).

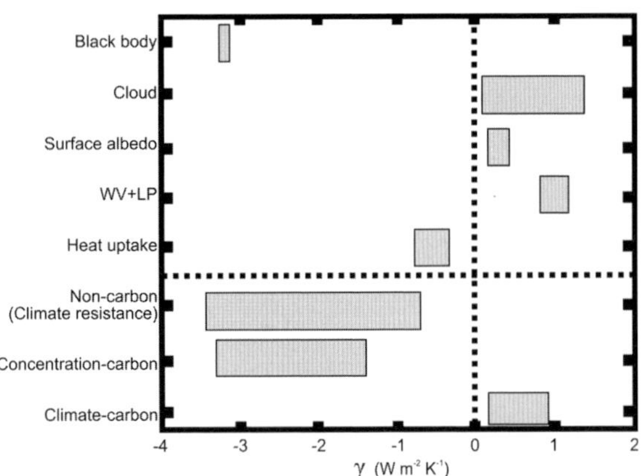

Figure 4.1 Comparison of physical feedbacks (black-body response, cloud, surface albedo, water vapour and lapse rate (WV + LP), and ocean heat uptake) with carbon-cycle feedbacks (the carbon-cycle response to change in atmospheric CO_2 and in change in climate). (Image first published in Gregory *et al.* (2009) *Journal of Climate*, **22**, doi:10.1175/2009JCLI2949.1; © American Meteorological Society. Reprinted with permission).

The Gregory *et al.* (2009) framework has recently been applied to other biogeochemical feedbacks (Arneth *et al.*, 2010a). Both nitrous oxide (N_2O) and methane (CH_4) are potent greenhouse gases with large terrestrial emissions, driven by biogenic processes. There is evidence that these emissions increase with temperature. Nitrous oxide is released by soils as a by-product of microbial remineralization of organic matter under suboxic conditions. A climate feedback due to increased N_2O emissions from natural soils has been described, based on observations and modelling (Xu-Ri *et al.*, submitted). Methane also has microbial sources, being produced by methanogenic decomposers under anoxic conditions. A growing number of studies have investigated the potential climate feedback through increased emissions of CH_4 from natural environments, mainly wetlands (Eliseev *et al.*, 2008; Ringeval *et al.*, 2011). Other potential biogeochemical feedbacks have been identified, but to date these have not been systematically quantified using comprehensive models. Later in this chapter, we consider some of these poorly quantified biogeochemical feedbacks that are nonetheless candidates for being significant in the context of climate change, including those involving marine N_2O emissions, CH_4 emissions linked to thawing of frozen soils or sediments and possibly clathrate destabilization, and pyrogenic (from fires) emissions of CO_2, trace gases, black carbon and other aerosol precursors.

4.1.2 Is there observational evidence for Earth system feedbacks?

It is impossible to measure the gain of any Earth system feedback directly. The strength of a feedback is the ratio between the change of a quantity (e.g. global temperature) accounting for the feedback and the change in the same quantity in the theoretical case where the feedback is inactive. Only the first term can be measured in the real world. What we can observe, however, is the strong correlation between observations of different components of the Earth system, giving hindsight at least on the sign of the feedbacks.

The ice-core record shows that polar temperatures (inferred from isotopic measurements in the ice) co-vary with the atmospheric concentrations of the long-lived greenhouse gases (CO_2, CH_4 and N_2O), which are directly measured in the air bubbles entrapped in the ice. Glacial periods have cold climates and low concentrations of these gases; interglacial periods have warm climates and higher concentrations. But this information does not allow us directly to estimate the strength of the feedback. The recorded change in temperature is not just due to the change in greenhouse gases, but also to changes in orbital parameters, planetary albedo, ice-sheet orography and atmospheric aerosols. To quantify the feedbacks we have to invoke our knowledge of the radiative properties of greenhouse gases. This has been done for CO_2, using the observed changes in atmospheric CO_2 both during the most recent glacial–interglacial cycle and the last millennium including the Little Ice Age.

One additional, highly promising approach to estimating the strength of Earth system feedbacks involves examining the sensitivity to climate of a given process at a timescale over which observations are available, and extrapolating this to longer timescales using an ensemble of models. Hall and Qu (2006) were the first to apply this method. Using both historical and future climate simulations from a series of models, they showed that in the 'model world' there is a close linear relationship between the present-day response of snow cover to the seasonal cycle of temperature (i.e. snow melting that takes place during the warming in spring), and the longer-term response of snow cover to global warming during the twenty-first century. Models with a large snow feedback tend to simulate both a rapid spring snowmelt and a strong response of snow cover to warming. Conversely, models with a small snow feedback tend to simulate both a slow spring snowmelt and a weak response of snow cover to warming. With the help of this relationship, Hall and Qu (2006) were able to use the *observed* seasonal cycle of snow cover to constrain the *future* snow feedback. A similar approach was proposed by Cox and Jones (2008) to estimate the climate–carbon feedback, using the observed interannual variability of the carbon cycle.

4.2 Land–atmosphere biogeophysical feedbacks

The land surface affects the atmosphere through biogeophysical feedbacks involving evapotranspiration and energy partitioning between latent and sensible heat fluxes, surface albedo and surface-roughness length. We present these three pathways and their consequences, and then consider their implications for the quantification of biogeochemical feedbacks.

4.2.1 Impacts on evapotranspiration

Evapotranspiration from the land is a major component of the global water and energy cycles (Seneviratne *et al.*, 2010). Evapotranspiration returns as much as 60% of total land precipitation to the atmosphere (Oki and Kanae, 2006), and the associated latent heat flux uses up more than half of the net radiation incoming on land surfaces (Trenberth *et al.*, 2009). Plant evapotranspiration and CO_2 uptake are closely coupled through photosynthesis, implying that changes in evapotranspiration have an impact on the carbon cycle while changes in the carbon cycle have an impact on the water and energy cycles (Section 4.2.4 below).

Figure 4.2 displays the (idealized) dependency of the evaporative fraction (latent heat flux normalized by net radiation) on soil moisture, following the classical framework of Budyko (1956). Under wet soil conditions (right side of Figure 4.2), actual evapotranspiration is close to potential evaporation (evaporation from a wet surface) and thus determined primarily by the available energy. Under those conditions, land-surface states and processes have limited control on land–atmosphere water exchanges, except through impacts on albedo and roughness length (discussed in Sections 4.2.2 and 4.2.3 below). Under dry soil conditions (left side of Figure 4.2), evapotranspiration is sensitive to soil moisture but too small to affect atmospheric conditions and regional climate. Land-surface conditions therefore mainly influence evapotranspiration in transitional zones between dry and wet climates (Koster

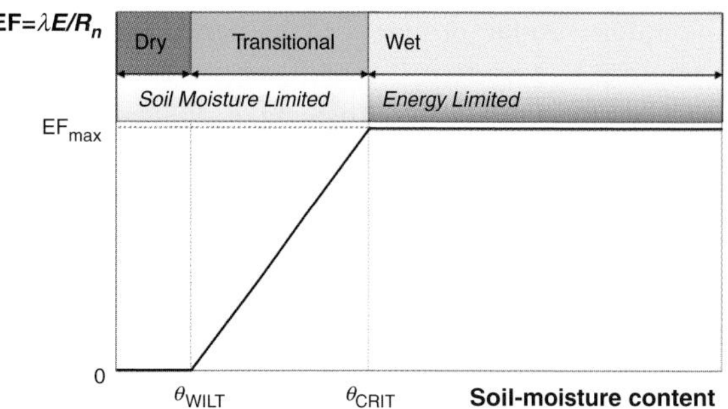

Figure 4.2 Soil moisture and evapotranspiration regimes following the classic framework of Budyko (1956). The term EF refers to the evaporative fraction (latent heat flux divided by available net radiation); EF_{max} is its maximum value, θ_{CRIT} the critical soil moisture content under which evapotranspiration becomes soil-moisture limited, and θ_{WILT} the plant wilting point. (Image first published in Seneviratne *et al.* (2010) *Earth-Science Reviews*, **99**, 125–161.)

et al., 2004; Guo *et al.*, 2006; Seneviratne *et al.*, 2010) where evaporation is substantial but limited by soil moisture. Within this regime, small spatial or temporal variations in soil-moisture content, land cover or vegetation state lead to variations in surface energy fluxes.

Several studies have highlighted the importance of soil-moisture control on evapotranspiration in transitional climate regions. These include contributions to soil-moisture–precipitation feedbacks, whereby enhanced soil-moisture content leads to either enhanced or diminished precipitation depending on boundary-layer conditions (e.g. Findell and Eltahir, 2003; Betts, 2004; Ek and Holtslag, 2004; see Seneviratne *et al.*, 2010 for an overview). Changes in evapotranspiration can also affect sensible heat flux and thus air temperature (reduced or enhanced evaporative cooling), with effects of up to several degrees, which can impact the strength or likelihood of a heat wave (e.g. Seneviratne *et al.*, 2006a; Fischer *et al.*, 2007; Vautard *et al.*, 2007; Jaeger and Seneviratne, 2011). In the context of climate change, it has been shown that modified land–atmosphere coupling characteristics linked with shifts in the location of transitional climate regions can explain projected changes in temperature variability and extremes (Seneviratne *et al.*, 2006a). Large differences between climate models are linked to differences in the representation of soil-moisture–evapotranspiration coupling in the models (Boé and Terray, 2008).

Evidence for the role of soil-moisture–evapotranspiration feedbacks in climate and atmospheric processes is mostly based on modelling, starting with the landmark study by Shukla and Mintz (1982) and expanded to the recent multi-model analyses within the international project Global Land–Atmosphere Coupling Experiment (GLACE) and those based on IPCC's AR4 simulations (Koster *et al.*, 2004; Dirmeyer *et al.*, 2006b; Guo *et al.*, 2006; Seneviratne *et al.*, 2006a,b; Koster *et al.*, 2010). Evidence directly based on observations is still scarce, but correspondences have been shown between modelling-identified 'hot spots' of land–atmosphere coupling and related observational diagnostics (Koster *et al.*, 2003; Teuling *et al.*, 2009; Seneviratne *et al.*, 2010), and observed impacts during heat waves have been documented using satellite data (Zaitchik *et al.*, 2006).

Figure 4.3 displays diagnostics related to land–atmosphere coupling for the boreal summer season (JJA) based on both modelling and observational investigations (Seneviratne *et al.*, 2010). The top two panels (adapted from Teuling *et al.*, 2009) are analyses of evapotranspiration regimes based on correlation measures applied to data from the second phase of the Global Soil Wetness Project (GSWP-2, Figure 4.3a: yearly correlation of evapotranspiration with radiation and precipitation) and to FLUXNET measurements (Figure 4.3b: daily correlations with radiation). The GSWP-2 data are a multi-model product driven with observation-based meteorological forcing (Dirmeyer *et al.*, 2006a). The FLUXNET data are eddy-covariance flux measurements of land–atmosphere exchanges made at more than 500 sites worldwide (Baldocchi *et al.*, 2001). The network sites are unevenly distributed with the greatest concentrations of sites in North America and Europe. The middle panels of the figure (redrawn from Koster *et al.*, 2004, 2006) display a measure of soil-moisture–atmosphere coupling ($\Delta\Omega$, as defined in Koster *et al.*, 2006) applied to simulations from the GLACE experiment and diagnosing the strength of coupling between soil moisture and temperature or precipitation (Figure 4.3c and 4.3d, respectively). The bottom two panels (Figure 4.3e and 4.3f) diagnose soil-moisture–evapotranspiration/temperature coupling in IPCC's AR4 simulations (using the GFDL, HadGEM1 and ECHAM5 models) for two time periods (1970–1989 and 2080–2099) from the correlation between evapotranspiration and temperature. Negative values (red shadings) indicate strong coupling, positive values (blue shadings) indicate weak coupling (Seneviratne *et al.*, 2006a).

Coherent patterns can be identified across data sets in Figure 4.3. The GLACE hot spot of soil-moisture–temperature and soil-moisture–precipitation coupling in the North American Great Plains is also identified in the IPCC's AR4 simulations, and corresponds to an area with a soil-moisture-limited evapotranspiration regime. The IPCC's AR4 twentieth-century simulations and the GSWP-2/FLUXNET analysis also all suggest a gradient of evapotranspiration regimes across Europe, with the Mediterranean region identified as a hot spot of soil-moisture–atmosphere coupling (projected to expand to central Europe in the twenty-first century, Figure 4.3f). This regional hot spot in the Mediterranean region is not captured in the GLACE simulations, possibly due to the use of prescribed sea-surface-temperature (SST) conditions from a single year in the experiment (Seneviratne *et al.*, 2006a). On the other hand, the GSWP-2, GLACE and IPCC's AR4 analyses agree on a soil-moisture-limited evapotranspiration regime on the Indian subcontinent.

These preliminary comparisons suggest some level of qualitative agreement between observations and models regarding soil-moisture–evapotranspiration interactions. But quantitative assessments remain difficult both due to lack of observations and methodological issues (Seneviratne *et al.*, 2010). Climate models still vary greatly in the way they represent land–atmosphere coupling (Koster *et al.*, 2004; Dirmeyer *et al.*, 2006a; Pitman *et al.*, 2009). Future research will need to focus on a more systematic evaluation of soil-moisture–evapotranspiration coupling in current climate models, in particular with the help of newly developed global evapotranspiration data sets (Jiménez *et al.*, 2011; Mueller *et al.*, 2011).

4.2.2 Impacts on surface albedo

Land-surface conditions affect the climate system through impacts on surface albedo and resulting modifications of the radiation balance. Lower albedo leads to a warming of the near-surface atmosphere due to enhanced shortwave-radiation absorption, and the converse also holds – higher albedo enhances reflection and results in a cooling. Albedo can be altered by changes in snow cover, soil-moisture content, vegetation cover, and dust levels of dry regions.

The literature on snow–albedo feedback mechanisms is extensive (e.g. Groisman *et al.*, 1994; Bony *et al.*, 2006; Hall and Qu, 2006; Brown and Mote, 2009). Snow cover leads to a cooling of the surface due to enhanced albedo, which in turn tends to maintain the snow cover, resulting in a positive feedback between low surface temperatures and snow cover. This feedback is very effective in regions of snow

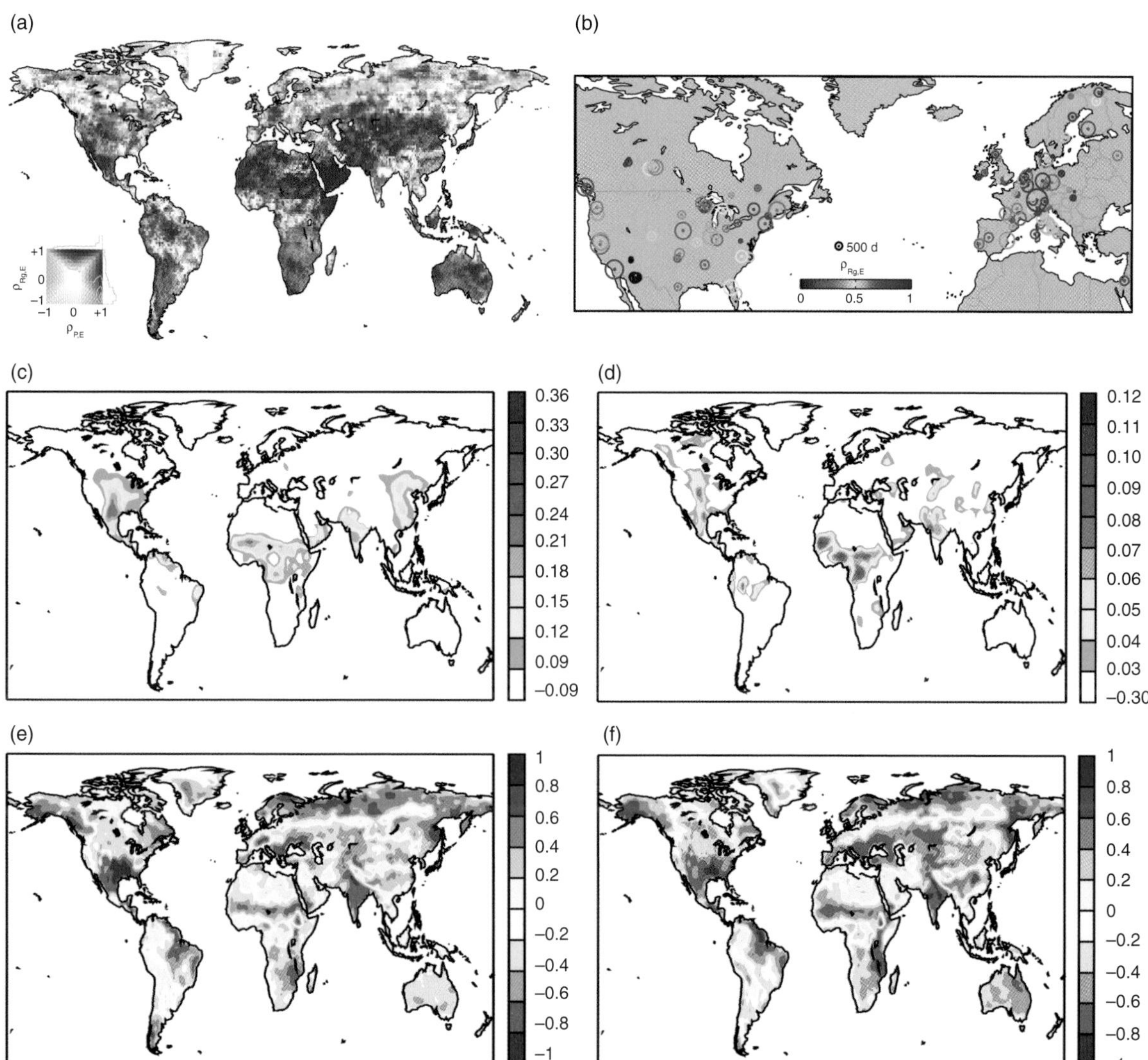

Figure 4.3 Diagnostics of land–atmosphere coupling, applied to observations and model results for the boreal summer. (Image first published in Seneviratne *et al.* (2010) *Earth-Science Reviews*, **99**, 125–161, adapted from (a,b) Teuling *et al.*, 2009; (c,d) Koster *et al.*, 2004, 2006; and (e,f) Seneviratne *et al.*, 2006a,b.) Top row: estimation of evapotranspiration drivers of moisture and radiation based on simulations from (a) the GSWP project (yearly correlations of evaporation with global radiation (R_g) and precipitation (P), redrawn for whole globe) and (b) FLUXNET observations of daily correlations between evaporation and global radiation (circle area is proportional to the number of days used). Middle row: estimates of (c) the soil-moisture–temperature and (d) soil-moisture–precipitation coupling, based on $\Delta\Omega$-coupling diagnostic and output of 12 general circulation models from the GLACE project. Bottom row: Estimates of soil moisture–temperature coupling for the climates of (e) 1970–1989 and (f) 2080–2099, based on three GCMs and diagnosed with the correlation between evapotranspiration and 2 m temperature.

vegetation such as tundra, but less so over boreal forest as the higher dark forest canopy masks the snow on the ground (Betts and Ball, 1997). A climate-induced northward extension of boreal forest could then produce further warming through a decrease of snow-covered surface albedo (Bonan *et al.* 1992). Quantifying these effects using Earth system models that include better representations of vegetation types is of relevance for assessing the net radiative forcing of boreal forests on climate. Snow aging and metamorphosis also influence snow albedo, with old (compacted, dirty) snowpacks being less reflective. Snow also affects climate through other pathways, in particular associated with snowmelt impacts on soil-moisture content and land-surface insulation by the snowpack (e.g. Lawrence and Slater, 2010).

Soil moisture has a three-fold impact on land-surface albedo (Seneviratne *et al.*, 2010) acting through changes in bare-soil albedo with soil-moisture content (soil has a lower albedo under wetter conditions), changes in vegetation albedo with soil-moisture content (yellowing of leaves occurring with soil drying) and changes in the exposed bare-soil fraction due to drought impacts on vegetation (which affect leaf orientation and leaf-area index). However, the effects of drought on surface albedo are not straightforward and have been shown to vary across the different spectral bands of the solar radiation (e.g. visible vs. near infrared) (Teuling and Seneviratne, 2008).

4.2.3 Impacts on roughness length

Climate models include a representation of surface roughness because frictional processes influence the atmospheric circulation. Several modelling studies have shown how an increase in surface roughness increases the convergence of the water vapour in the overlying atmosphere, thereby promoting rainfall. The impact on the Amazon rainforest has been studied extensively (see for example Pitman *et al.*, 1993; Lean *et al.*, 1996). The impact of this feedback between the land and the atmosphere is local (Pitman *et al.*, 2009) but it strongly affects the viability of large forest systems such as the Amazon (Cowling *et al.*, 2005).

Observational evidence of this biophysical feedback is hard to obtain. But a unique study of the Les Landes forest in France by Blyth *et al.* (1994) demonstrated that the observed patterns of cloud over the region were most likely caused by the presence of the forest, with its high roughness. Another example is from Los *et al.* (2006), who analyzed satellite data of vegetation cover and rainfall in the Sahel and demonstrated that there is a likely causal link between the presence of vegetation and rainfall, in addition to the more obvious causal link between rainfall and vegetation growth.

4.2.4 Implications of biogeophysical feedbacks for biogeochemical feedbacks

The implications of biogeophysical feedbacks for biogeochemical feedbacks can be considered from two main perspectives: first, the biosphere, where soil moisture impacts on plant function and CO_2 uptake; and secondly, the atmosphere, where feedbacks affect radiative forcing, which needs to be evaluated against those changes produced by biogeochemical feedbacks.

The interactions of soil moisture and evapotranspiration interactions have implications for carbon cycling, given the tight coupling between water loss and CO_2 uptake in plants. As an example, during the extreme drought and heat-wave summer of 2003 in Europe, it was found that plants became less productive; the European continent became a source of CO_2 in summer (Ciais *et al.*, 2005; Reichstein *et al.*, 2007). The 2005 drought in the Amazon rainforest also had major impacts on carbon uptake (Phillips *et al.*, 2009). Uncertainties associated with the simulation of soil-moisture–plant relationships in current climate models are likely to affect the representation of biogeochemical feedbacks in carbon–climate models. Assessments of carbon-cycle feedbacks in these models suggest large variations in land feedback characteristics (Friedlingstein *et al.*, 2006).

Quantification of the radiative forcing associated with biogeophysical versus biogeochemical feedbacks has a bearing on assessments such as the total climate impacts of afforestation (Betts, 2000; Claussen *et al.*, 2001; Bonan, 2008). Regional modifications of surface temperature associated with land-use and land-cover changes need to be taken into account alongside the first-order radiative forcing that they induce (Davin *et al.*, 2007).

4.3 Carbon-cycle feedbacks

The atmospheric content of CO_2 represents the balance between anthropogenic emissions and natural sources and sinks. (See Chapter 2 for a general description of the carbon cycle, and the mechanisms of CO_2 uptake.) From observations, we know that only about 55% of anthropogenic CO_2 emissions is taken up by land and ocean processes, and that the remaining 'airborne fraction' is responsible for the atmospheric CO_2 increase. One effect of this increase in atmospheric concentration is to set up a concentration gradient across the air–sea interface, which causes a net uptake of CO_2 into the surface ocean. The CO_2 is progressively exported to deeper layers through a combination of biological and physical processes. Increasing atmospheric CO_2 also progressively stimulates land-plant photosynthesis (the 'CO_2 fertilization' effect). If the system were in equilibrium, any increase in photosynthesis would be balanced by an increase in decomposition. But as CO_2 concentration is rising, the land biosphere is not in equilibrium. Instead, because of the 20–30-year residence time of carbon in terrestrial ecosystems, the increase in decomposition lags by 20–30 years behind the increase in photosynthesis. This imbalance provides a quantitative explanation for the continuing net uptake of CO_2 into vegetation and soils. The land and ocean together thus exert a major negative feedback on the carbon cycle, absorbing together more than half of the human-induced CO_2 currently being emitted. But there are good reasons to expect that this negative feedback is offset to some (increasing) extent by positive feedbacks due to the effect of climate on CO_2 uptake.

4.3.1 Climate–carbon-cycle interaction

Many Earth system modelling studies have concluded that future climate change will reduce the efficiency of land and ocean in removing CO_2 from the atmosphere, leading to more warming. But the size of this positive feedback has proved hard to pin down. The first studies, by Cox *et al.* (2000) and by Friedlingstein *et al.* (2001) and Dufresne *et al.* (2002), agreed on the sign (positive) of this feedback, but Cox *et al.* (2000) estimated a feedback three times larger than that of Friedlingstein *et al.* (2001) and Dufresne *et al.* (2002). The Coupled Climate–Carbon Cycle Model Intercomparison Project (C^4MIP), engaging 11 Earth system models, confirmed the original finding of a positive feedback, with both land and ocean showing a reduced uptake under a warming climate; but did nothing to reduce the uncertainty in its magnitude (Friedlingstein *et al.*, 2006).

The main processes leading to reduced uptake, at least, are common to the models. Increased stratification of the surface ocean due to warming at the surface reduces the export of carbon from the surface to the deep ocean, and hence limits the air–sea exchange of CO_2. A general, warming-induced increase of the rate of soil-carbon decomposition (heterotrophic respiration) per unit of decomposing material partially offsets the CO_2 fertilization effect on terrestrial ecosystems. But the responses of temperate ecosystems, especially, are highly variable among models.

The C^4MIP analysis provided a first estimate of the magnitude of the uncertainty in both the CO_2 concentration–carbon-cycle (negative) feedback and the climate–carbon-cycle (positive) feedback, revealing that the current uncertainty in these processes is larger than the uncertainty in many physical feedbacks (Figure 4.1, two bottom bars; Gregory *et al.*, 2009).

4.3.2 Carbon–nitrogen cycle interaction

Nitrogen is a limiting nutrient for most terrestrial ecosystems, in the sense that addition of reactive forms of nitrogen produces an increase in vegetation growth (Vitousek *et al.*, 1991). Terrestrial models that include carbon–nitrogen (C–N) cycle interactions (with fixed vegetation distributions) have existed since the early 1990s (e.g. Melillo *et al.*, 1993), but none of the C^4MIP models accounted for nitrogen. Several modelling groups have recently enhanced dynamic global vegetation models by including an explicit, prognostic nitrogen cycle coupled to the carbon cycle and vegetation dynamics (e.g. Thornton *et al.*, 2009; Xu-Ri and Prentice, 2008; Zaehle *et al.*, 2010). According to these modelling studies, nitrogen limitation on land has two main implications for the global-scale analysis of feedbacks. First, in regions where nitrogen is limiting to plant growth, the CO_2 fertilization effect is restricted (compared to a reference case with no active nitrogen cycle), reducing the magnitude of the negative feedback due to CO_2 fertilization. Secondly, the effect of warming on soil organic-matter decomposition becomes more complex: CO_2 is released from soil more quickly (a positive feedback) but this process also accelerates soil nitrogen mineralization (the conversion of nitrogen from organic to inorganic forms in the soil solution), hence increasing the amount of nitrogen available for plant growth (a negative feedback).

The introduction of C–N interactions into coupled models therefore alters the modelling of the climate–carbon-cycle feedback. All published C–N modelling studies agree that the strength of the climate–carbon-cycle feedback is reduced when accounting for nitrogen. However, the amplitude of that reduction varies considerably. The most extreme model (Thornton *et al.*, 2009) produces an increase in growth larger than the increase in decomposition, leading to an unrealistic net negative feedback. Other models, such as Zaehle *et al.* (2010), lead to the conclusion that nitrogen dynamics merely reduces the strength of the climate–carbon-cycle feedback. These studies agreed that nitrogen constraints reduce CO_2 fertilization, leaving more CO_2 in the atmosphere. Hence an apparent contradiction arises: although accounting for nitrogen in Earth system models reduces the strength of the climate–carbon-cycle feedback, the overall effect of accounting for nitrogen is to enhance the simulated atmospheric CO_2 growth rate, and hence to accelerate modelled global warming.

4.3.3 Observational constraints on climate–carbon-cycle feedback

An emerging approach for assessing the realism of modelled climate–carbon-cycle feedbacks involves the examination of independent global observations of CO_2 and climate. The CO_2 response to climate anomalies can provide constraints on the magnitude of the global carbon-cycle sensitivity to climate on several timescales (Friedlingstein and Prentice, 2010). The year-to-year variation in atmospheric CO_2 growth rate is strongly correlated with climate variability. In particular, El Niño years are associated with higher-than-average CO_2 growth rates. Many studies have analysed CO_2 variability on El Niño-Southern Oscillation (ENSO) timescales either using carbon-cycle models or top-down methods such as deconvolution or inversion of atmospheric CO_2 and $^{13}CO_2$ data. All evidence points to terrestrial carbon release during El Niño events, with the global ocean showing slightly enhanced uptake, but the terrestrial effect being dominant.

Interannual variations in the CO_2 growth rate have been used to constrain global climate–carbon-cycle models and to estimate the carbon-cycle sensitivity to climate (Cox and Jones, 2008). During El Niño events, tropical lands are anomalously warm, and are sources rather than sinks of CO_2, implying that the climate–land carbon-cycle feedback is positive on the ENSO timescale. Among models, there is a strong positive correlation between the magnitude of simulated interannual variability of atmospheric CO_2 and the tendency for tropical forests to lose carbon in global-warming scenarios. Recent analysis using the observed atmospheric CO_2 and tropical land-temperature variability has suggested that the response of tropical ecosystems amounts to about 40 Pg C loss per degree of warming (Cox *et al.*, unpublished work). This finding excludes the possibility that the climate–carbon-cycle feedback is negative as some recent modelling has suggested, but it also rules out 'high-end' estimates of the positive feedback.

On longer timescales, ice-core records allow us to investigate coupled CO_2 and climate variability and potentially to infer some properties of the coupled climate–carbon-cycle system. The last millennium is of special interest as the CO_2 record is at a high resolution, and time-series data for reconstructions of temperature are available. It includes periods of change over the same timescales of concern (decadal to centennial) as the ones in play over the coming century. For example, the low CO_2 and temperature excursion associated with the Little Ice Age (roughly AD 1600–1850) should be a relevant constraint. The larger CO_2 and climate oscillations over glacial–interglacial cycles may be less useful in the context of the twenty-first-century climate–carbon-cycle feedback, as the processes and timescales involved are different.

The main problems with using these data are the multiple sources (different methods for temperature reconstruction and different sets of ice-core measurements for atmospheric CO_2) and the necessary choices in the treatment of the data, including the analysis period, smoothing techniques, age calibration, consideration of time lags, and handling of uncertainties. Several studies have nevertheless used the last-millennium record to derive the carbon-cycle sensitivity to climate (Joos and Prentice, 2004; Scheffer *et al.*, 2006; Cox and Jones, 2008; Frank *et al.*, 2010). Frank *et al.* (2010), in the most comprehensive study so far, concluded that the carbon-cycle sensitivity to climate on centennial to millennial timescales is positive with a median amplitude of 7.7 ppm CO_2 K^{-1} and a likely (broad) range of 1.7 to 21.4 ppm CO_2 K^{-1}. Again, this analysis excludes both negative and very large positive feedbacks.

4.4 Nitrous oxide feedbacks

4.4.1 N_2O production pathways

Nitrous oxide is a potent greenhouse gas produced by nitrifying and denitrifying bacteria during the degradation of organic matter in soils and water. The specific formation pathway is determined by local environmental conditions, including oxygen level, substrate availability, pH and temperature (Ritchie and Nicholas, 1972; Baggs and Philippot, 2009 and references therein). In oxic environments, nitrifying bacteria produce N_2O as a by-product during the oxidation of ammonium to nitrate. Under anoxic and suboxic conditions more complex pathways dominate, often with a much higher yield of N_2O. These include N_2O production by denitrifiers (during the reduction of nitrate to nitrite and N_2); denitrification by nitrifiers (consisting of the oxidation of ammonium to nitrite and subsequent reduction to N_2O); and nitrate ammonification (reduction of nitrate to nitrite and ammonium, during which some ammonifying bacteria can produce N_2O). Kelso *et al.* (1997), Codispoti and Christensen (1985) and Naqvi *et al.* (2010) have reviewed the microbial pathways that produce N_2O.

The largest sources of N_2O are terrestrial (both natural and fertilized soils). Nitrous oxide production pathways in soils are highly sensitive to local environmental conditions, and can alternate between low-yield nitrification pathways and higher-yield pathways, primarily driven by changes in soil oxygen levels. The largest terrestrial sources of N_2O are in the wet and seasonal tropics. In aquatic environments N_2O is produced during the biologically mediated remineralization of organic matter. Current understanding is that the dominant source pathway in the oxygenated open ocean is via nitrification while combinations of higher-yield pathways dominate in suboxic marine environments, including coastal and estuarine zones and sediments (Bange, 2008).

4.4.2 Impact of climate change on N_2O production

Feedbacks between terrestrial N_2O fluxes and climate can arise due to climate-driven variations in soil temperature and soil-water content, both of which are important controls on the global pattern of soil N_2O emission. Both nitrification and denitrification respond positively to temperature over most of the physiological range. There is experimental evidence for the enhancement of soil N_2O emissions in response to temperature, soil moisture and CO_2 (e.g. Barnard *et al.*, 2005; Kamman *et al.*, 2008; Cantarel *et al.*, 2011; van Groenigen *et al.*, 2011) although some experiments have reported a decline in N_2O emission with increased CO_2 due to increased plant demand for nitrogen, which can deplete the soil pools of inorganic nitrogen from which N_2O is derived. Experiments on permafrost soils indicate a potential positive feedback mechanism due to increased N_2O emissions leaking from the thawing soil (Repo *et al.*, 2009). There are possible counteracting effects, such as greater denitrification to N_2 instead of N_2O if soil moisture increases (e.g. Kesik *et al.*, 2006). A global modelling study of N_2O terrestrial emissions during the twentieth century has suggested a dominant control by temperature, with a positive response of ~1 Tg N a^{-1} per degree of warming (Xu-Ri *et al.*, submitted).

Climatic influences on marine N_2O formation include changes in the extent of suboxic zones, and changes in local temperature. The potential expansion of oceanic suboxic regions under climate warming (Stramma *et al.*, 2008) could result in large increases in N_2O yield (Bange, 2008; Codispoti, 2010). *In situ* temperature is also reported as a possible influence on marine N_2O production (Butler *et al.*, 1989); this would imply potentially higher N_2O yields. The magnitude of the oceanic N_2O source is also determined by the rate of organic-matter production and subsurface remineralization. Factors influencing these flows include changes in vertical mixing and upper-ocean stratification (Bopp *et al.*, 2001; Behrenfeld *et al.*, 2006), and changes in nutrient supply via surface deposition and river outflow (e.g. iron and reactive inorganic nitrogen; see Jickells *et al.*, 2005; Duce *et al.*, 2008). Bopp *et al.* (2001) note possible reductions in marine export production under scenarios of future climate warming and increased ocean stratification; this could imply reduction in marine N_2O production. But such warming scenarios would be accompanied by increased ocean deoxygenation, and increased N_2O formation in suboxic waters. These separate effects could result in feedbacks of opposite sign. The overall impact of warming on ocean N_2O fluxes remains to be determined.

4.4.3 Observational constraints on climate–N_2O feedback

The recent atmospheric measurements and the ice-core records of N_2O concentration are underexploited resources for the analysis of climate feedbacks involving N_2O. Xu-Ri *et al.* and others (Prentice, personal communication, 2011) have shown that there is an empirical relationship between global mean land-temperature variations and the variable growth rates of N_2O concentration in the atmosphere during the twentieth century, as reconstructed from ice-core measurements. The growth-rate variations are of similar magnitude to variations in modelled N_2O emissions from soils. Xu-Ri *et al.* (submitted) also pointed out that the so-called anthropogenic source of N_2O in fertilized soils is likely to respond to climate in a similar way to natural soils, amplifying the feedback. This possibility has so far been neglected in global N_2O budgets and projections.

4.5 Methane feedbacks

4.5.1 CH_4 production pathways

Wetlands are environments where high water levels inhibit the diffusion of oxygen into the soil and produce anoxia. Anoxia slows down soil organic-matter decomposition, and allows a reduced compound such as CH_4 to be the terminal electron acceptor in the decomposition chain. Methane production (methanogenesis) is carried out mainly by archaea, which use either plant-derived acetate, or CO_2 and H_2, as substrates for growth. Natural wetlands cover about 7% of the global land surface area (Lehner and Döll, 2004) and are the largest source of CH_4 in the atmosphere (Hein *et al.*, 1997). The emission of methane from wetlands represents a balance between production by methanogens, and consumption by methane-oxidizing (methanotrophic) bacteria in the drier top layers (Frenzel, 2000).

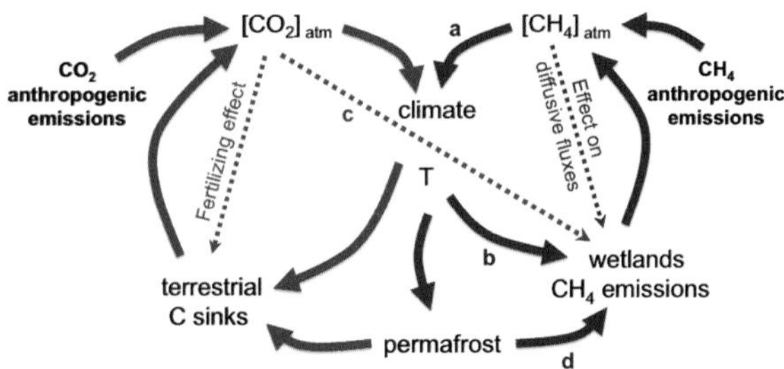

Figure 4.4 Methane and carbon dioxide feedback mechanisms. Red arrows represent the climate–carbon dioxide feedback and purple arrows the climate–methane feedback. The arrows labelled with letters a–d are discussed in the text: (a) the effect of methane on the climate as a greenhouse gas; (b) the effect of climate (temperature, T) on methane emissions by increasing rates of decomposition and influencing wetland extent; (c) the CO_2 fertilization effect on wetlands; and (d) the melting of permafrost, increasing the inundated area in northern areas. Figure modified with permission from Ringeval *et al.* (2011).

4.5.2 The carbon balance and CH_4 emissions of wetlands

The CH_4 concentration in the atmosphere and the climate are linked in a feedback loop (Figure 4.4): climate is the main control on wetland CH_4 emissions while a change in CH_4 concentration derived from those emissions must affect the climate, because CH_4 is a potent greenhouse gas. The climate–carbon-cycle feedback loop is connected to the CH_4 feedback loop, via the effect of CO_2 on climate, and the direct CO_2 fertilization effect on primary productivity in wetlands, which increases the substrate available for CH_4 production (Ringeval *et al.*, 2011). Some so-called anthropogenic sources are also affected by climate and atmospheric CO_2 concentration: for example, rice paddies react to climate and CO_2 concentration changes in a similar way to natural wetlands.

Northern wetlands have been extensively studied in terms of their involvement in the carbon cycle. Because of their low decomposition rates, roughly 20% of the world's soil carbon is stored in boreal and subarctic peatlands where it has been accumulating gradually under wet and cool climatic conditions since the end of the last ice age (Maltby and Immirzi, 1993). The most recent estimate of carbon stored in these ecosystems is about 550 Pg C (Yu *et al.*, 2010; Yu, 2011). Northern peatlands are a small sink for CO_2 but an important source of CH_4 and dissolved and particulate organic carbon which will eventually be oxidized to CO_2 (Billett *et al.*, 2004; Blodau *et al.*, 2004). Global warming, drainage and peat fires may be causing a decline of carbon accumulation in peatlands, and possibly even converting these ecosystems into a net source of carbon to the atmosphere (Gorham, 1991; Billett *et al.*, 2004). It has been speculated that increased summer droughts in mid-continental regions and increases in precipitation in oceanic regions will further reduce carbon uptake by peatlands (Belyea and Malmer, 2004).

The carbon budget of tropical wetlands is not as well understood as that of northern wetlands. Tropical wetlands are notable because their high productivity has led to the accumulation of deep peat deposits in certain areas of Amazonia and southeast Asia (Page *et al.*, 2002; Lähteenoja *et al.*, 2009). Amazonian peatlands have been estimated to cover 150,000 km^2. This area, combined with the great depth of peat deposits and high accumulation rates found by Lähteenoja *et al.* (2009), make these peatlands a potentially significant component of the global carbon budget.

4.5.3 Impacts of climate on CH_4 emissions

Methane emissions from wetlands depend on temperature, water table, pH, plant cover and type, time of year (i.e. the supply of metabolites), nutrient status, depth of the oxic zone and the presence of competing micro-organisms. The rate of methanogenic activity increases with temperature. Some models predict an increase of CH_4 emissions for a small temperature rise due to changes in net ecosystem production and/or increased methanogenesis (e.g. Cao *et al.*, 1998). On the other hand, a decrease of this flux has been predicted for an increase of temperature of more than 4 K because of soil-moisture depletion and a subsequent decrease in net ecosystem production (Cao *et al.*, 1998). It has been suggested that CH_4 and CO_2 emissions are only controlled directly by temperature in wetter sites, while in drier sites the position of the water table is a more

Table 4.1 A summary of methane models and their main features.

Model name	Dynamic wetland	Rice paddies	Permafrost	Methane increase	Carbon accum/ loss	Climate scenario	Reference
HadleyMOSESLSH	✓	✓	✗	×2	✗	IS92a	(Gedney *et al.*, 2004)
ORCHIDEE	✓	✗	✗	×1.5	✗	SRES A2	(Ringeval *et al.*, 2011)
ORCHIDEE	✓	✗	✓	×2	✗	SRES A2	(Koven *et al.*, 2011)
GISS	✓	✗	✗	×1.8	✗	2 x CO_2	(Shindell *et al.*, 2004)
LPJWHyMe	✗	✓	✓	×2.5	✓	SRES A2	(Wania *et al.*, 2009b)
IAP RAS CM	✗	✗	✓	×1.5	✗	All IPCC	(Eliseev *et al.*, 2008)

important control (Christensen, 1993; Bubier *et al.*, 2003). In general, lowering the water table increases the rate of CO_2 emission and decreases the emission rate of CH_4 in the long term (Crill *et al.*, 1991; Priemé, 1994; Daulat and Clymo, 1998; Blodau *et al.*, 2004; Strack and Waddington, 2007).

4.5.4 Climatic controls on wetland distribution

Climate affects not just the rates of carbon mineralization and methane fluxes in wetlands, but also the extent and distribution of these ecosystems. Wetland ecosystems exist within well-defined climatic boundaries (Wieder and Vitt, 2006) and are highly susceptible to changes in climate because they rely on the maintenance of high water tables. Models of peatland distribution for Canada suggest a retreat northwards of the southern boundary of peatlands under future warming (Gignac *et al.*, 1998). The climate envelope of British blanket bogs has been predicted to retreat towards the north and west when subjected to warming scenarios (Clark *et al.*, 2010; Gallego-Sala *et al.*, 2010). Globally, blanket bogs may be under threat from a changing climate (Gallego-Sala *et al.*, 2009). Palsa mires, which are marginal permafrost features and highly vulnerable to changes in the climate, are predicted to degrade rapidly due to warming (Luoto *et al.*, 2004). Temperature changes are the most important driver in all of these model projections. The importance of temperature as a factor controlling the destabilization of peatlands has been pointed out in many studies (Freeman *et al.*, 2001). Paludification, the positive feedback between accumulation of soil organic carbon and water-table rise due to the high water-holding capacity of peat and its low hydraulic conductivity, has been suggested to increase the sensitivity of peat decomposition to temperature when the climate is unfavourable to peat accumulation by accelerating carbon loss (Ise *et al.*, 2008).

The large area of peatlands currently under permafrost is especially vulnerable to climate change (Denman *et al.*, 2007). Thawing of the permafrost in Arctic and boreal peatlands may stimulate net primary productivity and hence the high-latitude carbon sink. However, permafrost thawing may also lead to an increase in soil organic-matter decomposition. Increased CH_4 production may counteract any potential increase in the carbon sink, in terms of radiative forcing (Turetsky *et al.*, 2007). A recent study, based on a terrestrial ecosystem model accounting for both permafrost and wetlands dynamics, predicts that, by the end of the twenty-first century, the land north of 60° N latitude could shift from a CO_2 sink to a source, once high-latitude processes such as soil insulation, cryoturbation and microbial heat release are accounted for. In addition, CH_4 emissions from these regions are predicted to experience up to a two-fold increase, although this rise will be partly compensated by a reduction in wetland extent (Koven *et al.*, 2011).

Several dynamic vegetation or land-surface models have added a module for wetland extent or methane emission or both (Gedney *et al.*, 2004; Shindell *et al.*, 2004; Wania *et al.*, 2009a,b; Ringeval *et al.*, 2010, 2011). These models differ in the way they calculate wetland extent (dynamic or fixed), whether they consider rice paddies and whether permafrost dynamics are modelled (Table 4.1). However, all models agree in predicting an increase in the methane fluxes from wetlands by the end of the twenty-first century (see the example in Figure 4.5 from the ORCHIDEE model). The only

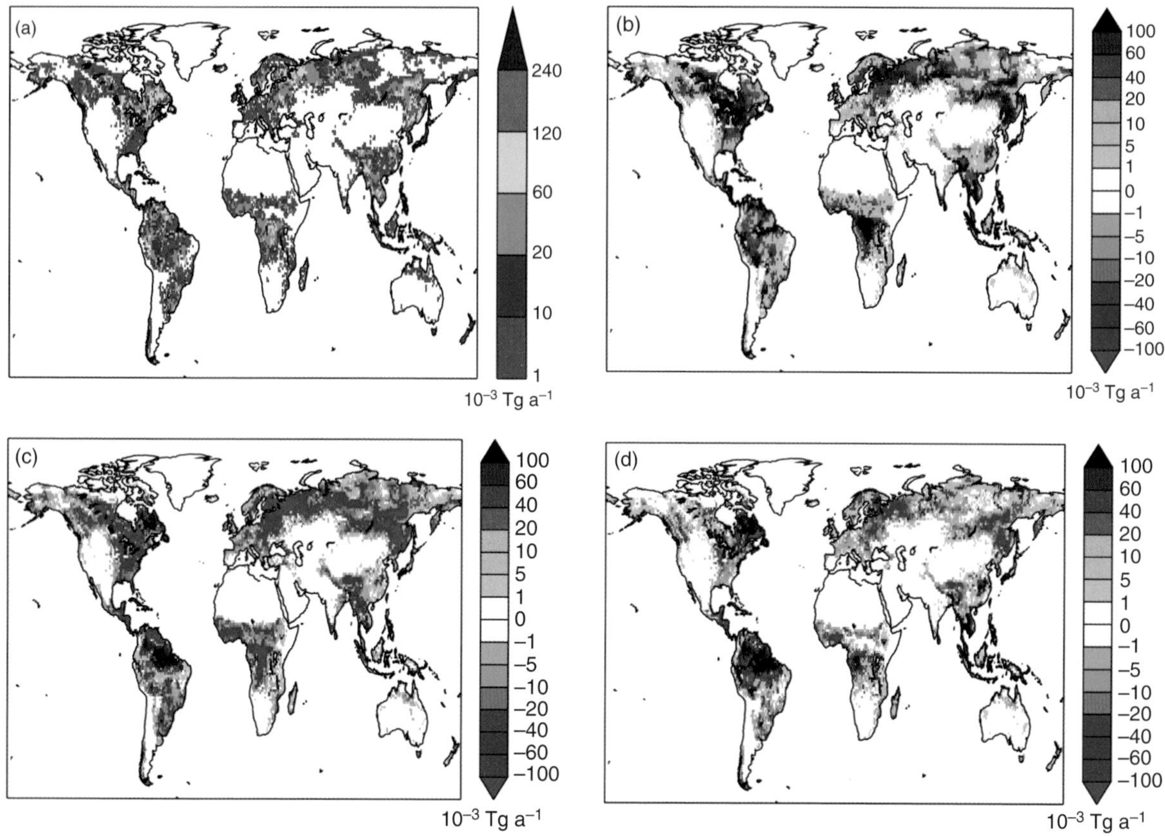

Figure 4.5 Methane emissions as simulated by the ORCHIDEE model. (Figure first published in Ringeval *et al.* (2011) *Biogeosciences Discussions*, **8**, 3203–3251). (a) Control emissions; (b) additional emissions by 2100 due to atmospheric CO_2 increase; (c) additional emissions by 2100 due to climate change; and (d) additional emissions by 2100 due to the combined effect of atmospheric CO_2 increase and climate change.

one of these models currently able to calculate global fluxes of CO_2 due to losses of carbon from peatlands is LPJWHyMe (Wania *et al.*, 2009b).

4.5.5 Climate change and CH_4 release from clathrates

Contemporary warming is greatest in the northern high latitudes and this high rate of warming is expected to continue, so these regions are liable to produce the strongest changes in natural emissions of CH_4. There is a large store of carbon in the form of CH_4, much of it bound in clathrate lattices, in permafrost soils on land and in marine sediments. There has been extensive speculation (not reviewed here) about the potential for warming, especially in the Arctic, to destabilize these CH_4 stores. Shakhova *et al.* (2010) drew attention to the East Siberian Shelf as a region whose sediments contain abundant CH_4. The water here is shallow enough that CH_4 released from the sediments can bubble up and be released at the sea surface without oxidation. Shakhova *et al.* (2010) showed that this shelf is indeed a source of CH_4. But the magnitude of the source is tiny compared to terrestrial sources, and there is no evidence that it has increased.

Chapter 3 discusses a key constraint on the potential total release of CH_4 from storage in a warming world. The circumpolar Arctic region was 3 to 5 K warmer during the last interglacial than it is today, yet CH_4 concentration (as recorded in ice cores) did not show a large increase.

However, this analogy does not help us in the case of an even greater warming. Sokolov *et al.* (2009) considered the case of a very large (12 K) temperature rise. The predicted increased methane release was 40 Tg CH_4 a^{-1}. This is still a small fraction of the anthropogenic source (currently 400 Tg CH_4 a^{-1}), and would contribute an additional temperature increase of 0.5 K.

Archer and Buffett (2005) estimated the total global store of deep-ocean methane clathrates to be ~ 5000 Pg C. They used a model to estimate the effect of global warming on this store. They found that a large global warming could eventually release thousands of Pg C from clathrates, but only very slowly – that is, over millennia or longer. Archer (2007) concluded that '*on the timescale of the coming century … most of the hydrate reservoir will be insulated from anthropogenic climate change. The exceptions are hydrate in permafrost soils, especially those [in] coastal areas, and in shallow ocean sediments where methane gas is focused by subsurface migration. The most likely response of these deposits to anthropogenic climate change is an increased background rate of chronic methane release, rather than an abrupt release.*'

4.5.6 Impact of biogenic volatile organic compound emissions on CH_4 lifetime and atmospheric chemistry

Various biogenic volatile organic compounds (BVOCs) are emitted from terrestrial sources and, although not greenhouse gases, are capable of impacting the lifetime of CH_4 in the atmosphere. The main process is via reaction with the hydroxyl radical, which is the main CH_4 sink; so increased BVOC emission can (indirectly) lead to an increase in the concentration of CH_4 (Folberth *et al.*, 2006). BVOCs can impact atmospheric chemistry in additional ways, producing or destroying tropospheric ozone (depending on the NO_x concentration), and promoting the production of secondary aerosols. Altogether, BVOCs have been estimated to represent a large emission source of carbon annually, on the order of 1000 Tg C a^{-1}, although this figure is not well constrained. Plants emit BVOCs in defence against high-temperature stress and toxic effects of oxidants (including ozone), and as signalling molecules. Among BVOCs, isoprene (C_5H_8) is the single most important compound in terms of its annual flux (estimated as 400 to 600 Tg C).

Models of isoprene and other BVOCs (mainly monoterpenes) are used to upscale observed measurements of leaf fluxes to whole ecosystems (Arneth *et al.*, 2010b) but the process of production and release of isoprene is complex and our knowledge is still very limited (Pacifico *et al.*, 2009). There remains a large uncertainty in the effect of a change in climate and CO_2 concentration on the emissions of isoprenes and other BVOCs (Arneth *et al.*, 2008; Peñuelas and Staudt, 2010; Pacifico *et al.*, in revision).

Different plant species produce and emit isoprene at greatly differing rates. Isoprene emissions have been observed to increase in the short term as a saturating function of solar radiation, and exponentially with temperature. There is also a CO_2–isoprene interaction, with emissions increasing at low CO_2 concentrations and decreasing at high CO_2 concentration (Arneth *et al.*, 2008). Some models of isoprene predict an increase of emissions in the future due to the temperature effect, the CO_2 fertilization of photosynthetic activity and land-use changes (Lathière *et al.*, 2006; Guenther, 2007; Andersson and Engardt, 2010) but others predict no significant change (Arneth *et al.*, 2007; Pacifico *et al.*, in revision) or a small reduction in emissions (Heald *et al.*, 2009) due to the impact of higher CO_2 concentration. These uncertainties in changes in future isoprene emissions translate into large uncertainties about the future changes in methane lifetime and tropospheric ozone. While Andersson and Engardt (2010) find that future isoprene emission would contribute significantly to a projected increase in tropospheric ozone, Pacifico *et al.* (in revision) find no significant isoprene-induced change in tropospheric ozone by the end of the century.

4.6 Fire feedbacks

Fire is a major influence on vegetation dynamics and community structure, an element of the global carbon cycle, and a mechanism by which CO, CH_4, NO_x and other trace gases, BVOCs and aerosol precursors are emitted to the atmosphere (Bowman *et al.*, 2009; Carslaw *et al.*, 2010; van der Werf *et al.*, 2010. General principles and recent developments in observing and understanding fire regimes are summarized in Chapter 2). Fires modify Earth surface characteristics via changes in vegetation cover and surface albedo, with consequences for the land–atmosphere exchanges as already discussed in Section 4.2. Smoke-borne aerosols from fires can disrupt radiative processes in the troposphere, as well as normal hydrological processes, potentially reducing rainfall (Andreae, 2007; Carslaw *et al.*, 2010). Fire is thus a cross-cutting process in the Earth system with the ability to perturb the Earth's vegetation and its carbon, hydrological and energy cycles, and to generate land–atmosphere feedbacks (Figure 4.6).

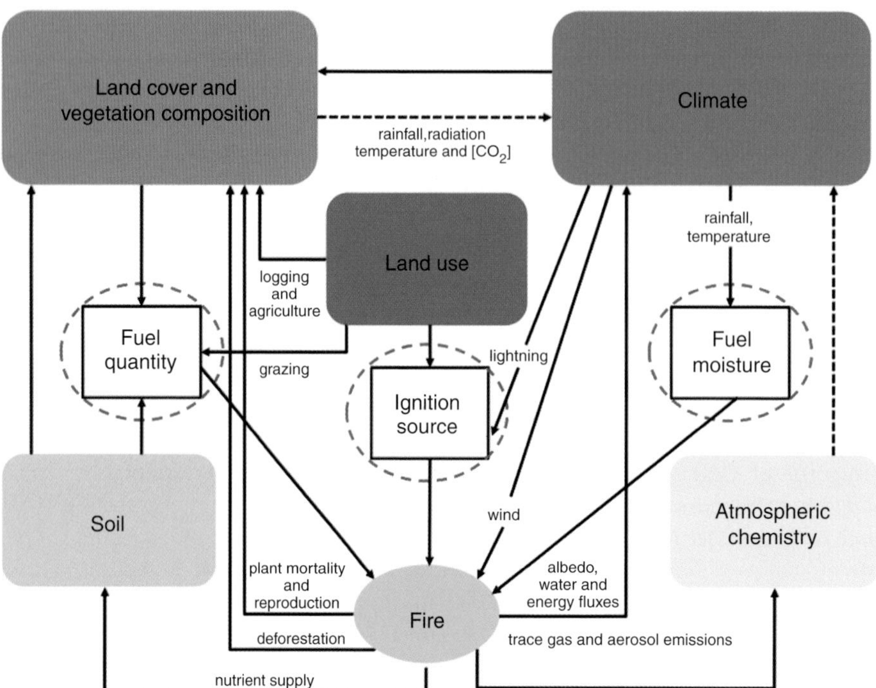

Figure 4.6 Fire functioning and feedbacks in the Earth system, illustrating the three requisites for fire to occur: a sufficient amount of fuel, sufficiently dry fuel, and an ignition source.

4.6.1 Climate change and land-use impacts on fire

Fire is extremely sensitive to climate. Analysis of remotely sensed burned-area data in relation to climate has shown that, globally, burned area increases with temperature across the full range of temperatures, whereas the relationship between burned area and precipitation is unimodal due to limitation by fuel availability in dry climates versus suppression by fuel moisture in wet climates (Prentice *et al.*, 2011; Daniau *et al.*, unpublished work). The dependence of fire on temperature and fuel availability has been confirmed in several regional studies. For example, Westerling *et al.* (2006) showed that fire activity in the western USA increased markedly in the mid 1980s, along with increased spring and summer temperatures, and earlier spring snowmelt, which produces an earlier, prolonged dry season, more opportunities for ignition and reduced fuel moisture. Spessa *et al.* (2005) and Harris *et al.* (2008) showed a strong positive relationship between antecedent rainfall and area burned in the tropical savannas of northern Australia, due to the positive effect of rainfall on biomass.

Many studies have linked fire activity to specific atmospheric circulation patterns, including an increase in fire activity during La Niña events in the southern United States and Central America (Kitzberger *et al.*, 2001; Le Page *et al.*, 2008; van der Werf *et al.*, 2010) and a marked increase in fire activity in tropical rainforests during El Niño events (Le Page *et al.*, 2008; van der Werf *et al.*, 2008, 2010). Anomalous drought and severe fire episodes in the Iberian Peninsula have been linked to the negative phase of the North Atlantic Oscillation (NAO) (Trigo *et al.*, 2006), while Balzter *et al.* (2005) reported a close correlation between the Arctic Oscillation (closely related to the NAO) and fire in the Russian boreal forests. However, all of these linkages are indirect; they come about through the effects of local climate on fuel supply and burning conditions.

In moisture-limited (high biomass) fire regimes, land-use practices can exacerbate the impacts of drought on fire. For example, active fire suppression is suggested to have led to fewer but much more intense fires during La Niña periods in the USA (Pyne *et al.*, 1996) and El Niño periods in southeastern Australia (Gill and Catling, 2002). Logging in closed forests produces more residual biomass, and can lead to more severe dry conditions because the canopy is fragmented and the surface exposed to increased radiation. The increased dry fuel increases the risk of fire, which can then further open up the canopy. Furthermore, logging

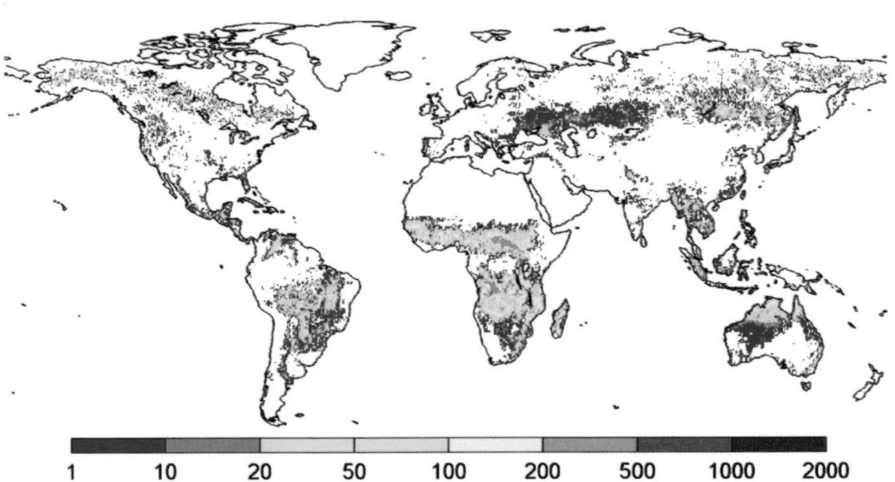

Figure 4.7 Average carbon releases from above- and below-ground wildfires, 1997–2008 (g C m^{-2} a^{-1}). Image first published in van der Werf *et al.* (2010), using data from www.falw.vu/~gwerf/GFED/index.html.

and fires in areas not usually subject to fires can create conditions that encourage future fires because of incursions by fire-tolerant grasses and shrubs which, in turn, lead to higher fine fuel loads and drier fuels surrounding the forest (Siegert *et al.*, 2001; Page *et al.*, 2002, 2010; Cochrane, 2003). Peat drainage for agricultural production can worsen impact of El Niño drought on peat moisture, thereby increasing the vulnerability of peats to fire (Page *et al.*, 2002; Ballhorn *et al.*, 2009).

Studies of climate impacts on future wildfires have used outputs from climate models forced with emission scenarios (Flannigan *et al.*, 2009). Climate-model outputs have been used to examine changes in fire due to weather-related climate variables such as temperature, rainfall and humidity. Alternatively, various climate fields can be combined to form a fire danger index as used by forest operations managers (e.g. the Canadian Fire Danger Index, Stocks *et al.*, 1989). Flannigan *et al.* (2009) list over 20 such studies in the last decade that have shown a general increase in fire risk under a future climate regime. Despite this consensus, studies relying on climate variables or fire danger index alone inform us only about the climatic risk of fire, and disregard the equally important effects of vegetation properties and land use.

4.6.2 Modelling fire–vegetation interaction

Modelling fire, burned area and emissions of trace gases under climate change calls for a process-based modelling approach. Prognostic fire models, embedded in process-based vegetation models, have been developed to simulate the effects of changes in climate and vegetation dynamics on fire activity and emissions (Lenihan and Neilson, 1998; Thonicke *et al.*, 2001; Venevsky *et al.*, 2002; Arora and Boer, 2006; Lehsten *et al.*, 2009; Thonicke *et al.*, 2010; Kloster *et al.*, 2010; Spessa and Fisher, 2010; Prentice *et al.*, 2011). Only a handful of studies have used climate-model outputs to drive coupled vegetation–fire models, in order to investigate fire activity under future climate change. All have reported a general increase in burned area (i.e. not just fire risk) under various scenarios (Arora and Boer *et al.*, 2006; Scholze *et al.*, 2006; Thonicke and Cramer, 2006; Lenihan *et al.*, 2008). But an important caveat remains: future fire regimes are not entirely predictable from climate, even when climate- and CO_2-induced vegetation changes are taken into account. Changes in fire regimes will necessarily depend on land-use decisions as well as on climate change (Archibald *et al.*, 2009; Aldersley *et al.*, 2011).

4.6.3 Fire effects on the emissions of CO_2 and other trace gases

A recent global assessment reports that wildfires (excluding peat fires) emit 2 to 4 Pg C a^{-1} to the atmosphere, with about half of this amount due to savanna fires in Africa (van der Werf *et al.*, 2010). Mediterranean forests and shrublands, tropical forests and boreal forests, where fuel loads (and therefore emissions) per unit area are much greater than in grasslands, are also globally significant for the carbon cycle even though the annual areas burned are much smaller (Figure 4.7).

Carbon dioxide emissions from these fires are expected to be balanced over decadal timescales by the additional CO_2 uptake due to regenerating vegetation. But many other trace atmospheric constituents are released by fires (Andreae and Merlet, 2001), including greenhouse gases and other reactive gases that influence the greenhouse effect through atmospheric chemistry. The emissions of these substances from fire amount to additional climate forcing agents, which are not balanced by uptake in the same way as for CO_2. Carbon dioxide released from deforestation and peatland fires, especially in South America and southeast Asia, also counts as an additional, unbalanced emission. These fires are thought to emit about 0.6 Pg C a^{-1} (Schultz *et al.*, 2008; van der Werf *et al.*, 2010) albeit with large uncertainty.

Fire is estimated to have been responsible for ~ 1 Pg C a^{-1} of the interannual variability of total land–atmosphere CO_2 fluxes during the past decade (van der Werf *et al.*, 2010). Short-term increases of 10% or more in the atmospheric growth rates of CO, NO_x and CH_4 are associated with El Niño-induced droughts in the tropics (van der A *et al.*, 2008; Yurganov *et al.*, 2008; van der Werf *et al.*, 2010). The influence of both wetlands and peatland fires on large global CH_4 pulses in 1997–1998, 2002–2003 and 2006 (all three coinciding with El Niño periods) has been confirmed by inverse modelling based on atmospheric CH_4 measurements (Bousquet *et al.*, 2011).

Peat has a high carbon content, and can burn when dry. Once ignited by the presence of a heat source (e.g. a wildfire penetrating the subsurface), it smoulders. Smouldering peat fires can burn undetected for weeks to months. Over the past 15 years or so, the burning of Indonesian peatlands and forests has contributed to detectable increases in global CO_2, CO, CH_4 and aerosol concentrations in the atmosphere (Page *et al.*, 2002; van der Werf *et al.*, 2008, 2010; Spessa *et al.*, 2010). During the El Niño-induced drought of 1997–1998, the most severe in recorded history, peat and forest fires in Indonesia released at least 1 Pg C (Page *et al.*, 2002). High-resolution analysis of fires in Borneo during 1997–2007 (Spessa *et al.*, 2010) has shown positive correlations between fire frequency and drought. The strength of these correlations increases with declining forest cover, implying a nexus between fire, climate and deforestation or forest degradation. Equatorial southeast Asia is projected to generally experience reduced rainfall in the next decades implying that the risks of drought-related fire will increase (Li *et al.*, 2007; Dai, 2011). At the same time, economic demands for establishing pulp paper and palm-oil plantations to replace native rainforests, especially on peatlands where tenure conflicts among landowners tend to be minimal, are expected to increase as well (Hooijer *et al.*, 2006).

Climate change projections show increased melting of permafrost and drought frequency in the northern high latitudes (IPCC, 2007). This suggests that peat fires in the boreal zone will also become more common, and the emissions from such fires will increase (Flannigan and de Groot, 2009).

4.6.4 Fire effects on aerosols and radiative forcing

Aerosols emitted from biomass burning are particularly important for land–climate interactions because of their light-absorbing properties, which can affect absorption of radiation in the atmosphere and at the Earth's surface, especially when the particles are deposited on snow and ice (principally in the form of black carbon); and because of their solar scattering properties (principally due to organic carbon), which can lead to cooling at the surface (Andreae, 2007; Carslaw *et al.*, 2010). However, estimates of aerosol production from fires vary widely (Schulz *et al.*, 2008; van der Werf *et al.*, 2010). And in contrast to trace-gas emissions, there is no agreement on the sign of the direct radiative forcing due to aerosols. Equivocal outcomes arise from several factors. Aerosols from wildfires contribute to either warming or cooling of the climate depending on the season and region, and from which ecosystems they are emitted. Also, model structure and complexity vary between different studies (Carslaw *et al.*, 2010). For example, Tosca *et al.* (2010) showed that biomass-burning aerosols from regional peat and forest fires can increase the intensity of future droughts in southeast Asia during El Niño events via an increase in tropospheric heating from black carbon, which leads to a reduction in shortwave-radiation forcing at the surface, which in turn leads to a further reduction in rainfall via increased surface cooling and hence reduced convection. On the other hand, positive forcing has been predicted to occur in areas with a very high surface reflectance, such as the desert regions in North Africa and the snow fields of the Himalayas, mainly due to increased absorption of radiation by black carbon (Forster *et al.*, 2007).

Aerosols have additional indirect effects on climate and ecosystems. Aerosol loading can increase photosynthesis (up to a point) by increasing the diffuse solar-radiation fraction (Mercado *et al.*, 2009). On the other hand, aerosols can disrupt droplet formation in clouds and thereby reduce rainfall (Andreae, 2007).

4.6.5 Fire effects on surface albedo

Surface albedo varies between ecosystems and interacts with climate and vegetation (Bonan, 2008). Fire regimes can influence climate through changes in surface albedo because of the interplay between how much vegetation cover is removed, how much black ash is produced, sensible and latent heat fluxes from fire scars, and available atmospheric water vapour.

In the Australian tropics, for example, the concentration of water vapour abruptly increases at the end of the dry season until increasing cloud cover results in the predominance of diffuse solar radiation. At this time, recently burned areas (relatively dark in colour with low albedo) can induce precipitation because of intense heating and deep moist convection. The development of such convective storms does not occur in the early dry season, because of lower humidity and solar radiation (Bowman *et al.*, 2007).

Throughout the North American boreal region, forest stands are being continuously renewed by stand-replacing fires. Fires of such severity not only kill the above-ground vegetation but also blacken the surface, influencing the albedo and the partitioning of sensible versus latent heat flux. An analysis of a single boreal forest fire by Randerson *et al.* (2006) showed a net effect of decreased radiative forcing, mainly from sustained multi-decadal increases in the surface albedo. Amiro *et al.* (2006) indicated that fire-induced lowering of albedo in these fire-prone regions has a net cooling effect. Randerson *et al.* (2006) concluded that this albedo effect exceeded the warming effects from fire-emitted greenhouse gases, but the generality of this conclusion is unclear.

Another albedo-related effect on climate arises when black-carbon particles are deposited on regions with summer snow cover at high latitudes or elevations (Forster *et al.*, 2007; Carslaw *et al.*, 2010). The global radiative forcing for black carbon is certainly positive, but its magnitude remains an open question.

4.7 Human feedbacks

Most of the physical and biogeochemical feedbacks discussed above are likely to be modified by human activities, especially land use, even though these interactions have not been studied in a systematic way. While observations of the global environment enable the intricate interactions of humans and the natural environment to be detected, monitored and mapped (Ellis and Ramankutty, 2008) existing modelling efforts struggle to represent the complexity of human feedbacks.

A particularly difficult challenge is to distinguish a socially mediated process as a climate feedback, rather than part of the wider societal context of climate change mitigation and adaptation (see also Chapters 1, 6, 7 and 8). Causality is complicated in human systems, especially where foresight is applied in decision-making. Further theoretical developments are needed, analogous to the work of Gregory *et al.* (2009) in integrating physical and biosphere-mediated feedbacks within the same conceptual framework. In the remainder of this discussion, we confine ourselves to situations where human responses to climate change alter the anthropogenic emissions of greenhouse gases.

To start with an example that has already been explored in detail from a biogeophysical angle, the incidence and distribution of fire is as sensitive to human decisions about land use as it is to climatic controls (Marlon *et al.*, 2008). Land use has the potential either to amplify or reduce climate feedbacks due to changes in the land–atmosphere fluxes of the biosphere-mediated greenhouse gases CO_2, N_2O and CH_4. A richer understanding of the natural processes involved in these feedbacks could enable society to target efforts to jointly optimize climatic and other benefits. For example, there may be situations where intensifying land use might reduce carbon losses and greenhouse-gas emissions from fires – a benefit to be set against possible losses of carbon associated with the land-use change.

We also need to consider another class of feedbacks whereby, as individuals, communities and a global population, people's reactions to changes in the climate have synergistic consequences for greenhouse-gas emissions. For example, deliberate, planned measures to reduce present energy use in response to society's perception of future climate change as a key political and economic issue could result in a (probably small)

Table 4.2 Examples of positive and negative human feedbacks: how changes in climate can cause people to increase or decrease greenhouse-gas emissions

Meteorological measure of climate change	Impact on human wellbeing	Human response	Consequence for greenhouse-gas emissions	Feedback
Higher overall temperatures	Thermal discomfort in buildings	Increased use of air-conditioning for cooling	Higher energy consumption	Raised emissions; positive feedback
Drier climate in populated coastal regions	Increased water shortage	Desalination	Higher energy consumption	Raised emissions; positive feedback
Milder spring, autumn	Extended period of thermal comfort in peripheral domestic areas (e.g. conservatory extensions, especially with double glazing)	Space heating added to conservatories for all year use	Higher energy consumption	Raised emissions; positive feedback
Milder winter climates	Improved thermal comfort in buildings	Reduced use of fossil-fuel heating and firewood	Reduced energy consumption and deforestation	Lowered emissions: negative feedback

negative feedback to climate change. However, Lenton and Watson (2011) have argued that a high-energy, high-recycling society is not incompatible with a stable climate; shifts to sustainable-energy production and enhanced recycling could have greater effects on future climate than changes in energy use. Another possible negative feedback is that milder winters in cold climates could reduce the requirement for space heating, currently a major use of fossil-fuel energy in developed countries. Table 4.2 above summarizes some other examples of potential positive and negative human feedbacks on the climate system.

The risk of 'maladaptation' (Barnett and O'Neill, 2010, is a concern, however. This term refers to rational responses to an immediate, local climate-related pressure that aggravate the global climate problem. For example, in some coastal regions where the climate is becoming drier, energy-intensive desalination technology has been adopted to partially make up for freshwater shortage. The net effect on climate of this response to a pressing societal need depends on the energy source. If increased greenhouse-gas emissions result from the energy used to desalinate seawater, a positive feedback ensues. The increased use of air-conditioning in hot years, using greater amounts of greenhouse-gas producing fossil fuels, is another example of a positive feedback. These two examples are direct feedbacks, but human behaviour changes also have unintended consequences. For example, a policy decision to encourage householders to use double glazing for windows in cold climates should in principle reduce household energy consumption and hence greenhouse-gas emissions. However, using double glazing in conservatory extensions has encouraged space heating in conservatories (e.g. Chu and Oreszczyn, 1991), which if single-glazed would be very difficult to heat to a comfortable temperature on a cold winter's day. The paradox is that increasing energy efficiency has often *increased* energy use, rather than reduced it.

The examples shown in Table 4.2 are simple and involve local actions. However, the linkages between humans and the Earth system occur at multiple temporal, spatial and socio-political scales (Turner *et al.*, 2003; Adger *et al.*, 2009; van der Leeuw *et al.*, 2011) through trade, migration, international policy and information exchange.

It is virtually impossible to assess the current sign and magnitude of human feedbacks because of the overwhelming impacts of non-climatic factors, such as poverty and access to natural resources. The IPCC Fourth Assessment Report (IPCC, 2007) made an effort to consider non-climatic factors from the perspective of human vulnerability to climate change. It concluded that decision-makers considering actions for economic development have available to them emissions-increasing or emissions-reducing options, and that it is difficult to determine the impact of development generally on emissions (Yohe *et al.*, 2007). It also concluded that climate change will make it harder to ensure global equity in development, through measures such as achieving the Millenium Development Goals, which embody basic human rights for all people on Earth (www.un.org/millenniumgoals/bkgd.shtml).

In the face of the known current economic uncertainties and development challenges, it is hard to imagine that our lack of quantitative understanding of human feedbacks constitutes a significant obstacle either to sustainable development in general, or to climate change mitigation in particular. However, a better grasp on human feedbacks is important. Human feedbacks can act against climate mitigation policies. Acknowledging the existence of human feedbacks may inform individual decisions, raise awareness of climate change mitigation and improve larger-scale decision-making by taking account of unplanned changes in behaviour.

References

Adger, W. N., Eakin, H., Winkels, A. (2009). Nested and teleconnected vulnerabilities to environmental change. *Frontiers in Ecology and the Environment*, 7, 150–157.

Aldersley, A., Murray, S. J. and Cornell, S. E. (2011). Global and regional analysis of climate and human drivers of wildfire. *Science of the Total Environment*, **409**, 3472–3481.

Amiro, B. D., Orchansky, A. L., Barr, A. G., *et al.* (2006). The effect of post fire stand age on the boreal forest energy balance. *Agricultural and Forest Meteorology*, **140**, 41–50.

Andersson, C. and Engardt, M. (2010). European ozone in a future climate: importance of changes in dry deposition and isoprene emissions. *Journal of Geophysical Research-Atmospheres*, **115**, 13.

Andreae, M. (2007). Atmospheric aerosols versus greenhouse gases in the twenty-first century. *Philosophical Transactions of the Royal Society*, **365**, 1915–1923.

Andreae, M. and Merlet, P. (2001). Emission of trace gases and aerosols from biomass burning. *Global Biogeochemical Cycles*, **15**, 955–966.

Archer, D. (2007). Methane hydrate stability and anthropogenic climate change. *Biogeosciences*, **4**, 521–544.

Archer, D. and Buffett, B. (2005) Time-dependent response of the global ocean clathrate reservoir to climatic and anthropogenic forcing. *Geochemistry, Geophysics, Geosystems*, **6**(3), doi:10.1029/2004GC000854.

Archibald, S., Roy, D. P., van Wilgen, B. W. and Scholes, R. J. (2009). What limits fire? An examination of burnt area in southern Africa. *Global Change Biology*, **15**, 613–630.

Arneth, A., Niinemets, U., Pressley, S. *et al.* (2007). Process-based estimates of terrestrial ecosystem isoprene emissions: incorporating the effects of a direct CO_2–isoprene interaction. *Atmospheric Chemistry and Physics*, **7**, 31–53.

Arneth, A., Monson, R. K., Schurgers, G., Niinemets, U. and Palmer, P. I. (2008). Why are estimates of global terrestrial isoprene emissions so similar (and why is this not so for monoterpenes)? *Atmospheric Chemistry and Physics*, **8**, 4605–4620.

Arneth, A., Harrison, S. P., Zaehle, S. *et al.* (2010a). Terrestrial biogeochemical feedbacks in the climate system. *Nature Geoscience*, **3**, 525–532.

Arneth, A., Sitch, S., Bondeau, A., *et al.* (2010b). From biota to chemistry and climate: towards a comprehensive description of trace gas exchange between the biosphere and atmosphere. *Biogeosciences*, **7**, 121–149.

Arora, V. and Boer, G. (2006). Fire as an interactive component of dynamic vegetation models. *Journal of Geophysical Research*, **110**, G02008.

Arrhenius, S. (1896). On the influence of carbonic acid in the air upon the temperature of the ground. *Philosophical Magazine of Science*, **41**, 237–276.

Baggs, E. M. and Philippot, L. (2009). Microbial terrestrial pathways to N_2O. In *Nitrous Oxide and Climate Change*, ed. K.A. Smith. London: Earthscan.

Baldocchi, D. D., Falge, E., Gu, L. *et al.* (2001). FLUXNET: a new tool to study the temporal and spatial variability of ecosystem-scale carbon dioxide, water vapor and energy flux densities. *Bulletin of the American Meteorological Society*, **82**, 2415–2435.

Ballhorn, U., Siegert, F., Mason, M., and Limin, S. (2009). Derivation of burn scar depths and estimation of carbon emissions with LIDAR in Indonesian peatlands. *Proceedings of the National Academy of Science USA*, **106**, 21213–21218.

Balzter, H., Gerard, F. F., George, C. T. *et al.* (2005). Impact of the Arctic Oscillation pattern on interannual forest fire variability in central Siberia. *Geophysical Research Letters*, **32**, L14709.1–L14709.4

Bange, H. W. (2008). Gaseous nitrogen compounds (NO, N_2O, N_2, NH_3) in the ocean. In *Nitrogen in the Marine Environment*, eds. D. G. Capone, D. A. Bronk, M. R. Mulholland and E.J. Carpenter. Amsterdam: Elsevier, pp. 51–94.

Barnard, R., Leadley, P. W. and Hungate, B. A. (2005). Global change, nitrification, and denitrification: a review. *Global Biogeochemical Cycles*, **19**, GB1007.

Barnett, J. and O'Neill, S. (2010). Maladaptation. *Global Environmental Change*, **20**, 211–213.

Behrenfeld, M., O'Malley, R. T., Siegel, D. A. *et al.* (2006). Climate driven trends in contemporary ocean productivity. *Nature*, **444**, 752–755.

Belyea, L. R. and Malmer, N. (2004). Carbon sequestration in peatland: patterns and mechanisms of response to climate change. *Global Change Biology*, **10**, 1043–1052.

Betts, A. K. (2004). Understanding hydrometeorology using global models. *Bulletin of the American Meteorological Society*, **85**, 1673–1688.

Betts, A. K. and Ball J. H. (1997). Albedo over the boreal forest. *Journal of Geophysical Research*, **102**, 28901–28909.

Betts, R. A. (2000). Offset of the potential carbon sink from boreal forestation by decreases in surface albedo. *Nature*, **408**, 187–190.

Billett, M. F., Palmer, S. M., Hope, D. *et al.* (2004). Linking land–atmosphere–stream carbon fluxes in a lowland peatland system. *Global Biogeochemical Cycles*, **18**, GB1024.

Blodau, C., Basiliko, N. and Moore, T. R. (2004). Carbon turnover in peatland mesocosms exposed to different water table levels. *Biogeochemistry*, **67**, 331–351.

Blyth, E. M., Dolman, A. J. and Noilhan, J. (1994). The effect of forest on mesoscale rainfall: an example from HAPEX-MOBILHY. *Journal of Applied Meteorology*, **33**, 445–454.

Boé, J. and Terray, L. (2008). Uncertainties in summer evapotranspiration changes over Europe and implications for regional climate change. *Geophysical Research Letters*, **35**, L05702.

Bonan, G. B. (2008). Forests and climate change: forcing, feedbacks, and the climate benefit of forests. *Science*, **320**, 1444–1449.

Bonan, G. B., Pollard, D. and Thompson, S.L. (1992). Effects of boreal forest vegetation on global climate. *Nature*, **359**, 716–718.

Bony, S., Colman, R., Kattsov, V. M. *et al.* (2006). How well do we understand and evaluate climate change feedback processes? *Journal of Climate*, **19**, 3445–3482.

Bopp L., Monfray, P., Aumont, O. *et al.* (2001). Potential impact of climate change on marine export production. *Global Biogeochemical Cycles*, **15**, 81–99.

Bousquet, P., Ringeval, B., Pison, I. *et al.* (2011). Source attribution of the changes in atmospheric methane for 2006–2008. *Atmospheric Chemistry and Physics*, **11**, 3689–3700.

Bowman, D., Dingle, J. K., and Johnston, F. H. (2007). Seasonal patterns in biomass smoke pollution and the mid 20th-century transition from Aboriginal to European fire management in northern Australia. *Global Ecology and Biogeography*, **16**, 246–256.

Bowman, D., Balch, J., Artaxo, P. *et al.* (2009). Fire in the Earth system. *Science*, **324**, 481–484.

Brown, R. D. and Mote, P. W. (2009). The response of northern hemisphere snow cover to a changing climate. *Journal of Climate*, **22**, 2124–2145.

Bubier, J. L., Bhatia, G., Moore, T. R., Roulet, N. T. and Lafleur, P. M. (2003). Spatial and temporal variability in growing-season net ecosystem carbon dioxide exchange at a large peatland in Ontario, Canada. *Ecosystems*, **6**, 353–367.

Budyko, M. I. (1956). *Teplovoi Balans Zemnoi Poverkhnosti*. Leningrad: Gidrometeoizdat. (*Heat Balance of the Earth's Surface*), translated by N. A. Stepanova. Washington, DC: US Weather Bureau, 1958.

Butler, J. H., Elkins, J. W., Thompson, T. M. and Egan, K. B. (1989). Tropospheric and dissolved N_2O of the West Pacific and East Indian oceans during the El Niño Southern Oscillation event of 1987. *Journal of Geophysical Research*, **94**(D12), 14865–14877.

Cantarel, A. M., Bloor, J. M. G., Deltroy, N. and Soussana, J. F. (2011). Effects of climate change drivers on nitrous oxide fluxes in an upland temperate grassland. *Ecosystems*, **14**, 223–233.

Cao, M., Gregson, K. and Marshall, S. (1998). Global methane emission from wetlands and its sensitivity to climate change. *Atmospheric Environment*, **32**, 3293–3299.

Carslaw, K. S., Boucher, O., Spracklen, D. V., *et al.* (2010). A review of natural aerosol interactions and feedbacks within the Earth system, *Atmospheric Chemistry and Physics*, **10**, 1701–1737.

Cess, R. D. (1975). Global climate change: an investigation of atmospheric feedback mechanisms. *Tellus*, **27**, 193–198.

Chu, W. L. and Oreszczyn, T. (1991).The energy duality of conservatories: should conservatories be exempt from the building regulations? *Building Technical File*, **32**, 9–13.

Ciais, P., Reichstein, M., Viovy, N. *et al.* (2005). Unprecedented European-level reduction in primary productivity caused by the 2003 heat and drought. *Nature*, **437**, 529–533.

Claussen, M., Brovkin, V. and Ganopolski, A. (2001). Biogeophysical versus biogeochemical feedbacks of large-scale land cover change. *Geophysical Research Letters*, **28**, 1011–1014.

Cochrane, M. (2003). Fire science for rainforests. *Nature*, **421**, 913–919.

Codispoti, L. A. (2010). Interesting times for marine N_2O. *Science*, **327**, 1339–1340.

Codispoti, L. A. and Christensen, J. P. (1985). Nitrification, denitrification and nitrous oxide cycling in the eastern tropical South Pacific Ocean. *Marine Chemistry*, **16**, 277–300.

Cowling, S. A., Betts, R. A., Cox, P. M. *et al.* (2005). Modelling the past and the future fate of the Amazonian forest. In *Tropical Forests and Global Atmospheric Change*, eds. Y. Malhi and O. L. Philips. Oxford: Oxford University Press, pp. 191–198.

Cox, P. M. and Jones, C. J. (2008). Illuminating the modern dance of climate and CO_2. *Science*, **321**, 1642–1644.

Cox, P. M., Betts, R. A., Jones, C. D., Spall, S. A. and Totterdell, I. J. (2000). Acceleration of global warming due to carbon-cycle feedbacks in a coupled climate model. *Nature*, **408**, 184–187.

Christensen, T. R. (1993). Methane emission from arctic tundra. *Biogeochemistry*, **21**, 117–139.

Clark, J. M., Gallego-Sala, A. V., Chapman, S. *et al.* (2010). Assessing the vulnerability of blanket peat in Great Britain to climate change using an ensemble of statistical bioclimatic envelope models. *Climate Research*, **41**, 131–150.

Crill, P. M., Harriss, R. C. and Bartlett, K. B. (1991). Methane fluxes from terrestrial wetland environments. In *Microbial Production and Consumption of Trace Gases*, ed. W. Whitman. Washington, DC: American Society of Microbiology, pp. 91–109.

Dai, A. (2011). Drought under global warming: a review. *WIREs Climate Change*, **2**, 45–65.

Daulat, W. E. and Clymo, R. S. (1998). Effects of temperature and watertable on the efflux of methane from peatland surface cores. *Atmospheric Environment*, **32**, 3207–3218.

Davin, E. L., de Noblet-Ducoudré, N. and Friedlingstein, P. (2007). Impact of land cover change on surface climate: relevance of the radiative forcing concept, *Geophysical Research Letters*, **34**, L13702.

Denman, K. L., Brasseur, G., Chidthaisong, A. *et al.* (2007). Couplings between changes in the climate system and biogeochemistry. In *Climate Change 2007: The Physical Science Basis. Contribution of Working Group I to the Fourth Assessment Report of the Intergovernmental Panel on Climate Change*, eds. S. Solomon, D. Qin, M. Manning *et al.* Cambridge: Cambridge University Press.

Dirmeyer, P. A., Gao, X., Zhao, M. *et al.* (2006a). GSWP-2: multimodel analysis and implications for our perception of the land surface. *Bulletin of the American Meteorological Society*, **87**, 1381–1397.

Dirmeyer, P. A., Koster, R. D. and Guo, Z. (2006b). Do global models properly represent the feedback between land and atmosphere? *Journal of Hydrometeorology*, **7**, 1177–1198.

Duce, R. A., LaRoche, J., Altieri, K. *et al.* (2008). Impacts of atmospheric anthropogenic nitrogen on the open ocean. *Science*, **320**, 893–897.

Dufresne, J.-L., Friedlingstein, P., Berthelot, M. *et al.* (2002). On the magnitude of positive feedback between future climate change and the carbon cycle. *Geophysical Research Letters*, **29**, 10.1029/2001GL013777.

Ek, M. B. and Holtslag, A. A. M. (2004). Influence of soil moisture on boundary layer cloud development. *Journal of Hydrometeorology*, **5**, 86–99.

Eliseev, A., Mokhov, I., Arzhanov, M., Demchenko, P. and Denisov, S. (2008). Interaction of the methane cycle and processes in wetland ecosystems in a climate model of intermediate complexity. *Izvestiya Atmospheric and Oceanic Physics*, **44**, 139–152.

Ellis, E. C. and N. Ramankutty (2008). Putting people in the map: anthropogenic biomes of the world. *Frontiers in Ecology and the Environment*, **6**, 439–447.

Findell, K. L. and Eltahir, E. A. B. (2003). Atmospheric controls on soil moisture–boundary layer interactions. Part I: framework development. *Journal of Hydrometeorology*, **4**, 552–569.

Fischer, E. M., Seneviratne, S. I., Vidale, P. L., Lüthi, D. and Schär, C. (2007). Soil moisture–atmosphere interactions during the 2003 European summer heatwave, *Journal of Climate*, **20**, 5081–5099.

Flannigan, M. and de Groot, W. (2009). Forest fires and climate change in the circumboreal forest. Proceedings, ESA-iLEAPS ALANIS Workshop, Austrian Academy of Sciences, Vienna, 20 April 2009.

Flannigan, M., Krawchuk, M., de Groot, W., Wotton, M. A. and Gowman, L. M. (2009). Implications of changing climate for global wildland fire. *International Journal of Wildland Fire*, **18**, 483–507.

Folberth, G. A., Hauglustaine, D. A., Lathiere, J. and Brocheton, F. (2006). Interactive chemistry in the Laboratoire de Meteorologie Dynamique general circulation model: model description and impact analysis of biogenic hydrocarbons on tropospheric chemistry. *Atmospheric Chemistry and Physics*, **6**, 2273–2319.

Forster, P., Ramaswamy, V., Artaxo, P. *et al.* (2007). Changes in atmospheric constituents and in radiative forcing. In *Climate Change 2007: The Physical Science Basis. Contribution of Working Group I to the Fourth Assessment Report of the Intergovernmental Panel on Climate Change*, eds. S. Solomon, D. Qin, M. Manning *et al.* Cambridge: Cambridge University Press.

Frank, D. C., Esper, J., Raible, C. C. *et al.* (2010). Ensemble reconstruction constraints on the global carbon cycle sensitivity to climate. *Nature*, **463**, 527–530.

Freeman, C., Ostle, N. and Kang, H. (2001). An enzymic 'latch' on a global carbon store – a shortage of oxygen locks up carbon in peatlands by restraining a single enzyme. *Nature*, **409**, 149.

Frenzel, P. (2000). Plant-associated methane oxidation in rice fields and wetlands. In *Advances in Microbial Ecology*, ed. B. Schink. New York: Kluwer Academic/Plenum Publications, vol. 16, pp. 85–114.

Friedlingstein, P. and Prentice, I. C. (2010). Carbon–climate feedbacks: a review of model and observation based estimates. *Current Opinion in Environmental Sustainability*, **2**, 251–257.

Friedlingstein, P., Bopp. L., Ciais, P. *et al.* (2001). Positive feedback between future climate change and the carbon cycle. *Geophysical Research Letters*, **28**, 1543–1546.

Friedlingstein, P., Dufresne, J.-L., Cox, P. M., and Rayner P. (2003). How positive is the feedback between climate change and the carbon cycle? *Tellus B*, **55**, 692–700.

Friedlingstein, P., Cox, P. M., Betts, R. *et al.* (2006). Climate–carbon cycle feedback analysis, results from the C^4MIP model intercomparison. *Journal of Climate*, **19**, 3337–3353.

Gallego-Sala, A. V., Clark, J., House, J. *et al.* (2009). Bioclimatic envelope modelling of the present global distribution of boreal peatlands. Proceedings, BIOGEOMON: 6th International Symposium on Ecosystem Behaviour, University of Helsinki, Finland, 29 June–3 July.

Gallego-Sala, A. V., Clark, J., House, J. I. *et al.* (2010). Bioclimatic envelope model of climate change impacts on blanket peatland distribution in Great Britain. *Climate Research*, **45**, 151–162.

Gedney, N., Cox, P. M. and Huntingford, C. (2004). Climate feedback from wetland methane emissions. *Geophysical Research Letters*, **31**, L20503.

Gignac, L. D., Nicholson, B. J. and Bayley, S. E. (1998). The utilization of bryophytes in bioclimatic modeling: predicted northward migration of peatlands in the Mackenzie River Basin, Canada, as a result of global warming. *The Bryologist*, **101**, 572–587.

Gill, A. M. and Catling, P. (2002). Fire regimes and biodiversity of forested landscapes of southern Australia. In *Flammable Australia: Fire Regimes and Biodiversity of a Continent*, eds. R. Bradstock, J. Williams and A. M. Gil. Cambridge: Cambridge University Press, pp. 351–369.

Gorham, E. (1991). Northern peatlands – role in the carbon-cycle and probable responses to climatic warming. *Ecological Applications*, **1**, 182–195.

Gregory, J. M., Jones, C. D., Cadule, P. and Friedlingstein, P. (2009). Quantifying carbon-cycle feedbacks. *Journal of Climate*, **22**, doi:10.1175/2009JCLI2949.1.

Groisman, P. Y., Karl, T. R. and Knight, R. W. (1994). Observed impact of snow cover on the heat balance and the rise of continental spring temperatures. *Science*, **263**, 198–200.

Guenther, A. (2007). Estimates of global terrestrial isoprene emissions using MEGAN (Model of Emissions of Gases and Aerosols from Nature) (Corrigendum to vol. 6, p. 3181–3186, 2006). *Atmospheric Chemistry and Physics*, **7**, 4327–4327.

Guo, Z. C., Dirmeyer, P. A., Koster, R. D., *et al.* (2006). GLACE: the Global Land–Atmosphere Coupling Experiment. Part II: analysis. *Journal of Hydrometeorology*, **7**, 611–625.

Hall, A. and Qu, X. (2006). Using the current seasonal cycle to constrain snow albedo feedback in future climate change. *Geophysical Research Letters*, **33**, L03502.

Hansen, J., Lacis, A., Rind, D. *et al.* (1984). Climate sensitivity: analysis of feedback mechanisms. In *Climate Processes and Climate Sensitivity*, eds. J. E. Hansen and T. Takahashi. AGU Geophysical Monograph 29, pp. 130–163.

Harris, S., Tapper, N., Packham, D., Orlove, B. and Nicholls, N. (2008). The relationship between the monsoonal summer rain and dryseason fire activity of northern Australia. *International Journal of Wildland Fire*, **17**, 674–684.

Heald, C. L., Wilkinson, M. J., Monson, R. K. *et al.* (2009). Response of isoprene emission to ambient CO_2 changes and implications for global budgets. *Global Change Biology*, **15**, 1127–1140.

Hein, R., Crutzen, P. J. and Heimann, M. (1997). An inverse modeling approach to investigate the global atmospheric methane cycle. *Global Biogeochemical Cycles*, **11**, 43–76.

Hooijer, A., Silvius, M., Wösten, H. and Page, S. (2006). PEAT-CO_2, Assessment of CO_2 emissions from drained SE Asia. Delft Hydraulics Report Q3943.

IPCC (2007). *Climate Change 2007: The Physical Science Basis. Contribution of Working Group I to the Fourth Assessment Report of the Intergovernmental Panel on Climate Change*, eds. S. Solomon, D. Qin, M. Manning *et al.* Cambridge: Cambridge University Press.

Ise, T., Dunn, A. L., Wofsy, S. C. and Moorcroft, P. R. (2008). High sensitivity of peat decomposition to climate change through water-table feedback. *Nature Geoscience*, **1**, 763–766.

Jaeger, E. B. and Seneviratne, S. I. (2011). Impact of soil moisture–atmosphere coupling on European climate extremes and trends in a regional climate model. *Climate Dynamics*, **36**, doi:10.1007/s00382-010-0780-8.

Jickells, T., An, Z. S., Anderson, K. K. *et al.* (2005). Global iron connections between desert dust, ocean biogeochemistry and climate. *Science*, **308**, 67–71.

Jiménez, C., Prigent, C., Mueller, B. *et al.* (2011). Global intercomparison of 12 land surface heat flux estimates. *Journal of Geophysical Research*, **116**, D02102.

Joos, F. and Prentice, I. C. (2004). A paleo-perspective on changes in atmospheric CO_2 and climate. In *The global carbon cycle: integrating humans, climate and the natural world, eds.* C. B. Field and M. R. Raupach. Washington, DC: Island Press, pp. 165–186.

Kammann, C., Müller, C., Grünhage, L. and Jäger, H. (2008). Elevated CO_2 stimulates N_2O emissions in permanent grassland. *Soil Biology and Biochemistry*, **40**, 2194–2205.

Kelso, B. H. L., Smith, R. V., Laughlin, R. J. and Lennox, S. D. (1997). Dissimilatory nitrate reduction in anaerobic sediments leading to river nitrite accumulation. *Applied and Environmental Microbiology*, **63**, 4679–4685.

Kesik, M., Brüggemann, N., Forkel, R. *et al.* (2006). Future scenarios of N_2O and NO emissions from European forest soils. *Journal of Geophysical Research*, **111**, G02018.

Kitzberger, T., Swetnam, T. W. and Veblen, T. T. (2001). Interhemispheric synchrony of forest fires and the El Niño-Southern Oscillation. *Global Ecology and Biogeography*, **10**, 315–326.

Kloster, S., Mahowald, N. M., Randerson, J. T. *et al.* (2010). Fire dynamics during the 20th century simulated by the Community Land Model. *Biogeosciences*, **7**, 1877–1902.

Koster, R. D., Suarez, M. J., Higgins, W. and Van den Dool, H. M. (2003). Observational evidence that soil moisture variations affect precipitation. *Geophysical Research Letters*, **30**, 1241.

Koster, R. D., Dirmeyer, P. A., Guo, Z. C. *et al.* (2004). Regions of strong coupling between soil moisture and precipitation. *Science*, **305**, 1138–1140.

Koster, R. D., Guo, Z. C., Dirmeyer, P. A. *et al.* (2006). GLACE: the Global Land– Atmosphere Coupling Experiment. Part I: overview. *Journal of Hydrometeorology*, **7**, 590–610.

Koster, R. D., Mahanama, S., Yamada, T. J. *et al.* (2010). The contribution of land initialization to subseasonal forecast skill: first results from the GLACE-2 project. *Geophysical Research Letters*, **37**, L02402.

Koven, C., Ringeval, B., Friedlingstein, P. *et al.* (2011). Permafrost carbon–climate feedbacks accelerate global warming. *Proceedings of the National Academy of Sciences USA*, **108**, 14769–14774.

Lähteenoja, O., Ruokolainen, K., Schulman, L. and Oinonen, M. (2009). Amazonian peatlands: an ignored C sink and potential source. *Global Change Biology*, **15**, 2311–2320.

Lathière, J., Hauglustaine, D. A., Friend, A. D. *et al.* (2006). Impact of climate variability and land use changes on global biogenic volatile organic compound emissions. *Atmospheric Chemistry and Physics*, **6**, 2129–2146.

Lawrence, D. and Slater, A. (2010). The contribution of snow condition trends to future ground climate. *Climate Dynamics*, **34**, 969–981.

Lean, J., Bunton, C. B., Nobre, C. A. and Rowntree, P. R. (1996). The simulated impact of Amazonian deforestation on climate using measured ABRACOS vegetation characteristics. In *Amazonian Deforestation and Climate*, eds. J. H. C. Gash, C. A. Nobre, J. M. Roberts and R. L. Victoria. Chichester: Wiley, pp. 549–576.

Lehner, B. and Döll, P. (2004). Development and validation of a global database of lakes, reservoirs and wetlands. *Journal of Hydrology*, **296**, 1–22.

Lehsten, V., Tansey, K., Balzter, H. *et al.* (2009). Estimating carbon emissions from African wildfires. *Biogeosciences*, **5**, 3091–3122.

Lenton, T. and Watson, A. (2011). *Revolutions that made the Earth*, Oxford: Oxford University Press.

Le Page, Y., Pereira, J. M. C., Trigo, R. *et al.* (2008). Global fire activity patterns (1996–2006) and climatic influence: an analysis using the World Fire Atlas. *Atmospheric Chemistry and Physics*, **8**, 1911–1924.

Lenihan, J. and Neilson, R. (1998). Simulating broad-scale fire severity in a dynamic global vegetation model. *Science*, **72**, 91–103.

Lenihan, J. M., Bachelet, D., Drapek, R. J. and Neilson, R. P. (2008). The response of vegetation distribution, ecosystem productivity, and fire to climate change scenarios for California. *Climate Change*, **87**, 215–230.

Li, W. H., Dickinson, R. E., Fu, R. *et al.* (2007). Future precipitation changes and their implications for tropical peatlands. *Geophysical Research Letters*, **34**, L01403.

Los, S. O., Weedon, G. P., North, P. R. J. *et al.* (2006). An observation-based estimate of the strength of rainfall–vegetation interactions in the Sahel. *Geophysical Research Letters*, **33**, L16402.

Luoto, M., Fronzek, S. and Zuidhoff, F. S. (2004). Spatial modelling of palsa mires in relation to climate in northern Europe. *Earth Surface Processes and Landforms*, **29**, 1373–1387.

Maltby, E. and Immirzi, P. (1993). Carbon dynamics in peatlands and other wetland soils: regional and global perspectives. *Chemosphere*, **27**, 999–1023.

Manabe, S. and Wetherald, R. T. (1967). Thermal equilibrium of the atmosphere with a given distribution of relative humidity. *Journal of Atmospheric Science*, **24**, 241–259.

Marlon, J. R., Bartlein, P. J., Carcaillet, C. *et al.* (2008). Climate and human influences on global biomass burning over the past two millennia. *Nature Geoscience*, **1**, 697–702.

Melillo, J. M., McGuire, A. D., Kicklighter, D.W. *et al.* (1993). Global climate change and terrestrial net primary production. *Nature*, **363**, 234–240.

Mercado, L., Bellouin, N., Sitch, S. *et al.* (2009). Impact of changes in diffuse radiation on the global land carbon sink. *Nature*, **458**, 1014–1017.

Mueller, B., Seneviratne, S. I., Jimenez, C. *et al.* (2011). Evaluation of global observations-based evapotranspiration datasets and IPCC AR4 simulations. *Geophysical Research Letters*, **38**, L06402.

Naqvi, S. W. A., Bange, H. W., Farías, L. *et al.* (2010). Marine hypoxia/anoxia as a source of CH_4 and N_2O. *Biogeosciences*, **7**, 2159–2190.

NRC – National Research Council (1979). *Carbon Dioxide and Climate: A Scientific Assessment.* Washington, DC: National Academy of Sciences, Climate Research Board.

Oki, T. and Kanae, S. (2006). Global hydrological cycles and world water resources. *Science*, **313**, 1068–1072.

Pacifico, F., Harrison, S. P., Jones, C. and Sitch, S. (2009). Isoprene emissions and climate. *Atmospheric Environment*, **43**, 6121–6135.

Pacifico, F., Folberth, G. A., Jones, C., Harrison, S. P. and Collins, W. J. (2011). Sensitivity of biogenic isoprene emissions to past, present and future environmental conditions and implications for atmospheric chemistry. *Journal of Geophysical Research*, in revision.

Page, S. E., Siegert, F., Rieley, J. O. *et al.* (2002). The amount of carbon released from peat and forest fires in Indonesia during 1997. *Nature*, **420**, 61–65.

Peñuelas, J. and Staudt, M. (2010). BVOCs and global change. *Trends in Plant Science*, **15**, 133–144.

Phillips, O. L., Aragão, L. E. O. C., Lewis, S. L. *et al.* (2009). Drought sensitivity of the Amazon rainforest. *Science*, **323**, 1344–1347.

Pitman, A. J., Durbidge, T. B. and Henderson-Sellers, A. (1993). Assessing climate model sensitivity to prescribed deforested landscapes. *International Journal of Climatology*, **13**, 879–898.

Pitman, A. J., de Noblet-Ducoudre, N., Cruz, F. T. *et al.* (2009). Uncertainties in climate responses to past land cover change: first results from the LUCID intercomparison study. *Geophysical Research Letters*, **36**, L14814.

Prentice, I. C., Kelley, D. I., Foster, P. N. *et al.* (2011). Modeling fire and the terrestrial carbon balance. *Global Biogeochemical Cycles*, **25**, GB3005.

Priemé, A. (1994). Production and emission of methane in a brackish and a fresh-water wetland. *Soil Biology and Biochemistry*, **26**, 7–18.

Pyne, S., Andrews, P. and Laven, R. (1996). *Introduction to Wildland Fire.* New York, NY: Wiley.

Randerson, J., Liu, T., Flanner, M. *et al.* (2006). The impact of boreal forest fire on climate warming. *Science*, **314**, 1130–1132.

Reichstein, M., Ciais, P., Papale, D. *et al.* (2007). Reduction of ecosystem productivity and respiration during the European summer 2003 climate anomaly: a joint flux tower, remote sensing and modeling analysis. *Global Change Biology*, **13**, 634–651.

Repo, M. E., Sanna, S., Lind, S. *et al.* (2009). Large N_2O emissions from cryoturbated peat soil in tundra. *Nature Geoscience*, **2**, 193–196.

Ringeval, B., de Noblet-Ducoudré, N., Ciais, P. *et al.* (2010). An attempt to quantify the impact of changes in wetland extent on methane emissions on the seasonal and interannual time scales. *Global Biogeochemical Cycles*, **24**, GB2003.

Ringeval, B., Friedlingstein, P., Koven, C., Ciais, P. and de Noblet-Ducoudré, N. (2011). Climate–CH_4 feedback from wetlands and its interaction with the climate–CO_2 feedback. *Biogeosciences Discussions*, **8**, 3203–3251.

Ritchie, G. A. F. and Nicholas, D. J. D. (1972). Identification of the sources of nitrous oxide produced by oxidative and reductive processes in *Nitrosomonas europaea. Biochemical Journal*, **126**, 1181–1191.

Scheffer, M., Brovkin, V. and Cox, P. (2006). Positive feedback between global warming and atmospheric CO_2 concentration inferred from past climate change. *Geophysical Research Letters*, **33**, L10702.

Schmittner, A., Urban, N. M., Shakun, J. D. *et al.* (2011). Climate sensitivity estimated from temperature reconstructions of the last glacial maximum. *Science*, **334**, 1385–1388.

Scholze, M., Knorr, W., Arnell, N. and Prentice, I. C. (2006). A climate-change risk analysis for world ecosystems. *Proceedings of the National Academy of Sciences USA*, **10**, 13116–13120.

Schultz, M., Heil, A., Hoelzemann, J. *et al.* (2008). Global emissions from wildland fires from 1960 to 2000. *Global Biogeochemical Cycles*, **22**, GB2002.

Sellers, W. D. (1969). A global climate model based on the energy balance of the Earth–atmosphere system. *Journal of Applied Meteorology*, **8**, 392–400.

Seneviratne, S. I., Lüthi, D., Litschi, M. and Schär, C. (2006a). Land–atmosphere coupling and climate change in Europe. *Nature*, **443**, 205–209.

Seneviratne, S. I., Koster, R. D., Guo, Z. *et al.* (2006b). Soil moisture memory in AGCM simulations: analysis of Global Land–Atmosphere Coupling Experiment (GLACE) data. *Journal of Hydrometeorology*, **7**, 1090–1112.

Seneviratne, S. I., Corti, T., Davin, E. L. *et al.* (2010). Investigating soil moisture–climate interactions in a changing climate: a review. *Earth-Science Reviews*, **99**, 125–161.

Shakhova, N., Semiletov, I., Salyuk, A. *et al.* (2010). Extensive methane venting to the atmosphere from sediments of the east Siberian Arctic shelf. *Science*, **237**, 1246–1250.

Shindell, D. T., Walter, B. P. and Faluvegi, G. (2004). Impacts of climate change on methane emissions from wetlands. *Geophysical Research Letters*, **31**, L21202.

Shukla, J. and Mintz, Y. (1982). The influence of land-surface evapotranspiration on the Earth's climate. *Science*, **215**, 1498–1501.

Siegert, F., Ruecker, G., Hinrichs, A. and Hoffmann, A. (2001). Increased fire impacts in logged forests during El Niño driven fires. *Nature*, **414**, 437–440.

Sokolov, A. P., Stone, P. H., Forest, C. E. *et al.* (2009). Probabilistic forecast for twenty-first century climate based on uncertainties in emissions (without policy) and climate parameters. *Journal of Climate*, **22**, 5175–5204.

Spessa, A. and Fisher, R. (2010). On the relative role of fire and rainfall in determining vegetation patterns in tropical savannas: a simulation study. *Geophysical Research Abstracts*, **12**, EGU2010–7142–6.

Spessa, A., McBeth, B. and Prentice, C. (2005). Relationships among fire frequency, rainfall and vegetation patterns in the wet–dry tropics of northern Australia: an analysis based on NOAA–AVHRR data. *Global Ecology and Biogeography*, **14**, 439–454.

Spessa, A., Weber, U., Langner, A. *et al.* (2010). Fire in the vegetation and peatlands of Borneo, 1997–2007: patterns, drivers and emissions from biomass burning. *Geophysical Research Abstracts*, **12**, EGU2010–7149.

Stocks, B. J., Lawson, B. D., Alexander, M. E. *et al.* (1989). The Canadian Forest Fire Danger Rating System: an overview. *Forest Chronicle*, **65**, 450–457.

Strack, M. and Waddington, J. M. (2007). Response of peatland carbon dioxide and methane fluxes to a water table drawdown experiment. *Global Biogeochemical Cycles*, **21**, GB1007.

Stramma, L., Johnson, G. C., Sprintall, J. and Mohrholz, V. (2008). Expanding oxygen-minimum zones in the tropical oceans. *Science*, **320**, 655–658.

Teuling, A. J. and Seneviratne, S. I. (2008). Contrasting spectral changes limit albedo impact on land–atmosphere coupling during the 2003 European heat wave. *Geophysical Research Letters*, **35**, L03401.

Teuling, A. J., Hirschi, M., Ohmura, A. *et al.* (2009). A regional perspective on trends in continental evaporation. *Geophysical Research Letters*, **36**, L02404.

Thonicke, K. and Cramer, W. (2006). Long-term trends in vegetation dynamics and forest fires in Brandenburg (Germany) under a changing climate. *Natural Hazards*, **38**, 283–300.

Thonicke, K., Venevsky, S., Sitch, S. and Cramer, W. (2001). The role of fire disturbance for global vegetation dynamics: coupling fire into a Dynamic Global Vegetation Model. *Global Ecology and Biogeography*, **10**, 661–678.

Thonicke, K., Spessa, A., Prentice, I. C., Harrison, S. and Carmona-Moreno, C. (2010). The influence of vegetation, fire spread and fire behaviour on global biomass burning and trace gas emissions. *Biogeosciences*, **7**, 697–743.

Thornton, P. E., Doney, S. C., Lindsay, K. *et al.* (2009). Carbon–nitrogen interactions regulate climate–carbon cycle feedbacks: results from an atmosphere–ocean general circulation model. *Biogeosciences*, **6**, 2099–2120.

Tosca, M. G., Randerson, J. T., Zender, C. S., Flanner, M. G. and Rasch, P. J. (2010). Do biomass burning aerosols intensify drought in equatorial Asia during El Nino? *Atmospheric Chemistry and Physics*, **10**, 3515–3528.

Trenberth, K. E., Fasullo, J. T. and Kiehl, J. (2009). Earth's global energy budget. *Bulletin of the American Meteorological Society*, **90**, 311–323.

Trigo, R. M., Pereira, J. M. C., Pereira, M. G. *et al.* (2006). Atmospheric conditions associated with the exceptional fire season of 2003 in Portugal. *International Journal of Climatology*, **26**, 1741–1757.

Turetsky, M. R., Wieder, R. K., Vitt, D. H., Evans, R. J. and Scott, K. D. (2007). The disappearance of relict permafrost in boreal north America: effects on peatland carbon storage and fluxes. *Global Change Biology*, **13**, 1922–1934.

Turner, B. L., Kasperson, R. E., Matson, P. A. *et al.* (2003). A framework for vulnerability analysis in sustainability science. *Proceedings of the National Academy of Sciences USA*, **100**, 8074–8079.

van der A, R. J., Eskes, H. J., Boersma, K. F. *et al.* (2008). Trends, seasonal variability and dominant NO_x source derived from a ten year record of NO_2 measured from space. *Journal of Geophysical Research*, **113**, D04302.

van der Leeuw, S., Costanza, R. Aulenbach, S. *et al.* (2011). Toward an integrated history to guide the future. *Ecology and Society*, **16**, 2.

van der Werf, G. R., Randerson, J. T., Giglio, L., Gobron, N. and Dolman, A. J. (2008). Climate controls on the variability of fires in the tropics and subtropics. *Global Biogeochemical Cycles*, **22**, GB3028.

van der Werf, G. R., Randerson, J., Giglio, L. *et al.* (2010). Global fire emissions and the contribution of deforestation, savanna, forest, agricultural, and peat fires (1997–2009). *Atmospheric Chemistry and Physics*, **8**, 7673–7696.

van Groenigen, K. J., Osenberg, C. W. and Hungate, B. A. (2011). Increased soil emissions of potent greenhouse gases under increased atmospheric CO_2. *Nature*, **475**, 214–218.

Vautard, R., Yiou, P., D'Andrea, F. *et al.* (2007). Summertime European heat and drought waves induced by wintertime Mediterranean rainfall deficit. *Geophysical Research Letters*, **34**, L07711.

Venevsky, S., Thonicke, K., Sitch, S. and Cramer, W. (2002). Simulating fire regimes in human-dominated ecosystems: Iberian Peninsula case study. *Global Change Biology*, **8**, 984–998.

Vitousek, P. M. and Howarth, R. W. (1991). Nitrogen limitation on land and in the sea: how can it occur? *Biogeochemistry*, **13**, 87–115.

Wania, R., Ross, I. and Prentice, I. C. (2009a). Integrating peatlands and permafrost into a dynamic global vegetation model: 1. Evaluation and sensitivity of physical land surface processes. *Global Biogeochemical Cycles*, **23**, GB3014.

Wania, R., Ross, I. and Prentice, I. C. (2009b). Integrating peatlands and permafrost into a dynamic global vegetation model: 2. Evaluation and sensitivity of vegetation and carbon cycle processes. *Global Biogeochemical Cycles*, **23**, GB3015.

Westerling, A. L., Hidalgo, H. G., Cayan, D. R. and Swetnam, T. W. (2006). Warming and earlier spring increase western US forest wildfire activity. *Science*, **313**, 940–943.

Wieder, R. K. and Vitt, D. H. (2006). *Boreal Peatland Ecosystems*. Berlin: Springer-Verlag.

Xu-Ri and Prentice, I. C. (2008). Terrestrial nitrogen cycle simulation with a dynamic global vegetation model. *Global Change Biology*, **14**, 1745–1764.

Xu-Ri, Prentice, I. C., Spahni, R. and Niu, H. S. (submitted). Modelling terrestrial nitrous oxide emissions and implications for climate feedback.

Yohe, G. W., Lasco, R. D., Ahmad, Q. K. *et al.* (2007). Perspectives on climate change and sustainability. In *Climate Change 2007: Impacts, Adaptation and Vulnerability. Contribution of Working Group II to the Fourth Assessment Report of the Intergovernmental Panel on Climate Change*, eds. M. L. Parry, O. F. Canziani, J. P. Palutikof, P. J. van der Linden and C. E. Hanson. Cambridge: Cambridge University Press, pp. 811–841.

Yu, Z. (2011). Holocene carbon flux histories of the world's peatlands: global carbon-cycle implications. *The Holocene*, **21**, 761–774.

Yu, Z., Loisel, J., Brosseau, D. P., Beilman, D. W. and Hunt, S. J. (2010). Global peatland dynamics since the last glacial maximum. *Geophysical Research Letters*, **37**, L13402.

Yurganov, L. N., McMillan, W. W., Dzhola, A. V. *et al.* (2008). Global AIRS and MOPITT CO measurements: validation, comparison, and links to biomass burning variations and carbon cycle. *Journal of Geophysical Research*, **113**, D09301.

Zaehle, S., Friedlingstein, P. and Friend, A. D. (2010). Terrestrial nitrogen feedbacks may accelerate future climate change. *Geophysical Research Letters*, **37**, L01401.

Zaitchik, B. F., Macaldy, A. K., Bonneau, L. R. and Smith, R. B. (2006). Europe's 2003 heat wave: a satellite view of impacts and land atmosphere feedbacks. *International Journal of Climatology*, **26**, 743–769.

Chapter

5 Earth system models: a tool to understand changes in the Earth system

Marko Scholze, J. Icarus Allen, William J. Collins, Sarah E. Cornell, Chris Huntingford, Manoj M. Joshi, Jason A. Lowe, Robin S. Smith and Oliver Wild

This chapter provides an overview of Earth system models, the various model 'flavours', their state of development including model evaluation, benchmarking and optimization against observational data and their application to climate change issues.

5.1 Introduction

The Earth system can be conceptualized as a suite of interacting physical, chemical, biological and anthropogenic processes that regulate the planet's flow of matter and energy. Earth system models (ESMs; Box 5.1) are built to mirror these processes. In fact, ESMs are the only tool available to the scientific community to investigate the system properties of the Earth, as we do not have an alternative planet to manipulate that could serve as a scientist's laboratory.

The term 'Earth system model' is commonly used to describe coupled land–ocean–atmosphere models that include interactive biogeochemical components. Such models have developed progressively from the physical climate models first created in the 1960s and 1970s. Conventional climate models apply physical laws to simulate the general circulation of atmosphere and ocean. As our understanding of the natural and anthropogenic controls on climate has grown, and given the steady advances in computing power, global climate models have been extended to include more comprehensive representations of biological and geochemical processes, involving the addition of the various interacting components of the Earth system with their own feedback mechanisms. Figure 5.1 shows the conceptual differences between a conventional global coupled atmosphere–ocean general circulation model (AOGCM) and an ESM. In terms of the coupling between components, ESMs are more complex, and they have correspondingly higher computational demands.

Earth system models are constantly being developed and are including progressively more interacting components of the geosphere, or the physical climate system, and the biosphere, or the living world. Figure 5.2 displays the historical evolution of physical-based climate models towards ESMs by the progressive introduction of new elements. In common with climate models, the core concept of Earth system models is the representation of transport and exchange of energy (heat and water), as laid out in Chapter 2. However, ESMs have been developed to include the dynamics of Earth's fluid components (the oceans and atmosphere); physical structures (continents, mountainous elevation); albedo (cloud, aerosol, land surface, ice); controls on natural greenhouse gases (land-cover types or biomes, the carbon and nitrogen cycles). They also include linked physical and biogeochemical processes (for example, the role of land vegetation in the formation of atmospheric aerosol, and the role of ocean circulation in marine biological productivity). In physical climate models, the only interactively computed greenhouse gas is water vapour. The additional components in ESMs allow the interactive computation of a whole range of greenhouse gases and other atmospheric constituents. They provide insights into many processes acting through the energy, water and carbon balance of land and ocean surfaces, changes in atmospheric chemistry, and the generation and fate of atmospheric aerosols as detailed in Chapter 4. Part of the underlying rationale for the construction and use of ESMs

Understanding the Earth System: Global Change Science for Application, eds. Sarah E. Cornell, I. Colin Prentice, Joanna I. House and Catherine J. Downy. Published by Cambridge University Press

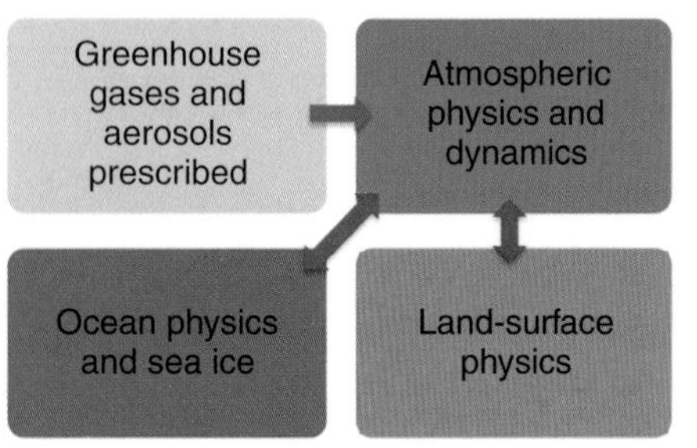

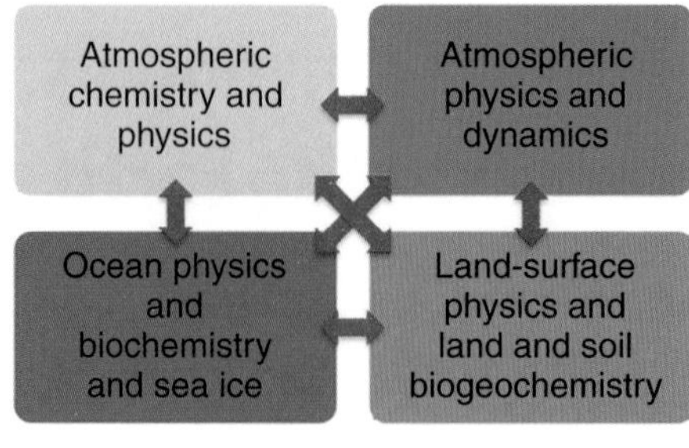

Figure 5.1 Conceptual schematic of an Earth system model compared with a 'traditional' atmosphere-ocean coupled general circulation model. Red arrows denote interactively coupled components, the blue arrow denotes prescribed input data.

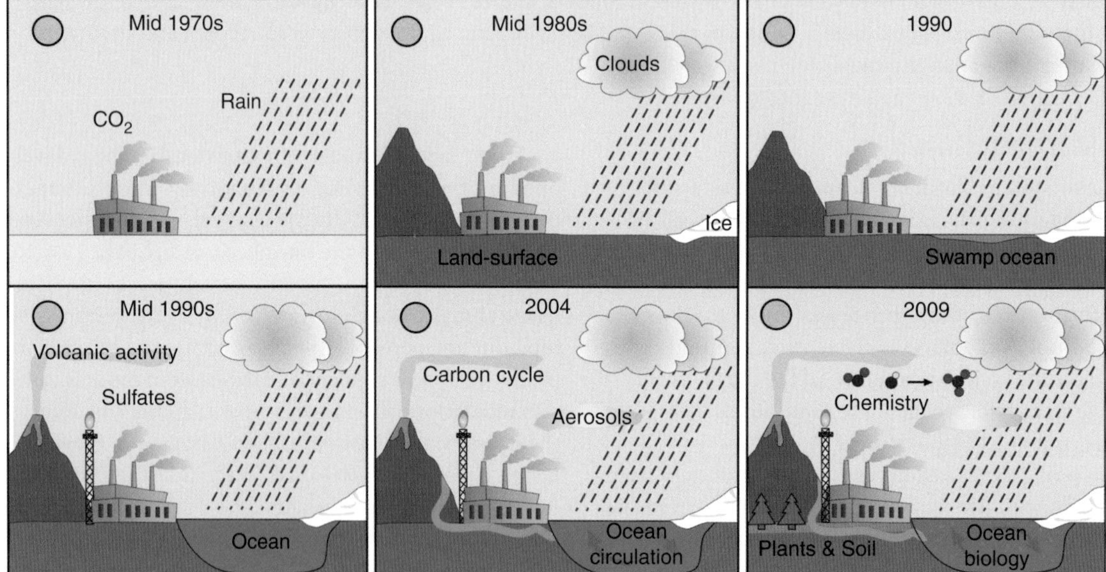

Figure 5.2 Historical evolution of climate models towards ESMs by addition of different biogeochemical components since the 2000s.

has been to address feedback processes, and thus to derive more realistic estimates of future climate change under different anthropogenic emission scenarios. The new generation of ESMs include a much larger range of climatic feedbacks, mediated through biophysical and biogeochemical processes, than could be addressed using the traditional physical climate models alone.

The internal feedbacks in the Earth system are important in providing the necessary conditions for the self-regulation of the system. Previous chapters have described the rationale and evidence for Earth as an integrated system. One prominent example of Earth system behaviour where feedbacks are paramount is the high correlation seen in the ice-core record between temperature and atmospheric greenhouse-gas concentrations over glacial–interglacial cycles. The more complete description of the Earth system in numerical models allows a better interpretation of observational Earth system records such as these, demonstrating that the atmospheric carbon dioxide (CO_2) concentration truly depends on the coupled physical–biogeochemical land–atmosphere–ocean system. Recent measurements from the EPICA ice core (Lüthi *et al.*, 2008; Loulergue *et al.*, 2008) show that the concordant variations of inferred local temperature and greenhouse-gas concentrations – here CO_2 and methane (CH_4) have been bounded within nearly constant upper and lower limits over the last 800,000 years. This observation highlights the self-regulating nature of Earth system prior to the industrial era. The climate variations themselves (the glacial–interglacial cycles), are paced by variations in insolation, but it is the internal dynamics and feedbacks of the Earth system that keep the variations within their observed bounds, and maintain the planet in a habitable state. Therefore, for future climate projections it is of paramount importance to understand the interactions of the various processes in the Earth system that are ultimately responsible for this observed self-regulatory behaviour.

In order to quantify the strength of a feedback analytically, the concept of gain, commonly used in electronics, has been introduced to the physical climate system (Hansen *et al.*, 1994), and more recently has been adapted for biogeochemical feedbacks (Friedlingstein *et al.*, 2006). The fundamentals of gain theory are laid out in Chapter 4. The main advantage of this linear feedback concept is that it allows an easy comparison of the strength of the different feedback processes. However, it has also its limitations in the complex system of the real Earth. Some feedbacks change in magnitude with the state of the Earth system (e.g. Hargreaves *et al.*, 2007; Senior and Mitchell, 2000). In this case, the concept of gain holds only for small perturbations around an equilibrium state. In addition, the processes responsible for a certain feedback mechanism may act on different timescales and thus include a dynamical component that is not captured within the gain concept. Hallegatte *et al.* (2006) demonstrated that neglecting these dynamical elements and using equilibrium states in the calculation of the feedback strength does not give a fully adequate characterization of the water-vapour feedback.

The current generation of ESMs can provide improved predictions of the climate response to both external and anthropogenic perturbations in the forcing. These predictions of climate change impacts are needed, as the impacts are often forcing agents themselves, at least on a regional scale. For example, changes in the hydrological cycle in the Sahel region impact the growth of vegetation, which itself feeds back on the hydrological cycle through changes in evapotranspiration (Zeng and Neelin, 2000). And since the beginning of anthropogenic fossil-fuel emissions, Earth's delicate self-regulating system, evident in the ice-core records and other palaeoclimatic observations, has been perturbed: human activities have created strong internal forcings on the system with their own feedback mechanisms (which are briefly discussed in Chapter 4). Earth System models need to incorporate the growing understanding of Earth's dynamic and interconnected processes, in order to enable these processes to be investigated more fully and, in turn, to provide new insights that may transform our assumptions about how the Earth works.

Box 5.1 Definition of Earth system models

Numerical models are a means of representing processes and embodying theories about the real world in systematic ways. Earth system models provide a method for collating and testing our current understanding of Earth's dynamic processes. The characteristic feature of ESMs, first proposed in the 1980s (Bretherton, 1985), is that they represent global environmental changes in terms of the physical climate system and Earth's biogeochemical cycles, including the various interactions between these two broad sub-systems.

By including multiple processes and multiple interactions, ESMs are increasingly able to provide robust information about possible feedbacks and system thresholds. While this may not mean that future climate can be predicted, this better understanding of the variability and interactions of the total Earth system does increase its predictability. Schellnhuber (1999) proposed that the Earth system should be defined as encompassing the climate system and the aspect of the environment that is influenced by human activity or modified for human use. A key aim for ESMs is therefore to address explicitly the interactions between the ecosphere and the 'anthroposphere'. This is essential if the models are to be used to create projections of possible futures for Earth, with different degrees of human perturbation of the natural environment.

The actual configuration of the models can vary enormously, depending on factors like the timescales of interest, the number of included processes, and so on. Structurally, ESMs often incorporate component models of each sub-system. These are often first developed 'offline', creating stand-alone modules that are then coupled together. The coupling involves ensuring that concurrent simulations by each module exchange relevant information throughout the model run. The coupling of the model components or modules is typically an iterative process, in which one sub-model's outputs are used as the inputs for calculation and interpolation in the next time-step of the model run. For instance, in land–atmosphere coupling, fluxes of biogenic emissions from land vegetation might be accumulated over the course of a modelled 'day', and then fed into the atmospheric model to drive its chemical reactions or aerosol processes, which then determine the inputs for the next 'day' of the land biosphere model.

From a technical perspective, coupling of models so they run concurrently and interactively can be difficult for many reasons. Randall and Wielicki (1997) classify models into four categories:

- elementary or mechanistic models, where physical laws are applied to environmental phenomena (for example, energy balance models);
- forecast models that predict the deterministic evolution of, say, the atmosphere;
- statistical models (higher-order closure models); and
- 'toy' models, which are stylized representations for use as educational or exploratory tools.

Earth system models include elements of all these types. Each sub-model may well have originally been developed for its own specific research purpose. Different research communities may operate with different software languages, global projections, grid types and scales, time-steps, protocols for input parameters and data formats, and also different requirements for the forcing data used to drive the model that does not originate from the other sub-models.

However, our ability to make better predictions of Earth's climate using models depends strongly on coupling the components efficiently and understanding what happens when these various components interact. Just as in the real world changes in one part of the environment influence other parts, the dynamics in one sub-model induce changes in the dynamics of the others. Many combinations of processes could result in the same observed change. As coupled models become structurally more complex, understanding the nature and sources of the variability in the final model output becomes a bigger challenge. It requires very extensive experimentation, with a degree of 'tuning' to reduce biases and rationalize the coupling structure. A complex model may also raise the demand for very complex observational data, together with new mathematical and statistical methods for comparing these observations with model outputs.

5.2 Horses for courses: no model is 'best'

There is no 'best' Earth system modelling framework for all applications. The appropriate structure and the degree of complexity of the model depend on the question being asked. The relative quality of the model's representations can only be judged or evaluated with respect to its uses. Hence, a range of modelling approaches exists. Deterministic, complex, process-based models designed for decade-to-century projections cannot be used for millennial-scale simulations due to computational constraints. These computational constraints force us to choose among complexity, resolution and timescale, and our choices about which to focus on govern which model approach to apply. However, there are certain attributes that an ESM ideally should possess:

1. It should include representations of all the processes and their interactions that are believed to affect climate, both natural and those perturbed by human activity. Such processes would be generally well depicted, with parameterizations known and modelled accurately.
2. It should reflect the recognition that some aspects of the climate system are inherently 'chaotic' (see Section 5.3). Hence, although average changes in 'weather' may be expected to be amenable to prediction, conditions at particular dates and places would ideally be described by probability distribution functions. To achieve probabilistic outputs, the model must be capable of operation in an ensemble capacity, where multiple simulations are made using slightly different initial conditions or parameter values that are all plausible given the past and current set of observations or measurements.
3. The ESM should be easy to refine and upgrade as new process understanding becomes available that affects Earth system simulations. For example, a current modelling challenge is the ability to assess methods of geoengineering the climate. Robust, evidenced insights into the functioning of the Earth system will be essential if society decides that geoengineering is a course of action it wishes to consider, or in the event that it becomes a necessity if society is faced with potentially dangerous levels of climate change but an inability to rapidly de-carbonize its economy.
4. The model should be able to entrain observations of the Earth system to provide 'real-world' constraints on the simulations, and ensure the model state is consistent with the observed conditions (see Section 5.3). Such a model is then much more likely to increase our process understanding of the Earth system and more likely to make accurate predictions. This improved accuracy is particularly important for shorter timescales (i.e. decadal predictions rather than

those with century or millennial time horizons) because these timescales in some respects are becoming increasingly important to policy-makers planning adaptive and mitigation responses to climate change.

The question therefore becomes, 'What is preventing such a model becoming available?' With regard to point (1) above, there remains considerable uncertainty surrounding fundamental aspects of how the Earth system functions, including its interactions with human activity. The technical summary of the IPCC's Fourth Assessment Report (Solomon *et al.*, 2007, pp. 81–91) lists a range of outstanding uncertainties, with examples drawn from many components of the climate system. These include aerosol–cloud interactions, land–atmosphere interactions and the likelihood of future extreme events. It can be expected that highly targeted ongoing modelling and measurement campaigns will narrow such gaps in understanding.

Addressing chaotic behaviour (point 2) through ensembles is largely an aspect of the availability of computing power and relates also to choices made on model resolution. For future climate change projections we aspire to high spatial resolution, so that projections can provide highly regional assessments of expected meteorological responses to altered levels of radiatively active atmospheric gases. Despite continuous increases in processing power during the last few decades, this additional computational speed must be balanced against the ever-increasing number of environmental processes described in climate models. For this reason, the spatial resolution of climate models has not decreased as much, nor ensemble sizes increased as quickly as might have been anticipated by capitalizing on higher processor speeds.

An important dimension of point (3), the ease of making model adjustments, is associated with the accessibility of ESMs, which require considerable scientific and programming expertise to build and operate them. ESMs are clearly of a profoundly transdisciplinary nature, thus requiring different and traditionally separate disciplines to work closely together (as discussed in Chapter 1). This also relates to the complaint that only specialists can perform simulations to predict the future. This is seen by some as a problematic lack of transparency for an issue of such global importance; it potentially prevents people with relevant expertise working in different disciplines (social scientists and economists as well as other natural scientists) from participating fully in debates about how best to manage or respond to future climate. This objective in some ways lies in tension with the final point, namely the ability to confront ESMs with data – and vice versa – which requires the ability to perform a sophisticated form of data assimilation, drawing together multiple streams of available environmental data. The formidable tasks of data–model comparison are discussed more fully in Section 5.3. Nevertheless, we suggest that complexity in itself need not preclude greater transparency in the modelling process. There are several exemplars of the wide scope of options for better engagement and scrutiny in model development. The climateprediction.net initiative engaged members of the public in a major ensemble modelling project, both improving public understanding of important concepts in global change science (climate sensitivity) and demonstrating something of the technical and intellectual investment required for climate modelling. The community of scientists using the programming language and environment R (http://r-project.org) make the codes and documentation used for model visualization and statistical analysis available. The Bern climate model was implemented in Java (Matthews, 2005), with internationalized code, so that researchers and members of the public could interact with a web version of the model and have a better understanding of its workings. And, increasingly, the opportunities that open-source software presents for public transparency, scientific critique and accelerated development are shifting the 'business model' in Earth system modelling itself away from single institutional ownership of the code, with the costs and scientific inertia that this brings. Early examples of community models include the US Earth System Modelling Framework (www.earthsystemmodeling.org) and the freely available UCAR Community Earth System Model (www.cesm.ucar.edu). The MITgcm (http://mitgcm.org) is an open-source and publicly documented ocean GCM. All of the components in the Planet Simulator (Fraedrich, 2005 *et al.*; www.mi.uni-hamburg.de/plasim) are open source.

The (non-exhaustive) catalogue above of desirable or aspirational features of an ESM and the associated problems with attaining this should not be interpreted as a reason to either distrust the broad predictions of current models, or to give up on Earth system modelling as being too difficult or impossible with present-day computing capability. What is emerging

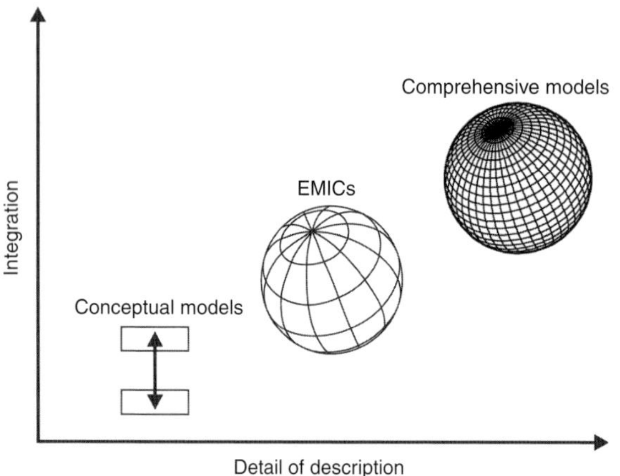

Figure 5.3 Model hierarchy as a function of detail of description (number of interactive components and processes represented, as well as spatial and temporal resolution) and resources (computing power, staff resource to maintain and develop the model).

now is a set of intermediate and quite pragmatic solutions or 'work-arounds' that address each of the above points individually. There is no single modelling solution to predicting future climate change, but different modelling structures can be adopted to answer specific questions. This is sometimes referred to as the 'hierarchy' or 'spectrum' of climate models (Figure 5.3). What is perhaps surprising is that although such differing and effective solutions are available, there has been little formal comparison between the merits of each. One such example is an attempt by Ferro *et al.* (2012) to consider, for a given increase in computational resource, whether the biggest gains in prediction capability come from raising ensemble size, increasing complexity or increasing resolution. In the rest of this section we present the various modelling approaches to Earth system modelling and refer to examples from models developed and used in the UK.

5.2.1 Full-complexity Earth system models

Climate-modelling centres are generally required to simulate climate changes, under different scenario assumptions, on timescales from a decade to centuries. The range of requirements is often met by different climate models or different configurations of the same modelling system. Most current model configurations for 10–15-year-timescale predictions have not included biogeochemical Earth system components, in the expectation that these processes have less relative importance on timescales of up to a decade, although it could be argued that atmospheric chemistry and aerosol processes interact with the climate on precisely these scales. So far, the models used for centennial climate projections have also been predominantly physical models, that is, global coupled AOGCMs. The IPCC's Fourth Assessment Report (AR4) included physical climate models defined by the CMIP3 protocol (Meehl *et al.*, 2007a, 2007b), i.e. global coupled AOGCMs. Only a few of the models featured in AR4 explored carbon-cycle feedbacks by computing atmospheric CO_2 concentrations interactively. This model configuration is now becoming much more widely used for centennial predictions and many modelling groups are starting to include biogeochemical Earth system components (e.g. Collins *et al.*, 2011) in order to allow quantification of some of the biogeochemical feedbacks that might be important on these timescales. Such processes include those coupling the climate, ecosystems and atmospheric composition (Figure 5.2).

For an 'assessment-type' or 'production' ESM (i.e. one designed to provide future scenario simulations for IPCC assessment reports and to be run for intercomparison studies under many different scenarios), it is important to have a well-defined frozen configuration, and one that is not prohibitively demanding in computational resources to run for a total of many centuries. This latter criterion means that not all the complexity desired can necessarily be afforded, and that compromises and trade-offs have to be made between the different components. The first criterion is in some ways more problematic for Earth system research. The need for a frozen configuration often means that the individual components are less likely to be updated even if the model is used for many years. This means that the components are well understood and evaluated but may not be 'cutting-edge' in their science. Since ESMs have started to include chemical and biological components, this limitation has become more pronounced. In the early years of coupled modelling, the atmosphere and ocean components were both developed by physical-process modellers who knew the state of the art in their (closely related) scientific fields. Now, climate modellers draw on a much wider range of scientific knowledge, with a greater risk that the various sub-models may not adequately reflect the current state of knowledge. Addressing this risk, by ensuring much closer working relationships between people with technical modelling skills and specialist scientific expertise, presents an opportunity to make much better models. The UK's research programme for Earth system science, QUEST, explicitly encouraged this process. There is still some way to go, but the development of the Quest Earth System Model (QESM) showed that there is considerable willingness for the needed co-operation, and it represents a valuable first step towards greater integration of the many disciplines contributing to the modelling enterprise.

The issue of the necessary trade-offs in Earth system modelling can be illustrated by reference to the Met Office Hadley Centre's HadGEM2 Earth system model (Collins *et al.*, 2011). Since global carbon-cycle and chemistry–aerosol feedbacks are likely to be important for centennial-scale climate change, existing components for the terrestrial carbon cycle (TRIFFID (Cox, 2001) and RothC (Coleman and Jenkinson, 1999)), the ocean carbon cycle (diat-HadOCC (Palmer and Totterdell, 2011)) and atmospheric chemistry (UKCA (Morgenstern *et al.*, 2009)) have been coupled to a physical climate model. In this coupling process compromises were made. This was particularly evident in the scope of the atmospheric chemistry, for which a simple troposphere-only scheme was used; it is sufficient to represent the oxidation of sulfate (important in aerosol formation) and the evolution of the ozone radiative forcing. In contrast, the full complexity of the diat-HadOCC ocean biogeochemistry was required to represent emissions of dimethyl sulfide (again important for aerosol formation) and the fertilization of marine biological productivity by iron inputs from the atmosphere.

The advantage of the 'assessment' model configurations is that the most important biogeochemical feedbacks are included in predictions of future climate under a range of scenarios. This information is important for the evaluation of mitigation options (including geoengineering), adaptation requirements, and the identification of potential tipping points (a point at which a sudden transition from one stable state to another stable state in the Earth system occurs). A disadvantage of these well-established configurations is that the most up-to-date process knowledge is not included, and some known processes are excluded due to the computational cost. However, although individual sub-components in these models are not cutting-edge science, the coupling together of a range of physical and biogeochemical processes in a production configuration such as HadGEM2 means that important Earth system feedbacks are simulated that standalone cutting-edge process models cannot capture.

Complementary to these assessment-type ESMs are models composed of state-of-the-art sub-models. Such models are pushing forward the ESM development and stand at the forefront of Earth system modelling complexity. They are, however, only designed for investigations of decadal- to centennial-scale climate and Earth system change. Depending on the resolution, multidecadal timescale integrations are computationally feasible with this type of model, thus enabling the study of continental-scale feedbacks and climate impacts in detail, despite the model complexity.

The QESM is one such cutting-edge ESM and one of the two models directly supported by QUEST (the other being FAMOUS, see below). Although similar to the HadGEM2 ESM (as outlined above) in the physical complexity and spatial resolution of its physical atmospheric and oceanic components, QESM has a more detailed treatment of biological and chemical complexity, and also includes stratospheric processes. All the interactive components outlined in Figure 5.1 for an ESM are included in QESM. Oceanic biogeochemistry is modelled using a 'plankton functional type' approach with different functional classes of phytoplankton and zooplankton. This is a step change from the simpler 'nutrient–phytoplankton–zooplankton–detritus' (NPZD) approach of HadGEM2, which has only one type of phytoplankton and one of zooplankton. This change enables a better study of problems such as ocean acidification and marine ecosystem change in general. The land surface in QESM has three main advances: vegetation-cover changes are modelled using a more realistic approach by including different age classes in the vegetation cover rather than the simpler TRIFFID model in HadGEM2; fire, which is important for vegetation succession and cover – especially in semi-arid areas – is included. A representation of the nitrogen cycle is also included in QESM. The QESM atmosphere has 60 layers in the vertical dimension, extending from the surface to 80 km in altitude. The inclusion of the stratosphere and lower mesosphere enables better simulation of the radiative and dynamical perturbations that result from changes in stratospheric ozone, which QESM can model self-consistently.

The chemical and biological sub-models of QESM are some of the most advanced available: UKCA is an atmospheric-composition–climate module providing a range of chemistry schemes, including tropospheric, stratospheric and whole-atmosphere (Morgenstern *et al.*, 2009). Biogenic emissions use the MEGAN scheme (Guenther *et al.*, 2006), while surface deposition is calculated interactively, depending on surface type. The new aerosol microphysics module included in UKCA resolves the size, particle number concentration and composition of the aerosol. This module is coupled to the modelled terrestrial and oceanic aerosol and precursor emissions (from JULES and PlankTOM10, see below) and UKCA's chemistry module.

The land-surface component of QESM, JULES, includes all processes critical for modelling the full atmosphere–surface exchanges (Clark *et al.*, 2011). It represents plant succession following disturbance with a canopy gap-type modelling approach (ecosystem demography; Moorcroft *et al.*, 2001; discussion in Fisher *et al.*, 2010a), thus providing a more realistic representation of vegetation dynamics than the earlier TRIFFID scheme. Natural wildfires, an important agent of vegetation disturbance, are captured by inclusion of a fire model, which simulates fire activity and fire-induced vegetation mortality and emissions from biomass burning (SPITFIRE, Thonicke *et al.*, 2010). The land nitrogen and carbon cycles are simulated through the inclusion of a soil carbon and nitrogen model (ECOSSE; Smith *et al.*, 2010) and a vegetation nitrogen uptake model (FUN; Fisher *et al.*, 2010b). Furthermore, the biogenic emissions model allows coupling between vegetation and atmospheric chemistry via fluxes of isoprene.

The ocean biogeochemistry module in QESM, PlankTOM10, represents lower-trophic marine ecosystems through the data-based parameterization of the ten most important types of plankton needed to represent the interactions between marine ecosystems and climate (Le Quéré *et al.*, 2005). These include bacteria, six classes of phytoplankton and three of zooplankton, as well as size-categorized organic material with sinking rates based on particle density. Including the ballast effect of plankton shells is important because it is a vital part of the cycling of carbon (via burial in sediments), and is likely to be impacted by increased atmospheric CO_2 concentrations. PlankTOM10 represents full prognostic cycles of several elements: carbon, nitrogen, phosphorus, silicon, sulfur, calcium and oxygen. It also incorporates a simple representation of the biogeochemical cycle of iron (an essential nutrient), as well as the associated air–sea fluxes of CO_2 and dimethyl sulfide.

5.2.2 Fast Earth system models

While modern full-complexity ESMs are capable of addressing scientific questions involving a wide range of Earth system feedbacks, the computational expense required limits their application to timescales of decades or centuries. Until recently, longer-timescale simulations (covering millennia and beyond) were carried out with fundamentally less sophisticated models, based on simplified dynamical equations and less complex parameterizations of physical and chemical processes (EMICs, discussed below). Some modelling groups are now developing ESMs based around previous generations of AOGCMs retaining the full complexity of the primitive-equation dynamics and using representations of many of the processes traditionally left out of simplified physics models. These new models are often run at lower spatial resolutions than the current state-of-the-art ESMs to reduce their computational overheads. They represent a step-change in the range of atmospheric feedbacks available to millennial-scale ESMs, crucial for representing short-term variability, geographical teleconnections and ocean and ice feedbacks which can influence the mean climate state (as an example see Figure 5.4).

The FAMOUS model (Smith *et al.*, 2008) is one such example that occupies this new niche. The fact that FAMOUS's basic code and climate simulation are similar to the higher-resolution HadCM3 AOGCM (Gordon *et al.*, 2000) means that it can also be used to directly inform more detailed studies, as well as take advantage of all that is already known about the parent model.

Typically for this new class of model, FAMOUS uses a quasi-hydrostatic, primitive-equation atmosphere with a coarse horizontal resolution (~ 6°) and 11 vertical layers. The rigid-lid ocean model (a numerical approximation that does not allow any fluid volume transfer at the surface to eliminate fast ocean-surface waves, such as tides and tsunamis) has a horizontal resolution of around 3°, with 20 vertical layers, and includes isopycnal mixing and a zero-layer thermodynamic sea-ice model. Interactive models of the carbon cycle are also included. Ocean biogeochemistry is modelled by a simple NPZD scheme. Land and soil processes are modelled at the same spatial resolution as the atmosphere with the capability to represent nine different surface types within each gridbox and which can evolve dynamically with the climate. The FAMOUS model can also be coupled to a land ice-sheet model, allowing fully interactive glacial simulations.

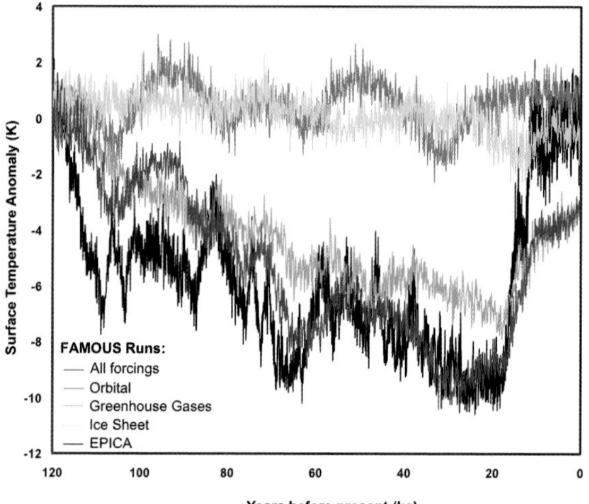

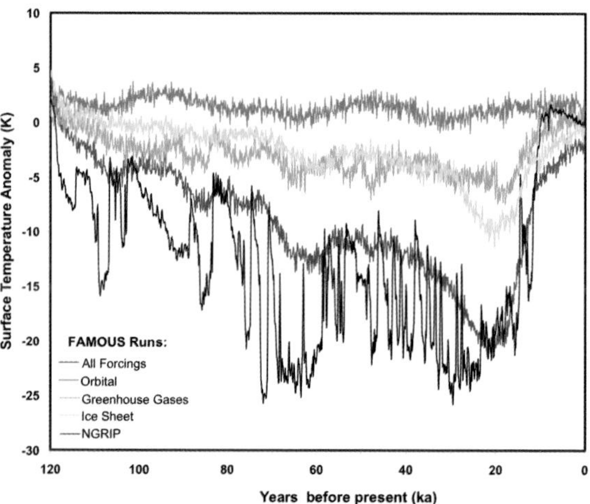

Figure 5.4 Results from FAMOUS simulation on glacial cycles (Smith and Gregory, 2012), demonstrating effectiveness of a fast ESM. The graphs show surface temperature anomalies from present day over the last glacial cycle at (a) Dome C, Antarctica, and (b) the NGRIP station in northern Greenland. The black line shows temperature reconstructions from ice core: (a) from the EPICA ice-core data (Jouzel *et al.*, 2007); (b) from the North-Grip ice core data (Masson-Delmotte *et al.*, 2005). The coloured lines are taken from a suite of FAMOUS simulations covering this period. The blue experiment was forced with solar insolation changes due to orbital forcing, the yellow one with atmospheric greenhouse gases as recorded in the EPICA ice-core (Lüthi *et al.*, 2008), and the blue one with the northern-hemisphere ice-sheet reconstruction of Zweck and Huybrechts (2005). The red line shows a FAMOUS simulation with all of these forcings applied together.

5.2.3 Earth system models of intermediate complexity

Some Earth system processes operate on very long-term timescales. For instance, reorganizations of ocean circulation and nutrient cycling as well as changes in biological productivity and surface temperatures all modulate the marine uptake of CO_2 from the atmosphere and are likely to have been central in past climate change such as the glacial–interglacial cycles of the past ~800,000 years (Kohfeld and Ridgwell, 2009). On longer timescales, interactions between the ocean, deep-sea sediments, and weathering on land exert further controls on atmospheric CO_2 (Archer *et al.*, 1997; Ridgwell and Hargreaves, 2007). These longer-timescale processes (see Section 5.4) are suspected to have dominated the response and recovery of the Earth system from catastrophic CO_2 release in the geological past (Zachos *et al.*, 2005).

Coupled climate–carbon and Earth system models are designed to resolve many of the key biological processes on ocean and land that link biogeochemistry and climate. However, their computational demands make them unsuitable for investigating timescales much beyond 1,000 years, and they generally omit processes characterized by a timescale longer than this as mentioned above. At the other extreme, Earth system models based on a 'box-model' representation of the ocean (e.g. Lenton and Britton, 2006; Munhoven, 2007; Shaffer *et al.*, 2008; Zeebe, 2011) are extremely computationally efficient and traditionally have been the tools of choice for very-long-timescale questions. But they have limitations. For instance, validation against marine observations and sediment records may be problematic because large volumes of the ocean are homogenized in creating each 'box'. The tension between resolved-but-slow, and abstracted-but-fast models, each with their own limitations, stimulated the development of a new class of model characterized by reduced spatial resolution and/or more highly parameterized 'physics' in one or more elements of the climate system – often known as Earth system models of intermediate complexity (EMICs; e.g. Claussen *et al.*, 2002). Archer *et al.* (2009) give an overview of the different flavours of EMICs.

Because, for a comparable resolution, an atmospheric GCM must be run with a slower time-step than an ocean GCM, the first compromise that EMICs generally make is to reduce the complexity of the atmospheric component. At one extreme, atmospheres in EMICs may consist of nothing more than a simple zero-dimensional energy-balance calculation, applied to the same grid as the ocean and land surface and often including a parameterization of lateral transport of heat and moisture to create a quasi-two-dimensional model (e.g. GENIE-1, Edwards and Marsh, 2005). Other, slightly less simplified, approaches have been taken, resolving something of the vertical structure of the atmosphere by including a statistical–dynamical approach (CLIMBER-2, Petoukhov *et al.*, 2000) or even employing a full atmospheric GCM at reduced resolution (GENIE-2, Lenton *et al.*, 2007). The ocean in EMICs is generally three-dimensional although reduced complexity physics may be employed here, such as the frictional–geostrophic approximation used in GENIE (Edwards and Shepherd, 2002). Further computational efficiency, albeit at the expense of further degraded resolution, can be achieved by making a zonally averaged '2.5-dimensional' approximation of the ocean (Stocker *et al.*, 1992). Sea-ice, as necessitated by both the often low resolution of the ocean grid as well as lack of an adequately realistic atmospheric dynamics, is often relatively simplified. Because EMICs can access the timescale relevant to the growth and decay of ice sheets, ice-sheet components may also be coupled in some manner (e.g. CLIMBER-2).

For the land carbon cycle, use of highly simplified representations of heat and moisture transport will necessitate the use of a comparably simple and robust terrestrial scheme, as employed in GENIE-1 (Williamson *et al.*, 2006). However, even simple terrestrial schemes may include plant functional types (e.g. CLIMBER-2). The marine carbon and other biogeochemical cycles are characterized by timescales of adjustment of order 1,000 years and longer. This is where EMICs come into their own compared to ESMs, which generally cannot access the *c.* 2,000–5,000-year timescale necessary to fully spin up the ocean and carbon cycle from 'cold'. The complexity of the overall biogeochemical cycling hence tends to be greater in EMICs than in ESMs, and because EMICs are often applied to past/geological questions, trace-element and isotopic tracers may also be included (Ridgwell *et al.*, 2007). However, EMICs, being more limited in resolution and time-step, tend to include a less complex plankton ecosystem scheme (if any). Many EMICs incorporate the ocean–sediment carbon-cycle component necessary to address carbon–climate perturbations on timescales longer than 1,000 years. The GENIE model (Ridgwell and Hargreaves, 2007) is one of the few EMICs that is based on a dynamical ocean and which incorporates the sea-floor and terrestrial calcium carbonate ($CaCO_3$) neutralization sinks for CO_2 as well as the weathering feedback.

5.2.4 Other modelling approaches

Because the land surface plays such an important role for society, and alteration through climate change could markedly affect food and water security, an impact modelling system to address these specific issues has emerged. This system, IMOGEN (Integrated Model Of Global Effects of climatic aNomalies; Huntingford *et al.*, 2010), is an intermediate-complexity model except that its structure retains the full complexity of the underlying land-surface model. The land surface in IMOGEN is gridded to the same resolution as the Hadley Centre's HadCM3 GCM. For computational speed, the meteorological drivers are derived by 'pattern-scaling' (Huntingford and Cox, 2000). In this approach, changes in surface conditions, such as temperature, humidity, surface radiation and rainfall, are assumed for different geographical positions and months to scale linearly with the amount of global warming. For HadCM3 outputs, this assumption has been tested and shown to work reasonably well. This system, currently focused on the land surface, could potentially be used for detailed analysis of other constituents of the Earth system. For instance, in IMOGEN, the global-warming amount that scales the patterns of change is from an energy-box model (again calibrated against HadCM3), forced by different concentrations of atmospheric greenhouse gases. Thus users can prescribe different future trajectories of CO_2 and non-CO_2 greenhouse gases, rather than operating a full GCM, to assess expected climate impacts on terrestrial ecosystems. The model can also be used as an intermediate testbed for implementation of new process understanding before inclusion in a full ESM. For instance, Huntingford *et al.* (2011) have used IMOGEN to demonstrate that comparing changes in concentrations of atmospheric constituents only in terms of their alteration to radiative forcing is insufficient to represent the impacts of climate change on ecosystem services.

Integrated assessment models (IAMs) are designed to account for the interaction between natural and socio-economic systems, providing quantitative detail about qualitative-scenario narratives of possible futures (as explained in Section 5.4). In the context of global change research, the term is usually applied to models that link simplified representations of climate and the global economy, with modules representing factors such as energy, land use and demographics. A brief overview of the approach taken in these models was given in Chapter 1. Chapter 7 provides some insights in the use of IAMs for defining and assessing climate policies, particularly on mitigation.

There are many IAMs. They vary greatly in their level of geographic and socio-economic detail, and in policy scope. A few integrate spatial and societal dimensions, allowing for the calculation of geographic changes in land use and impacts. From an Earth system perspective, important features of IAMs are that they are global in their scale, and that they are actually 'integrated', in the sense that they include both human and natural system components within the model structure. In other words, they deal dynamically with the interactions between human and natural system variables rather than having one or the other as externally prescribed inputs (Kelly and Kolstad, 1999; Costanza *et al.*, 2007).

One inherent challenge that arises in the task of 'quantifying the qualitative' using IAMs is the sensitivity of the models to changes in the underlying parameter values and assumptions. While this is true of all models, history tells us that predicting societal futures is particularly problematic. IAMs require strongly simplifying assumptions to be made about such things as technological change and societal responses to climatic or economic changes. Schneider (1997) drew attention to the risks that these '*assumptions [are] not universally lauded, not always transparent, and plausible alternatives could radically change the "answer"*.' The models can provide insights into how the socio-environmental system works, and make conditional forecasts about potential consequences of given policy decisions. However, the ultimate predictive power of IAMs is open to question.

Another challenge is created by the fact that IAMs combine sub-models of different types or ontologies (see Box 5.1). Sub-models can be based on statistical treatments of empirical data (typically general equilibrium economic models), mechanistic simulations (components of the climate models) and heuristics (e.g. the description of the relationship between environmental impact and wealth). There is also great diversity in the approaches used for meshing the different sub-models. For instance, the downscaling of climate-model output from global to regional scenarios can use simple proportionality rules, or empirical statistical techniques. As 'hybridized' modelling systems, the evaluation of IAMs needs particularly careful attention. As with any other modelling approach, IAMs need to include clear analyses of uncertainty.

The computational advances of recent years have opened up the vistas for integrated assessment modelling, as they have for Earth system models. The most advanced (spatially explicit, multi-component) IAMs are now converging towards intermediate-complexity ESMs, rather than using simplified (parameterized or reduced-form) climate-model output as one of their driving elements. In the not-too-distant future, it is likely that full-complexity Earth system models will be usable to address the questions currently the remit of integrated assessment. Schneider (1997) anticipated this trend; Tol (2006) has reviewed progress in the increasing integration of natural and social science models.

5.3 Understanding observations

So far in this chapter we have focused on modelling, which is a vital activity in Earth system science. However, just as many Earth system scientists are engaged in the activity of making observations and obtaining measurements. This work can involve, for example, the development of Earth-observation instruments and algorithms to generate remote-sensing data of Earth's properties and processes from satellites, or the meticulous counting and identification of grains of pollen from lake sediments or foraminiferal shells from ocean sediments. Science at the global scale now involves a vast effort of data synthesis, data analysis and data stewardship. Unfortunately, to some extent, scientists involved in modelling and observation have tended to form distinct communities of expertise.

Improving our understanding of the mechanisms and processes of the Earth system as a whole demands above all that we bring observations and modelling close together. First of all, models are needed to understand many of the large and/or complex observational data sets that exist for global processes; but at the same time we need observations to evaluate models. Without data, the models are just abstractions. In order to obtain understanding and inform real-world decision-making, we need our models to tell us 'the truth' about the world, and observational data are a critically important part of that verification. Moss and Schneider (2000) proposed that while models and data need to agree, scientific confidence also requires them to accord with theory, with a high degree of expert consensus. Increasingly, approaches are also required for optimizing models by assimilating observations. Assimilation can be thought of as a dynamic process, rather than the static comparison of models with data. Here, we discuss each of these challenges.

5.3.1 Interpreting observations

Observations can only be interpreted with the help of an underlying model. Simple, conceptual models are useful for a more qualitative description and understanding of the major features of the Earth system, while ESMs with a sufficient level of completeness are needed for a detailed quantitative analysis and direct interpretation of the full set of available observations. To assess the consistency of a model simulation with observations, it is most effective to provide model output that is as directly comparable to the observations, as possible. In general, it is easier to work in a forward sense and include the key processes in the model for simulating the observations directly (that is, aim to model what you can see), rather than going the inverse way and using the observational data to infer reconstructed information about the state and evolution of the Earth system. This is especially true if the processes involved show highly non-linear behaviour; but in many instances – notably in studies of past climate – there is no option other than to make inferences, and test them with the models.

For example, pollen and plant macrofossil data derived from sediment profiles can be used to reconstruct the vegetation cover for past times. By applying empirical statistical methods (transfer functions) these data can be used for the reconstruction of climate variables (Bartlein *et al.*, 2011). But it is also possible to interpret the data at a more basic level in terms of biomes (Prentice *et al.*, 1996) or abundances of plant functional types. Then the data can be compared with model outputs in various ways: by using a process-based biogeography model to convert GCM-simulated climate into biome distributions (one-way coupling); using such a model to update land-surface properties for the GCM, and iterating until a stable climate and biome distribution are obtained (asynchronous coupling); or including a dynamic vegetation model as an interactive component of an ESM (full coupling). One-way coupling has been widely used as a diagnostic for palaeoclimate simulations with GCMs. The Palaeoclimate Model Intercomparison Project (PMIP) has recommended this approach using the BIOME4 biogeography model as standard. Harrison and Prentice (2003), for example, used this approach to demonstrate the need to include physiological CO_2 effects in the model in order to explain the observed huge expansion of forests after the Last Glacial Maximum. Asynchronous coupling was used to explore the possible role of biogeophysical vegetation–climate feedbacks in amplifying orbital forcing during the initiation of glaciations (de Noblet *et al.*, 1996) and the maintenance of a 'green' Sahara (Claussen, 1997; de Noblet-Ducoudré *et al.*, 2000) and northward-extended boreal forests (Texier *et al.*, 1997) during the mid Holocene. Some ESMs that include a carbon cycle (Friedlingstein *et al.*, 2006) now routinely use full coupling of vegetation and climate dynamics through the inclusion of a dynamic global vegetation model. The PMIP has recently developed schemes that allow the outputs of several dynamic global vegetation models to be compared directly to palaeovegetation data expressed as biomes.

A major use of ESMs is to improve understanding of contemporary climate change, and relate the observed changes now to what is known about past variation in the Earth system, in terms of atmospheric chemical composition and land cover. In this context, two key questions addressed are the following: 'What is the balance of impact on global temperature between increased greenhouse gases, increased aerosols, manipulating the land surface and natural phenomena such as variation in solar output and volcanic eruptions?' and 'To what extent are human activities altering the climate system?' While atmospheric carbon dioxide is still important in this context, ESMs give important insights also for important non-CO_2 greenhouse gases like CH_4, and the cooling effect of raised aerosol concentrations. Earth system models can be forced by conditions corresponding to each individual climate forcing to determine their unique response, and these signatures can be searched for in the gridded global temperature records. The statistical method to do this is sometimes called 'optimal fingerprinting', which is an advanced form of regression analysis (Hasselmann, 1997). Many studies have now found that the greenhouse-gas signal can be observed in the temperature record with greater than 95% confidence, thus leading to the statement in the latest IPCC reports that humans are 'very likely' affecting the climate system. From a growing list of research papers confirming this finding, studies leading to this attribution statement include those of Stott *et al.* (2000), Allen and Stott (2003), Stott *et al.* (2003) and a combined multi-model analysis, Huntingford *et al.* (2006).

As ESMs evolve, the new developments are extending the use of ESMs as an interpretative tool to a wider range of Earth system observations. For instance, Ridgwell (2007) has coupled a mechanistic model for simulating deep-sea sediment cores to the GENIE-I EMIC to help interpret the excursion in the abundance of calcium carbonate compensation depth as recorded in deep-sea sediment cores from the South Atlantic during the Palaeocene–Eocene Thermal Maximum

(PETM, see Chapter 3). The PETM was a global warming event caused by a massive release of carbon to the atmosphere and ocean (Zachos *et al.*, 2008), discussed further in Chapter 3. The data show that the event was accompanied by a large excursion of the compensation depth, as expected due to ocean acidification. Ridgwell simulated the biogeochemical response of the ocean sediments to a massive release of CO_2 and by comparison with the available sediment data could show that a CO_2 release over just a thousand years would have been sufficient to account for the sediment data.

5.3.2 Model evaluation and benchmarking

The above examples demonstrate the usefulness of ESMs in interpreting observations of the Earth system. But how can we have any confidence in these complex models? Model intercomparison projects (MIPs) have shown that any given model 'experiment' can evince a range of responses from different models. This should perhaps not be surprising given that the climate system is intrinsically affected by uncertainty.

Lorenz (1975) defined two kinds of predictive uncertainties. The first kind deals with the initial state of the system. Because of the non-linearity and instability of the primitive equations of atmospheric dynamics (which are at the core of an ESM) any prediction depends on the initial values, which are not very well known or practically specifiable. Numerical weather prediction suffers from predictive uncertainty of this first kind. Predictive uncertainty of the second kind is determined by the response of the climate system to external perturbations, that is, how the state of the climate system will change as some of the external forcing is altered. It is believed to be less dependent on the initial conditions. Because of the processes involved, uncertainty in climate predictions is mainly of the second kind and arises from model errors and uncertainties in representing the processes of the Earth system.

Formal methods are needed to assess the abilities of ESMs to represent processes and feedbacks. Identifying and reducing uncertainty in the model internal feedback loops (see Chapter 4) is a top priority, because these can strongly influence the relation between a temperature-based policy target and the compatible emissions of anthropogenic greenhouse gases. But the hierarchy of scales and processes represented in ESMs presents a challenge in developing a model-evaluation strategy. Model intercomparison projects define common protocols to deliver a controlled evaluation of ESMs or model components (see www.clivar.org/organization/wgcm/projects.php for a list of MIPs). These projects have proved useful in identifying structural model errors (see Box 5.2 on model uncertainty). They have confirmed, for example, that the representation of cloud-radiation processes and the cloud feedback is one of the largest uncertainties in climate prediction with AOGCMs (Le Treut *et al.*, 2007). In ESMs with an interactive carbon cycle, the C[4]MIP showed that considerable uncertainties in simulated global temperatures at 2100 can also be caused by uncertainties in the climate–carbon-cycle feedback (Friedlingstein *et al.*, 2006). However, model intercomparison studies have their limits: not least that all models are treated as equal and independent, which in general is not realistic (Masson and Knutti, 2011; Box 5.1).

A necessary complement to model intercomparisons is provided by model evaluation and benchmarking against observations. Quantitative model evaluation based on observations is a prerequisite to understanding and improving the process representation of feedbacks in ESMs. A formal quantitative evaluation of forecasts is routinely performed for numerical weather prediction. The statistical verification scores are intercompared among the weather-forecasting centres; these historical comparisons of the scores prove the increasing reliability of weather forecasts (e.g. Simmons and Hollingsworth, 2002). While for numerical weather prediction, the evaluation of a forecast is straightforward (we only have to wait a few days to compare the forecast against the actual weather), it is impossible on climate timescales. However, what is possible is the evaluation of climate predictions in hindcast. That is, the models are run to predict the climate for the timescales on which it has been observed.

In a rigorous evaluation of ESMs, we require not only to demonstrate that modelled climates in the different regions agree with current observations, but also to demonstrate that we can successfully reproduce past changes in climate. Only by considering past observations can we be confident that we can represent climate states that are radically different from those of today as well as the transient behaviour of the system, and thus have confidence that our treatment of the controlling processes is realistic. Glecker *et al.* (2008) have confronted model simulations of the twentieth century from the Coupled Model Intercomparison Project (CMIP3) archive with 'quasi-observed' reference data from two 40-year reanalysis products, namely NCEP/NCAR (Kalnay *et al.*, 1996) and ERA40 (Simmons and Gibson, 2000), and ranked the models using a performance index. They showed that an index-measuring model performance on interannual timescales is only weakly correlated with an index of the mean climate performance, illustrating the importance of evaluating a wide range of processes and scales. PMIP has extended the concept to assess our ability to simulate much larger changes of climate that are documented for the geologically recent past (Braconnot *et al.*, 2007).

Coherent frameworks have recently been developed for process-oriented evaluation of ESMs focusing on terrestrial biogeochemistry (Cadule *et al.*, 2010) and atmospheric chemistry (Eyring *et al.*, 2005). Each process is associated with one or more model diagnostics, and with observational data sets that can be used for their evaluation. The diagnostics used in these studies go beyond climatological means and include measures of spatial patterns, trends, and characteristics of variability on different timescales to reflect the complexity of the processes. Cadule *et al.* (2010) used observed atmospheric CO_2 concentrations from several measurement locations and extracted the long-term trend, seasonal cycle and interannual variability of the CO_2 signal to compare with model results. They also used performance metrics, both for different versions of individual ESMs and for different generations of the ensembles of models that are being used in international assessments. Randerson *et al.* (2009) and Blyth *et al.* (2011) have provided more comprehensive sets of benchmark tests for land–atmosphere carbon and water cycling.

5.3.3 Data assimilation in Earth system science

Systematic model evaluation as described above can greatly help to identify inadequate process representations in ESMs as shown by a large mismatch between model results and observations. But a mismatch between model results and observations does not necessarily imply a structural error in the model; it could also be caused by poorly chosen values of one or more key parameters. This is likely to be particularly an issue for the biogeochemical components of an ESM, as biogeochemical process-parameter values are often estimated from studies at the scale of individual organisms or ecosystems, or from short-term experiments which may not necessarily 'scale up'. Parameter-optimization methods are useful in this context.

Some attempts to quantify parametric uncertainty in atmospheric GCMs have been performed using perturbed physics ensembles, i.e. model parameters have manually been perturbed to assess the impact on model diagnostics (Murphy *et al.*, 2004). Stainforth *et al.* (2005) have taken this further by using a distributed-computing method to increase the ensemble size massively.

More computationally efficient and rigorous parameter optimization methods rely on data assimilation. Data assimilation is a formal technique to produce the best simulation that the combination of model and observations allow. The observational information is accumulated into the model and propagated to all parts of the model. The advantage of using data assimilation is that the constraints on the model state or model parameters are consistent with the governing physical laws. Data assimilation was first introduced in numerical weather prediction to improve estimates of the initial state of the forecasting model. It is now an emerging field in Earth system science.

A data-assimilation system consists of three ingredients: observations, a dynamical model and an assimilation method. Central to the concept of data assimilation are the concepts of errors, error estimation and error modelling. Observations have errors arising from various sources such as instrumental measurement error, sampling and the representation of measurements. Dynamical models have errors arising from the parameterizations, and the discretization of analytical dynamics into a numerical model. There are two basic methods for data assimilation: sequential assimilation, which considers observations at discrete time-steps and thus evolves over time; and variational assimilation, which considers all relevant observations at once over a defined period. Various flavours of these approaches have been developed and are now routinely used at operational meteorological agencies and centres around the world. They differ in their numerical efficiency and optimality for their specific use. Figure 5.5 shows a general data-assimilation scheme for state/parameter estimation. A more complete introduction to the concepts of data assimilation is given in the textbooks by Daley (1991) and Tarantola (2005) as well as in the lecture notes of the European Centre for Medium-range Weather Forecast on data assimilation and the use of satellite data (ECMWF, 2001).

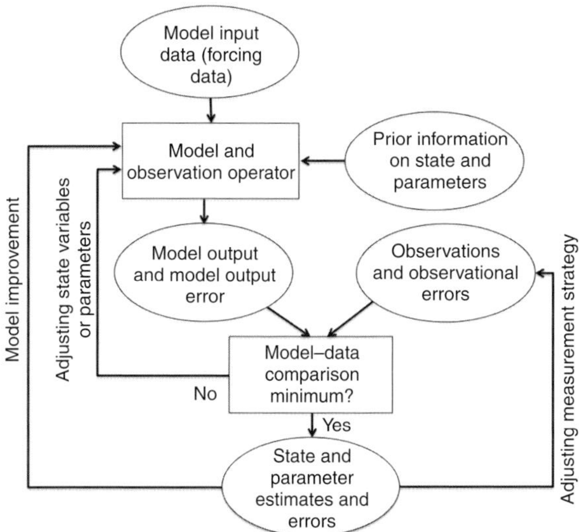

Figure 5.5 Schematic of a data assimilation system. Boxes represent computations, ovals represent data/observations and arrows represent the flows of information. The inner loop ('Model-data comparison' box to 'Model and observation operator' box) indicates the iterative assimilation process. Often, the analysis of residuals in model–data comparison lead to either model improvements or adjustment of the measurement strategies (outer loops).

Prominent products that are used in Earth system science primarily for model evaluation and that heavily rely on the assimilation of observations are the reanalysis products of the atmosphere (see above and Chapter 2). For the oceans, similar reanalysis or state estimation products exist as well. The ECCO/GECCO (Stammer *et al.*, 2003) and the Mercator Ocean (Bahurel *et al.*, 2006) products are just two examples; they are a powerful tool for detecting climate signals in physical oceanography (e.g. Lee *et al.*, 2009).

Data assimilation using ESMs is still in its infancy. The most advanced applications so far have been on the estimation of process parameters of both the marine and the terrestrial biogeochemical components of ESMs. For marine ecosystem models, the advent of remotely sensed ocean-colour information has stimulated an increasing number of data-assimilation applications (Gregg *et al.*, 2009; Ward *et al.*, 2010) to estimate poorly known model parameters by minimizing the misfit between model simulations and observed data. For example, Friedrichs *et al.* (2007) assessed 12 lower-trophic-level models of varying complexity objectively in two distinct regions (equatorial Pacific and Arabian Sea) using variational assimilation of chlorophyll-a, nitrate, and export and primary productivity. The same misfit metric was used to assess model skill. The authors demonstrated that within a region, the simplest models fitted the data as well as those with multiple phytoplankton functional groups, but cross-validation experiments revealed that as long as only a few key biogeochemical parameters were optimized, the models with greater phytoplankton complexity were generally better able to perform well in diverse regions and physical settings. However, while complex models may be more portable across regions, it remains unclear whether these models are mechanistically more realistic. Improved ocean-state estimation, as outlined above, is important in this context because the behaviour of marine ecosystems is critically dependent upon their physical environment.

For terrestrial ESM components, the most advanced data-assimilation system is the Carbon Cycle Data Assimilation System (CCDAS: Rayner *et al.*, 2005), which applies the variational approach. This system uses observations of the carbon cycle (currently atmospheric CO_2 concentration and remotely sensed greenness) to optimize process parameters in a prognostic terrestrial-biosphere model by iterative evaluation of the cost function and its gradient (the first derivative) with respect to the parameters. In addition, posterior uncertainty of the estimated parameter set, which is consistent with assumed observational and model uncertainties, is approximated by the inverse of the cost function's Hessian (second-derivative) matrix, evaluated for the optimal parameter set. The advantage of optimizing process parameters is that the calibrated model can subsequently be used for both diagnostic and prognostic simulations. Hence, uncertainties in model predictions can substantially be reduced by constraining the model with observations as demonstrated by Scholze *et al.* (2007) in a hindcast experiment with CCDAS. Ziehn *et al.* (2011) used a different assimilation method (Markov Chain Monte Carlo) but the same terrestrial ecosystem model as in CCDAS to demonstrate that assimilating plant-trait data at the leaf level substantially reduces the uncertainty range in simulated plant productivity for a future climate change scenario. Other attempts to constrain the parameters of the terrestrial ESM component have been made, based on the assimilation of eddy-covariance measurements of CO_2 and energy fluxes using a variety of methods (e.g. Wang *et al.*, 2001; Santaren *et al.*, 2007).

The QUEST programme has been in the vanguard of these developments. The QUEST Climate-Carbon Modelling, Assimilation and Prediction (CCMAP) project has taken the CCDAS variational approach further, to investigate the potential of parameter optimization and uncertainty reduction in a computationally efficient ESM (the Planet Simulator, Fraedrich *et al.*, 2005). A first feasibility study showed that the method could deliver important information for parameter optimization. However, the application of this approach on climate timescales is still a challenging problem due to the chaotic nature of the primitive equations (Kaminski *et al.*, 2007). Some earlier studies have successfully optimized a limited number of parameters (~10) in an EMIC with an energy–moisture-balance model for the atmosphere, thus avoiding the chaotic behaviour induced by the primitive equations (Hargreaves *et al.*, 2004; Ridgwell *et al.*, 2007). A 'grand challenge' in data assimilation in the context of Earth system modelling is the initialization of coupled models to achieve physical consistency across the coupled-model components, in this case mainly ocean and atmosphere models, especially because of the huge range of timescales involved.

Box 5.2 Uncertainty in Earth system modelling

The IPCC's Fourth Assessment Report (IPCC, 2007) showed that future climate change projections yield a wide range in simulated global surface warming for the period between 2000 and 2100 (see Figure 5.8), and recognized that large uncertainties exist in climate change simulations. Different types of uncertainty are involved in Earth system modelling. At the most general level, uncertainty can be distinguished into two types: ***aleatory*** uncertainty, which is a property of the system related to its intrinsic variability; and ***epistemic*** uncertainty, which arises due to a lack of knowledge about the system. An informative essay on this distinction is given by Der Kiureghian and Ditlevsen (2009). In the context of Earth system modelling, it is helpful to further categorize uncertainty so that it can be better characterized and reduced.

Parametric uncertainty

Parametric uncertainty arises because the 'best' values for the parameterizations (that is, the mathematical descriptions) of the individual processes being modelled are not usually well known. This situation arises for several reasons: uncertainty exists in the original empirical measurements used to parameterize a process; scaling issues arise when a parameterization derived from point measurements (as in local field studies) is used at the larger scales at which ESMs operate; and many processes rely on semi-empirically derived parameterizations, for which parameter values are not readily measurable. Formally, the time evolution of the state s of the system is a given by the model M and its parameters p:

$$\frac{\partial s(t)}{\partial t} = M(s(t), p) \tag{5.1}$$

Uncertainty in the parameter values p as described by probability density functions propagates to uncertainty in the model state and model diagnostics. In principle, this uncertainty can be quantified, for example by Monte Carlo studies. However, the very large number of model runs that these approaches require mean that they are usually not feasible in practice due to the high computational demands of ESMs. Instead, ensemble simulations using perturbed physics ensembles are often used to constrain the parameters and quantify the impact of this uncertainty on the resulting model projections. Parametric uncertainty can be reduced through parameter optimization against observations, using techniques of data assimilation as outlined in Section 5.3.3.

Structural uncertainty

Structural uncertainty arises from uncertainty as to the correct mathematical representation of processes within the models. It can relate to incomplete or imprecise understanding of a process, or the need to parameterize a process rather than represent it dynamically in the model due to computational costs. The uncertainty from unrepresented but relevant process descriptions also belongs in this category. A prominent example in Earth system science is the lack of an interactive carbon cycle in 'traditional' AOGCMs. Equation (5.1) can be generalized to include uncertainty in the functional form f of the parameterizations:

$$\frac{\partial s(t)}{\partial t} = M_f(s(t), p_f) \tag{5.2}$$

where f represents the parameterizations describing the involved processes. Here again, uncertainty in the process formulation will propagate to uncertainty in the model state and model diagnostics. Structural uncertainty can best be quantified by model intercomparisons or by testing different process formulations in one modelling framework. Comparison of ESMs with more detailed process-specific and process-resolving models from a specific discipline can help to reduce structural uncertainty.

Residual uncertainty

Model outputs depend strongly on the initial and boundary conditions, or the 'external forcing' of the system. The residual uncertainty in modelling arises from uncertainty in this external forcing. The most obvious example in Earth system modelling now is the uncertainty in the anthropogenic forcing, but this category also includes uncertainty in solar and volcanic forcing and, for instance, the state of the deep oceans as an initial condition. If *e(t)* represents the external forcing then Equation (5.2) can be expanded to:

$$\frac{\partial s}{\partial t} = M_f(s(t), p_f, e(t)) \qquad (5.3)$$

The uncertainty in model results due to residual uncertainty can be quantified by ensemble simulations using different realizations of the external forcing. The different realizations for the anthropogenic forcing, for example, is provided by the various emission scenarios as described in Section 5.4.2. Even here, further distinctions can be made: the uncertainty in solar forcing arises from the lack of knowledge about its variability, while the uncertainty in the anthropogenic forcing has an element of subjectivity in that choices can be made by society about emissions.

Intrinsic uncertainty

The intrinsic uncertainty in Earth system modelling is a consequence of the stochastic nature of the dynamical equations. It is an irreducible element of climate variability. The numerical representation and discretization of analytical, continuous equations form an additional part of the intrinsic uncertainty.

The way forward

The increasing number of model components in ESMs provides further progress towards completeness in the representation of the elements, interactions and feedbacks that constitute the Earth system but, in itself, this provides no additional confidence that the model results obtained are better. In fact, the variability in responses may increase, as process addition increases the number of degrees of freedom. The greater complexity can also sometimes increase model bias due to the omission of other interacting processes. Therefore a need for careful attention to the quantification and reduction of uncertainty in Earth system modelling is inevitable. Necessary ingredients are observations, together with their uncertainties, along with a wide range of diagnostics and appropriate metrics. Discipline-based, process-resolving models can be deployed to stringently evaluate and benchmark models. However, that in turn introduces a challenge: traceability has to be ensured from these process models, which generally include greater complexity, down to the simpler versions present in ESMs.

Last, but not least, we still need a common standard for quantification and terminology for reporting uncertainties. The IPCC made some steps towards providing a common approach for expressing uncertainty in its synthesis assessments (IPCC, 2005). However, as Hargreaves (2010) notes, this guidance was not consistently adopted across all of the IPCC's 2007 Fourth Assessment Report. This suggests that the guidance does not yet fit the requirements of the diverse and interdisciplinary research community involved in climate science, so further refinement and research dialogue are needed to reduce the risks of misperceptions by the wider, interested public about uncertainty in climate modelling.

5.4 Predicting future global change

Earth system models provide one important means of better understanding the climate system. Where there is demonstrated skill in reproducing observable aspects of the past and present they can be used to provide a guide to the future. However, this requires carefully specifying the questions to be addressed, and construction of a suitable experimental design. Part of the experimental design is to set out the inputs to the model as a scenario. A typical input scenario for a climate model could consist of estimates of the atmospheric greenhouse-gas concentrations and aerosol emissions over the remainder of the twenty-first century. But how are these obtained? In this section we discuss different types of scenario, how they are constructed, and some results from models driven by these scenarios. It is useful to recognize that in many modelling studies the preferred approach is to use several different climate models in order to learn more about the uncertainty in the response to a given input scenario.

5.4.1 Concentration- versus emission-driven scenarios

The first choice to make is whether the inputs to an ESM are to be provided as atmospheric concentrations of greenhouse gases and aerosols, emissions of greenhouse gases and aerosol precursors, or a mix of the two. Until recently, the most complex climate models were always driven by concentrations of greenhouse gases. Historically, this is because few of these models have contained the necessary biogeochemistry to translate the emissions of carbon dioxide and other gases into atmospheric concentrations. Offline biogeochemistry models (such as the Bern model, Joos *et al.*, 1996) were used for this step. So the standard model set-up in IPCC assessments to date has been as follows: (1) define a scenario of future greenhouse-gas emissions; (2) run this through the Bern model, to yield a corresponding scenario of concentrations; (3) use these concentrations to drive an AOGCM.

The concentration-driven approach has several advantages. First, it allows modelling teams that have not yet added the Earth system components necessary for an emissions-driven approach to their climate models to take part in intercomparison studies. Secondly, the results of multi-model studies are often easier to interpret because one component of the spread in climate response – that due to uncertainty in the relationship between emissions and concentrations – is eliminated. However, if the results of the projection exercise are to be used for planning purposes, for instance to examine the need for different adaptation options, then a concentration-driven approach will underestimate the spread in the climate response to a given emissions scenario.

Unlike AOGCMs, ESMs including a complete (land–ocean–atmosphere) carbon cycle can be driven directly with emissions. Since controls on greenhouse-gas and aerosol emissions are currently seen as the major lever for reducing climate change, a growing number of models used for policy advice are now set up to perform at least some emission-driven experiments. One of the first emission-driven studies with a complex climate model, HadCM3-LC (Cox *et al.*, 2000), indicated that the inclusion of biogeochemical feedbacks in an emission-driven experiment could substantially alter the predictions of future climate. The effects of including carbon-cycle feedbacks in this model for a 'business-as-usual' scenario caused an additional increase in future temperatures at 2100 by around 1.5 K, implicitly reducing the allowed emissions for a stabilization of climate by as much as 3 Pg C a^{-1} (Jones *et al.*, 2006). But this particular ESM was found to show a much stronger climate–carbon-cycle feedback than others in the C^{4}MIP intercomparison (Friedlingstein *et al.*, 2006), and a recent study combining ensemble model runs with temperature reconstructions (Frank *et al.*, 2010) indicated that the observed amplification of the climate–carbon-cycle feedback is not consistent with this very strong feedback. This issue is further elaborated in Chapter 4.

Booth *et al.* (unpublished work) found that the spread in climate predictions due to uncertainty in the carbon-cycle modelling in emission-driven experiments is comparable to the spread due to uncertainty in the physical parameters. Thus, a key challenge remains: by including an important process (the carbon cycle) in ESMs, we have unearthed a new and large source of uncertainty! More work is evidently needed on the carbon-cycle component of ESMs, but we suggest that rapid progress will be made if carbon-cycle models adopt common observational benchmarks, which up to now has not been the case. Meanwhile, the climate modelling community has adopted a pragmatic approach to scenario development. Models used in the IPCC Fifth Assessment Report will be explicitly concentration-driven, with concentrations following 'Representative Concentration Pathways'; but the uncertainties in 'compatible emissions' for any given pathway will be assessed using either offline models or the biogeochemical components of the ESM, depending on each model's current capabilities.

5.4.2 How are the input scenarios of emissions and concentrations constructed?

Where the goal is to understand the response of the climate system to a given applied forcing then simple changes in greenhouse-gas concentrations can be specified *a priori*. Typically, climate models are forced with concentrations of atmospheric CO_2 increasing from pre-industrial levels at a compound rate of 1% per annum until the atmospheric concentration has doubled. The concentrations then level out and the climate models continue to run for several decades, or even several hundred years to reach a quasi-equilibrium state. Other types of idealized scenarios involve instantaneously increasing the CO_2 concentration from pre-industrial levels to either twice or four times this value, then again holding the concentration steady whilst the climate models adjust. These types of idealized simulations can be useful for model intercomparison experiments and for understanding the emergent timescales coming from the climate models for warming over land or ocean heat uptake. They also provide some guidance on the linearity of the response of the models to different levels of radiative forcing. They are not supposed to be in any sense realistic representations of the future.

To provide projections of the future for real-world planning of mitigation and adaptation policy, more plausible input scenarios are required. Here, the scenario usually starts at pre-industrial observed values, followed by actual historic changes in greenhouse gases and aerosols, with the greenhouse-gas emissions resulting from a 'what if' type question being applied for the future. Some types of 'what if' questions considered include: 'What if there is no mitigation policy and economic growth leads to an increased demand for energy?' and 'What if nuclear power generation, renewable energy and carbon capture and storage are deployed at rapid rates in the future?' These are specified not in terms of the physical quantities of emissions or atmospheric greenhouse-gas concentrations needed by ESMs. They are expressed in terms of how we think the socio-economics of the Earth might evolve. In particular, they rely on assumptions of population growth, demand for energy and other resources, and the development of new technologies as the world economy moves forward. The current tool of choice for converting the socio-economic scenarios into the inputs needed by ESMs is the integrated assessment models (IAMs, see Section 5.2). The 'what if' nature of the scenarios allows different views, often informed by the past, to be tested. This can include applying the constraints resulting from potential international climate agreements.

An area of current debate is whether concentration-overshoot scenarios should be considered (e.g. Huntingford and Lowe, 2007). These pathways examine a future where there is a target level for greenhouse-gas concentrations, for instance, consistent with Article 2 of the UNFCCC (1992), but where the concentrations of some greenhouse gases temporarily rise above the final target level before coming back down. It has been shown that this type of approach may have a lower mitigation cost than simply approaching the target concentrations from below (van Vuuren *et al.*, 2008). However, the approach relies on the inertia of the climate system to prevent climate change reaching unacceptable levels before the atmospheric concentrations can be brought back down from their peak to the eventual target. Furthermore, some scenarios of this type imply the possibility of globally negative emissions of CO_2 before the end of the twenty-first century, an outcome which most commentators consider extremely unlikely.

5.4.3 The IPCC experimental design

The IPCC's Third (Cubasch *et al.*, 2001) and Fourth (Meehl *et al.*, 2007a) Assessment Reports used scenarios from the Special Report on Emission Scenarios (SRES; IPCC, 2000) to provide input to the climate models. In total there are around 40 SRES scenarios for the twenty-first century, each providing a different view of how the world will evolve socio-economically and technologically. These are divided into four main groups, each with one or more marker scenarios:

- A1 group consists of a world with rapid economic growth, and the rapid spread of new technologies. There is significant convergence in standards of living between regions. The population rises to a peak of around 9 billion in 2050 and then gradually declines. Marker scenarios include: A1FI, which has a continued emphasis on fossil fuels; A1T, which places more emphasis on non-fossil-fuel energy; and an intermediate scenario, A1B, with a balance between energy sources.
- A2 group consists of a world with continuously increasing population and less economic convergence between the different regions. The increases in personal wealth and introduction of new technologies are slower than A1.
- B1 group consists of an ecologically friendly world, which includes the rapid introduction of clean and resource-efficient technologies. However, like the other SRES scenarios it does not have explicit mitigation policy. The evolution of global population growth is similar to that in the A1 group.
- B2 group consists of a world with a continuously increasing population, but with growth slower than in the A2 group. There are intermediate levels of economic development and technological change is slower than in the A1 and B1 groups.

Only a subset of the marker scenarios has been used as input to complex ESMs, because of the computational expense of simulating climate over periods of a century or more. In both the IPCC Third and Fourth Assessment Reports, most experiments were driven by concentration of greenhouse gases rather than emissions. However, a few experiments were driven by emissions (Friedlingstein *et al.*, 2006; Meehl *et al.*, 2007b). In some ways, those early, emission-driven studies were still proof-of-concept studies, but they have paved the way for a larger number of ESMs, incorporating biogeochemical as well as physical processes, to be part of future synthesis assessments.

While the SRES scenarios are still used in some studies, some aspects of their design have been considered unrealistic (Castles and Henderson, 2003; Höök *et al.*, 2009; Patzek and Croft, 2010). Several authors have also suggested that available reserves of fossil fuels may not be sufficient for the SRES scenarios with the largest amounts of future emissions (reviewed in Vernon *et al.*, 2011), but this point continues to be debated. Furthermore, the lack of any explicit treatment of mitigation policy in the SRES scenarios severely limits their value for policy.

For the IPCC Fifth Assessment Report, a new set of scenarios, the Representative Concentration Pathways (RCPs), are being used in place of the SRES scenarios (Moss *et al.*, 2010). Each RCP is named after the total change in radiative forcing to the year 2100. The four main pathways (Figure 5.6) are:

- RCP 3-PD (also often called RCP 2.6): a concentration-overshoot scenario, with radiative forcing peaking at approximately 3 W m^{-2} before returning to 2.6 W m^{-2} by 2100.
- RCP 4.5: a stabilization scenario where total radiative forcing is stabilized before 2100 by employment of a range of technologies and strategies for reducing greenhouse-gas emissions.
- RCP 6.0: a stabilization scenario where total radiative forcing is stabilized after 2100 without concentration overshoot.
- RCP 8.5: a scenario characterized by increasing greenhouse-gas emissions over time.

It is understood that the first three scenarios are unlikely to be realistic in the absence of mitigation policy. Each pathway was originally generated using one of four of the world's leading IAMs (IMAGE; MiniCAM; AIM; and MESSAGE), and ESM users are provided with either concentrations or emissions of greenhouse gases and aerosols as input for their experiments. A novel feature of the RCP approach is that a range of alternative socio-economic storylines are being developed for each of the physical sets of emissions and concentrations. This is important because impacts of climate change depend strongly on both the socio-economic background (including the population growth and the degree of protection they can afford) and the physical climate change projected to occur. Most of the ESMs taking part in the IPCC Fifth Assessment are being driven by concentration of greenhouse gases, but some are also running emission-driven experiments. Even where the ESMs are driven by atmospheric greenhouse-gas concentrations it is possible to extract from some of them the implied emissions of carbon dioxide that would have led to the concentration pathway, as previously demonstrated in the ENSEMBLES E1 experiments (Johns *et al.*, 2011; Figure 5.7).

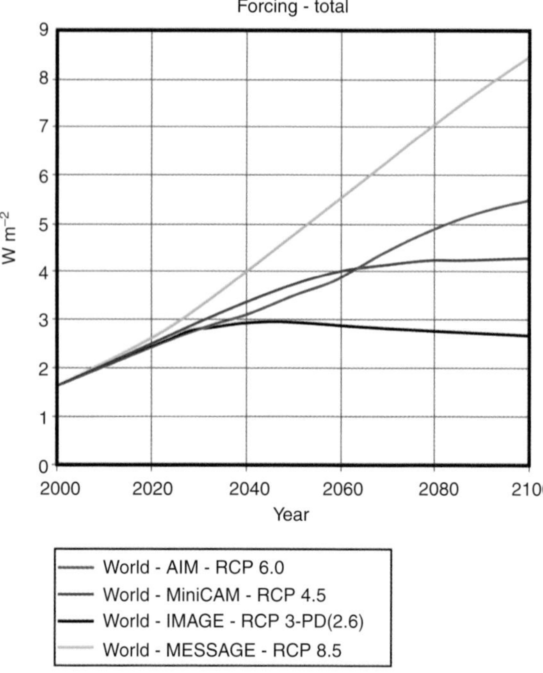

Figure 5.6 Total forcing during the twenty-first century for four RCP scenarios.

5.4.4 Projections of climate change

The IPCC's Fourth Assessment Report projected global average near-surface temperatures for a range of SRES scenarios between the years 1980–1999 and 2080–2099. Figure 5.8 shows the warmings ranging from 1.1 K (B1 scenario) to 5.4 K (A2 scenario), using a mix of complex and simple climate models; the highest projected warming of 6.4 K occurs under the A1FI scenario, which is not shown in the figure. Concomitant with this warming are large changes in patterns of precipitation, loss of sea ice, weakening of the Atlantic Meridional Overturning Circulation and an increase in sea level in the range of 18 cm to 59 cm (excluding any increase in dynamic effects associated with acceleration of outlet glaciers and ice streams on the great

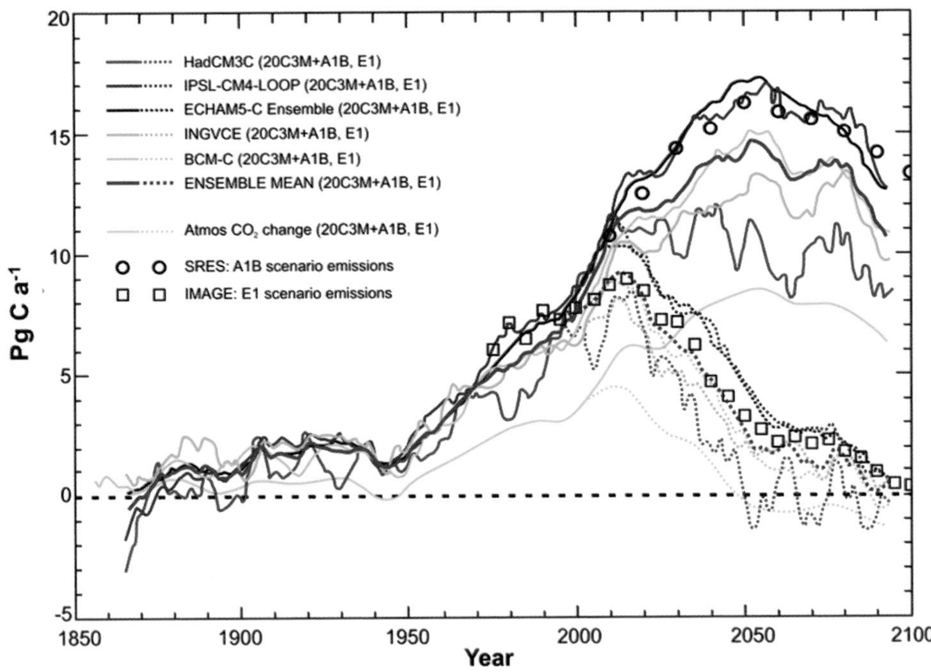

Figure 5.7 Implied ('allowable') anthropogenic net CO_2 emissions to the atmosphere (Pg C a^{-1}), diagnosed from the imposed change in atmospheric CO_2 concentrations and the modelled net carbon exchange between the atmosphere and land surface and ocean. An 11-year running average is applied to all curves, including concentration changes. ECHAM5-C results show an ensemble mean of six (20C3M + A1B) and three (E1) independent simulations, tending to smooth those results compared with other models. The ENSEMBLE MEAN curve weights each independent model equally. Corresponding SRES A1B and IMAGE E1 scenario values (the sum of fossil-fuel plus land-use-change emissions; every 5 years from 1970 to 2100 for E1, and every 10 years from 2000 to 2100 for A1B) are shown as symbols for comparison. From Johns *et al.*, 2011, with permission.

ice sheets). While global average increases in warming, sea level and a more intense hydrological cycle are considered robust results, the regional variations are much more uncertain. The agreement across models is greater for regional temperature than it is for changes in precipitation. The regional variations in future sea level also remain uncertain across models.

The multi-model ensemble of simulations for the IPCC Fifth Assessment Report, known as the CMIP5 ensemble, is not complete at the time of writing. But a simple climate-model study (Meinshausen *et al.*, 2011) suggests global average near-surface warming will peak in the range 1.3 K to 2 K (90% confidence interval range) above pre-industrial levels for RCP 3-PD, and reach between 3.6 K and 6.3 K for RCP 8.5. The first results have begun to appear for the runs using Earth system models of intermediate complexity (Schewe *et al.*, 2011) and complex ESM results (e.g. Arora *et al.*, 2011).

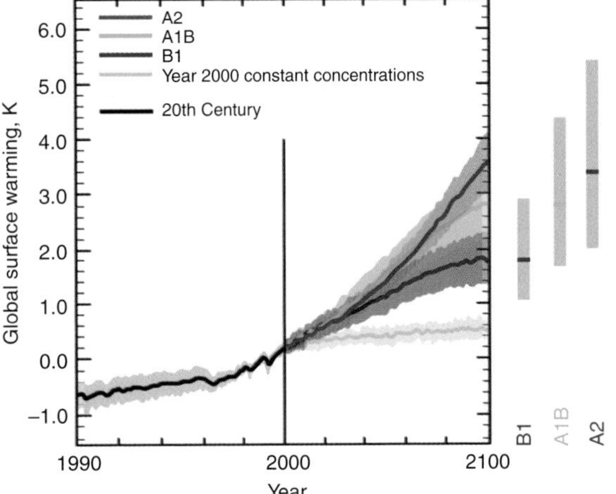

Figure 5.8 Past and future warming trends. Solid lines are multi-model global averages of surface warming (relative to 1980–1999) for the IPCC scenarios A2, A1B and B1, shown as continuations of the twentieth-century simulations. Shading denotes the ±1 standard-deviation range of individual model annual averages. The orange line shows an experiment where concentrations were held constant at year 2000 values. The grey bars at right indicate the best estimate (solid line within each bar) and the likely range assessed for the six SRES marker scenarios. The assessment of the best estimate and likely ranges in the grey bars is based on models and observational constraints. (Image used with permission from IPCC's AR4, SPM.5.)

5.4.5 How long is 'forever'?

An important application of ESMs is in clarifying the 'lifetime' of CO_2 added to the atmosphere, and hence the peak atmospheric CO_2 concentration (and thus warming) achieved for any given emissions scenario. This field includes consideration of CO_2 release associated with geological events (e.g. Panchuk *et al.*, 2008; Archer *et al.*, 2009; Zeebe *et al.*, 2009) as well as future fossil-fuel consumption. Chapter 2 discusses and debunks the still rather common misconception that there is a single, century-scale lifetime for CO_2.

The timescales of CO_2 removal by different processes can be explored and quantified with Earth system models (e.g. Ridgwell and Hargreaves, 2007) as illustrated in Figure 5.9. It is found that a *proportion* of CO_2 added to the atmosphere is indeed initially removed quickly: years to decades (for the CO_2-fertilization response of plant productivity) to centuries (soil-carbon inventory adjustment). At the same time, CO_2 quickly dissolves in seawater at the ocean surface and is transported in intermediate and deep waters to depth (Sabine *et al.*, 2004) making the ocean also an important sink of CO_2 on a timescale of years to hundreds of years (Archer *et al.*, 1997). However, neither the terrestrial nor the ocean sink has unlimited capacity for CO_2 sequestration, and on a timescale of a few thousand years a new equilibrium would be established between the atmosphere and ocean and terrestrial biosphere (Ridgwell and Hargreaves, 2007). At this point, 15–40% of total emissions will remain in the atmosphere (Archer, 2005).

On timescales longer than a millennium, additional components of the global carbon cycle come into play (Ridgwell and Edwards, 2007). Of the calcium carbonate ($CaCO_3$) produced in the ocean surface by calcifying plankton, about 10–15% is preserved in deep-sea sediments (Feely *et al.*, 2004). But the addition of CO_2 to seawater reduces the stability of $CaCO_3$, driving greater dissolution of $CaCO_3$ and thereby neutralizing the CO_2 (Ridgwell and Zeebe, 2005). A further geological buffering of atmospheric CO_2 arises through imbalances induced between the weathering of carbonate and silicate rocks on land and the burial of sediment in the ocean. These processes operate on timescales in the range of up to 10,000 years and once balance has been re-established between inputs (weathering) and sinks (sedimentary burial), some 7–8% of total emissions would remain in the atmosphere (Archer *et al.*, 1997; Ridgwell and Hargreaves, 2007). On even longer timescales (order 100,000 years, Colbourn, 2011), enhanced rates of silicate weathering resulting from the elevated atmospheric CO_2 concentrations and higher land-surface temperatures are expected to draw down the final fraction of excess CO_2 from the atmosphere. But on any conceivable human-relevant timescale, a significant fraction of emitted fossil-fuel CO_2 remains in the atmosphere and is effectively there 'for ever'.

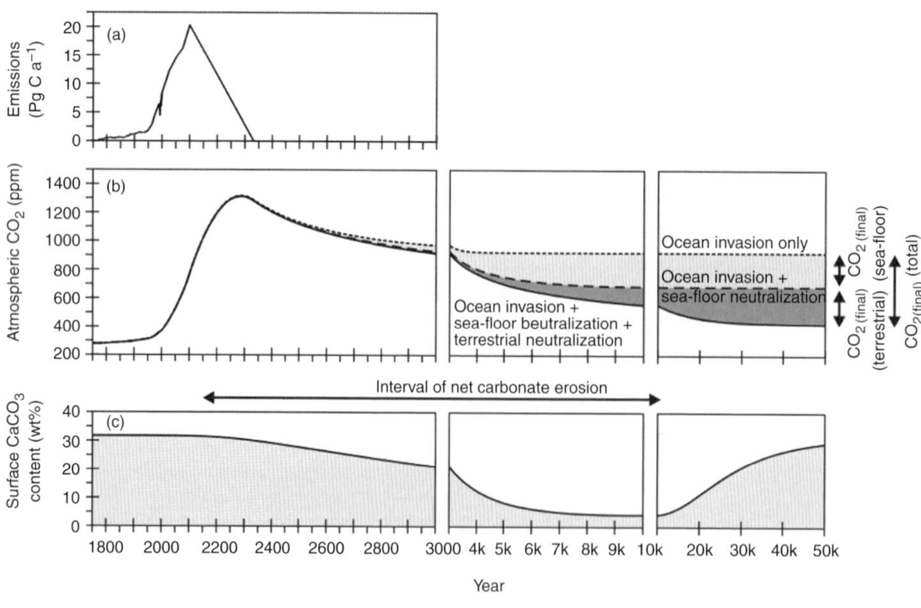

Figure 5.9 Earth system model analysis of the timescales of fossil-fuel CO_2 removal from the atmosphere. (Adapted from Ridgwell and Hargreaves, 2007). (a) Time-history of the rate of CO_2 emission to the atmosphere prescribed in the GENIE Earth system model (Ridgwell and Hargreaves, 2007). (b) Predicted trajectory of atmospheric CO_2. The light shaded region indicates the reduction of atmospheric CO_2 that is due to sea-floor $CaCO_3$ neutralization alone ($CO_{2(final)}$ sea-floor), while the dark shaded region shows that due to terrestrial $CaCO_3$ neutralization alone ($CO_{2(final)}$ terrestrial). (c) Evolution of global surface sediment composition (% $CaCO_3$). The minimum in sedimentary carbonate content corresponds to the transition of $CaCO_3$ accumulation from net erosion to re-deposition and replenishment of the surface sedimentary $CaCO_3$ reservoir.

5.5 A perspective on future model developments

5.5.1 Increasing complexity to cover a wider range of environmental problems

The current generation of ESMs already include an unprecedentedly wide range of Earth system components and are starting to account for the linkages and interactions between them. The first generation of models can account for the key transfers of matter and energy between components in an interactive way, allowing major system feedbacks to be identified. An example of this is the inclusion of the terrestrial and oceanic components of the carbon cycle, allowing changes in atmospheric CO_2 concentrations to affect plant photosynthesis and growth, vegetation distribution and hydrology, and thus affecting climate.

The second generation of models, now under development, need to account for more subtle interactions between components that arise from added detail or complexity in each Earth system compartment, thus building on the framework of current first-generation models. These involve processes that have been studied in some detail within particular scientific disciplines because of their importance in that field, but which have not previously been placed in a wider Earth system context. The availability of first-generation models allows these wider interactions to be explored, and the inclusion of these new components allows quantification both of new feedbacks arising and of changes to existing feedbacks.

The extension of biosphere–atmosphere interactions beyond the basic links associated with transfer of heat, moisture and carbon provides a number of examples of this. Vegetation constitutes a major source of reactive hydrocarbons to the atmosphere, altering atmospheric composition through formation of oxidants such as ozone and secondary organic aerosol particles. Ozone is a greenhouse gas, but it is also detrimental to plant health, damaging leaf surfaces and stomata and hence affecting photosynthesis and fluxes of moisture and carbon. Aerosol particles scatter shortwave radiation, altering light reaching plants at the Earth's surface, but they also affect cloud formation and thus both radiation balance and the water cycle. The importance of these interactions of atmospheric chemistry, ecosystems and climate is not yet fully understood, and their wider importance in the Earth system can only be clearly evaluated by including them in large-scale models where the key processes can be resolved.

Fire provides another important example of this type of process. Fire plays a vital role in vegetation clearance and renewal, providing one mechanism for climate to influence vegetation distribution through changes in temperature, soil moisture and lightning ignition. However, fire is an important source of oxidants and aerosol particles to the atmosphere, influencing greenhouse gases and cloud formation and thus impacting climate. While the effects of fire have long been considered within disciplines dealing with vegetation and atmospheric composition, the opportunity to link these fields and quantify the feedbacks that arise through vegetation–atmosphere–climate interactions involving fire is only now becoming available (see QESM in Section 5.2.1).

Improved understanding of biogeochemical and ecosystem processes in the terrestrial and oceanic environment and their response to climate change will necessitate the inclusion of additional nutrient cycles in large-scale models. While carbon cycling is now included in most climate models, essential nutrients, notably nitrogen and phosphorus, are generally still dealt with in a rudimentary way, using fixed ratios or simple assumptions. The effects of soluble iron in dust on growth of phytoplankton in the ocean are now widely appreciated and can be explored in an elementary way in current models. However, in practice, both crustal composition and plankton nutrient limitation vary greatly over different parts of the globe, but current models do not reflect this complexity. A proper assessment of the feedbacks between the biosphere and climate requires that the controls on the global cycles of these key nutrients be better understood, included in the models, and the results evaluated against relevant observations.

What types of problem will this new generation of models allow us to address? A more complete treatment of physical, chemical and biological processes and their interactions is needed to understand how the Earth system responds to perturbations from both natural and anthropogenic sources. The effects of the eruption of Mount Pinatubo in the Philippines in 1991, further discussed in Chapter 3, provide an excellent illustration of the complexity of Earth system responses. This major eruption released large volumes of reactive gases and particulate matter high into the atmosphere, altering atmospheric composition and dynamics. Studies of the aftermath of Pinatubo have largely focused on the effects in one Earth system component, for example the effects on surface temperature (McCormick *et al.*, 1995), on stratospheric circulation (Thomas *et al.*, 2009) and on vegetation growth (Lucht *et al.*, 2002). QUEST research has further explored interactions among domains: for example, how changes in climate and vegetation affected atmospheric composition through changes in emissions of biogenic hydrocarbons, contributing to the observed reduction in the atmospheric abundance of methane (Telford *et al.*, 2010).

5.5.2 Modelling priorities

As ESMs evolve, new processes are progressively added to the models. Priorities for model development include fire, nitrogen and phosphorus cycling, emissions of biogenic trace gases from the land and marine biosphere, stratospheric chemistry, ocean biogeochemistry including trophic effects, land–ocean exchanges and land-use change. Here we discuss a few examples at the frontier of Earth system modelling.

Fire

Fire engages both the terrestrial biosphere and climate. Recent developments in process-based vegetation–fire modelling (Thonicke *et al.*, 2010; Prentice *et al.*, 2011) make it possible to include prognostic fire modelling as an interactive component in ESMs. In contrast to a still fairly common assumption that the present extent of biomass burning is mostly due to human ignitions (e.g. Ito and Penner, 2005) more recent studies (discussed in Chapters 2 and 4) have shown that this is a gross oversimplification, and that present global biomass burning is at an historic low point, probably because of land use. Thus, modelling the effect of land use on fire has emerged as a higher priority than modelling human ignitions. Nevertheless, future climate change is bound to influence fire occurrence (Scholze *et al.*, 2006; Pechony and Shindell, 2010), with potential feedbacks on climate. Quantifying this feedback will require the inclusion of fire as an interactive component of ESMs, as has been done in QESM (see Section 5.2.1).

Ocean biogeochemistry and biology

Understanding how marine-plankton communities change under varying environmental conditions is crucial to assessing the effects of global change on marine ecosystems. Marine-plankton community dynamics are generated from the non-linear combination of biotic and abiotic ecosystem forcing. The traditional approach has been to couple bulk biomass plankton functional type (PFT) models to hydrodynamic models. The resulting model ecosystems are commonly used to quantify the interactions between climate and ocean biogeochemical cycles (e.g. Moore *et al.*, 2004; Bopp *et al.*, 2005; Le Quéré *et al.*, 2005). However, this approach means that the model biologies are incapable of adapting to new states, limiting the models' ability properly to predict ecosystem changes under changing environmental forcing. In order to capture ecosystem dynamics, we also need to describe the complex web of trophic interactions, which link all species together and feed back on each other. In the context of marine modelling, food webs are particularly important (Hannah *et al.*, 2010), yet a detailed understanding of the structure of some parts of the marine ecosystem and their interactions is currently lacking. Key processes that require further emphasis in model development and process inclusion are physiological processes, including those at the cellular level, which impact on ecosystem structure and function, and population dynamics and trophic interactions, such that the model ocean biology can adapt (and possibly evolve) in response to environmental change.

Land–ocean exchanges

The land–ocean interface embraces the biogeochemical pathways that link river catchments via estuaries and shelf seas to the open ocean. These areas play a particularly important but largely unquantified role in the Earth system. Estuaries act to mediate riverine inputs of nutrients, carbon and other land-derived material. Along with shelf and coastal seas, their significance is due to their exceptionally high biological productivity and close interaction with human activity. That this class of geographical regime deserves special attention is apparent in three important respects: their role in global biogeochemical cycles, their role in global ocean dynamics and their socio-economic significance. Coastal seas comprise only 18% of the Earth's surface, yet support half the world's fish and marine mammal biodiversity, account for more than half of global primary production, global denitrification and carbonate deposition, and provide most of the global fisheries catch (Walsh *et al.*, 1991; Longhurst *et al.*, 1995; Sloan *et al.*, 2007). Accurately simulating estuaries, coastal zones, the continental shelves and the ocean margins represents a particular challenge for ESMs, owing to the complex biophysical interactions in these regions that occur on scales orders of magnitude smaller than in the open ocean. It is well established that shelf seas are regions of exceptionally high biological productivity. This arises from the re-supply of nutrients to otherwise depleted surface waters by various processes, including heterotrophic nutrient recycling (by zooplankton and bacteria in pelagic and benthic ecosystems), coastal upwelling, cross-frontal transport and riverine inputs.

5.6 The outlook for Earth system science

To date, ESMs have been mainly used to predict the consequences of anthropogenic emissions of CO_2 and other trace atmospheric constituents for climate. The constant evolution of ESMs, due both to advances in our understanding of the Earth system and growth in computing resources, will be critical for evaluating the risks of future climate change but also, more broadly, for evaluating and improving strategies to mitigate and adapt to them. The idea of geoengineering the climate will pose important and potentially urgent challenges to ESMs, to estimate not only the effects on the target of global mean temperature but also the ramifications of any such measures through rainfall regimes and terrestrial and marine ecosystem function. These challenges also bring a responsibility: to match increasingly complex modelling with the creative use of observations to establish the validity of ESMs and quantify their limitations. We have identified several risks and potential pitfalls in the modelling enterprise. We believe they can be avoided. But we also note that to be effective and credible, the Earth system science enterprise will require a continuation of the process of breaking down barriers (between model development and data communities; field-data collection and remote-sensing communities; biologists, chemists and physicists; and, perhaps above all, between model developers and climate change stakeholders) that was promoted by QUEST.

References

Allen, M. R. and Stott, P. A. (2003). Estimating signal amplitudes in optimal fingerprinting, Part 1: theory. *Climate Dynamics*, **21**, 477–491.

Archer, D. (2005). Fate of fossil fuel CO_2 in geologic time. *Journal of Geophysical Research*, **110**, doi:10.1029/2004JC002625.

Archer, D., Kheshgi, H. and Maier-Reimer, E. (1997). Multiple timescales for neutralization of fossil fuel CO_2. *Geophysical Research Letters*, **24**, 405–408.

Archer, D., Eby, M., Brovkin, V. *et al.* (2009). Atmospheric lifetime of fossil-fuel carbon dioxide. *Annual Reviews of Earth and Planetary Sciences*, **37**, 117–134.

Arora, V. K., Scinocca, J. F., Boer, G. J. *et al.* (2011). Carbon emission limits required to satisfy future representative concentration pathways of greenhouse gases. *Geophysical Research Letters*, **38**, doi:10.1029/2010GL046270.

Bahurel, P. and the MERCATOR Project Team (2006). Mercator Ocean global to regional ocean monitoring and forecasting. In *Ocean Weather Forecasting, an Integrated View of Oceanography*, eds. E. P. Chassignet and J. Verron. Berlin: Springer-Verlag.

Bartlein, P. J., Harrison, S. P., Brewer, S. *et al.* (2011). Pollen-based continental climate reconstructions at 6 and 21 ka: a global synthesis. *Climate Dynamics*, **37**, 775–802.

Blyth, E., Clark, D. B., Ellis, R., *et al.* (2011). A comprehensive set of benchmark tests for a land surface model of simultaneous fluxes of water and carbon at both the global and seasonal scale. *Geoscientific Model Development*, **4**, 255–269.

Bopp, L., Aumont, O., Cadule, P., Alvain, S. and Gehlen, M. (2005). Response of diatoms distribution to global warming and potential implications: a global model study. *Geophysical Research Letters*, **32**, doi: 10.1029/2005GL019606.

Braconnot, P., Otto-Bliesner, B., Harrison, S. *et al.* (2007). Results of PMIP2 coupled simulations of the mid-Holocene and last glacial maximum – Part 1: experiments and large-scale features. *Climate of the Past*, **3**, 261–277.

Bretherton, F. P. (1985). Earth system science and remote sensing. *Proceedings of the IEEE*, **73**, 1118–1127.

Cadule, P., Friedlingstein, P., Bopp. L., *et al.* (2010). Benchmarking coupled climate–carbon models against long-term atmospheric CO_2 measurements. *Global Biogeochemical Cycles*, **24**, doi:10.1029/2009GB003556.

Castles, I. and Henderson, D. (2003). The IPCC Emission Scenarios: an economic-statistical critique. *Energy and Environment*, **14**, 159–185.

Clark, D. B., Mercado, L. M., Sitch, S., *et al.* (2011). The Joint UK Land Environment Simulator (JULES), model description – Part 2: carbon fluxes and vegetation. *Geoscientific Model Development Discussions*, **4**, 641–688.

Claussen, M. (1997). Modelling biogeophysical feedback in the African and Indian Monsoon region. *Climate Dynamics*, **13**, 247–257.

Claussen, M., Mysak, L. A., Weaver, A. J. *et al.* (2002). Earth system models of intermediate complexity: closing the gap in the spectrum of climate system models. *Climate Dynamics*, **18**, 579–586.

Colbourn, C. (2011). Weathering effects on the carbon cycle in an Earth system model. PhD thesis, University of East Anglia, Norwich.

Coleman, K. and Jenkinson, D. S. (1999). *RothC 26.3 – A model for the turnover of carbon in soil: model description and Windows Users Guide.* Harpenden: IACR-Rothamsted Research.

Collins, W. J., Bellouin, N., Doutriaux-Boucher, M. *et al.* (2011). Development and evaluation of an Earth-system model – HadGEM2. *Geoscientific Model Development Discussions*, **4**, 997–1062.

Costanza, R., Leemans, R., Boumans, R. and Gaddis, E. (2007). Integrated global models. In *Sustainability or Collapse: An Integrated History and Future of People on Earth*, eds. R. Costanza, L. J. Graumlich and W. Steffen. (Dahlem Workshop Report 96.) Cambridge, MA: MIT Press, pp. 417–446.

Cox, P. M. (2001). Description of the TRIFFID Dynamic Global Vegetation Model. Exeter: Hadley Centre, Met Office, Technical Note 24.

Cox, P. M., Betts, R. A., Jones, C. D., Spall, S. A. and Totterdell, I. J. (2000). Acceleration of global warming due to carbon-cycle feedbacks in a coupled climate model. *Nature* **408**, 184–187.

Cubasch, U., Meehl, G. A., Boer, G. J. *et al.* (2001). Projections of future climate change. In *Climate Change 2001: The Scientific Basis. The Scientific Basis. Contribution of Working Group I to the Third Assessment Report of the Intergovernmental Panel on Climate Change*, eds. J. T. Houghton, Y. Ding, D. J. Griggs *et al.* Cambridge: Cambridge University Press.

Daley, R. (1991). *Atmospheric Data Analysis*. Cambridge: Cambridge University Press.

de Noblet, N., Prentice, I. C., Joussaume, S. *et al.* (1996). Possible role of atmosphere–biosphere interactions in triggering the last glaciation. *Geophysical Research Letters*, **23**, 3191–3194.

de Noblet-Ducoudré, N., Claussen, M. and Prentice, I. C. (2000). Mid-Holocene greening of the Sahara: first results of the GAIM 6000 yr BP experiment with two asynchronously coupled atmosphere/biome models. *Climate Dynamics*, **16**, 643–659.

Der Kiureghian, A. and Ditlevsen, O. (2009). Aleatory or epistemic? Does it matter? *Structural Safety*, **31**, 105–112.

ECMWF – European Centre for Medium-range Weather Forecasts (2001). Lecture notes – Data assimilations and the use of satellite data. See: www.ecmwf.int/newsevents/training/rcourse_notes/DATA_ASSIMILATION/index.html

Edwards, N. R. and Shepherd, J. G. (2002). Bifurcations of the thermohaline circulation in a simplified three-dimensional model of the world ocean and the effects of inter-basin connectivity. *Climate Dynamics*, **19**, 31–42.

Edwards, N. R. and Marsh, R. (2005). Uncertainties due to transport-parameter sensitivity in an efficient 3-D ocean–climate model. *Climate Dynamics*, **24**, 415–433.

Eyring, V., Harris, N. R. P., Rex, M. *et al.* (2005). A strategy for process-oriented validation of coupled chemistry–climate models. *Bulletin of the American Meteorological Society*, **86**, 1117–1133.

Feely, R. A., Sabine, C. L., Lee, K. *et al.* (2004). Impact of anthropogenic CO_2 on the $CaCO_3$ system in the oceans. *Science*, **305**, 362–366.

Ferro, C. A. T., Jupp. T. E., Lambert, F. H., Huntingford, C. and Cox, P. M. (2012). Model complexity versus ensemble size: allocating resources for climate prediction. *Philosophical Transactions of the Royal Society A*, **370**, 1087–1099.

Fisher, J. B., Sitch, S., Malhi, Y. *et al.* (2010b). Carbon cost of plant nitrogen acquisition: a mechanistic, globally applicable model of plant nitrogen uptake, retranslocation, and fixation. *Global Biogeochemical Cycles*, **24**, doi:10.1029/2009GB003621.

Fisher, R., McDowell, N., Purves, D. *et al.* (2010a). Assessing uncertainties in a second-generation dynamic vegetation model caused by ecological scale limitations. *New Phytologist*, **187**, 666–681.

Fraedrich, K., Jansen, H., Kirk, E., Luksch, U. and Lunkeit, F. (2005). The Planet Simulator: towards a user friendly model. *Meteorologische Zeitschrift*, **14**(3), 299–304.

Frank, D. C., Esper, J., Raible, C. C. *et al.* (2010). Ensemble reconstruction constraints on the global carbon cycle sensitivity to climate. *Nature*, **463**, 527–532.

Friedlingstein, P., Cox, P., Betts, R. *et al.* (2006). Climate–carbon cycle feedback analysis, results from the C^4MIP model intercomparison. *Journal of Climate*, **19**, 3337–3353.

Friedrichs, M. A. M., Dusenberry, J. Anderson, L. A. *et al.* (2007). Assessment of skill and portability in regional marine biogeochemical models: the role of multiple plankton groups. *Journal of Geophysical Research*, **112**, doi:10.1029/2006JC003852.

Glecker, P., Taylor, K. E. and Doutriaux, C. (2008). Performance metrics for climate models. *Journal of Geophysical Research*, **113**, doi:10.1029/2007JD008972.

Gordon, C., Cooper, C., Senior, C.A. *et al.* (2000). The simulation of SST, sea ice extents and ocean heat transports in a version of the Hadley Centre coupled model without flux adjustments. *Climate Dynamics*, **16**, 147–168.

Gregg, W. W., Friedrichs, M. A. M., Robinson, A. R. *et al.* (2009). Skill assessment in ocean biological data assimilation. *Journal of Marine Systems*, **76**, 16–33.

Guenther, A., Karl, T., Harley, P. *et al.* (2006). Estimates of global terrestrial isoprene emissions using MEGAN (Model of Emissions of Gases and Aerosols from Nature). *Atmospheric Chemistry and Physics*, **6**, 3181–3210.

Hallegatte, S., Lahellec, A. and Grandpeix, J.-Y. (2006). An elicitation of the dynamic nature of water vapor feedback in climate change using a 1D model. *Journal of Atmospheric Science*, **63**, 1878–1894.

Hannah, C., Vezina, A. and St John, M. (2010). The case for marine ecosystem models of intermediate complexity. *Progress in Oceanography*, **84**, 121–128.

Hansen, J., Lacis, A., Rind, D. *et al.* (1984). Climate sensitivity: analysis of feedback mechanisms. In *Climate Processes and Climate Sensitivity*, eds. J. E. Hansen and T. Takahashi. AGU Geophysical Monographs 29, American Geophysical Union, pp. 130–163.

Hargreaves, J. C. (2010). Skill and uncertainty in climate models. *WIREs Climate Change*, **1**, 556–564.

Hargreaves, J. C., Annan, J. D., Edwards, N. R. and Marsh, R. (2004). An efficient climate forecasting method using an intermediate complexity Earth system model and the ensemble Kalman filter. *Climate Dynamics*, **23**, 745–760.

Hargreaves, J. C., Abe-Ouchi, A. and Annan, J. D. (2007). Linking glacial and future climates through an ensemble of GCM simulations. *Climate of the Past*, **3**, 77–87.

Harrison, S. P. and Prentice, I. C. (2003). Climate and CO_2 controls on global vegetation distribution at the last glacial maximum: analysis based on palaeovegetation data, biome modeling and palaeoclimate simulations. *Global Change Biology*, **9**, 983–1004.

Hasselmann, K. (1997). On multifingerprint detection and attribution of anthropogenic climate change. *Climate Dynamics*, **13**, 601–611.

Höök, M., Sivertsson, A. and Aleklett, K. (2009). Validity of the fossil fuel production outlooks in the IPCC emission scenarios. *Natural Resources Research*, **19**, 63–81.

Huntingford, C. and Cox, P. M. (2000). An analogue model to derive additional climate change scenarios from existing GCM simulations. *Climate Dynamics*, **16**, 575–586.

Huntingford, C. and Lowe, J. (2007). 'Overshoot' scenarios and climate change. *Science*, **316**, 829.

Huntingford, C., Stott, P. A., Allen, M. R. and Lambert, F. H. (2006). Incorporating model uncertainty into attribution of observed temperature change. *Geophysical Research Letters*, **33**, doi:10.1029/2005GL024831.

Huntingford, C., Booth, B. B. B., Sitch, S. *et al.* (2010). IMOGEN: an intermediate complexity model to evaluate terrestrial impacts of a changing climate. *Geoscientific Model Development*, **3**, 679–687.

Huntingford, C., Cox, P. M., Mercado, L. M. *et al.* (2011). Highly contrasting effects of different climate forcing agents on terrestrial ecosystem services. *Philosophical Transactions of the Royal Society A*, **369**, 2026–2037.

IPCC (2000). *Special Report on Emissions Scenarios: a Special Report of Working Group III of the Intergovernmental Panel on Climate Change*. Cambridge: Cambridge University Press.

IPCC (2005). *Guidance notes for Lead Authors of the IPCC Fourth Assessment Report on addressing uncertainties.* Geneva: Intergovernmental Panel on Climate Change.

IPCC (2007). Summary for policy makers. In *Climate Change 2007: The Physical Science Basis. Contribution of Working Group I to the Fourth Assessment Report of the Intergovernmental Panel on Climate Change*, eds. S. Solomon, D. Qin, M. Manning *et al.* Cambridge: Cambridge University Press.

Ito, A. and Penner, J. E. (2005). Historical emissions of carbonaceous aerosols from biomass and fossil fuel burning for the period 1870–2000. *Global Biogeochemical Cycles*, **19**, doi:10.1029/2004GB002374.

Johns, T. C., Royer, J.-F., Höschel, I. *et al.* (2011). Climate change under aggressive mitigation: the ENSEMBLES multi-model experiment. *Climate Dynamics*, **37**, doi: 10.1007/s00382-011-1005-5.

Jones, C. D., Cox, P. M. and Huntingford, C. (2006). Climate–carbon cycle feedbacks under stabilization: uncertainty and observational constraints. *Tellus B*, **58**, 603–613.

Joos, F., Bruno, M., Fink, R. *et al.* (1996). An efficient and accurate representation of complex oceanic and biospheric models of anthropogenic carbon uptake. *Tellus B*, **48**, 397–417.

Jouzel, J., Masson-Delmotte, V., Cattani, O. *et al.* (2007). Orbital and millennial Antarctic climate variability over the past 800,000 years. *Science*, **317**, 793–796.

Kalnay, E. M., Kanamitsu, M., Kistler, R., Collins, W. and Deaven, D. (1996). The NCEP/NCAR 40-year reanalysis project. *Bulletin of the American Meteorological Society*, **77**, 437–471.

Kaminski, T., Blessing, S., Giering, R., Scholze, M. and Voßbeck, M. (2007). Testing the use of adjoints for estimation of GCM parameters on climate time-scales. *Meteorologische Zeitschrift*, **16**, 643–652.

Kelly, D. L. and Kolstad, C. D. (1999). Integrated assessment models for climate change control. In *International Yearbook of Environmental and Resource Economics*

1999/2000: A Survey of Current Issues, eds. H. Folmer and T. Tietenberg. Cheltenham: Edward Elgar.

Kohfeld, K. E. and Ridgwell, A. (2009). Glacial–interglacial variability in atmospheric CO_2. In *Surface Ocean-Lower Atmosphere Processes*, eds. C. Le Quéré and E.S. Saltzman. AGU Geophysical Monograph 187, pp. 251–286.

Le Quéré, C., Harrison, S. P., Prentice, I. C. *et al.* (2005). Ecosystem dynamics based on plankton functional types for global ocean biogeochemistry models. *Global Change Biology*, **11**, 2016–2040.

Le Treut, H., Somerville, R., Cubasch, U. *et al.* (2007). Historical overview of climate change. In *Climate Change 2007: The Physical Science Basis. Contribution of Working Group I to the Fourth Assessment Report of the Intergovernmental Panel on Climate Change*, eds. S. Solomon, D. Qin, M. Manning *et al.*. Cambridge: Cambridge University Press.

Lee, T., Awaji, T., Balmaseda, M.A., Greiner, E. and Stammer, D. (2009). Ocean state estimation for climate research. *Oceanography*, **22**, 160–167.

Lenton, T. and Britton, C. (2006). Enhanced carbonate and silicate weathering accelerates recovery from fossil fuel CO_2 perturbations. *Global Biogeochemical Cycles*, **20**, doi:10.1029/2005GB002678.

Lenton, T. M., Marsh, R., Price, A. R. *et al.* (2007). A modular, scalable, Grid ENabled Integrated Earth system modelling (GENIE) framework: effects of atmospheric dynamics and ocean resolution on bi-stability of the thermohaline circulation. *Climate Dynamics*, **29**, 591–613.

Longhurst, A. R., Sathyendranath, S., Platt, T. and Caverhill, C. (1995). An estimate of global primary production in the ocean from satellite radiometer data. *Journal of Plankton Research*, **17**, 1245–1271.

Lorenz, E. N. (1975). The physical bases of climate and climate modelling. In *Climate Predictability*. GARP Publication Series 16. Geneva: World Meteorological Association, pp. 132–136.

Loulergue, L., Schilt, A., Spahni, R. *et al.* (2008). Orbital and millennial-scale features of atmospheric CH_4 over the past 800,000 years. *Nature*, **453**, 383–386.

Lucht, W., Prentice, I. C., Myneni, R. B. *et al.* (2002). Climatic control of the high-latitude vegetation greening trend and Pinatubo effect. *Science*, **296**, 1687–1689.

Lüthi, D., Le Floch, M., Bereiter, B. *et al.* (2008). High-resolution carbon dioxide concentration record 650,000–800,000 years before present. *Nature*, **453**, 379–382.

Masson-Delmotte, V., Jouzel, J., Landais, A. *et al.* (2005). GRIP deuterium excess reveals rapid and orbital-scale changes in Greenland moisture origin. *Science*, **309**, 118–121.

Masson, D. and Knutti, R. (2011). Climate model genealogy. *Geophysical Research Letters*, **38**, doi:10.1029/2011GL046864.

Matthews, B. (2005). The Java climate model. See: www.chooseclimate.org/jcm/index.html.

McCormick, M., Thomason, L. and Trepte, C. (1995). Atmospheric effects of the Mount Pinatubo eruption. *Nature*, **373**, 399–405.

Meehl, G. A., Stocker, T. F., Collins, W. D. *et al.* (2007a). Global climate projections. In *Climate Change 2007: The Physical Science Basis. Contribution of Working Group I to the Fourth Assessment Report of the Intergovernmental Panel on Climate Change*, eds. S. Solomon, D. Qin, M. Manning *et al.* Cambridge: Cambridge University Press.

Meehl, G. A., Covey, C., Delworth, T. *et al.* (2007b). The WCRP CMIP3 multi-model dataset: a new era in climate change research. *Bulletin of the American Meteorological Society*, **88**, 1383–1394.

Meinshausen, M., Wigley, T. M. L. and Raper, S. C. B. (2011). Emulating atmosphere–ocean and carbon cycle models with a simpler model, MAGICC6 – Part 2: applications. *Atmospheric Chemistry and Physics*, **11**, 1–15.

Moorcroft, P. R., Hurtt, G. C. and Pacala, S. W. (2001). A method for scaling vegetation dynamics: the ecosystem demography model (ED). *Ecological Monographs*, **71**, 557–586.

Moore, J. K., Doney, S. C. and Lindsay, K. (2004). Upper ocean ecosystem dynamics and iron cycling in a global three dimensional model. *Global Biogeochemical Cycles*, **18**, doi:10.1029/2004GB002220.

Morgenstern, O., Braesicke, P., O'Connor, F. M. *et al.* (2009). Evaluation of the new UKCA climate-composition model – Part 1: The stratosphere. *Geoscientific Model Development*, **2**, 43–57.

Moss, R. H. and Schneider, S. H. (2000). Uncertainties in the IPCC TAR: recommendations to lead authors for more consistent assessment and reporting. In *Guidance Papers on the Cross-Cutting Issues of the Third Assessment Report of the IPCC*, eds. R. Pachauri, T. Taniguchi and K. Tanaka. Geneva: World Meteorological Organization, pp. 33–51.

Moss, R. H., Edmonds, J. A., Hibbard, K. A. *et al.* (2010). The next generation of scenarios for climate change research and assessment. *Nature*, **463**, 747–756.

Munhoven, G. (2007). Glacial–interglacial rain ratio changes: implications for atmospheric CO_2 and ocean–sediment interaction. *Deep-Sea Research II*, **54**, 722–746.

Murphy, J. M., Sexton, D. M. H., Barnett, D. N. *et al.* (2004). Quantification of modelling uncertainties in a large ensemble of climate change simulations. *Nature*, **430**, 768–772.

Palmer, J. R. and Totterdell, I. J. (2011). Production and export in a global ocean ecosystem model. *Deep Sea Research I*, **48**, 1169–1198.

Panchuk, K., Ridgwell, A. and Kump, L. R. (2008). Sedimentary response to Paleocene Eocene Thermal Maximum carbon release: a model–data comparison. *Geology*, **36**, 315–318.

Patzek, T. W. and Croft, G. D. (2010). A global coal production forecast with multi-Hubbert cycle analysis. *Energy*, **35**, 3109–3122.

Pechony, O. and Shindell, D. T. (2010). Driving forces of global wildfires over the past millennium and the forthcoming century. *Proceedings of the National Academy of Sciences USA*, doi:10.1073/pnas.1003669107.

Petoukhov, V., Ganopolski, A., Brovkin, V. *et al.* (2000). CLIMBER-2: a climate system model of intermediate complexity. Part I: model description and performance for present climate. *Climate Dynamics*, **16**, 1–17.

Prentice, I. C., Guiot, J., Huntley, B., Jolly D. and Cheddadi, R. (1996). Reconstructing biomes from palaeoecological data: a general method and its application to European pollen data at 0 and 6 ka. *Climate Dynamics*, **12**, 185–194.

Prentice, I. C., Kelley, D. I., Foster, P. N. *et al.* (2011). Modeling fire and the terrestrial carbon balance. *Global Biogeochemical Cycles*, **25**, doi:10.1029/2010GB003906.

Randall, D. A. and Wielicki, B. A. (1997). Measurements, models, and hypotheses in the atmospheric sciences. *Bulletin of the American Meteorological Society*, **78**, 399.

Randerson, J. T., Hoffman, F. M., Thornton, P. E. *et al.* (2009). Systematic assessment of terrestrial biogeochemistry in coupled climate–carbon models. *Global Change Biology*, **15**, 2462–2484.

Rayner, P. J., Scholze, M., Knorr, W. *et al.* (2005). Two decades of terrestrial carbon fluxes from a carbon cycle data assimilation system (CCDAS). *Global Biogeochemical Cycles*, **19**, doi:10.1029/2004GB002254.

Ridgwell, A. (2007). Interpreting transient carbonate compensation depth changes by marine sediment core modeling. *Paleoceanography*, **22**, doi:10.1029.2006PA001372.

Ridgwell, A. and Zeebe, R. E. (2005). The role of the global carbonate cycle in the regulation and evolution of the Earth system. *Earth and Planetary Science Letters* **234**, 299–315.

Ridgwell, A. and Edwards, U. (2007). Geological carbon sinks. In *Greenhouse Gas Sinks*, eds. D. Reay, N. Hewitt, J. Grace and K. Smith. Wallingford: CABI.

Ridgwell, A. and Hargreaves, J. (2007). Regulation of atmospheric CO_2 by deep-sea sediments in an Earth system model. *Global Biogeochemical Cycles*, **21**, doi:10.1029/2006GB002764.

Ridgwell, A., Hargreaves, J. C., Edwards, N. R. *et al.* (2007). Marine geochemical data assimilation in an efficient Earth system model of global biogeochemical cycling. *Biogeosciences*, **4**, 87–104.

Sabine, C. L., Feely, R. A., Gruber, N. *et al.* (2004). The oceanic sink for atmospheric carbon. *Science*, **305**, 367–371.

Santaren, D., Peylin, P., Viovy, N. and Ciais, P. (2007). Optimizing a process-based ecosystem model with eddy-covariance flux measurements: a pine forest in southern France. *Global Biogeochemical Cycles*, **21**, doi:10.1029/2006GB002834.

Schellnhuber, H. J. (1999). 'Earth system' analysis and the second Copernican revolution. *Nature*, **402**, Millennium Supp., C19–C23.

Schewe, J., Levermann, A. and Meinshausen, M. (2011). Climate change under a scenario near 1.5 °C of global warming: monsoon intensification, ocean warming and steric sea level rise. *Earth System Dynamics*, **2**, 25–35.

Schneider, S. (1997) Integrated assessment modeling of global climate change: transparent rational tool for policy making or opaque screen hiding value-laden assumptions? *Environmental Modeling and Assessment*, **2**, 229–249.

Scholze, M., Knorr, W., Arnell, N. W. and Prentice, I. C. (2006). A climate change risk analysis for world ecosystems. *Proceedings of the National Academy of Sciences USA*, **103**, 13116–13120.

Scholze, M., Kaminski, T., Rayner, P., Knorr, W. and Giering, R. (2007). Propagating uncertainty through prognostic carbon cycle data assimilation system simulations. *Journal of Geophysical Research*, **112**, doi:10.1029/2007JD008642.

Senior, C. A. and Mitchell, J. F. B. (2000). The time-dependence of climate sensitivity. *Geophysical Research Letters*, **27**, 2685–2688.

Shaffer, G., Malskær Olsen, S. and Pepke Pedersen, J. O. (2008). Presentation, calibration and validation of the low-order, DCESS Earth System Model (Version 1). *Geoscientific Model Development*, **1**, 17–51.

Simmons, A. J. and Gibson, J. K. (2000). The ERA-40 project plan. Reading: ERA-40 Project Report Series No. 1.

Simmons, A. J. and Hollingsworth, A. (2002). Some aspects of the improvement in skill of numerical weather prediction. *Quarterly Journal of the Royal Meteorological Society*, **128**, 647–677.

Sloan, N. A., Vance-Borland, K. and Ray, G. C. (2007). Fallen between the cracks: conservation linking land and sea. *Conservation Biology*, **21**, 897–898.

Smith, J., Gottschalk, P., Bellarby, J. *et al.* (2010). Estimating changes in Scottish soil carbon stocks using ECOSSE. I. Model description and uncertainties. *Climate Research*, **45**, 179–192.

Smith, R. S., Gregory, J. M. and Osprey, A. (2008). A description of the FAMOUS (version XDBUA) climate model and control run. *Geoscientific Model Development*, **1**, 53–68.

Smith, R. S. and Gregory, J. M. (2012). The last glacial cycle: transient stimulations with an AOGCM. *Climate Dynamics*, **38**, 1545–1559.

Solomon, S., Qin, D., Manning, M. *et al.* (2007). Technical summary. In *Climate Change 2007: The physical science basis. Contribution of Working Group I to the Fourth Assessment Report of the Intergovernmental Panel on Climate Change*, eds. S. Solomon, D. Qin, M. Manning *et al.* Cambridge: Cambridge University Press.

Stainforth, D. A., Aina, T., Christensen, C. *et al.* (2005). Uncertainty in predictions of the climate response to rising levels of greenhouse gases. *Nature*, **433**, 403–406.

Stammer, D., Wunsch, C., Giering, R. *et al.* (2003). Volume, heat, and fresh-water transports of the global ocean circulation 1993–2000, estimated from a general circulation model constrained by World Ocean Circulation Experiment (WOCE) data. *Journal of Geophysical Research*, **108**, doi:10.1029/2001JC001115.

Stocker, T. F., Wright, D. G. and Mysak, L. A. (1992). A zonally averaged, coupled ocean–atmosphere model for paleoclimate studies. *Journal of Climate*, **5**, 773–797.

Stott, P. A., Tett, S. F. B., Jones, G. S. *et al.* (2000). External control of 20th century temperature change to natural and anthropogenic forcings. *Science*, **290**, 2133–2137.

Stott, P. A., Allen, M. R. and Jones, G. S. (2003). Estimating signal amplitudes in optimal fingerprinting, Part II: application to general circulation models. *Climate Dynamics*, **21**, doi:10.1007/s00382-003-0314-8.

Tarantola, A. (2005). *Inverse Problem Theory and Methods for Model Parameter Estimation*. Philadelphia, PA: Society for Industrial and Applied Mathematics.

Telford, P. J., Lathiere, J., Abraham, N. L. *et al.* (2010). Effects of climate-induced changes in isoprene emissions after the eruption of Mount Pinatubo. *Atmospheric Chemistry and Physics*, **10**, 7117–7125.

Texier, D., de Noblet, N., Harrison, S. P. *et al.* (1997). Quantifying the role of biosphere–atmosphere feedbacks in climate change: coupled model simulations for 6000 yr BP and comparison with palaeodata for northern Eurasia and northern Africa. *Climate Dynamics*, **13**, 865–882.

Thomas, M. A., Giorgetta, M. A., Timmreck, C., Graf, H.-F. and Stenchikov, G. (2009). Simulation of the climate impact of Mt. Pinatubo eruption using ECHAM5 – Part 2: sensitivity to the phase of the QBO and ENSO. *Atmospheric Chemistry and Physics*, **9**, 3001–3009.

Thonicke, K., Spessa, A., Prentice, I. C. *et al.* (2010). The influence of vegetation, fire spread and fire behaviour on global biomass burning and trace gas emissions: results from a process-based model. *Biogeosciences*, **7**, 1991–2011.

Tol, R. S. J. (2006). *Integrated Assessment Modelling*. Working Paper FNU-102. Research Unit Sustainability and Global Change, Hamburg University and Centre for Marine and Atmospheric Science. See: http://ideas.repec.org/p/sgc/wpaper/102.html.

UNFCCC (1992). United Nations Framework Convention on Climate Change, United Nations, Rio de Janeiro. Bonn: United Nations Framework Convention on Climate Change. See: www.unfccc.int/resource/ccsites/senegal/conven.htm.

Van Vuuren, D. P., Meinshausen, M., Plattner, G.-K. *et al.* (2008). Temperature increase of 21st century mitigation scenarios. *Proceedings of the National Academy of Sciences USA*, **105**, 15258–15262.

Vernon, C., Thompson, E. and Cornell, S. (2011). Carbon dioxide emission scenarios: limitations of the fossil fuel resource. *Procedia Environmental Sciences*, **64**, 206–215.

Walsh, J. J., Biscaye, P. E. and Csanady, G. T. (1991). Importance of continental margins in the marine biogeochemical cycling of carbon and nitrogen. *Nature*, **359**, 53–59.

Wang, Y. P., Leuning, R., Cleugh, H. A. and Coppin, P. A. (2001). Parameter estimation in surface exchange models using nonlinear inversion: how many parameters can we estimate and which measurements are most useful? *Global Change Biology*, **7**, 495–510.

Ward, B. A., Friedrichs, M. A. M., Anderson, T. R. and Oschlies, A. (2010). Parameter optimisation techniques and the problem of under-determination in marine biogeochemical models. *Journal of Marine Systems*, **81**, 34–43.

Williamson, M. S., Lenton, T. M., Shepherd, J. G. and Edwards, N. R. (2006). An efficient numerical terrestrial scheme (ENTS) for Earth system modeling. *Ecological Modelling*, **198**, 362–374.

Zachos, J. C., Röhl, U., Schellenberg, S. A. *et al.* (2005). Rapid acidification of the ocean during the Paleocene–Eocene thermal maximum. *Science*, **308**, 1611–1615.

Zachos, J. C., Dickens, G. R. and Zeebe, R. E. (2008). An early Cenozoic perspective on greenhouse warming and carbon-cycle dynamics. *Nature*, **451**, 279–283.

Zeebe, R. E. (2011). LOSCAR: Long-term Ocean-atmosphere-Sediment CArbon cycle Reservoir Model. *Geoscientific Model Development Discussions*, **4**, 1435–1476.

Zeebe, R. E., Zachos, J. C. and Dickens, G. R. (2009). Carbon dioxide forcing alone insufficient to explain Palaeocene–Eocene thermal maximum warming. *Nature Geosciences*, **2**, 576–580.

Zeng, N. and Neelin, J. D. (2000). The role of vegetation–climate interaction and interannual variability in shaping the African savanna. *Journal of Climate*, **13**, 2665–2670.

Ziehn, T., Kattge, J., Knorr, W. and Scholze, M. (2011). Improving the predictability of global CO_2 assimilation rates under climate change. *Geophysical Research Letters*, **38**, doi:10.1029/2011GL047182.

Zweck, C. and Huybrechts, P. (2005). Modeling of the northern hemisphere ice sheets during the last glacial cycle and glaciological sensitivity. *Journal of Geophysical Research*, **110**, doi: 10.1029/2004JD005489.

Chapter

6 Climate change impacts and adaptation: an Earth system view

Richard A. Betts, Nigel W. Arnell, Penelope M. Boorman, Sarah E. Cornell, Joanna I. House, Neil R. Kaye, Mark P. McCarthy, Douglas J. McNeall, Michael G. Sanderson and Andrew J. Wiltshire

In this chapter, we address the biophysical impacts of climate change, and the consequent impacts on socio-economic systems. Modelling the impacts associated with future climate change provides important information for society's mitigation and adaptation responses. It also presents significant challenges for Earth system science. We discuss the ways in which uncertainty in impact modelling arises and how it can be managed.

6.1 Introduction

6.1.1 Key concepts

Changes in climate, including those arising as a consequence of anthropogenic perturbations of the climate system, can result in a wide variety of impacts on Earth's ecosystems and the human activities that depend on them. There are two good practical reasons why it is important to understand the processes involved and assess the possible magnitudes of impacts.

First, an assessment of the extent to which continued anthropogenic climate change could inflict damage is needed in order that well-informed decisions can be made about the reduction of human influences on climate. Our understanding of Earth system behaviour alerts us to the fact that action to mitigate climate change through reductions in greenhouse-gas emissions is not without consequences; so decisions to pursue mitigation options need to be weighed up on the basis of reliable estimates of the costs, risks and benefits of different courses of action.

Secondly, the increase in atmospheric greenhouse-gas concentrations since the Industrial Revolution means that further climate change is inevitable even if greenhouse-gas emissions were to be reduced soon (Figure 6.1). It is therefore necessary for society to adapt to unavoidable changes. Since adaptation action is also not without consequences, it is important that adaptive action addresses credible risks, and represents an efficient allocation of resources.

Understanding the impacts of climate changes to which past human activities have already committed us also provides important information for assessing the effectiveness of action to mitigate climate change. Given that adaptation is necessary, assessing the impacts of different rates of climate change above that which is already unavoidable (Figure 6.1) may help inform decisions on how best to minimize risks through a combination of mitigation and adaptation measures.

Mitigation and adaptation decisions call for different kinds of input from climate and ecosystem science (Figure 6.2). For mitigation, the fundamental question is: 'What are the consequences of different emissions pathways?' Reducing future climate change requires a full-system assessment of the consequences of emissions, through their effects on the climate and also on the social and biophysical dimensions of the Earth system. Since CO_2 and other well-mixed greenhouse gases exert effects across the globe, the globally aggregated impacts of emissions are of the greatest relevance. Furthermore, since emissions reductions will take several decades to take effect, longer (e.g. century) timescales are of the greatest interest.

For adaptation, the fundamental question is: 'What conditions are expected in a particular place?' Information is needed both on the average state of the

Understanding the Earth System: Global Change Science for Application, eds. Sarah E. Cornell, I. Colin Prentice, Joanna I. House and Catherine J. Downy. Published by Cambridge University Press © Cambridge University Press 2012.

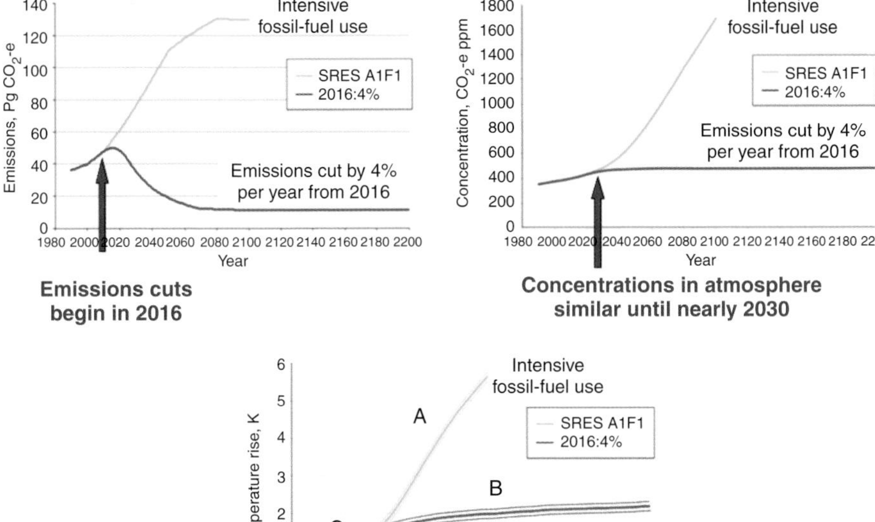

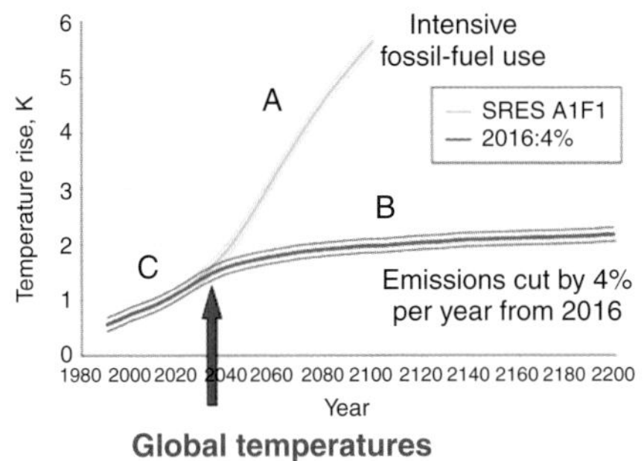

Figure 6.1 Avoidable and unavoidable climate change. Emissions are in carbon dioxide equivalent, CO_2-e. In the bottom panel, A: Impacts avoidable through mitigation; B: residual impacts after mitigation; C: committed near-term impacts. The difference (B–A) indicates the benefits of mitigation; C and B require adaptation.

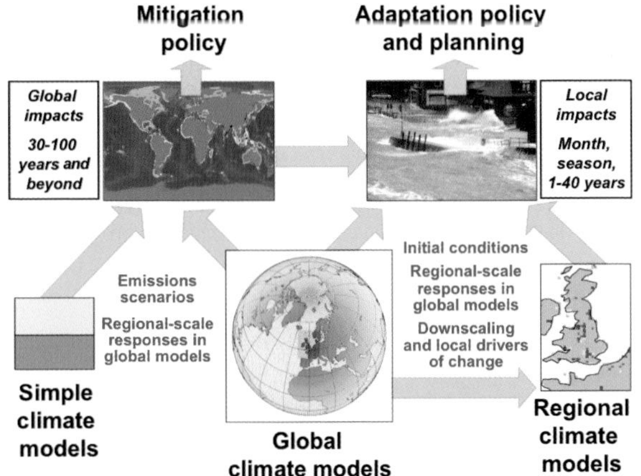

Figure 6.2 Aspects of climate change science relevant to mitigation and adaptation advice

climate and on the likelihood and frequency of extreme climatic events. In this context it is local to regional scales, and nearer-term timescales (up to a few decades), that are most salient.

'Climate' refers to the long-term set of weather conditions over a given area, over a range of spatial scales from a few kilometres to the entire Earth's surface, defined in terms of physical quantities such as temperature, precipitation, humidity and windspeed. The variability and extremes of these quantities are important, as well as the mean state. 'Climate change' is a change in the statistics of these quantities. The impacts of climate change include anything that occurs as a consequence of these changes. The link between climate change and its impacts is commonly viewed as the final stage in a linear chain of causes and effects, starting with human industrial and agricultural activities and moving through greenhouse-gas emissions to changes in climate and on to impacts (Figure 6.3(a)).

However, as has been demonstrated in earlier chapters, the Earth operates as a tightly coupled system of which humans are only a part. The representation of climate change and its impacts as a linear chain of causes and effects may thus fail to capture the true processes of change, including interactions and synergies. For example, as well as leading to greenhouse-gas emissions, industrial and agricultural activities influence the climate through changes in the physical properties of the land surface. These influences need to be considered alongside the impacts of changes in greenhouse-gas concentrations. Some greenhouse gases also exert impacts aside from their influence on radiative forcing. Changing atmospheric concentrations of CO_2 and O_3, for example, both directly affect the physiology of plants and other life forms. Some impacts of climate

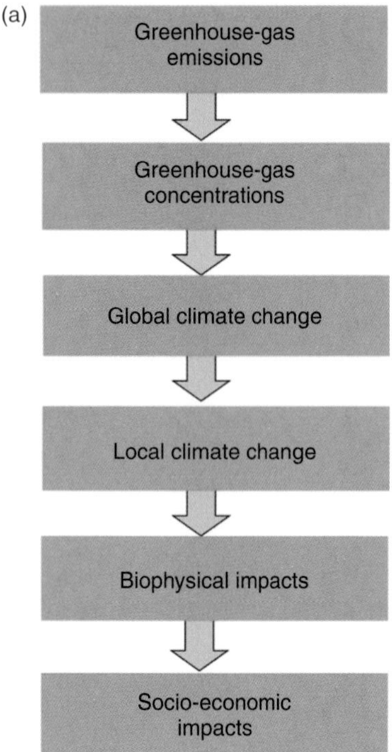

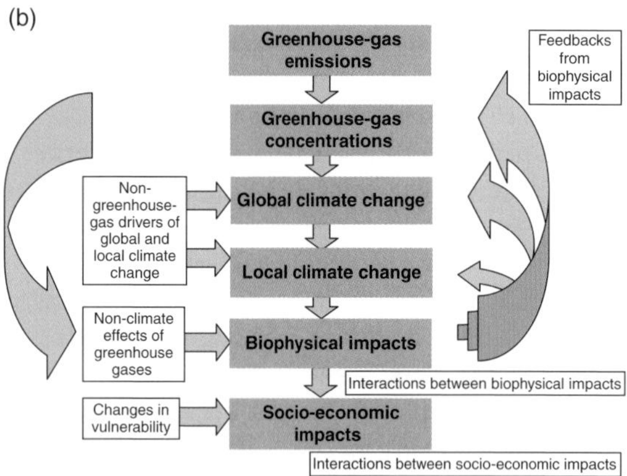

Figure 6.3 Impacts of anthropogenic climate change: (a) a common linear view of causes and effects focusing on radiatively forced climate change, and (b) a wider view considering additional human drivers of change, interactions between impacts, and feedbacks to climate change.

change have the potential to influence climate change itself, either directly through the natural system (for example through ecosystem responses affecting the carbon cycle) or indirectly through the human system (for example through energy-use patterns, land-use practices or the design of urban areas), as discussed in Chapter 4. Moreover, different impacts can interact. For instance, changes in water demand for agriculture can affect the availability of water for other purposes; the physiological responses of plants to changes in climate and atmospheric composition affect the terrestrial water cycle through changes in evapotranspiration; and so on (Figure 6.3(b)).

It is necessary also to consider non-human-driven climatic processes, including both natural forcing and internal variability of the climate system. Although these may be viewed as peripheral to the issue of mitigation – it can be argued that they merely provide an external context to the outcomes of different scenarios of human influence – they are nevertheless part of the 'baseline' against which anthropogenic impacts are to be assessed.

It makes sense to distinguish between biophysical and socio-economic impacts of climate change. Biophysical impacts are components of the physical, biological and chemical Earth system which are affected by the state of the climate – changes in the mass of ice sheets and glaciers; sea levels; the quantity of freshwater on or under the land; the productivity and character of vegetation. Socio-economic impacts are the next step – the knock-on effect on human society, which tends to be adapted to local conditions and hence will probably need to adjust, possibly to a very large extent, as conditions change. For example, coastal and riverside cities are built to align with historical fluctuations in water levels; agricultural systems use crops and management techniques appropriate to the local climate; and buildings are designed to cope with and exploit expected weather conditions.

It is especially useful to make this distinction because socio-economic impacts depend on biophysical impacts, but are also directly affected by human factors. Biophysical impacts alter the local terms of reference for human settlements and lifestyles, with potential consequences for societies and economies, but the actual consequences depend on the resilience and flexibility of the human systems: the nature of the physical infrastructure, the economic and political situation, and the ability to find alternatives. This introduces additional complexities in distinguishing the impacts of climate change from other changes in the world. Indeed, it may not be possible to identify a 'socio-economic impact of climate change' that is genuinely separate from other changes in society. Nevertheless, it is important to establish some level of understanding of how climate change affects society, and where links to biophysical impacts of climate change are unequivocal.

The problem of climate impacts can be cast in terms of risk, where:

$$\text{risk} = \text{hazard} \times \text{vulnerability} \quad (6.1)$$

Risk is defined with respect to a particular 'recipient', such as a species, a business, a country or a person. Hazard is an external factor (e.g. the nature and magnitude of climate change), while vulnerability is a property of the recipient – different recipients may have different levels of vulnerability to the same hazard, and may be exposed to different levels of risk. So for river flooding, the hazard is the extreme rainfall and the resulting increase in river levels, while vulnerability is a function of the location of buildings, existence of flood defences, preparedness of the residents for coping with flooding, and so on.

Adaptation can be considered to be largely about reducing vulnerability, while mitigation is about reducing hazard.

6.1.2 The contribution of Earth system science

A full Earth system approach is relevant to both mitigation and adaptation questions (Table 6.1).

Mitigation policy can benefit from information on the potential impacts of different scenarios of energy production and land use, including the additional climatic effects of different mitigation options within these sectors (such as biofuels and avoided deforestation). The non-climatic effects of different emissions scenarios, such as the physiological effects of changes in CO_2 and O_3 concentrations, are also important. This situation calls for projections of impacts on the timescales over which the different options would exert distinguishable impacts. Adaptation policy also requires projections of future changes at local to regional scales including all drivers of change, not just greenhouse-gas emissions, but taking into account the natural climate variability baseline as well. Some adaptation requires identification and understanding of impacts that are already occurring, so it is important to be able to explain current observed changes and attribute them to their causes (or at least assess the change in risk arising from different influences), whether these are changes in climate or other factors. Finally, given the difficulty in making well-constrained predictions of future change, some adaptation decisions simply require an assessment of vulnerability to a spectrum of potential changes, so that, where possible, actions may be targeted to reduce this vulnerability.

This chapter deals with issues surrounding the assessment of the impacts of climate change from the perspective of informing both mitigation and adaptation. In Section 6.2, evidence for past and current impacts of climate change is discussed, including a commentary on observed changes which may be at least partly due to climate change and a discussion on the extent to which changes such as these can be formally attributed to anthropogenic climate change. Section 6.3 gives an overview of some of the methods used to assess future impacts of climate change on different sectors. Section 6.4 summarizes the kinds of things that are known, and not known, about future climate change impacts and takes a critical look at some of the ways in which climate change mitigation has been quantified. This section briefly considers impacts at different levels of global warming, impacts of different emissions scenarios, which include time-dependency and additional drivers of change, and an approach based on thresholds of impact. Section 6.5 addresses the issue of vulnerability to climate change and the reduction of vulnerability through adaptation in the real world.

6.2 Measuring and modelling potential impacts of climate change

Explaining the present impacts of climate change and projecting future impacts requires quantification of the link between climate change and impacts. Many aspects of the natural environment are routinely monitored, not only meteorological quantities but also other quantities such as river flows, snow and ice cover, sea level, and the presence and phenology (that is, the timing of seasonal and cyclic events) of numerous life forms, providing a wealth of data for analysis. Understanding of impacts also comes from experiments carried out in controlled environments. Mathematical expressions of responses established through observations and experiments form the basis of numerical models, which can then be used together with climate projections to assess future hazards.

Models are used to estimate the changes in the longer-term character of the climate, under different assumptions concerning the drivers of change (increases in greenhouse-gas concentrations and other

Table 6.1 Issues for scientific studies informing climate change mitigation and adaptation

Issues	Mitigation	Adaptation
Timescale	30 years and beyond (mitigation cannot influence outcome in nearer term)	Next few years – adaptation to variability and current climate change Next few decades – adapting to changing mean climate state Maybe longer for major infrastructure and policy
Spatial scale	Global Key countries (because of influence on climate or on decision-making)	Local to national scale Global if considering reach of particular institution
Role of non-greenhouse-gas drivers of climate change (e.g. land-cover change, urban heat islands)	Full-system comparison of different mitigation options	Improved projections of local change
Role of non-climate anthropogenic drivers of change (e.g. impacts of changes in atmospheric composition on physiological processes; ecosystem disturbance)	Unintended consequences of mitigation action Full-system comparison of different forcing agents	More complete picture of changes in risk
Role of interactions between impacts	More complete and robust picture of consequences of different mitigation options	More complete picture of changes in risk
Role of non-anthropogenic drivers of change	Sets context (relative magnitude of influence and impacts) Possible synergistic influence on magnitude of anthropogenic impacts	Require complete assessment of all changes from all causes – adaptation is to 'change' not just 'climate change' in isolation
Role of observations and attribution of existing impacts	Testing and improving of models used for future projections	Establishing whether perceived changes are real, to avoid maladaption
Modelling issues and tools	Feedbacks Emissions scenarios (differences between them in impacts and drivers) Natural forcing as systematic boundary condition Integrated assessment models General circulation models (multi-decadal projections) Regional climate models (multi-decadal projections)	Emissions scenarios not so relevant – need to know likely/ plausible future changes Initialized forecasts and natural forcing to account for natural variability General circulation models (multi-decadal projections and near-term initialized forecasts) Regional climate models (multi-decadal projections and near-term initialized forecasts) Weather generators, statistical downscaling

factors). The outputs of climate models are increasingly used to assess the potential future impacts of climate change, both through linking them with other process-based models and by using statistical relationships between atmospheric quantities and 'impacts quantities'. Alternatively, instead of using a chain of models to predict future impacts, projected future climate changes can be examined in the context of metrics relating to vulnerability. For example, if history shows that a particular threshold of river streamflow is important for a particular societal impact, a combination of models and past data can be used to assess whether this threshold may be crossed more or less frequently in future.

Explaining the causes of past global changes is challenging, given the lack of controlled experimental conditions. Statistical analysis of quantities changing in the context of each other can provide circumstantial evidence of a relationship, and often such evidence can be strong, but conclusions may be vulnerable to unidentified confounding factors. A particular difficulty is that relationships with climate change do not allow separate identification of the relative contributions of anthropogenic and natural contributions to climate change. One 'work-around' is to use models as a virtual reality: the real world is assumed to be reasonably well replicated within the model, which is then subject to controlled experiments. This approach is already used for the detection and attribution of anthropogenic effects on aspects of climate change, such as changes in temperature. An inherent difficulty is that climate change impacts are likely to be experienced through extreme events, and a single weather event cannot be attributed to a particular cause. However, estimates can be made of the contribution of past anthropogenic climate change to the relative probability of particular kinds of events, which can then be related to impacts associated with those events.

6.2.1 Observing and monitoring change and potential impacts

To understand an impact of climate change, we need to know something about what the baseline conditions are, both in terms of the current state of the system and in terms of its usual dynamics. For that, we need long records, and we need good high-resolution coverage of the whole world. In reality, impacts research faces constraints on both those requirements. We start our discussion by considering biophysical impacts.

Many properties of ecosystems are studied systematically, providing a rich information base from which evidence of change may be obtained. Site-specific phenology studies have provided a substantial portion of the direct evidence for changes in terrestrial ecosystems over several decades – generally speaking, it is this aspect of ecosystems that appears to have been systematically monitored for longest, with some records dating back over 250 years (Puppi, 2007). Phenological changes have been documented for plant, animal, bird and insect species; phenological studies of multiple species and trophic levels at the same site are particularly valuable, such as an 80+-year record of phenology of a number of plant and animal species in New York state (Cook *et al.*, 2008). More recently, satellite imagery has allowed for a more global monitoring of the vegetative phenology of land plants.

Satellite observations of the land surface now allow for wider and more complete geographical coverage of many ecosystem and landscape properties, although data are only available for the last 20–30 years. Earth observation can in principle be used to monitor aspects of vegetation cover including the fraction of absorbed photosynthetically active radiation (fAPAR) and fire regimes (discussed in Chapter 2). However, a difficulty arises when using remote sensing for long-term monitoring because the instruments making the observations have changed, reflecting rapid developments in technology. Furthermore, the records for some instruments are not long. Advanced Very High Resolution Radiometer (AVHRR) data are available from 1982 onwards, giving observations of land cover, ocean state, clouds and other climate variables, which have underpinned many recent studies (e.g. Potter *et al.*, 2007; and other studies contributing to the Fourth Assessment Report of the IPCC). The Moderate Resolution Imaging Spectroradiometer (MODIS) became operational in 1999 (on the Terra satellite) and 2002 (on the Aqua satellite), delivering widely used observation products for land, atmosphere, cryosphere and ocean processes. Comparison of data from one instrument against another relies on certain assumptions about the behaviour of the sensors, and involves complex algorithms for data transformation. Considerable efforts have been made to eliminate spurious trends or shifts in long-term remotely sensed data sets.

Many ecosystem characteristics that are not (or not currently) observed from space, such as biomass, productivity and species composition of vegetation, only began to be monitored when concern about climate change became prominent. These programmes tend to focus on key ecosystems, notably forests. Several dozen sites across the moist tropics have been monitored intermittently to estimate changes in forest biomass. In a major effort, the international RAINFOR project (e.g. Phillips *et al.*, 2009) measured the girth of trees across the tropics to estimate biomass, on occasions several years apart. This effort now provides a ground-based record that can be used to calibrate remote-sensing data, and from which changes in tropical forest biomass can be estimated.

In addition to direct observations, information about baseline conditions and variability can be obtained by studying aspects of life forms that retain information on their growth and functioning in the past, such as tree rings and stable carbon isotope composition. Tree-ring records can be an indicator of ecosystem change and variability, potentially extending well back into the past (before human record-keeping). Ring width and wood density can indicate annual tree growth, and stable isotopes within the rings can also indicate changes in physiological functioning. Although tree rings are widely used as an indirect measure of past climatic changes, they can also be used in impacts studies as indicators of vegetation growth and functioning.

Mean sea level is monitored using records from tide gauges and, more recently, satellite altimetry. The data set of the Permanent Service for Mean Sea Level (Woodworth and Player, 2003) allows trends to be estimated, although the data contain a geographical bias towards the northern hemisphere and a strong bias towards coastlines rather than ocean interiors (Bindhoff *et al.*, 2007). Satellite altimetry provides global coverage, although precise calibration against tide gauges is required.

All these information sources have their own methodological challenges affecting the level of confidence in observed changes. A persistent difficulty for global change science is that the coverage of site-specific studies is far from globally uniform. In many cases, the land-based records are mainly in Europe and North America. Research and policy interest in specific ecosystems means there there are some in high latitudes and the tropics too, but data gaps remain in large areas, notably much of Africa and central and southern Asia. Also, some sites have records extending back several decades, or in exceptional cases centuries, while others have only recent or temporally sparse data. Thus, the various types of site-specific record share the common challenge of trying to establish a large-scale signal from a patchy evidence base.

When it comes to 'observing' socio-economic impacts, on the one hand there is a long-standing and very well-administered process in almost all countries of tracking many sorts of social and economic measures. On the other hand, as noted earlier, we are concerned with socio-economic impacts arising from environmental change, so they have a biophysical component nested within them – but they also depend on human factors. We illustrate the challenge that this presents with the example of impacts linked to drought.

Human society depends fundamentally on its agricultural systems for its food security, so climate impacts in this sector are an area of very great interest. If drought becomes more frequent, or drought events more extreme, the viability of agriculture can be affected; with implications for national economies, public health, and migration pressures (when communities depending on subsistence agriculture seek to move elsewhere). Holton *et al.* (2003) pointed out that '*the importance of drought lies in its impacts. Thus definitions should be region-specific and impact- or application-specific in order to be used in an operational mode by decision makers*'. It is common to distinguish between meteorological drought (low precipitation), agricultural drought (low soil moisture and increased water stress on plants), hydrological drought (reduced streamflow) and socio-economic drought (a deteriorating balance of supply of and demand for fresh water).

For large scales, river flows provide an indicator of the net water balance of the river basin. However, many rivers are affected directly by human intervention (such as extraction by irrigation, or introduction of dams) so many long-term data sets are not recommended for use as indicators of climate change or variability (Dai and Trenberth, 2002). As an indicator of drought specific to local agriculture, soil moisture would probably be the most relevant measure, but there are very few long-term records of soil moisture (Solomon *et al.*, 2007). Estimates of past changes in soil moisture, going back some 50 years, have been made using land-surface models driven by meteorological observations and reanalyses (Sheffield and Wood, 2008; Dai, 2011), and used to reconstruct patterns of drought.

Meteorological measurements, particularly temperature and precipitation, are much more widespread and observational records are relatively long, so drought tends to be quantified in terms of metrics calculated using these variables. One such metric is the Standard Precipitation Index (SPI), which relates precipitation over a given time period to that expected from the past climatology for the locality. Another widely applied metric is the Palmer Drought Severity Index (PDSI; Palmer, 1965), which is used to estimate the difference in moisture supply from normal, on the basis of precipitation, temperature and sometimes other variables such as humidity. As with the SPI, the PDSI is locally defined taking account of conditions expected from local climatology. It has been noted that PDSI appears to become systematically biased at higher temperatures (Sheffield and Wood, 2008), so despite its extensive application to date in development and water-resources contexts, it may not be suitable for studies of future climate change.

Since drought metrics tend to be locally defined, this must be borne in mind when analyzing long-term and global changes. Just as 'drought' depends on local experience, so a given level of rainfall or soil moisture may be regarded as drought in one location but not in another (e.g. Fraser, 2006, 2007). The perception of drought may therefore also evolve over time, as it depends on conditions to which the local human population are accustomed. However, the timescales that apply to becoming 'accustomed' to a change in climatic conditions are also variable (Brooks *et al.*, 2009; Berrang-Ford *et al.*, 2011). Drought is therefore a relative rather than absolute term and there are important unresolved issues in how drought-related climate impacts should be quantified.

6.2.2 Experiments to understand impacts-related processes

It is not enough simply to monitor changes that may or may not be a consequence of climate change – it is also important to understand the underlying physical, chemical and biological processes, so that past changes can be properly explained in terms of their causes. Ecological and agricultural processes of importance to climate change are studied experimentally by methods in which environmental conditions are manipulated under controlled conditions and the ecosystem responses studied. For example, 'dry-down' experiments are used to study ecosystem responses to drought (Brando *et al.*, 2008). Free Air Carbon Dioxide Enrichment (FACE) experiments are used to study the response of plant species (including many crops) to rising CO_2 concentrations in conditions of a free atmosphere as opposed to in a laboratory (Long *et al.*, 2006). Similar techniques are used to examine the impacts of increasing ozone pollution (Morgan *et al.*, 2003).

6.2.3 Modelling the impacts of climate change

In modelling studies of climate change, the climate system is represented by interacting sub-systems including the atmosphere, the ocean, the land surface (i.e. terrestrial ecosystems), marine ecosystems and the cryosphere. The interactions between sub-systems can be handled in two ways. The traditional method is to assume a linear cause–effect chain in which one sub-system is simulated first, and the resulting state is then applied as input to a second sub-system model, and so on. For example, projections of radiatively forced climate change and its impacts on ecosystems and hydrology generally consist of such a multi-stage linear sequence of simulations with models of different components of the climate system. Scenarios of greenhouse-gas and aerosol emissions are derived by models that follow scenarios of future changes in population, technology and economic state. These emissions scenarios are then translated into scenarios of greenhouse-gas concentrations in the atmosphere, using models that consider the processes acting to reduce concentrations through chemical reactions in the atmosphere or uptake at the land and ocean surfaces. These scenarios are then applied as inputs to climate models, which compute the radiative forcing due to the scenarios and the consequent changes in climate. The resulting climate scenarios are then used to drive sector-specific impacts models, which simulate the response of ecosystems, hydrology and other aspects of the Earth system, including agriculture, energy demands and other human activities.

The alternative to this linear cascade approach is Earth system modelling. Much effort is being channelled into developing fully coupled Earth system models where many of these biophysical sub-systems are modelled interactively rather than in sequence, as discussed in Chapter 5. However, it remains useful to keep a distinction between the biophysical components and the socio-economic sub-system. This is in part because the impacts of climate change in any given location depend on many factors: changes in mean climate, changes in extreme events, other drivers of local climate change, and non-climatic effects. Remote effects are important as well as local ones, through both physical links (rivers, etc.) and links through the human system such as international trade links. When impacts include non-climate factors, these aspects of the projected change should not be interpreted as a consequence of climate change. Nevertheless, these factors do need to be taken into account as part of the context to the change – vulnerability may change over time due to increasing population or changing economic situation, as well as climate changes, and ignoring this would give an unrealistic estimation of impacts.

For studies informing mitigation, impacts should be expressed as a change under one scenario relative to another. For example, flood risk may change over time as a result of both climate change and increased numbers of people living on flood plains. Both these factors should be considered in estimating the impact of climate change. The relevant information for informing mitigation would be an estimate of the proportion of the population at risk of flooding under one emissions scenario compared to another emissions scenario, with a future estimate of the number of floodplain-dwellers included in each.

In the following sections we discuss the present state of the modelling that underpins impact studies, and some key issues surrounding this modelling. It is a large field and so the examples cited are illustrative rather than comprehensive.

Sea-level rise

Large-scale changes in mean sea level are simulated with process-based ocean general circulation models (ECMs) taking account of ocean thermodynamics and circulation, with inputs of water from land (including ice melt) and outputs via evaporation (Randall *et al.*, 2007). This procedure therefore provides simulations of changes in sea level arising from changes in the total mass of water both globally and regionally, due to changes in the water budget and ocean circulation (Lowe and Gregory, 2006) and changes in global and regional water volume arising from changes in the temperature and mass of the oceans. In addition, short-lived, localized changes due to atmospheric-pressure variations (storm surges) are simulated using storm-surge models of local sea areas driven by regional-scale climate models, driven in turn by larger-scale changes simulated by global GCMs (Lowe *et al.*, 2009). The impacts of these sea-level changes on coastal flooding are then simulated with models that combine the projections of large-scale sea level changes with regionally specific information. Such information includes both the physical factors that modify the local response to large-scale sea-level changes (such as the morphology of the coastal land and sea floor); and socio-economic factors that affect the extent of actual flooding (such as sea defences) and the size and adaptive capacity of populations exposed to flooding (Nicholls *et al.*, 2011).

The use of GCMs that couple atmospheric and ocean components for the assessment of coastal impacts ensures that the simulated large-scale sea-level changes are consistent with the overlying atmospheric changes, in terms of the thermodynamic contributions and changes in circulation affected by wind stress. Land-ice models, however, are not currently coupled to atmospheric GCMs and instead are driven by meteorological outputs of atmospheric GCMs to simulate changes in ice mass. Changes in meltwater flows are used as inputs to ocean models for estimating sea-level rise, but the changes in large ice bodies may not necessarily be consistent with the overlying atmospheric state due to the lack of two-way coupling. This may affect the timescale of projected melting (Ridley *et al.*, 2005). Changes in large-scale rapid ice dynamics were highlighted as a major uncertainty in sea-level-rise projections in the IPCC's Fourth Assessment Report (AR4; Solomon *et al.*, 2007; Parry *et al.*, 2007).

Land hydrology

Hydrological models simulate the land-surface components of the water cycle (evapotranspiration and runoff) in response to meteorological inputs. The boundary conditions of hydrological models include topography (which determines the surface flow of water), soil characteristics (such as grain size and other properties that control water infiltration, sometimes varying vertically through different soil layers) and vegetation characteristics (such as rooting depth, leaf area and stomatal and aerodynamic conductance). The domains of hydrological models range from individual catchments to the globe. A key distinction between types of model is whether the models are 'lumped', with their domain considered as a single entity with single net inflows and outflows (Moore, 2007) or 'spatially distributed', consisting of an array of grid cells each with their own precipitation and evaporation input and output, connected laterally to neighbouring cells (Bell *et al.*, 2007). Some approaches simulate the lateral flow routing separately using a discrete model (Oki and Sud, 1998), taking as input the runoff simulated by a model of the surface–atmosphere hydrological balance. In practice, given that horizontal resolutions of distributed models typically decrease with domain size, it can be considered that the individual grid cells of spatially distributed models are lumped models in themselves.

Usually, hydrological models are used 'offline' (i.e. using outputs of a climate model, but not feeding back to the climate model) in studies of the impacts of climate change on surface hydrology (e.g. Arnell *et al.*, 2001). However, climate models and ecosystem models generally include their own surface hydrology sub-models as a necessary component of the climate or ecological system (e.g. Cox *et al.*, 1999). Such models are typically simpler than offline hydrological models in terms of their representation of landscape- and soil-related processes such as horizontal transport and water-table depth. Nevertheless, hydrological sub-models within climate and ecosystem models allow for interactions and feedbacks between hydrology, ecosystem processes and climate.

Both rainstorm (pluvial) and river (fluvial) flooding events depend mostly on extremely intense rainfall or high antecedent levels of rainfall, so a major limiting factor in the projection of flood impacts is the capability of the climate models in being able to simulate precipitation. However, flood-inundation risk for a given extreme precipitation event is dependent on local factors such as the shape of the river channel and floodplain, and the existence and nature of flood defences.

Land ecosystems

Equilibrium biogeography models may be used to assess consequences of climate change for the distribution of potential natural vegetation, with implications for changes in the kind of landscapes and species that are viable in any one place. They can also be used in conjunction with GCMs, regional climate models or mesoscale models for idealized studies of feedbacks between ecosystems and climate. But they do not include time-dependent responses of ecosystems to environmental changes, so they cannot predict how rapidly ecosystems will change under different scenarios. Dynamic global vegetation models (DGVMs) combine the large-scale applicability of biogeography models with time-dependent vegetation dynamics. Terrestrial vegetation is modelled at the global scale, including the dynamics of competition and succession. Plant physiological processes are modelled in a process-based way, allowing, for example, CO_2 effects on carbon and water cycling to be considered. Therefore, DGVMs are the tool of choice for modelling vegetation and terrestrial carbon-storage responses to transient climate change.

All current DGVMs represent vegetation in terms of plant functional types (PFTs) but the number, definitions and properties of PFTs vary among models. For instance, the VECODE model (Brovkin *et al.*, 1997) identifies only evergreen and deciduous trees and herbaceous plants, whereas some other models (e.g. LPJ, Sitch *et al.*, 2003; IBIS, Foley *et al.*, 1996) distinguish dry and cold deciduousness; tropical, temperate and boreal trees; and C_3 and C_4 herbaceous plants. Competition between PFTs is either represented in terms of individuals over several patches and scaled up to the grid cell (HYBRID; Friend *et al.*, 1997) or in terms of competition between populations (TRIFFID; Cox, 2001).

Dynamic global vegetation models are concerned with the general structure and functional character of vegetation rather than the behaviour and distribution of particular species, so they cannot directly model climate change impacts on biodiversity. However, the models can be used to investigate processes of relevance to climate change and biodiversity. Changes in the general viability of a PFT under a climate change will imply potential impacts on species within that PFT. Furthermore, changes in the abundance of species making up a particular PFT can impinge on the resources available to the species of another PFT, so changing competitive balances between PFTs suggests changes in ecosystem structure which have implications at the species level. If a PFT is simulated to disappear in a particular location, this may be taken to imply a threat of local extinction of all species within that PFT. However, it should be noted that the PFTs by definition do not represent the diversity within a group of similar species, some of which may well be more resilient than others. More realistically accounting for biodiversity in DGVMs is an active research goal.

The basis of DGVMs in plant-level processes means that they can be used to model large-scale ecosystem properties in a physically consistent way. They are not confined to simulating potential natural vegetation; indeed, one major field of application of DGVMs is the estimation of the separate and combined effects of CO_2, climate and land use on the global terrestrial carbon balance (Prentice *et al.*, 2001). Also, DGVMs can be used as global hydrological models, showing similar skill to purpose-built models (Gerten *et al.*, 2004; Murray *et al.*, 2011), while unlike conventional hydrological models they can enable consequences of plant-level processes, including effects of land-use change, fire and changing atmospheric CO_2 concentration, to be included in global water-resource assessments.

Crop suitability and productivity

As with land ecosystems, impacts of climate change on crops can be assessed either with empirical representations of the suitability of the local climate, or with process-based crop models. There is a vast literature on modelling specific crops, but most models require too much local information (for example on the crop variety, and planting and harvest dates) to allow the models to be used in large-scale assessments of climate impacts. Recently, some efforts have been made to develop simpler and more generally applicable modelling systems for crop suitability and yield. Crop models generally operate on similar principles to process-based ecosystem models as described above, simulating photosynthesis, respiration and transpiration, with the additional process of translating net productivity into harvestable yield. Challinor *et al.* (2004) noted that the strongest correlation between weather and crop yield could be seen at larger scales, so developed a generalized large-scale crop model specifically designed to operate at those scales. Ramankutty *et al.* (2002) developed a crop-suitability index that provided an indication of the suitability of local climates and soil conditions for major crops. The key features of the climate for this index are temperature and the availability of water for plants. The effect of changes in atmospheric CO_2 concentrations on crop water-use efficiency by plants can be included in the index.

Like ecosystem models, crop models are commonly driven offline by climate data and/or models. However, if process-based crop models are coupled to soil-moisture models, this allows for feedbacks from agricultural land use to land hydrology changes to be quantified.

Socio-economic impacts

Increasingly, there are calls for impacts of climate change to be expressed in terms of their economic implications (for specific sectors or entire national economies) and other human implications, such as rates of mortality and morbidity (incidence of disease) due to various causes, some of which (e.g. heatstroke, diarrhoea, insect vector-borne diseases) are closely linked to climate. Although in principle some aspects of these could be approached in a process-based manner, as with some of the biophysical impacts described above, in practice a more empirical approach is usually taken. Often this involves the derivation of 'response functions' (Toth *et al.*, 2000) relating specific impacts indicators to climate quantities, calibrating these over periods of observed data and then applying them to future projections.

Lloyd *et al.* (2011) developed a novel process-based modelling approach for estimating future climate impacts on child health, including both climate-determined food supply (using a crop model) and the more general socio-economic causes of under-nutrition. A challenge they faced was the very limited availability of data for future projections of impact. Despite the importance of socio-economic influences on health and many other aspects of human wellbeing, to date the available future global scenarios (created with the primary objective of projecting greenhouse-gas emissions) include little more in socio-economic terms than population and gross domestic product (GDP). Nevertheless, even with a very simple treatment of economic development, their model was able to quantify the climatic and socio-economic contributions to under-nutrition in scenarios of future climate change.

6.2.4 Using impacts models to simulate past and future changes

To provide insights into past changes, impacts models are driven by data on the observed climate. However, to understand the present climate situation, and enable the currently observed changes to be attributed to their human and natural causes, impacts models need to be driven by the outputs of climate models. Similarly, to provide future projections of the consequences of climate change, impacts models need to be driven by models that project future climate.

The biophysical component of impacts would ideally be assessed using full Earth system models, to ensure physical consistency with the projections of climate change. In reality, this objective is thwarted both by the limited number of Earth system model simulations that can be performed with available computing power, and by the existence of systematic biases in Earth system models. As discussed in Chapter 5, these problems require opposing solutions – computing-power constraints demand faster, simpler models, while addressing the shortcomings and biases caused by simplifications in process representation demands slower, more realistic, higher-resolution models. It is therefore unlikely that both limitations will be solved in the near future, so we will continue to need methods to work around them.

To minimize the effects of biases in Earth system models, it is usual to take the simulated climate anomalies from the Earth system model and apply these to observed climatologies. To perform more simulations (for instance, to enable the exploration of a wider range of emissions scenarios), simpler, more efficient climate models are used, often including pattern-scaling techniques, which again take an anomaly plus climatology approach.

The following sections discuss some of the strengths and limitations of these models and techniques, including some limitations that are not yet widely acknowledged.

Uncertainties in simulating climate change and its impacts

A central theme in this book is the substantial progress being made towards prediction in many areas of climate and ecosystem science, by combining process knowledge, real-time Earth observation and computer models. However, as projections of impacts become more detailed and time- and location-specific, they have the potential to inform decision-making in the real world. Attention thus has shifted to the problems of taking action under conditions of uncertainty.

There are many dimensions of uncertainty. It is helpful to distinguish between uncertainty in a model, and uncertainty that arises from the way in which the model is used. The main approaches for quantitative analysis of the uncertainty in Earth system models (e.g. Moss and Schneider, 2000; Allen *et al.*, 2000) have been described more fully in previous chapters but, briefly, they relate to the quality and quantity of input data; the selection and definition of model parameters; and the way in which the model defines the functional relations between the inputs. Together, these add up to a measure of the extent to which an Earth system model is a simplified abstraction of reality.

By bridging biophysical and socio-economic dimensions, impacts assessment introduces new sources of model uncertainty and faces further challenges with uncertainty relating to model use. Various suggestions have been made for classifying uncertainty in linked socio-environmental systems (e.g. van der Sluijs, 1996; van Asselt and Rotmans, 2002; Regan *et al.*, 2002; Meijer *et al.*, 2006; Maier *et al.*, 2008; Lo and Mueller, 2010). They all group the kinds of uncertainty into two main strands. Epistemic uncertainty arises from a lack of knowledge about the system, so, in principle, it can be reduced by obtaining more information. The monitoring programmes described above aim to address this kind of uncertainty. Stochastic uncertainty arises from the inherent randomness and variability of the system, so for this aspect of uncertainty, obtaining more data about the system will not reduce the envelope of possibility. Biophysical and socio-economic components alike possess both kinds of uncertainty, with additional complexity affecting impacts studies because of the interlinked biophysical and socio-economic impacts of climate.

Quantification of uncertainty in complex socio-environmental systems is challenging. Van der Sluijs (1996) and Batty and Torrens (2005) discuss the lack

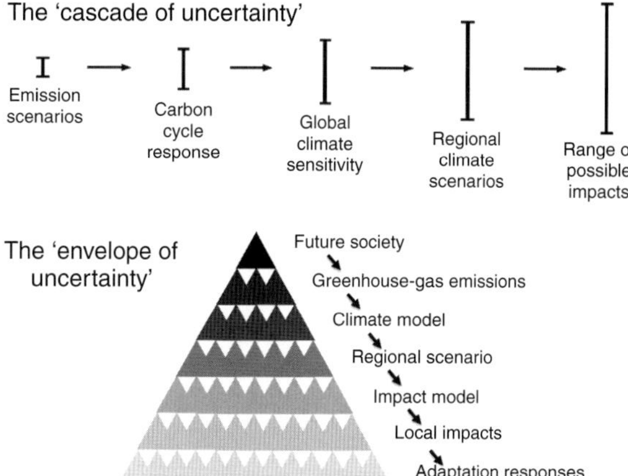

Figure 6.4 Cascading uncertainty in impacts modelling – the classic view, in which the magnitude of uncertainty at each step increases.

of methodologies for handling surprises in the climate system (e.g. discontinuities in climatic trends, and discrete random events). As discussed in Chapter 1, some social scientists have argued that the future of human society is intrinsically unpredictable, because there are features – like community life and innovative behaviour – that shape how society responds to climate change (and to predictions of change) that cannot be robustly quantified (Ackoff and Emery, 1972).

Several scientists have drawn attention to a 'cascade' or 'explosion' of uncertainty in climate impact assessments (Figure 6.4; Henderson-Sellers, 1993; Jones, 2000; Moss and Schneider, 2000; Mearns *et al.*, 2001; Wilby and Dessai, 2010). This idea is useful in demonstrating the many stages in the climate risk-assessment process where attention must be paid to the degree of confidence in the scientific knowledge. However, the cascade idea is often presented as a problem for climate science. Echoing the earlier studies, Wilby and Dessai (2010) state, '*the range (or envelope) of uncertainty expands at each step of the process to the extent that potential impacts and their implied adaptation responses span such a wide range as to be practically unhelpful*'.

If the best available science about global change were really of no practical use in informing society's responses, this would present a serious problem. But that is not the situation. The uncertainty cascades or pyramids described in the literature oversimplify the different 'flavours' of uncertainty in impacts analysis, and merge them with concepts of climate risk management that extend beyond the modelling and assessment of climate impacts. They draw on concepts of numerical error aggregation (appropriate for quantitative assessment of uncertainty) and decision-tree analysis (appropriate for the practical management of uncertainty), but sometimes mix them together inappropriately. For example, Jones (2000) refers to the error aggregation approach when he states that '*ranges are multiplied to encompass a comprehensive range of future possibilities*', but not all of the uncertainty stages in his cascade are multiplicative. Some relate to societal choices. Other studies use the language of decision-tree approaches, but describe simple aggregation (as in Dessai and Wilby, 2010): '*information is cascaded from one step to the next, with the number of permutations of emission scenario, climate model, downscaling method, and so on, proliferating at each stage*'. In real-world decision-making, choices at a given level can reduce the number of available permutations; they do not proliferate endlessly. Cornell and Jackson (in press) address in more depth these and other issues relating to socio-environmental uncertainty.

A key value of the uncertainty cascade concept lies in illuminating the practical steps needed to address the different kinds of uncertainty at the various stages of the climate risk-analysis process. Policy-makers have long experience in decision-making under uncertainty, and they do not require all aspects of uncertainty to be quantified in the same units. Targeted observation and monitoring programmes address epistemic knowledge limitations relating to the input data and fundamental processes represented in models; methods for confronting models with data (including palaeodata) help identify the possible states of the system and define better the margins of stochastic uncertainty. The careful design of scenarios also is important in this latter regard. However, when science cannot give absolute certainty – which is the usual situation in decisions about future societal adaptation to climate change – responses need to be designed to be robust against uncertainty (van der Sluijs, 1996; Mearns, 2010). In identifying robust options, Moss and Schneider (2000) emphasized the essential role of collective judgements, involving dialogue between scientists and the diverse stakeholders in decision-making about responses to climate change.

Many uncertainties in future projections of the impacts of climate change are well recognized, but some are not (Figure 6.5). Widely recognized uncertainties include climate sensitivity and emissions

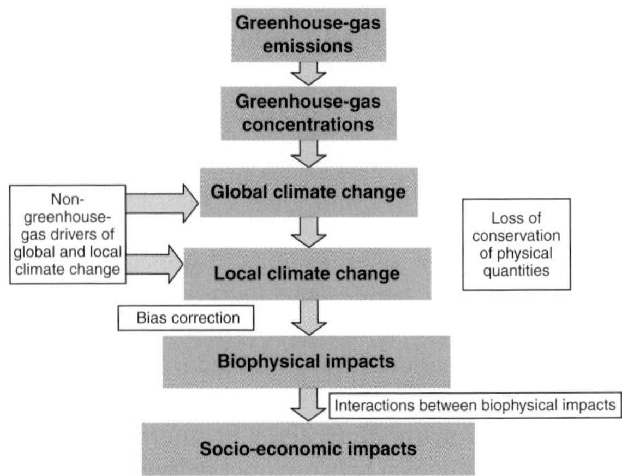

Figure 6.5 Additional uncertainties in assessing the long-term impacts of anthropogenic climate change. The central column (orange text-boxes) shows the classic conceptual model of the 'cascade of uncertainty' with its well-recognized propagation of uncertainty along the chain of causes and effects. Side boxes show additional sources of uncertainty that are currently not widely recognized in impacts modelling. Note that for near-term changes, the predominant source of uncertainty is the internal variability of the climate system. The implications of uncertainty in emissions and concentrations only become significant for impacts assessment over longer timeframes.

scenarios. Uncertainties in regional climate change are reasonably well recognized, but usually less emphasized than emissions and climate sensitivity, and frequently ignored. It is still common to see impacts studies that consider different emissions scenarios yet use a single climate model, apparently assuming that the difference between the emissions scenarios is the main uncertainty in regional climate change. This is not generally true. Different climate models can vary hugely in their projections of e.g. seasonal precipitation changes at regional scales, and this can lead to differences even in the sign of the impact. Figure 6.6 illustrates this in the context of simulated runoff changes by the climate models used for the IPCC's AR4.

A poorly recognized yet significant uncertainty relates to the translation of emissions scenarios to concentrations. It is very common to see climate modelling and impacts papers stating that a particular emissions scenario was used, when in reality the climate model was driven by a prescribed scenario of greenhouse-gas *concentrations*, without recognizing that this is only one interpretation of the original emissions scenario. For example, the CMIP3 climate-model projections used in the IPCC's AR4 (Solomon *et al.*, 2007) were driven by prescribed concentrations previously simulated by

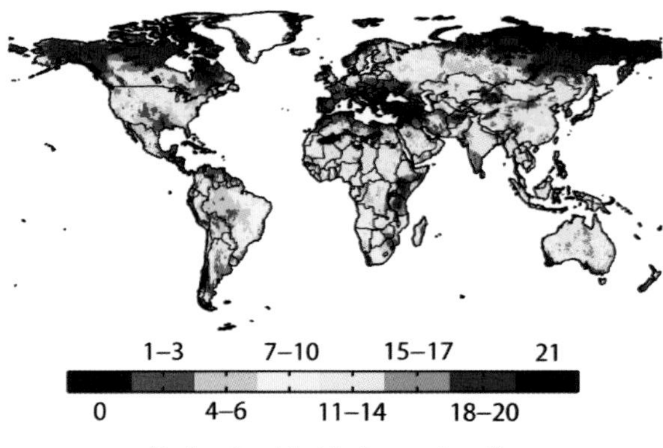

Figure 6.6 Implications of uncertainties in climate projections for global runoff. The number of climate models under which an increase in runoff is projected for each land grid cell, using runoff simulations by the MacPDM hydrological model driven by the IPCC AR4 climate model outputs.

simple climate models. They therefore did not capture the uncertainty in concentrations arising from uncertainties in the strength of climate–carbon-cycle feedbacks (Friedlingstein *et al.*, 2006). For regional climate change, however, this may still be a secondary uncertainty in comparison with that arising from regional climate responses as noted above. Murphy *et al.* (2009) made a systematic assessment of the relative contributions of emissions scenarios, climate–carbon-cycle feedbacks, and regional climate response to uncertainty in climate change.

Some sources of uncertainty in projections of climate change and its impacts seem to be largely unrecognized. Some relate to the omission in models of additional forcing factors such as land-cover change and urban effects; some to interactions within the climate system in response to external forcings; others simply to methods to deal with imperfections in the models themselves. These issues are discussed below.

Uncertainties in addressing biases in climate models

General circulation models are developed with the aim of representing present-day climate as accurately as possible. The vast majority of model testing and evaluation is against observed present-day data with the aim of further improving the accuracy of the simulation. Yet climate models still feature systematic biases in comparison with observations, especially at regional scales. When applying these models to impacts studies, these systematic biases may be of concern if they result in unrealistic simulations of impacts processes and quantities at the present day – the obvious implication being that if the present-day simulations of impacts are unrealistic, the future projections will also be unrealistic. This is particularly concerning when the aim is to make quantitative projections of the absolute state of impacts in the future, as opposed to merely the change (e.g. the absolute crop yield or water stress in the future, as opposed to simply the direction of change relative to present-day).

To reduce these systematic biases, '*bias-correction*' techniques are often employed (Figure 6.7). Generally, bias correction involves using the climate model as a source of information on the *change* in climate, and applying the projected changes to the observed present-day climate state in order to obtain a future climate state which is – theoretically, at least – less affected by systematic biases.

In improving the representation of variability of a climatic variable, the modeller can use the absolute anomalies around the baseline trend or proportional 'change factors'. Using absolute values risks artificially introducing meaningless negative values at the lower end of a trend; and using percentage-change factors risks introducing disproportionate changes. There is also a risk of inconsistency, since the real processes may be non-linear. Another often overlooked issue is whether it matters if the 'adjusted' climate drivers then give rise to impacts that are inconsistent with the overlying climate. For example, projections could be made of evaporation and runoff that do not give closure of the surface-water budget, if the original climate model simulated changes in atmospheric moisture content and surface fluxes from land using different schemes.

Looking more closely at a concrete example, changing rainfall in Amazonia, most of the IPCC AR4 models simulated less annual mean precipitation than is observed, and the magnitude of this bias varies considerably between models (Figure 6.8(a)). Typically, this issue is addressed by assuming that these are systematic biases that can be accounted for by taking the simulated changes in climate relative to the present day, and applying these to an observational climatology (Scholze *et al.*, 2006; Salazar *et al.*, 2007; Lapola *et al.*, 2009; Malhi *et al.*, 2009). However, such bias correction has its own difficulties, especially for precipitation and especially when the model biases are large (Figure 6.8(b)). For example, if changes in precipitation are expressed as a percentage of the baseline state and then applied to a bias-corrected baseline, and the baseline precipitation is too low, small changes in the

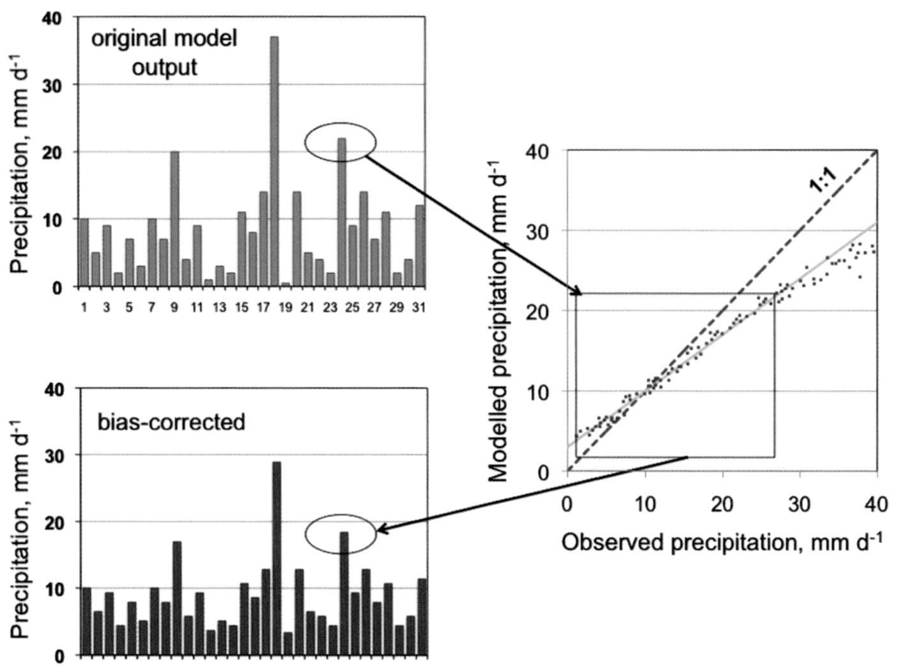

Figure 6.7 A simple example of bias correction of model output

original projections (Figure 6.8(a)) can be magnified and result in large changes in the bias-corrected projection (Figure 6.8(b)). Such changes can become physically inconsistent with other aspects of the climate projections, as quantities such as moisture cease to be conserved.

One possible approach to correcting biases whilst avoiding inconsistencies between the impact and the overlying climate change is to take the anomaly of the physical impacts variable (e.g. river flow) and add this to an observed climatology of that variable. However, while this approach may be viable for continuous quantities such as river flows, it may not be appropriate for threshold-sensitive processes such as crop growth and yield.

Additional human drivers of climate change

As well as changing in response to climate change, land cover is being modified directly by human activities. Here we consider some of the ways in which land-use change affects climate.

Land-use change is a direct consequence of human action, and generally not a result of climate change. Its effects on climate are therefore correctly treated as a forcing as opposed to a feedback. From a global perspective, the dominant aspect of historical land-cover change has been the extensive deforestation in temperate regions during historical times. A number of modelling studies (e.g. Betts, 2001; Govindasamy *et al.*, 2001; Myhre and Myhre, 2003) suggest that this has exerted a radiative forcing which is most likely negative (i.e. a global cooling effect) as a result of increased surface albedo, especially in high-latitude winter and spring when snow is lying. Most estimates of the global mean radiative forcing relative to the pre-industrial state are in the region of -0.2 to -0.3 W m^{-2}. This global forcing is small in comparison to the positive radiative forcing of ~ 3 W m^{-2} estimated to arise from increased greenhouse-gas concentrations, but the forcing due to greenhouse gases is spread relatively evenly across the globe whereas the effects of anthropogenic surface-albedo change are concentrated in temperate regions. At the regional scale, these radiative forcings are estimated to be -5 W m^{-2} or larger, leading to a regional cooling of 1–2 K in the absence of any other influences. This calculation implies that the greenhouse warming in temperate regions has been partly offset regionally by anthropogenic changes in land cover.

Land-cover change can also modify the surface energy budget through changes in the fluxes of latent and sensible heat. While such processes act as anthropogenic perturbations to the climate system, they do not involve direct perturbations to the Earth's radiation budget and therefore cannot be considered as radiative

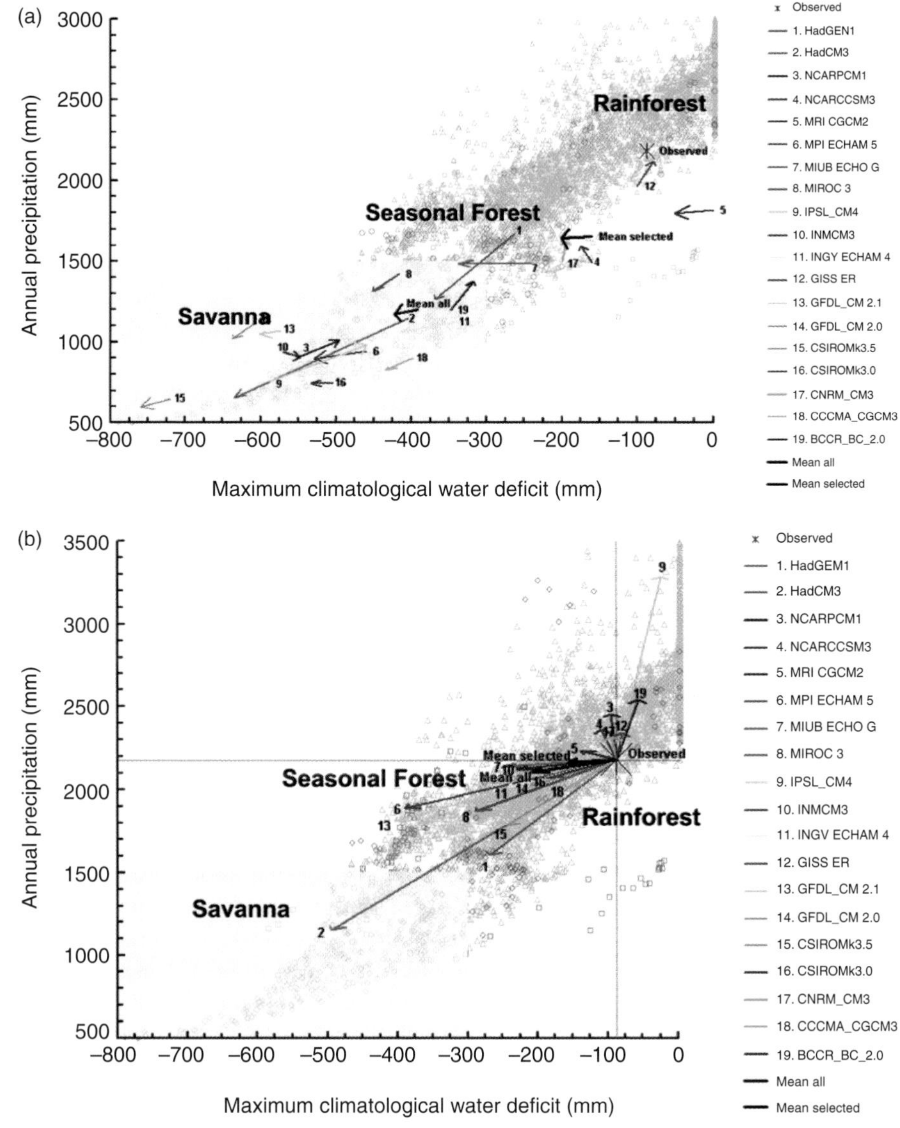

Figure 6.8 Difficulties with bias correction. (a) Simulated and observed rainfall regimes in Amazonia, shown as precipitation against maximum water deficit. The grey markers show observed present-day precipitation and water deficit inferred from observations. The coloured arrows show the change in simulated rainfall regime between the late twentieth century and the end of the twenty-first century. The shaded area shows the regime where savanna vegetation dominates, rather than forest. (b) Recalculated precipitation and water deficit derived by applying a 'bias correction' to the simulated changes shown in (a). Reproduced with permission from Malhi *et al.*, 2009.

forcing. Although the term 'non-radiative forcing' has been proposed (Jacob *et al.*, 2005), this cannot be quantified in terms comparable with radiative forcing, so at present the effects of these processes are best discussed in terms of temperature changes. Model results suggest that the combined effects of past tropical deforestation may have exerted regional warmings of ~ 0.2 K. Such changes may also have perturbed the global atmospheric circulation, thus affecting regional climates at a distance (Chase *et al.*, 2000; Pielke *et al.*, 2002).

Many scenarios of future greenhouse-gas emissions include contributions from land-cover change, but the biophysical effects of such changes are not usually included as inputs to climate models. Feddema *et al.* (2005) examined the importance of land-cover changes, simulating climate changes induced by changes in the physical properties of the land surface. These simulations suggested that the land-cover effects on regional climate change are significant in comparison with those of greenhouse-gas-induced climate change. For example, the extensive deforestation of Amazonia by 2100 in the A2 scenario simulation leads to a warming of that region of approximately 2 K due to a reduction in evapotranspiration. This warming would occur in addition to the warming caused by increased greenhouse-gas concentrations.

Feedbacks from the oceans may enhance the effects of land-cover change on climate. For example, Delire *et al.* (2001) found that in the Indonesian Archipelago, the impacts of deforestation on simulated wind speeds may be sufficient to modify ocean upwelling and cause warming over the surrounding ocean surfaces, in addition to the warming caused over land by reduced evaporation.

To explore these kinds of issues more systematically, land-cover change is included in many of the current generation of Earth system models being used for climate projections for the IPCC Fifth Assessment Report (e.g. HadGEM2-ES, Jones *et al.*, 2011; Arora and Montenegro, 2011), using a standardized set of land-cover reconstructions and future scenarios (Hurtt *et al.*, 2011).

Attention is now directed towards human changes in the water cycle. For example, irrigation can increase the surface-moisture flux resulting in a cooling influence on local near-surface temperatures. Lobell and Bonfils (2008) found that daily maximum temperatures were reduced by up to 5 K in highly irrigated areas (compared to surrounding non-irrigated regions) in California.

Plans for adaptation to climate change require specific, local detail of climate change. Although this can be provided by regional climate models, these have typically only been used for downscaling of radiatively forced global climate change. Climate change adaptation planning may therefore be misdirected if based on projections that ignore the climatic effects of land-use change and water-system change, both of which have strong human controls.

The other direct effect of human activities on climate which we consider here is the effect of urbanization and anthropogenic heat release. Built-up areas exert significant influences on their local climates, with an 'urban heat island' being observed in many cities. This effect is due partly to the influence of the urban landscape on the surface energy budget and local meteorology, and partly from sources of heat arising from human activities ('human energy production'). The nature of the land surface is a key factor influencing the sensitivity of near-surface climates to radiative forcing by increasing greenhouse-gas concentrations, so the responses of urban climates to radiative forcing may be different to those of non-urban climates. Moreover, increases in anthropogenic heat sources may exert an additional direct forcing of local climates. The energy consumption in urban areas in 2006 was estimated as 335 million million million joules (International Energy Agency, 2008), or approximately 0.02 W m^{-2} as a global average energy flux expressed in terms that can be compared with global radiative forcing. With urban areas covering approximately 0.4% of the global land surface (Hurtt *et al.*, 2011) this implies an average heat flux of approximately 18 W m^{-2} in urban areas. This average value masks a great deal of variation; for example, daytime values in central Tokyo typically exceed 400 W m^{-2} with a maximum of 1590 W m^{-2} in winter (Ichinose *et al.*, 1999). Thus, although human energy production is a very small influence at the global scale, it may be critically important for local climate changes in cities (Crutzen, 2004).

Previously, climate-model simulations of climate change did not consider the influences of urban areas on their own local climates. Consequently, it was not clear whether it was appropriate to apply such scenarios to climate change impacts studies in the context of built-up areas. While this problem may be partially addressed by combining climate change scenarios with representations of present-day urban heat islands, this approach does not address the interactions between urban landscapes and radiative forcing nor account for increasing anthropogenic heat sources. These processes can only be accounted for by the inclusion of urban meteorological processes and human energy production within high-resolution climate models.

McCarthy *et al.* (2010) included an urban-heat-island representation into a GCM and examined the effects of the urban landscape and a tripling of the global anthropogenic heat release, which could plausibly be estimated to occur at approximately the same time as a doubling of CO_2 concentration (Figure 6.9). Although there was no discernible effect on global mean temperatures, the additional warming effect of the tripled heat source on minimum temperatures in cities was approximately 13% of that due to doubling CO_2. This suggests that plausible increases in anthropogenic heat release could exert a significant additional influence on urban climates, with particular implications for impacts on human health through unrelieved heat stress overnight. Urban effects are also now being included in some Earth system models being used for climate projections for the IPCC Fifth Assessment Report (e.g. HadGEM2-ES; Jones *et al.*, 2011).

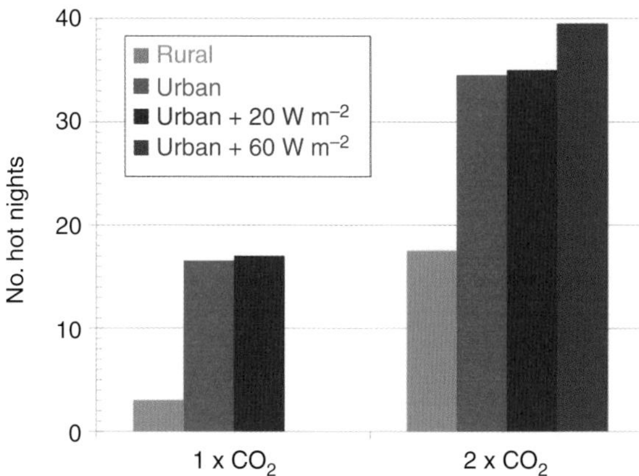

Figure 6.9 Climate change in cities. Comparison of effects of doubled-CO_2 climate change, urban-landscape effects and anthropogenic heat release on the number of nights in Delhi defined as 'hot', using the hottest 10% in the baseline simulation. (After McCarthy *et al.*, 2010).

Interactions between impacts and synergies with other anthropogenic drivers of change

Most studies of impacts of climate change are carried out for selected individual impacts separately. However, in many cases there can be a strong interdependency between impacts, both biophysical and socio-economic. For example, physical impacts on ecosystems, agriculture, irrigation, river flows and glacier melt can be closely linked. The supply of water for irrigation provides an illustration of how impacts in one sector and region can modify impacts in another sector and region. Irrigated agricultural land comprises less than one-fifth of all cropped area but produces 40–45% of the world's food (Döll and Siebert, 2002), and the water for irrigation is often extracted from rivers, which can depend on climatic conditions far from the point of extraction (Figure 6.10). Some major rivers are fed by mountain glaciers; about one-sixth of the world's population lives in glacier-fed river basins. Populations are projected to rise considerably in major glacier-fed basins such as on the Indo-Gangetic plain, and other major irrigated areas such as the Yangtze and Nile. Changes in remote rainfall and the magnitude and seasonality of glacial meltwater could therefore impact food production for many people.

Changes in atmospheric composition, particularly CO_2 and O_3 concentrations, can affect ecosystems through physiological processes. Although these are frequently taken into account in ecosystem and crop impacts studies, their knock-on effects on other sectors are often not considered. For example, increased plant water-use efficiency in response to elevated CO_2 has implications for the wider hydrological cycle as reduced transpiration necessarily results in increased soil moisture and/or runoff (Figure 6.11). An integrated Earth system approach to modelling is needed to allow such interactions to be accounted for.

Linkages between climate models and impacts models

One advantage of using a chain of loosely coupled models in impacts analysis is that it allows each sub-system model to be developed to best represent the particular sub-system, such as with appropriate spatial resolutions and time-steps, without any constraint imposed by a need for compatibility with the other models. For example, coarse-resolution output from a global-scale climate model can be used in combination with statistics of fine-scale climatic features in order to provide input to a higher-resolution hydrology model capturing catchment-scale topographic detail. A second advantage is that any biases in the output of one model can potentially be accounted for by adjustment of data prior to input to the next model. Such adjustment is frequently used when applying climate-model output to climate impacts models, in order to remove systematic errors from the climate simulations as discussed previously. A third advantage is that analysis of causes and effects is relatively straightforward, as processes can be broken down into components.

However, the use of a linear cause–effect chain of sub-system models neglects the possibility of feedbacks between the sub-systems. For example, if an ecosystem model simulates major changes in terrestrial carbon storage in response to a simulated climate change, this might imply a potential feedback to atmospheric CO_2 and hence climate, which was not accounted for in the original climate simulation. Furthermore, some sub-system models represent the same processes at the interface between the sub-systems, but with inconsistent methods. Atmospheric models, for instance, include models of surface evaporation in order to simulate a conservative water cycle. Hydrological models also simulate surface evaporation, but often at a higher spatial resolution and with different treatments of aspects such as surface saturation. If the evaporation simulated by a hydrological model is different from that simulated by the atmospheric model from which its input data are derived, the input data are inconsistent with the simulation. This may be critical because if

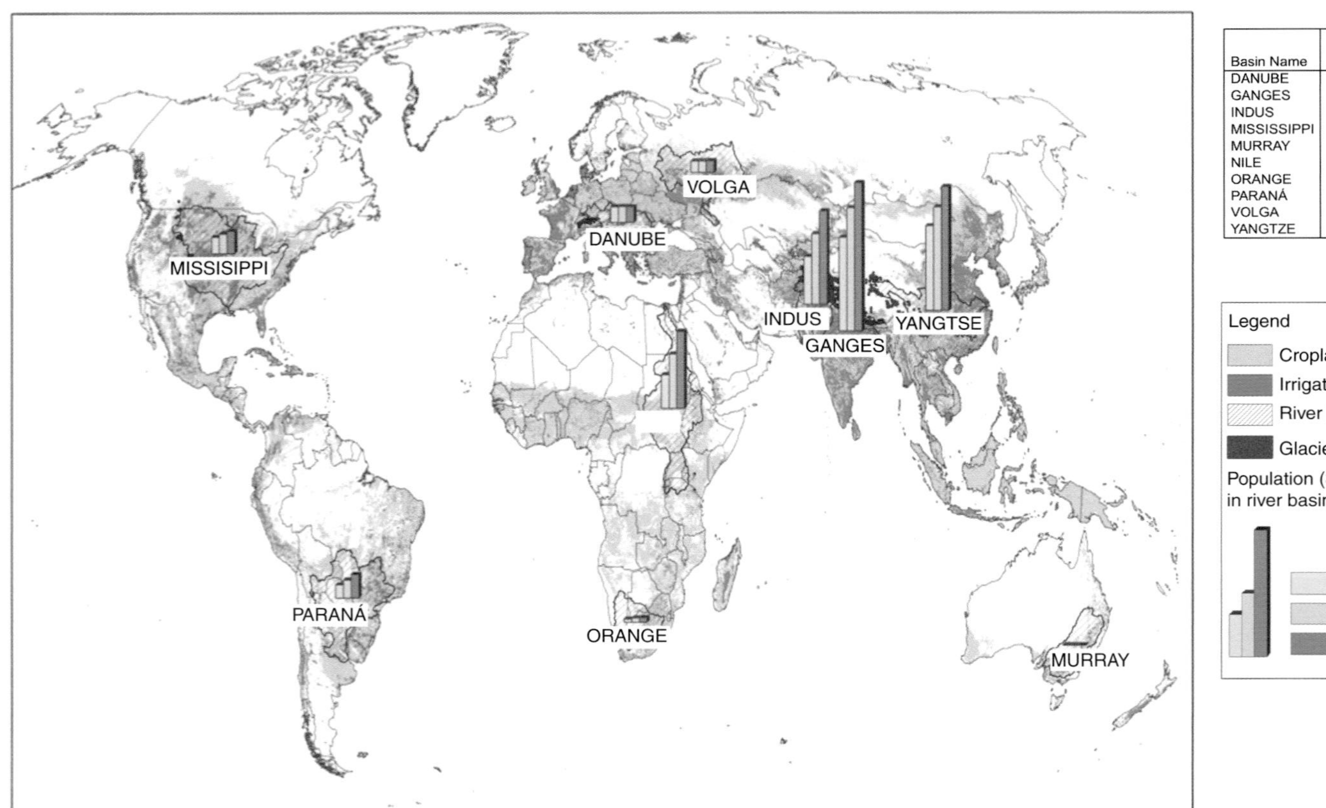

Basin Name	Population (Millions) 2005	2025	2045
DANUBE	82.8	84.1	83.8
GANGES	478.8	633.3	765.7
INDUS	244.8	361.0	479.1
MISSISSIPPI	84.6	100.3	114.9
MURRAY	2.1	2.2	2.2
NILE	174.3	281.2	396.5
ORANGE	16.7	20.6	22.6
PARANÁ	67.8	92.5	120.4
VOLGA	58.8	58.9	61.4
YANGTZE	432.2	531.2	639.3

Legend
Cropland
Irrigated cropland
River basin
Glaciers in river basin
Population (SRES A2) in river basin
2005
2025
2045

Figure 6.10 Interactions between land impacts of climate change. Most assessments consider only local climate changes which are important for rain-fed croplands (brown). However, irrigated croplands (green) can depend on river systems (blue) affected by precipitation and evaporation in remote regions, and in some cases may also be fed by melt-water from glaciers (red). Population growth (bar charts) is projected to be large in many regions where such synergistic impacts may be a key consideration.

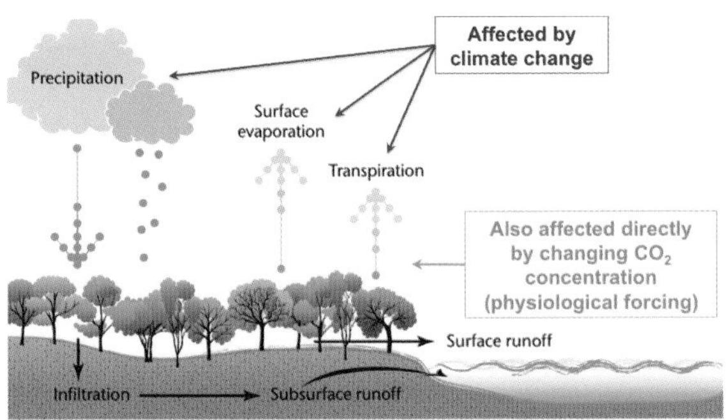

Figure 6.11 Impacts of climate change and CO_2 physiological forcing on hydrological cycle. Land hydrology is affected by flows of water to and from the land surface, through precipitation, evaporation and transpiration. All three processes are affected by climate change, and the latter is additionally affected by changes in vegetation water-use efficiency in response to changing atmospheric CO_2 concentrations.

the overall simulated water budget does not conserve water, the results by definition must be unrealistic.

In the light of these drawbacks, an alternative modelling approach has come into use in the last decade, involving the coupling of models of the climate sub-systems to allow reciprocal interactions. While this means that model features such as resolution may be an additional constraint on some of the sub-system models, and makes adjustment of biases and understanding of model behaviour more difficult, it does ensure consistency between the models and allows the simulations to include feedbacks within the climate system. Indeed, in some cases, climate models already simulate the quantities required for impacts studies as a necessary part of the climate simulation, but often without these being used for impacts assessments. For example, runoff (and sometimes river flow) is simulated by climate models as part of the hydrological cycle, but with a few exceptions (e.g. Nohara *et al.*, 2006) this information is rarely used.

The importance of the offline versus online differences can be illustrated by comparing the impacts of climate change on river runoff as simulated directly within a climate model with those simulated by applying the meteorological outputs of the climate model to an offline hydrological model. Here we used the HadCM3 climate model, which incorporates the MOSES land-surface hydrology scheme (Cox *et al.*, 1999), and the MacPDM global hydrological model driven offline by meteorological outputs of HadCM3. The two approaches give different projections of runoff change in many areas of the world, sometimes

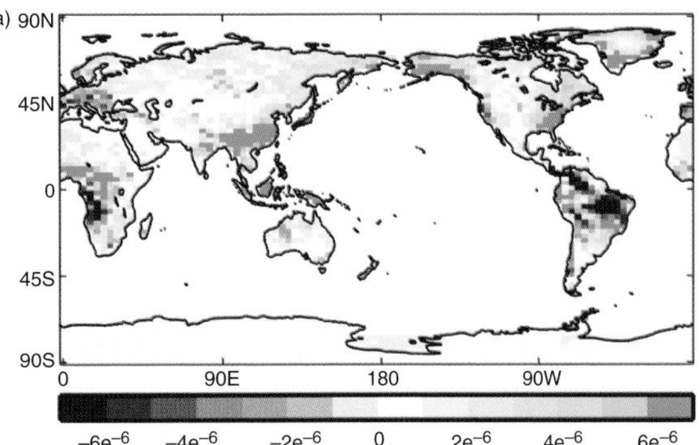

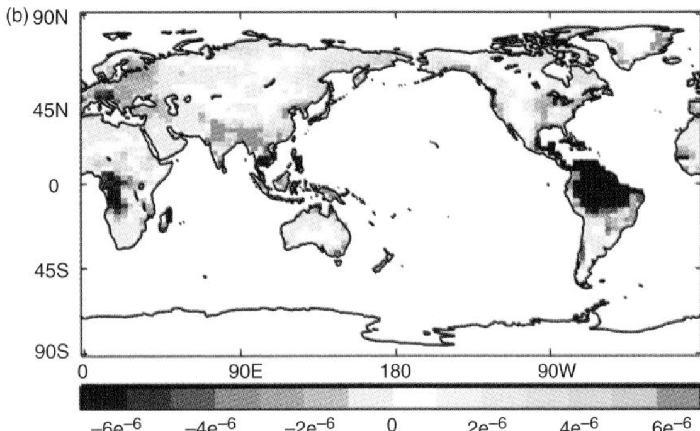

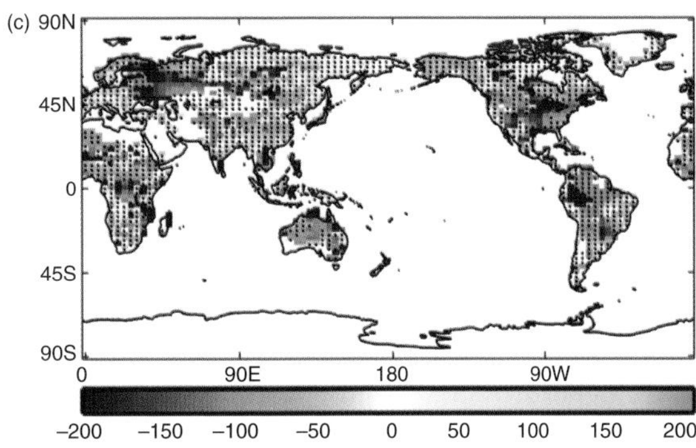

Figure 6.12 Changes in runoff by the end of the twenty-first century, (a) simulated by the HadCM3 model directly, and (b) by the MacPDM offline hydrological model driven by HadCM3 meteorological output; (c) shows the difference between the two. Stipples show where the two approaches agree on the sign of the runoff change.

differing in sign as well as magnitude (Figure 6.12). More crucially, the offline model MacPDM simulates changes in evapotranspiration that were markedly different, and very often opposite in sign, to those simulated by the land-surface sub-model within HadCM3. As a result, the offline modelling approach leads to major losses of conservation of water, and hence it may, for example, simulate much larger decreases in runoff in Amazonia under the drying climate simulated by HadCM3. If the increases in evapotranspiration simulated by offline MacPDM had been replicated within HadCM3, its regional precipitation reduction might not have been so severe, so a less dry climate would have been used to drive MacPDM. In other words, these offline simulations of hydrological impacts must be wrong; either because the offline hydrological model itself is wrong, or because it has been driven by a wrongly simulated climate that neglects feedbacks from the land surface.

It is not yet clear whether the inconsistencies between the online and offline analyses arise from different approaches used by the two land-surface models, or simply from the fact that the land-surface sub-model is coupled to the atmospheric model while MacPDM is not. Nevertheless, this analysis shows the importance of maintaining consistency between the hydrological model and climate model – neglecting physical consistency may lead to over- or under-estimation of projected impacts.

These differences between the original climate model's own simulation of runoff and that simulated by an external hydrological model driven with the same meteorology can be an important component of the uncertainty in regional impacts projections, even compared to uncertainty in the regional climate change. The apparent 'consensus' between climate models on either increasing or decreasing runoff can depend on whether this is examined using the runoff simulated by the climate models themselves or by applying their meteorological outputs to an external hydrological model (Figure 6.13).

Whether an online or offline approach is inherently more useful than the other remains an open question. The answer may depend on the application – whether for informing mitigation or adaptation. For adaptation, the best possible estimate of absolute future conditions is important. For mitigation, since the question is more about relative change ('What are the consequences of different emissions policies?'), it may be possible to consider some uncertainties or non-climate drivers simply as systematic biases which are cancelled by comparing the impacts of two emissions scenarios in which the same systematic biases exist. The difficulties with this assumption are that the biases may not be equivalent if they involve non-linear feedbacks, and that thresholds may be important (e.g. where mitigation action may only be justified if absolute impacts exceed some critical threshold).

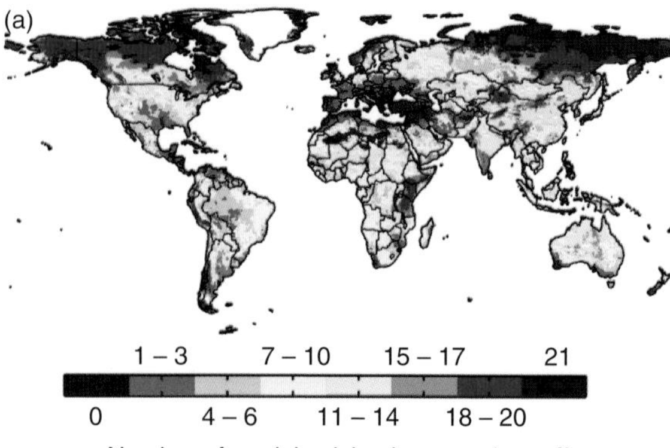

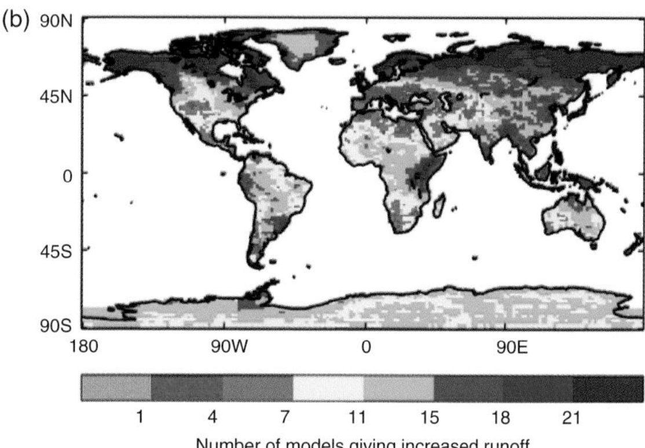

Figure 6.13 Differences in model consensus on increasing or decreasing runoff. (a) Runoff change assessed using a hydrological model (MacPDM) driven by meteorological outputs of the IPCC AR4 climate models. (b) Runoff directly simulated by the climate models.

6.3 Evidence for impacts of climate change in the recent past

6.3.1 Detection and attribution of impacts

Anthropogenic climate change is already occurring and it is reasonable to assume that it may already be exerting impacts both in the natural world and in human affairs. But such impacts are hard to identify unequivocally against a background of natural climate variability and social and economic change. It is now commonplace that both extreme weather events and ecological shifts are reported in the media to be 'changing due to climate change' (e.g. Magrath, 2007) but there is very rarely any scientific evidence that could unambiguously attribute the supposed impact to climate change.

The issue of attribution of impacts has been regarded as part of the evidence base for anthropogenic climate change, with ecological and hydrological changes possibly being viewed as part of the picture of a 'warming world' (Parry *et al.*, 2007). On the other hand, it has been argued that attribution is a lower priority than simply obtaining direct evidence of warming itself, such as from temperature measurements; ecological changes are so complex that they may provide only weak evidence for effects of anthropogenic warming (Parmesan *et al.*, 2011). Nevertheless, 'proof' of anthropogenic climate change is not the only reason to establish whether observed changes are impacts of climate change or not. It is also important to understand the causes of changes in order to inform adaptation, and in particular to avoid maladaptation (e.g. Barnett and O'Neill, 2010) or adaptation to a trend or change which may not continue as expected because it has been poorly understood.

A key issue is whether change is significant (or not) in the context of past variability. This can be difficult to ascertain for many impacts which have not been monitored until recently, although, as outlined in Section 6.2.1, evidence for some biophysical impacts can (at least in principle) be obtained from sources such as pollen records, tree rings, geochemical data and geomorphic traces. Identifying the signal of a change outside the noise of variability is known as 'detection'.

Once a change has been detected, the next step is to 'attribute' it to causes. The first task in attribution is to identify climate signals using statistical techniques. Most 'impacts-attribution' studies have focused on this step. But attribution to *anthropogenic* climate change is more difficult. Furthermore, anthropogenic influences on climate are not restricted to global-scale changes in atmospheric composition (Pielke *et al.*, 2011; McCarthy *et al.*, 2010), and some non-climatic drivers may be directly associated with the causes of human influence on climate – examples of this are land-cover change (which affects ecosystems directly as well as influencing climate) and the effects of changes in atmospheric composition on plants. Detection and attribution of biophysical impacts may nevertheless be easier than for socio-economic impacts, where the direct human drivers are likely to be dominant.

6.3.2 Detection of observed changes

Impacts of recent climate change may already be detectable in ecosystems. Regions where the presence or functioning of life forms is limited by temperature, particularly high-latitude and mountainous regions, are strong candidates for featuring ecological changes associated with the observed general warming trend, which has been attributed to anthropogenic influences, at least at continental scales (Hegerl *et al.*, 2007).

General shifts in phenology have been observed. The predominant picture appears to be of earlier onset of spring events, such as budburst and first observation of migratory or hibernating species. However, not all species in all sites show such trends – some show no significant trend, others show a trend towards later dates. For example, a relatively long time series of observations in Spain (Gordo and Sanz, 2009), from 1943–2003, show some consistent changes occurring since the 1970s, with leaf unfolding, flowering and fruiting dates advancing by around half a day each year. Less consistent changes can be seen in the autumn leaf-falling dates, with some species advancing, and some delaying. During the observation period, the growing season lengthened overall by 18 days. In Switzerland, 38 years of pollen data (1969–2006) indicate that birch trees now start flowering 15 days earlier (Frei and Gassner, 2008). Satellite imagery also suggests general shifts in high-latitude vegetation phenology at large scales. Changes in abundance, distribution and phenology of woodland bird species in temperate regions have been observed (Leech and Crick, 2007). The earliest appearances of butterfly species have been documented in several UK records since the 1970s, generally showing a trend for the earliest appearance to occur earlier in the year by the 1990s (Sparks and Yates, 1997). In North America, phenological trends have been seen in a number of plant and animal species, although they differ in magnitude and even direction (Cook *et al.*, 2008).

Widespread 'greening' trends in satellite observations of the normalized difference vegetation index (NDVI) have been reported in temperate and boreal forest regions (Nemani *et al.*, 2003). However, more recent evidence suggests that such trends have not continued at the same rates nor been geographically uniform (Goetz *et al.*, 2007). More than 25% of North American forest areas (excluding areas recently disturbed by fire) showed a decline in productivity and no systematic change in growing season length, particularly after 2000 (for records extending to 2005 as measured by AVHRR data). In the boreal forests there is evidence of migration of 'keystone ecosystems' in upland and lowland treeline of mountainous regions across southern Siberia (Soja *et al.*, 2007). In mountainous regions, more species appear to be surviving at high altitudes: for example, an increase of ~ 10%

in species richness was seen in measurement plots in Austrian Tyrol between 1994 and 2004 (Pauli *et al.*, 2007). In the maritime Antarctic, two native vascular plants, antarctic pearlwort (*Colobanthus quitensis*) and antarctic hair grass (*Deschampsia antarctica*) have become more prolific over recent decades, possibly benefiting from the warming climate more than mosses due to their efficient acquisition of nutrients released in the faster decomposition of soil organic matter (Hill *et al.*, 2011).

Some large-scale changes are being seen outside the cold regions where the temperature changes have been most pronounced. For example, woody vegetation is encroaching into shortgrass steppe, savanna and cerrado landscapes worldwide (Mitchard *et al.*, 2009; Golubiewski and Hall-Beyer, 2008, and references therein). However this 'woody thickening' is not necessarily due to climate change; changes in grazing pressure, and direct effects of increasing CO_2 concentration on tree-grass competition in the tropics, have been proposed as primary causes. Chapter 3 discussed evidence for a strong effect of CO_2, also required to fully explain the worldwide increase of tree cover during the last transition from ice age to interglacial climate.

A general trend of increasing forest biomass across the tropics today is seen in plot-scale measurements (Phillips *et al.*, 1998), although with episodes of declining biomass at regional scales. For example, a fall in biomass in Amazonia was measured after the severe 2005 drought (Phillips *et al.*, 2009). Preliminary analysis of satellite measurements suggests there was an even more widespread decline in vegetation productivity after the 2010 Amazon drought than in 2005.

For much of the twentieth century, a number of tree-ring records have shown a general ongoing increase in tree growth (e.g. Briffa *et al.*, 2008), which often correlates with temperature. Direct CO_2 effects (increased growth especially under drought conditions) have been identified in approximately 20% of the sites in the International Tree Ring DataBase (ITRDB; Gedalof and Berg, 2010), and studied in detail at some sites (e.g. Koutavas, 2007). However, since the 1980s, a number of records show a decline in tree growth, for which a number of possible causes have been suggested, including increasing water stress and ozone damage (e.g. D'Arrigo *et al.*, 2008). Most tree-ring chronologies only extend to the 1980s (Gedalof and Berg, 2010) so conclusions for more recent dates are based on less evidence. Tree-ring studies may not be representative of forests in general, as most studies have been specifically designed to examine growth in response to environmental changes (Gedalof and Berg, 2010). Sites that are not sensitive to environmental changes may be under-represented.

Species extinctions are ongoing, and growing numbers of species are becoming increasingly endangered (e.g. Butchart *et al.*, 2010) but these appear predominantly to be due to direct human intervention through habitat loss or hunting. Attribution of individual extinctions specifically to climate change currently is not possible, or at least not clear. With new isotope evidence from tropical trees in the region (Anchukaitis and Evans, 2010), the extinction in the 1980s of the Monteverde golden toad, which became iconic as a reputed climate change-driven extinction, now appears to be linked to climate variability, and the extreme El Niño of 1986–87, rather than climate change *per se*.

Changes in hydrological quantities vary in sign and magnitude across the globe. River-flow reconstructions from gauge observations gap-filled with a land-surface model indicate that of the world's top 200 rivers, about one-third show statistically significant trends between 1948 and 2004, with 45 rivers showing decreasing trends in flow and 19 showing increasing trends in flow (Dai *et al.*, 2009). Broadly speaking, decreasing flows were seen at lower latitudes and increasing flows at higher latitudes, but in both cases some basins showed changes of opposite sign to the general case.

Reconstructions of soil moisture using a model driven by meteorological observations suggest that mean drought duration, number and spatial extent all decreased between 1950 and 2000 at the global scale (Sheffield and Wood, 2008). Local changes varied considerably; generally, Europe, North and South America and Australia have seen decreases in drought since the 1950s, although the trend appears to have shifted towards increasing drought since the 1980s. Most of Asia and Africa have generally seen increased drought (Sheffield and Wood, 2008), although again with multi-decadal variability. A major drought in the Sahel peaked in the late 1980s and declined thereafter.

The frequency and magnitude of 'extreme' climatic events (floods, droughts, storms and extreme temperatures) may be changing. The IPCC recognizes that these changes increase environmental risks, and accordingly has supported the preparation of a special report, '*Managing the risks of extreme events and disasters to advance climate change adaptation*', which includes a review of the present state and trends in climate-related extreme events in addition to an assessment of the risks these events present and the risk-management options.

6.3.3 Attribution of observed impacts to their causes

Even when environmental changes have been detected, the attribution of changes to their causes is challenging. The established scientific method for explaining a phenomenon is to carry out a controlled experiment, in which two (or more) samples are examined, with one sample being subject to a deliberate change while another is held unchanged. Clearly, this method cannot be applied to the Earth, since we only have one! However, as a next-best alternative, it is possible to use models as a 'virtual' controlled experiment in which different drivers of change can be included or excluded. This method is widely used with climate models to provide evidence for the causes of past climate change (Tett *et al.*, 2002; Stott *et al.*, 2006; Hegerl *et al.*, 2007), and the same approach is now being used for the impacts of climate change.

In the IPCC AR4, few studies undertaking joint attribution were available. Working Group II of the AR4 (Parry *et al.*, 2007) was able to synthesize the quantitative evidence for impacts trends and events that are consistent with a trend of climatic warming, thus providing a basis from which attribution work could begin. Much of this concerns correlations of temperature with ecological indicators such as phenology, as discussed above. However, the critical stage in the joint attribution problem – establishing the fraction of change attributable to anthropogenic climate change and explaining the causal mechanism – is still rarely carried out.

In an early effort at impact attribution, Gillett *et al.* (2004) claimed detection of the effect of anthropogenic climate change on Canadian forest fires, using a simple relationship between burned area and simulated 5-year mean summer temperature from a climate model driven by anthropogenic forcings. This study was not able to explicitly examine the contribution of naturally forced climate change as the necessary simulations were not available to the authors, but previous work had suggested that natural forcings have made no significant contributions to observed climate change in the region. In Indonesia, human land-use activity has been reported to influence the size of fires occurring in closely comparable drought conditions, resulting in larger fire events (Field *et al.*, 2009). Interactions between land-use and climatic conditions form a key component of landscape fire activity, so this area is a rising priority for climate research.

Records of hydrological quantities such as river flows (Dai *et al.*, 2009) may also provide evidence of changes that may be affected by climate change, although other environmental and anthropogenic drivers are also highly likely to play a part. Climate change may affect river flows and large-scale runoff through changes in both precipitation and evaporation, and the latter may also be affected by other environmental changes, such as changes in water-use efficiency by plants and changes in incoming solar radiation due to atmospheric pollution. Direct anthropogenic influences include damming of rivers, extraction of river water for irrigation and introduction of additional water into river systems due to the extraction of deep groundwater.

Gedney *et al.* (2006) used established detection and attribution techniques in a first attempt to assess the contributions to changes in continental river runoff of climate change and variability, CO_2 physiological forcing, land-use change and solar dimming. They used the MOSES2 land-surface scheme (Essery *et al.*, 2003), which is a coupled ecosystem–hydrology model with mechanistic representations of plant physiology with a close coupling between photosynthesis and transpiration. The model was driven by different combinations of observational climate data and measured CO_2 concentrations, reconstructed land-use change and modelled changes in shortwave radiation. Its outputs were compared with observationally based continental-scale river-flow records (Labat *et al.*, 2004). Using optimal fingerprinting techniques, Gedney *et al.* (2006) found that suppression of transpiration by CO_2 physiological forcing was apparently detectable in the historical records. The model reproduced the 3% rise in global river flow over the twentieth century that was shown in the data set, with the simulated increase arising from CO_2 physiological forcing enhancing river flows by 5% with 2% being offset by climate change.

The attribution study of Gedney *et al.* (2006) assumed that the global runoff reconstructions of Labat *et al.* (2004) were sufficiently representative of past trends, but more recent data sets show different trends in some areas (e.g. Dai *et al.*, 2009). These recent data sets still appear to show a direct CO_2 contribution, although these are smaller than the precipitation contribution (Gerten *et al.*, 2008). However, the anthropogenic component of the precipitation contribution has not yet been established, and other direct anthropogenic contributions to runoff also need to be considered (Dai *et al.*, 2009), so a full-system attribution of anthropogenic CO_2 emissions (and other greenhouse gases and further anthropogenic effects) has yet to be provided. Nevertheless, these results indicate that a complete assessment of the consequences of the anthropogenic CO_2 rise for water resources requires the interactions with ecosystem impacts also to be considered.

Most large-scale changes in vegetation appear to be due to human land use or land abandonment, rather than climate. The likely exceptions are generally in colder regions: boreal forests, tundra, mountains and polar regions (including Antarctica), both because they are more sparsely populated and because the temperature change signal in some of these regions has been more pronounced. Land-cover change (including urbanization and the development of urban heat islands) is likely to be an important anthropogenic driver of local to regional climate change that will need to be taken into consideration in attribution efforts for more 'downstream' impacts, such as crop production and food security, as well as human health.

Many impacts relate to extreme events, and while the actual occurrence of individual events cannot be attributed to specific causes, such as anthropogenic climate change, changes in the probability of their occurrence arising from different causes can be estimated. This has previously been done for meteorological events, notably the European summer heatwave of 2003 (Stott *et al.*, 2004), but similar logic has now been applied to climate impacts. Pall *et al.* (2011) used a large climate ensemble to assess the probability that climate change was a cause of the extreme river flooding in England in the autumn of 2000, demonstrating that anthropogenic emissions contributed significantly to the risk of floods of this type.

Some claims have been made linking an apparent increase in the severity of weather-related disasters to climate change. For example, the Global Humanitarian Forum (2009) stated that '*every year climate change leaves over 300,000 people dead, 325 million people seriously affected, and economic losses of US $125 billion*'. This assessment is based on a simplistic attribution methodology, which compares the rate of increase in weather-related disasters to non-weather disasters (such as earthquakes), notes a steeper rate of increase in weather-related disasters and assumes the difference is due to anthropogenic climate change. However, this is based on 25 years of data from 1980–2005, a short period when dealing with changes in return periods and magnitudes of extreme events. It also does not address the joint attribution problem of distinguishing between anthropogenic climate change and natural climate variability, which may be very significant especially at local scales. Also, it does not account for any influence of climate change on the impact following non-weather disasters, such as earthquakes. Survival and wellbeing of large populations exposed to the elements after such events is dependent on the weather and climate. To the extent that attribution requires an explanation of the causal mechanisms, not just the manifestation of an impact, the assessment of the human and financial impact of climate change quoted above is poorly founded.

6.4 Global-scale impacts of future climate change

6.4.1 Difficulties of assessing 'dangerous' levels of global warming

The objective of the United Nations Framework Convention on Climate Change is to achieve '*stabilization of greenhouse gas concentrations in the atmosphere at a level that would prevent dangerous anthropogenic interference with the climate system*' (UNFCCC, 1992). The identification of this level of greenhouse-gas concentrations has therefore been perceived as a key requirement of scientific advice for informing action under the convention. Assessing the global-scale impacts of climate change is relevant to this process. The UNFCCC does also aim to prepare for adaptation to the impacts of climate change, stating that action under the convention should minimize '*adverse effects on the economy, on public health and on the quality of the environment*'.

Recently, scientific research and policy relating to 'dangerous' change has focused more narrowly on levels of global mean temperature change, rather than greenhouse-gas concentrations. The 2009 Copenhagen Accord cited a '*scientific view that the increase in global temperature should be below 2 degrees Celsius*'.

Emissions and concentrations of the various greenhouse gases are generally compared with each other on the basis of their effect on global mean radiative forcing. To the first order, radiative forcing appears to determine the global mean temperature rise, and so it is assumed to be a useful metric of the impacts of greenhouse-gas emissions. However, identification of the impacts manifest at specific levels of global warming (or radiative forcing) is problematic, as discussed further below, making it difficult both to identify 'dangerous' impacts arising from specific greenhouse-gas concentrations and to provide an informed view of different mitigation strategies, including their potential adverse effects. Moreover, with adaptation to climate change also now necessary, it is important that climate information presented to inform mitigation policy is not simplified so much that it becomes misleading for adaptation. Presenting impacts aligned to specific levels of global warming, such as the two-degree policy goal, without accounting for uncertainties in regional climate change, rates of change, or non-climatic drivers, may lead to inappropriate decisions being made on adaptation.

Assessing the impacts of climate change at specific levels of global warming presents a number of difficulties (Figure 6.14):

- relating regional climate change and its impacts to global mean temperature rise;
- the timing and rates of change, especially in relation to socio-economic changes which affect vulnerability;
- dealing with the drivers of change that are connected to climate change but with a large uncertainty in how they relate to global mean temperature change.

A full-system view may be required, using Earth system models to assess the impacts of different emissions scenarios through a wider range of processes.

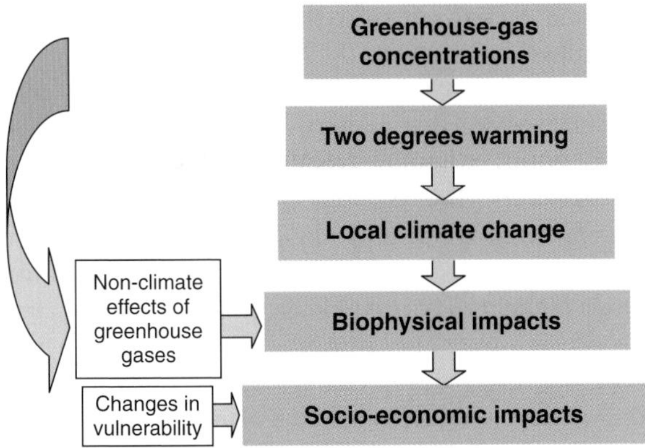

Figure 6.14 Difficulties with aligning specific levels of global warming with impacts: uncertainties in regional climate change, additional drivers of change associated with greenhouse gases, and changes in vulnerability.

6.4.2 Relationships between global warming and regional climate change

The patterns of regional climate change associated with a given increase in global mean temperature are highly uncertain. This is not merely an issue of uncertainties in the ability of models to represent the real world – many different patterns of regional temperature and precipitation change could be associated with a single global mean temperature rise, either because of different spatial patterns of forcing or because of local feedback processes and internal variability in climate. As well as the uncertainty in the regional climate response to a global mean radiative forcing arising from well-mixed greenhouse gases, it is important to recognize that a number of anthropogenic forcings of climate change have considerable spatial heterogeneity, and hence lead to very different patterns of climate change across the globe, even if the global mean forcing is the same.

Analysis of the set of climate-model projections used for IPCC AR4 shows that, in some regions, a very wide range of climate changes is associated with a given increase in global mean temperature. For example, in eastern Amazonia, some models project annual mean precipitation to decline to 1,000 mm at a global warming of 4 K while others project it to remain high at 3,000 mm (Figure 6.15). This means that it is not possible to make confident statements about the climate of the Amazon in a world that is four degrees warmer; only qualitative statements about potential risk.

In the IPCC AR4, Fischlin *et al.* (2007) made a preliminary systematic assessment of ecosystem impacts at different levels of global warming, using results from the literature and aligning the associated regional climate states with global mean temperature rise on the basis of 17 climate-model projections. Although this established a methodology for synthesizing multiple information sources against a common framework of global-warming levels, the robustness was limited by the availability of information relating global climate changes to the regional changes used in the original literature.

The potential impacts on the Amazon forest again provide a good example of the difficulties. The IPCC's 2007 assessment included a projection of major loss of forest cover in the Amazon due to climate change later in the twenty-first century, occurring at a global warming of 2 K to 3 K, based on simulations made with the HadCM3–LC coupled climate–carbon-cycle model (Cox *et al.*, 2000; Betts *et al.*, 2004). However, as shown in Figure 6.15, climate models produce

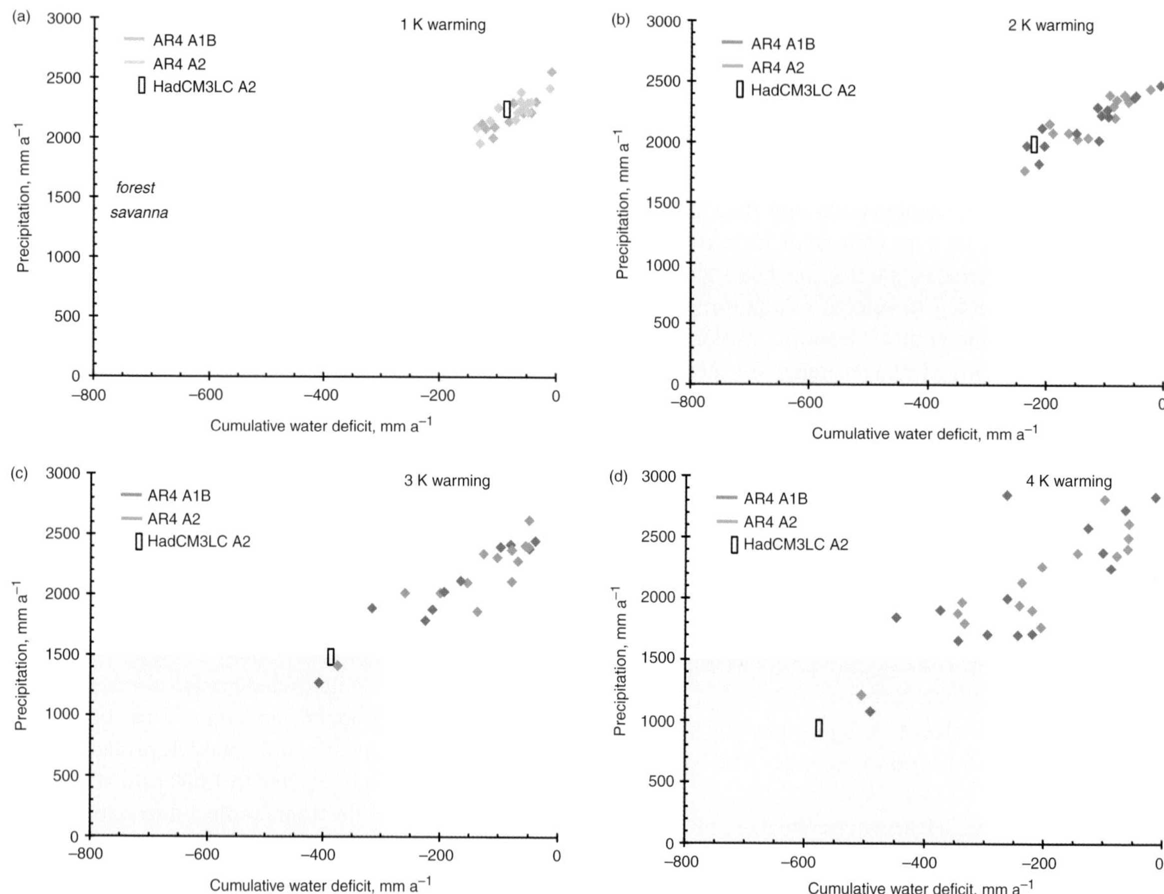

Figure 6.15 The rainfall regime of eastern Amazonia associated with different levels of global warming. Global mean temperature increases of (a) 1 K, (b) 2 K, (c) 3 K and (d) 4 K relative to pre-industrial, projected by the HadCM3–LC coupled climate–carbon-cycle model and the IPCC AR4 models under the SRES A2 and A1B emissions scenarios (linearly extrapolating the climate response to the higher levels of warming if these levels were not reached within the original simulations). Rainfall regime for savanna vegetation rather than forest (shaded area) is quantified following Malhi *et al.* (2008).

widely varying projections of change for Amazonia. Many of the AR4 models do not simulate changes that approach the critical threshold for rainforest (namely, a precipitation threshold of 1,500 mm and cumulative water deficit of −300 mm a^{-1}) even at four degrees of global warming.

6.4.3 Non-climatic drivers of change related to emissions

There is not a one-to-one correspondence between CO_2 concentration and global warming (this is the climate sensitivity issue discussed in earlier chapters). In cases where the direct effects of CO_2 can offset or enhance the impacts of climate change, the overall effect of a given change in temperature on the impacts depends on the CO_2 concentrations that actually produced that warming.

The CO_2 fertilization effect on vegetation is one instance where this matters. If climate sensitivity is high, the policy limit of two degrees of warming would be reached with a relatively small CO_2 increase, and the offsetting of warming impacts by CO_2 fertilization would be small. Conversely, if climate sensitivity is low, a higher CO_2 rise would be necessary to reach the two-degree warming, with a larger CO_2 fertilization offsetting more of the climate change impact. These differences could be critical in determining whether there are overall increases or decreases in crop yield, water resources (since vegetation responses to CO_2 also impact on surface hydrological processes) and flood risk. Furthermore, the actual mix of greenhouse gases that leads to the warming will influence the overall ecosystem and hydrological response, because non-CO_2 greenhouses gases do not impact directly on ecosystems in the way CO_2 does.

Climate sensitivity also determines the relationship between warming and ocean acidification. For high climate sensitivity, two degrees of warming caused by a small CO_2 rise would result in a small decrease in ocean pH with relatively lower potentially detrimental impacts on calcifying organisms. The relative impact of warming and acidification could be crucial.

Climate sensitivity is commonly expressed as a range although more sophisticated analyses have expressed the uncertainty in terms of probabilities. In impacts studies, the inverse problem is of interest: what is the probability that different levels of CO_2 concentration would result in, say, a two-degree warming? Because many biological, ecological and agricultural impacts of climate change are also affected directly by the CO_2 in the atmosphere, establishing the relationship between CO_2 concentrations and climate can better inform assessments of the possible implications. Figure 6.16 illustrates the uncertainty in the CO_2 concentration that would be associated with two degrees of global warming, for emissions scenarios both with and without global emissions reductions, using an ensemble of 58 variants of the HadCM3C Earth system model. These preliminary results suggest that in the A1B scenario (no climate policy, no early emissions reduction, and hence ongoing CO_2 rise), global warming could pass through two degrees warming relative to pre-industrial temperatures at CO_2 concentrations ranging from 430 to 620 ppm, with a mode of 460 ppm. However, in the RCP2.6 scenario which has an early peak and decline in emissions, and hence a near-stabilization of CO_2 concentrations, the two degrees of global warming could be reached at CO_2 concentrations ranging from 410 to 500 ppm. This stabilization scenario tends to give two degrees of warming at lower CO_2 concentrations because the slower CO_2 rise allows more time for the temperature to fully respond. In the ongoing emissions scenario, the warming seen at the time of any given CO_2 concentration is not the full warming that would ultimately result from that CO_2 concentration. In both scenarios, however, it is clear that the CO_2 concentration associated with a 2 K change in global mean temperature is subject to high uncertainty, and hence the extent of CO_2 physiological effects either offsetting or adding to climate effects on ecosystems is also highly uncertain, and depends on the climate sensitivity.

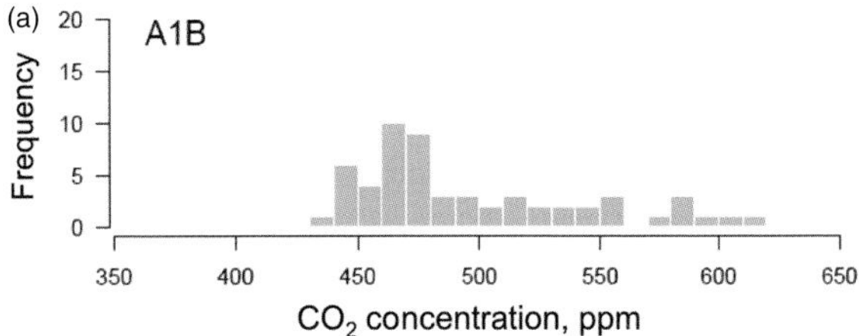

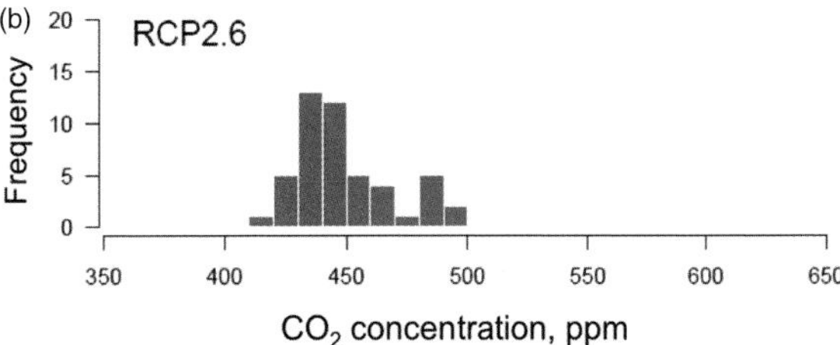

Figure 6.16 Frequency distributions for the atmospheric CO_2 concentration associated with a simulated global warming of 2 K above pre-industrial in an ensemble of 58 variants of the HadCM3C Earth system model, for (a) the SRES A1B scenario of emissions with no climate mitigation policy, and (b) the RCP2.6 scenario in which global emissions peak then decline with the aim of limiting warming to two degrees. 'Frequency' refers to the number of model variants first reaching 2 K global warming at the same time that their simulated atmospheric CO_2 concentration reaches the values shown, after smoothing of temperature variability.

6.4.4 Time-dependent impacts of climate change

Rather than attempting to align any particular level of global mean temperature rise to the policy goal of avoiding 'dangerous climate change', and hence risk detrimental unintended consequences, a more complete picture would be gained by addressing the question 'What are the full-system impacts of different perturbations to the Earth system?' This is a much more complex issue than simply plotting impacts against global warming – it depends on the timing of when that warming is reached, the importance of additional non-climatic effects of greenhouse gases, and whether adaptation happens or not. Presumptions of steady linear change may not always be realistic for most impacts, so informing social responses to future climate change requires information about the way that impacts are expected to unfold over time.

We illustrate this idea with two examples: the implications of climate change projections on thresholds of water stress, and on land suitability for crop production.

Water stress

Widely used metrics of water stress at the global and regional scale are based on the volume of surface-water runoff per capita. Using runoff simulated by an ensemble of 17 variants of the HadCM3 climate model driven by the A1B scenario, along with the population projection associated with this scenario, we have analysed the simulated changes in water stress for each of the 17 model variants. To avoid the problematic inconsistencies in the usual approach of applying bias-corrected, pattern-scaled meteorological outputs of climate models to an offline hydrological model, this analysis used the runoff output generated by the climate model, ensuring the hydrology is consistent with the overlying climate change over time. The systematic biases in the climate-model runoff output have been corrected directly. Climatological observations of runoff are added to the absolute projected runoff change in order to project runoff volumes in the future, and these are then compared with the water-stress thresholds. Three combinations of drivers of change in water stress were considered:

(i) population change alone;

(ii) population change and climate change, from 17 variants of the climate model, with no influence of CO_2 rise on vegetation water-use efficiency;

(iii) population change, climate change and increases in vegetation water-use efficiency due to CO_2 physiological forcing, from 17 variants of the climate model.

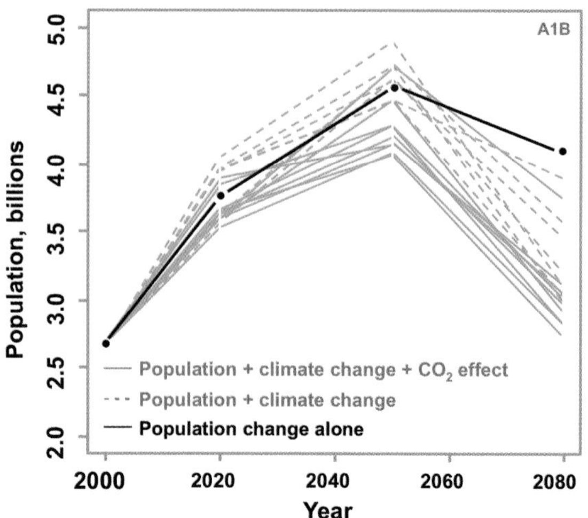

Figure 6.17 Simulated changes in global water stress due to changes in population, climate and CO_2 effects on vegetation projected by an ensemble of variants of the HadCM3 climate model for the SRES A1B scenario. The y-axis shows the population subject to high or extreme water stress (defined as annual mean runoff less than 1000 m^3 of water per capita). All model variants were driven by the standard greenhouse-gas concentrations corresponding to the A1B scenario as used in CMIP3; uncertainty in translating emissions to concentrations was not represented. The population projection was also from the A1B scenario. The black line shows changes in water stress due to population change alone; dashed green line shows changes due to both population and climate change (without changes in CO_2 physiological forcing); and solid green lines show changes due to population, climate change and CO_2 physiological forcing.

Figure 6.17 shows the projected changes in global population under high or extreme water stress (defined as less than 1000 m^3 of runoff per capita) for each of these combinations of drivers. In this analysis, the principal driver of increased water stress at global scales is population change: the solid black line shows the changes in global water stress due to population change alone, and this rises until 2050 due to population increases and then declines thereafter as the population declines under this scenario. The dashed green lines show the projected changes in water stress when climate change is also considered, for each model variant – in some model variants, water stress increases more rapidly than the population-only projection until the 2050s, while in other model variants the increase in water stress is less rapid when climate change is considered compared to population change alone. However, by the 2080s all model variants indicate a smaller fraction of global population being water-stressed when climate change is considered. When CO_2 physiological effects are also included (solid green lines) most model variants show less rapid increases in global water stress than the population-only projection until the 2050s, and all show less water stress by the 2080s – indeed for some model variants, global water stress is simulated to be nearly reduced back to present-day levels when both climate change and CO_2 effects are included.

At the scale of individual basins, there are some regions where water stress increases further due to climate change, while in others it increases less (Figure 6.18). Further work is still needed to refine this analysis. Using annual total runoff may not be the most appropriate measure of impact, as this assumes that seasonality is not important. If seasonality becomes more pronounced under future climate change then adaptive responses, such as increased water storage, may become necessary.

Changes in the suitability of cropland climates

The implications of climate change for croplands have been assessed in a joint study by the QUEST and AVOID

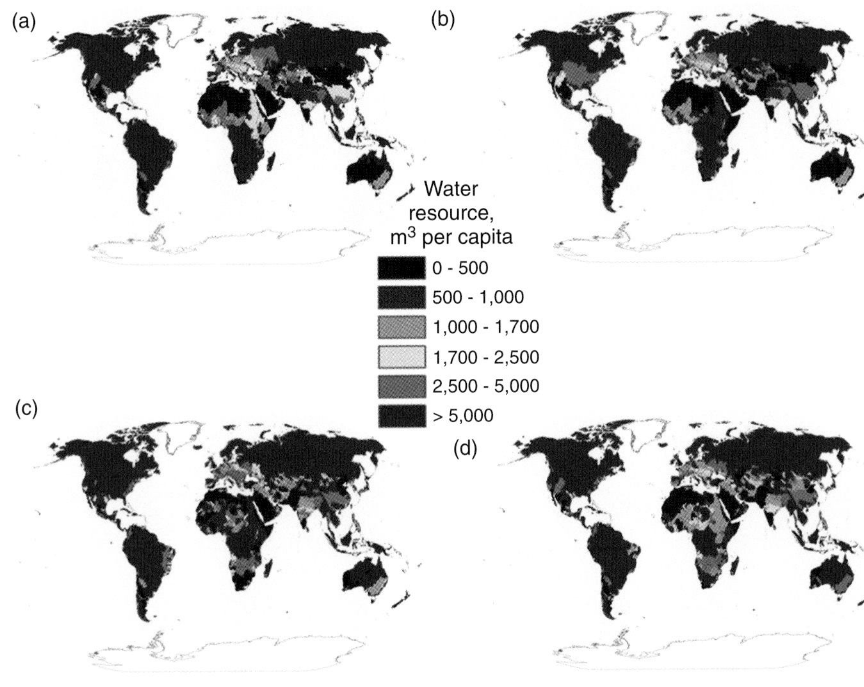

Figure 6.18 Simulated changes in regional water stress due to changes in population, climate and CO_2 effects on vegetation. (a) Present-day water stress; (b) water stress in the 2080s with population change alone; (c) water stress in the 2080s with population and climate; (d) water stress in the 2080s with population, climate change and physiological effects of CO_2 on vegetation water-use efficiency. Climate and CO_2 effects obtained from the mean of the 17-member HadCM3 ensemble. Water stress is defined in terms of annual mean runoff per capita. Moderate water stress is experienced at 1,000–1,700 m^3 per capita; high water stress at 500–1000 m^3 per capita, and extreme water stress at < 500 m^3 per capita.

programmes (Met Office, 2010). This assessment used an index of the suitability of an area for cultivation based on climatic quantities (temperature and water availability) and soil characteristics (Ramankutty *et al.*, 2002). Patterns of climate change from 21 model projections were used, for two emissions scenarios (Gohar and Lowe, 2009): first, the IPCC's SRES A1B scenario, which features ongoing emissions; and second, a 'stabilization' scenario (A1B-2016–5-L) in which global emissions peak at 2016 and then decline by 5% per year thereafter. The climate impact was quantified in terms of the proportion of croplands becoming more or less suitable for cultivation compared with current cropland areas, according to the crop suitability index. The crop suitability index varies significantly for current croplands across the world (Ramankutty *et al.*, 2002), with some current cropland areas occupying low-suitability lands according to this index. Also, some areas that are not currently used as croplands may already be suitable for cultivation. It is assumed that, to first order, these limitations of the approach tend to cancel each other out in this analysis. This study did not include the known effects of changing atmospheric CO_2 concentration on evapotranspiration (discussed in Chapters 2 and 4), nor its effect on photosynthesis and crop yield, both of which could offset part of the calculated decreases in crop suitability.

Under all the climate projections, some existing cropland areas become less suitable for cultivation while other existing cropland areas become more suitable (Figures 6.19, 6.20). The areas of increased and decreased suitability differ considerably according to the climate model used, but at the global scale these differences between models do not increase very much into the future (so the greater uncertainty about climate change further into the future is not reflected in increasing uncertainty in future cropland suitability).

For the A1B scenario, the area of current global croplands becoming more suitable initially increases by ~10–30%, but is not projected to increase further beyond 2030. In contrast, by 2030 under A1B, between 15% and 50% of current global croplands are projected to become less suitable for cropping, increasing to 30–70% by 2080.

Until 2030, the changes projected by the stabilization scenario are similar to those in A1B. The key difference is seen beyond 2030, in the expansion of areas of decreased suitability. Impacts under the stabilization scenario are lower than for A1B – by 2080 between 20% and 60% of current cropland have become less suitable for cultivation. At the global scale, then, the stabilization scenario implies changes in the suitability of current global croplands which are either less detrimental or slightly more beneficial than those implied by the A1B scenario.

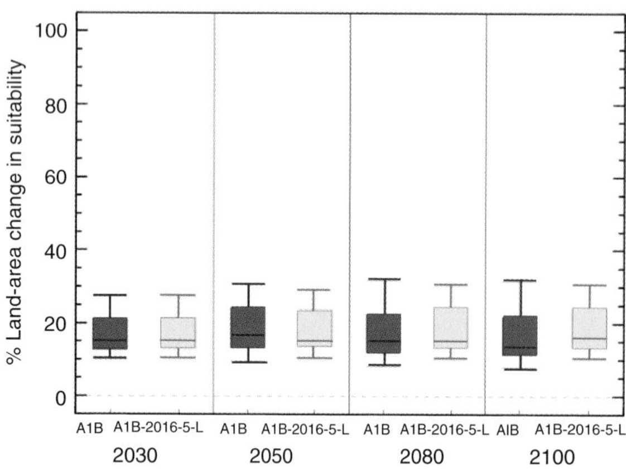

Figure 6.19 Areas of current cropland projected to *increase* in climatic suitability for two emissions scenarios by 2030, 2050, 2080 and 2100. Blue: A1B scenario of ongoing emissions; yellow: A1B-2016–5-L scenario of global emissions peaking in 2016 and declining by 5% per year thereafter. Boxes show 25%–75% range of model results, whiskers show full range.

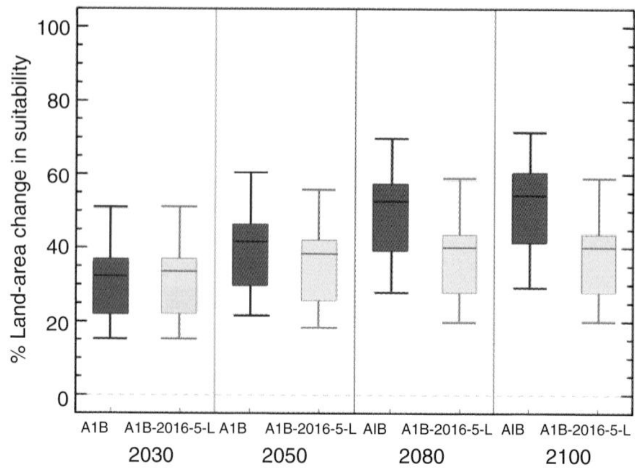

Figure 6.20 Areas of current cropland projected to *decrease* in climatic suitability for two emissions scenarios by 2030, 2050, 2080 and 2100. Blue: A1B scenario of ongoing emissions; yellow: A1B-2016–5-L scenario of global emissions peaking in 2016 and declining by 5% per year thereafter. Boxes show 25%–75% range of model results, whiskers show full range.

Although this study gives an indication of the global impacts of increased emissions, it makes no indication of whether the suitability of a particular area will continue to increase (or decrease) over time, and it only examined the direction of change in climatic suitability, not the magnitude of change nor any further indicators such as crop productivity.

At present, the aggregate impacts of climate change on global-scale agricultural productivity cannot be reliably quantified. In the QUEST/AVOID study, only local changes in climate could be addressed, but they relate only to rain-fed agriculture (some 58% of global food production; Rosegrant *et al.*, 2002). In areas dependent on irrigation, other climate-related factors come into play, including the behaviour of remote rainfall, snowmelt and glaciers, and groundwater recharge (Figure 6.10). Indirect impacts via sea-level rise (including salinization of groundwater), storms and diseases may also be important (Gornall *et al.*, 2010).

6.4.5 An impacts-focused metric for comparing different greenhouse gases?

To avoid 'dangerous' climate change, climate policy-makers need to know the total level of atmospheric greenhouse gases to avoid exceeding, and how one means of cutting emissions compares with another. This is routinely done in terms of quantifying the 'CO_2 equivalency' of a greenhouse gas or emissions source.

There are two forms of 'CO_2 equivalent' in common use. One focuses on the concentrations of greenhouse gases at a particular time compared to, say, the Industrial Revolution, and measures the contribution of each greenhouse gas to global warming in terms of radiative forcing (Forster *et al.*, 2007). This concept can be extended to several other changes to the climate system, for example changes in the concentration of atmospheric aerosols, such as dust, soot and sulfate, which can have net warming or cooling effects, or changes in surface albedo resulting from modified forest cover and the deposition of soot on snow.

Another 'CO_2 equivalent' focuses not on concentrations but on emissions. The usual metric, used by the UNFCCC and the Kyoto Protocol, is the global-warming potential (Forster *et al.*, 2007), which integrates the radiative forcing of an emitted quantity over a specified timescale, and compares that with CO_2. To some extent this measure incorporates the fact that different greenhouse gases are naturally removed from the atmosphere over different timescales. However, this approach assumes that radiative forcing is the only mechanism through which greenhouse gases influence climate.

Although emissions of greenhouse gases from their various sources are the dominant cause of global warming, the overall impacts of each greenhouse gas from each individual source are not in proportion to their relative contribution to global warming. For example, methane has a 100-year global-warming potential (GWP) of 23, suggesting that, as a molecule, it is 23 times 'more potent' than CO_2 in influencing climate. However, methane does not exert a physiological effect, so its relative effect on climate impacts (such as water resources, agricultural yields and biodiversity) may not necessarily be 23 times that of the equivalent mass of CO_2, despite what would be expected on the basis of the GWP.

The relative importance of various emissions sources can be illustrated by considering what the current impacts of climate change would be if all of the observed warming to date had resulted from CO_2 from fossil-fuel burning, with no deforestation and no other greenhouse gases being emitted. The impacts may well have been different from those now observed. Ocean acidification would have proceeded further, plant growth and carbon uptake would probably have increased further, and global average river runoff could be higher, all because of the higher CO_2 rise exerting direct impacts. Moreover, the lack of forest-cover change would also have had an effect; temperate regions would be warmer, due to their lower surface albedo with intact forest cover, while parts of the tropics would be cooler.

In reality, fossil-fuel CO_2 has only caused about half of the current anthropogenic 'greenhouse' warming, defined as global mean radiative forcing (Forster *et al.*, 2007). The relative contribution of fossil-fuel CO_2 emissions is increasing, but land-use change and other greenhouse gases still contribute approximately two-fifths of the CO_2-equivalent emissions defined in terms of GWPs (Metz *et al.*, 2007). If society is aiming to avoid greenhouse-gas concentrations rising above, say, 550 parts per million of 'CO_2 equivalent', are the overall impacts the same whether most of this rise is CO_2 itself, or CO_2 is limited to much lower levels at the expense of other greenhouse-gas concentrations? Would avoiding deforestation of an area which would have emitted a million tonnes of CO_2 have the same effect as reducing fossil-fuel emissions by that amount? Are biofuels and nuclear power equivalent in terms of their contribution to avoided damages? In all of these cases, the evidence suggests not. Determining the relative benefit of cutting emissions of different gases from different sources remains vital.

Given that the current concept of CO_2 equivalent is incomplete and cannot deal adequately with these kinds of questions, what is the alternative? While radiative forcing allows comparison in terms of a well-defined quantity, it does not allow for comparison with other drivers of climate change quantified in different units. Since the ultimate aim of climate change mitigation is to reduce the (economic and other) cost of human interference with the climate system, the ideal option would be to compare emissions sources in terms of their final impact, such as on human lives lost, species driven to extinction, and economic damages. In particular, since water resources and flood risk are among the key issues of concern over climate change, and are strongly influenced by the interactions of physical climate and the biosphere, metrics allowing the comparison of greenhouse gases in terms of hydrological impacts would be desirable.

At this stage, it is difficult to say whether such 'impact-based' CO_2 equivalents for non-CO_2 greenhouse gases would be greater or less than those indicated by standard CO_2 equivalents based on radiative forcing. However, in the case of comparing the effects of deforestation with fossil-fuel burning, there is good reason to believe that an end-to-end climate-impact metric would give a greater relative weight to the overall impact of deforestation than does the current metric of radiative forcing by emitted greenhouse gases. Decreasing deforestation rates in the tropics would not only contribute to climate change mitigation through reducing CO_2 emissions, it could also help to maintain relatively cooler, moister climates in these regions and conserve biodiversity and ecosystem services, the natural processes and resources that ecosystems supply and from which humankind benefits (Millennium Ecosystem Assessment, 2005). Several of the aims of stabilizing greenhouse-gas concentrations would be served more effectively by a more complete consideration of ecosystem services.

The existing definitions of CO_2 equivalent are vital for the current policy process on climate change mitigation, so should not be discarded in haste. They are, however, limited in their ability to assess the real impact of climate change drivers. In time, more far-reaching decisions will need to be taken in the context of conflicting priorities. Current CO_2 equivalents will be supplemented or supplanted by more comprehensive metrics.

6.5 Adaptation in practice

6.5.1 Linking scientific advice to adaptation practice

Much of the need for forecasting to inform forward planning requires information on relatively near-term timescales: multi-annual to decadal, or a few decades. Some adaptation planning, like major infrastructure, needs to look further ahead. Also, adaptation information is usually needed at the local scale rather than the global. The high uncertainties in future projections and attribution, together with the socio-political requirements of governance, mean that much adaptation practice needs to focus on reducing vulnerability to a range of possible changes (Mearns, 2010). Realistic projections or scenarios are also needed in order to inform vulnerability reduction. With many possible areas requiring action, there is a need to prioritize efforts and resources, informed by assessment of the magnitude of possible impacts and the estimated likelihood of events.

There are several general issues that arise in adaptation efforts:

- ***The adaptation gap*** – in many cases, nations and communities are inadequately equipped to adapt to current vulnerability, because of problems such as poverty and resource scarcity (Parry *et al.*, 2007). Adaptive responses need to cope with the current baseline climate variability, as well as future changes.
- ***Problem definition*** – responses depend fundamentally on what kind of things need to be adapted to. In most contexts, adaptation is needed to changes in mean climate conditions and also to extreme events, but the specific climatic threats vary from context to context.
- ***Response options*** – what does adaptation mean in practice, in a given context? It may involve institutional and infrastructure modifications, such as changes in business practices, building design or land planning, but it also involves other measures to improve society's preparedness for the increased likelihood of adverse conditions.
- ***The timing of adaptation*** – decisions about when to adapt require the identification of appropriate lead times to allow planning over correct timescale. Ideally, planning also includes flexibility to allow for the modification of adaptation as more information (e.g. better forecasts) becomes available.
- ***The integration of adaptation*** – decisions are needed about the extent to which adaptation is addressed as a specific policy issue in its own right, compared with efforts at mainstreaming adaptation into business-as-usual planning.

A key question for climate science today is whether the available models and other techniques are useful in meeting these various requirements. One priority research area is in providing information at the required temporal and spatial scales. Mid-term (seasonal to decadal) forecasting, in which model projections are initialized at the correct state within current cycles of variability, is a rapidly moving area of research and has demonstrated success in improving forecasts of global mean temperature over the timescale of a few years (Smith *et al.*, 2007). However, achieving forecast skill at regional scales, especially for precipitation, is a considerable challenge.

In the following section, we describe current efforts to link scientific advice with adaptation policy and practice, allowing decisions to be as informed as possible even in the context of large uncertainties in regional climate changes.

6.5.2 Case studies in adaptation

Adapting food systems to climate change

Food security in the context of climate variability and climate change depends on the rapidity of change, the lag in adaptation (social and technological) and the vulnerability of food crops, particularly to climate extremes. Forecasting ability is severely limited for many aspects of climate change impacts at the scale required for planning, so adaptation measures in the short to medium term need to focus on reducing vulnerability.

The vulnerability of food systems can be reduced through a number of approaches (Gregory *et al.*, 2005). Increased crop productivity may be achieved through the selection of cultivars with greater resilience to climate change and variability (Porter and Semenov, 2005), traits such as greater efficiency of water use, and tolerance to extremes of temperature and to pests and diseases. Changes in management tools and techniques such as irrigation may also assist with adaptation to shifting mean climate states. Adaptation of post-production aspects of food systems may be needed, such as planning to improve the resilience of food distribution infrastructure, or the introduction of improved communication for food producers.

The implementation of such adaptation will either require climatic changes to be discernible enough to prompt action but gradual enough to allow time to respond, or forward planning on the basis of future risk assessment. Improved food security in the face of climate extremes may therefore require intervention at the scales of local or national government. Some extreme events can be relatively localized, so food security can be improved by maintaining and improving food distribution networks, mitigating impacts on local productivity by access to production elsewhere. Reducing the vulnerability of individual farms' productivity to extreme climatic events is likely to be particularly challenging. For example, Brondizio and Moran (2008) suggested that small farmers in Amazonia often lack the knowledge and information on climatic extremes that would allow them to be resilient to unusual events such as the 2005 drought.

Given that resilience is already poor in many areas, improving people's ability to cope with existing climate variability is important. However, development that aims to increase resilience against current climate variability should take account of potential future climate changes in order to maintain this resilience in the longer term.

Using seasonal climate forecasting as a climate change adaptation tool can facilitate early warnings of impacts on food production that may allow preparations to be made. Seasonal forecasts in some regions such as Ghana and northeast Brazil are already demonstrating predictive skill due to improved understanding and modelling of links between regional weather and sea-surface temperatures. While this approach is in its infancy, and there are still many improvements to be made in the scientific capability of seasonal and decadal forecasting, there is also considerable scope for improved pull-through of such forecasts to a wider range of actions on the ground.

Flood defences in the Thames Estuary

In 2004, a UK Government-commissioned Foresight study published its report (DTI 2004; Thorne *et al.*, 2007) on the prospects for Britain's river and coastal flooding and coastal erosion, in the context of future scenarios of climate change. It combined hydrological, geomorphological and economic mapping to highlight areas of greatest national significance in terms of future risk over an 80-year timeframe. As might be expected because of its pivotal role in UK and international economies, London emerged as a priority area for flood-risk responses. This study was influential in shaping the methodology for the pre-emptive adaptive responses that are currently under way in London.

In terms of the current and future drivers of flood risk, the population of London is exposed to land-level isostatic readjustments due to the last deglaciation some 10,000 years ago, sea-level rise, storm surges, rainfall runoff, and the consequences of water-level management in surrounding areas, including the use of the tidal barrier downstream of central London that was built in the 1980s. This list includes long-term, fairly predictable geological change, climate-related changes, and political decisions. The tidal reach of the Thames Estuary is great, extending to the west side of London, so the risks associated with sea-level rise have potential impacts on a large proportion of the inhabitants of the estuary area. In economic terms, the costs of flood management in the UK have risen steadily, but so also have the costs of flood events. The commissioning of the Foresight Future Flooding report marked a national policy shift away from flood defence to more multi-pronged flood-risk management, including improving community preparedness, and incorporating landscape-scale planning changes to manage the distribution and flows of water.

Like most major cities, the Thames Estuary exists within a complicated policy context with many stakeholders, including several local authorities, diverse government agencies and non-governmental organizations, and influential business interests. It has a complex legislative context, with many regulations, directives and acts, which have different levels of influence. A number of tensions exist, including pressure for London to expand, involving the redevelopment of low-lying lands, with associated shifts in governance between public and private responsibility. Navigating these constraints while finding opportunities for mutual benefits is a typical challenge for real-world adaptive responses to climate.

In this context, a detailed assessment of the risks of flooding and options for managing these risks was carried out in order to develop the Thames Estuary 2100 plan (Environment Agency, 2009). Global and regional climate change and sea-level projections together with storm-surge and river-flow models were

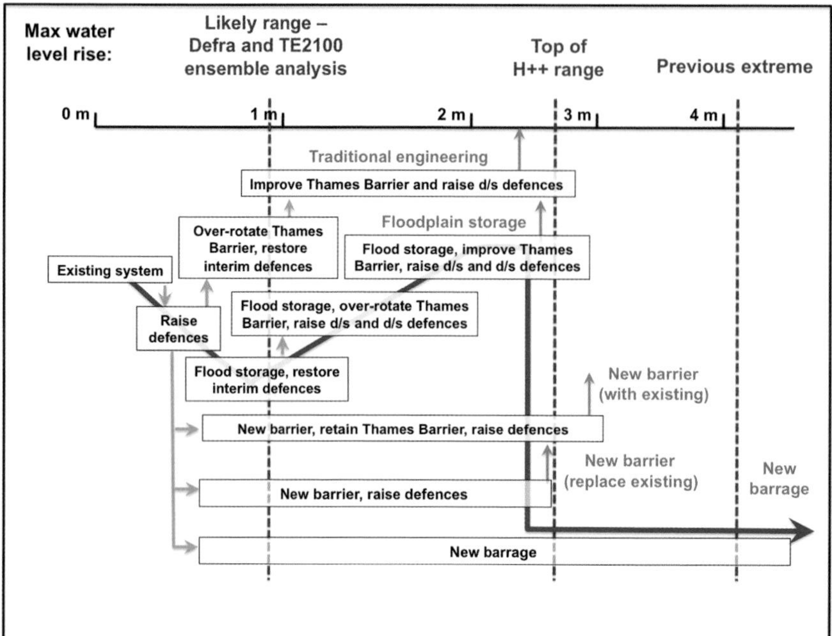

Figure 6.21 Conceptual tool for flexible planning for adaptation of Thames Estuary flood defences. Text boxes describe different options for flood defences, shown against amounts of mean sea-level rise that each would protect against. Red dashed lines show estimates of sea-level rise considered in the Thames Estuary 2100 adaptation planning project. H++ is an extreme vulnerability scenario that considers the highest physically feasible values for all components of twenty-first century sea-level rise. The blue curve indicates the changing level of flood risk with the implementation of options; d/s denotes drainage system. In the face of large uncertainty about future sea-level rise and the need to be able to plan upgrades to defences potentially decades in advance, it is necessary to identify pathways (turquoise arrows) through different options so that choices can be made and changed if necessary as improved information becomes available.

used to generate scenarios of coastal and river flooding, and to estimate the likely ranges of impact along with worst-case scenarios. These were then used to inform assessments of the potential need for different flood-defence options (including drainage systems, structural defences and even a full barrage across the estuary), and of the potential windows of opportunity for revising plans if and when improved advice on risks becomes available.

A key feature of the Thames Estuary 2100 process is its focus on adaptive management (Figure 6.21). For example, some physical interventions, such as permanent flood defences and embankments, have been designed for incremental construction. The additional cost now of building a defence with a wider-than-normal footprint that allows for subsequent heightening if needed is minor compared with the costs of complete reconstruction of defences that prove to be inadequate later.

6.6 Key messages

An Earth system view of global change is crucial for informing strategies for adaptation to anthropogenic climate change and the mitigation of its causes.

Quantification of 'dangerous climate change' in terms of global mean temperature rise is problematic due to the large uncertainties in the regional climate change and impacts associated with different levels of global warming, the additional influences of CO_2 beyond its role as a greenhouse gas, and the importance of the rate of warming in comparison with socio-economic development pathways. Comparison of the full Earth system impacts of different time-dependent emissions scenarios would provide a clearer picture of the consequences of different future pathways of greenhouse-gas emissions.

Mitigation decisions that rely on the comparison of the effects of different greenhouse gases should take account of the climatic effects of these climate change drivers that do not act through the enhanced greenhouse effect. The current concept of 'CO_2 equivalent' based on radiative forcing does not provide a meaningful comparison of the impacts of different greenhouse gases on the climate system. If mitigation policies require a prioritization of actions based on the relative impacts of different greenhouse gases and other forcing agents, the full range of effects of these agents should be considered.

Strategies for mitigating climate change that involve land-cover change, such as plantations for carbon sequestration or biofuels, will exert additional effects on climate change and its impacts. Merely accounting for the net reduction in carbon emissions will not provide an accurate measure of the full effect of these mitigation strategies. To fulfil the UNFCCC aim of avoiding 'dangerous' climate warming, the full effects of changes in land cover should also be considered.

Adaptation strategies require robust assessments of future climate change and its biophysical and socio-economic impacts at global and regional scales. These depend on ecosystem feedbacks on radiatively forced climate change, additional anthropogenic forcings through land-cover change, and plant physiological responses to changes in CO_2 concentration and other atmospheric constituents. A general implication of these issues is that adaptation strategies should regard existing projections of climate change impacts as preliminary. To date, most climate research has focused on assessing whether anthropogenic radiatively forced climate change may be serious enough to warrant action. This question arguably implies a relatively low requirement for regional detail and process representation. However, adaptation plans require a more complete picture of climate change in the context of other aspects of global change, and more precise information on individual places.

Any adaptation strategy or decision that is sensitive to the rate of climate change will be influenced by consideration of feedbacks on global climate change through the carbon cycle and other biogeochemical cycles. Decisions based on climate scenarios that do not take these feedbacks into account may not allow for a sufficiently rapid response or rate of adaptation. For example, increases in the frequency of extreme weather events may occur more rapidly than projected by the simpler models, so there may be a more urgent need to 'future-proof' infrastructure and put measures in place to deal with humanitarian crises in vulnerable regions.

Strategies relating to land use, agriculture and water availability will require regional climate change scenarios that consider the effects of land-use change and biological processes on climate. They will need to consider climate impacts acting in synergy with each other, and with changes in atmospheric composition. Issues relating to food security, food supply, health risks and migration pressures are rising priorities for national governments, international bodies such as the World Health Organization and the Food and Agriculture Organization, and bodies responsible for dealing with humanitarian crises.

At city scales, issues relating to building design, urban planning and health-service provision will need to consider the localized effects of urban heat islands in addition to larger-scale climate change. This will affect strategic decision-making by local authorities and other organizations responsible for aspects of urban development, as well as the growing number of individuals living in cities.

Most current climate change and impacts assessments still fail to provide internally consistent projections of future change, due to the separate consideration of different components of the Earth system. This is manifest in the loss of conservation in model budgets of water and energy. The consequent loss of realism creates further uncertainty in future projections in addition to the better-known uncertainties in future emissions, regional climate responses and natural internal variability in the climate system.

In the face of this incomplete understanding of climate change and its future impacts, adaptation decisions need to be robust against uncertainties, and flexible enough to be able to be modified in the light of improved scientific advice.

References

Ackoff, R. L. and Emery, F. E. (1972). *On Purposeful Systems: An Interdisciplinary Analysis of Individual and Social Behaviour as a System of Purposeful Events*. London: Tavistock Publications.

Allen, M. R., Stott, P. A., Mitchell, J. F. B., Schnur, R. and Delworth, T. L. (2000). Quantifying the uncertainty in forecasts of anthropogenic climate change. *Nature*, **407**, 617–620.

Anchukaitis, K. J. and Evans, M. N. (2010). Tropical cloud forest climate variability and the demise of the Monteverde golden toad. *Proceedings of the National Academy of Sciences*, **107**, 11, 5036–5040.

Arnell, N., Liu, C. Z., Compagnucci, R. *et al.* (2001). Hydrology and water resources. In *Climate Change 2001: Impacts, Adaptation, and Vulnerability*, eds. J. J. McCarthy, O.F. Canziani, N.A. Leary, D. J. Dokken and K. S. White. Cambridge: Cambridge University Press, pp. 192–234.

Arora, V. K. and Montenegro, A. (2011). Small temperature benefits provided by realistic afforestation efforts. *Nature Geoscience*, **4**, 514–518.

Barnett, J. and O'Neill, S. (2010). Maladaptation. *Global Environmental Change*, **20**, 211–213.

Batty, M. and Torrens, P. M. (2005). Modelling and prediction in a complex world. *Futures*, **37**, 745–766.

Bell, V. A., Kay, A. L., Jones, R. G. and Moore, R. J. (2007). Use of grid-based hydrological and regional climate model outputs to assess changing flood risk. *International Journal of Climatology*, **27**(12), 1657–1671.

Berrang-Ford, L., Ford, J. D. and Paterson, J. (2011). Are we adapting to climate change? *Global Environmental Change*, **21**, 25–33.

Betts, R. A. (2001). Biogeophysical impacts of land use on present-day climate: near-surface temperature and radiative forcing. *Atmospheric Science Letters*, doi:10.1006/asle.2000.0023.

Betts, R. A., Cox, P. M., Collins, M. *et al.* (2004). The role of ecosystem-atmosphere interactions in simulated Amazonian precipitation decrease and forest dieback under global climate warming. *Theoretical and Applied Climatology*, **78**, 157–175.

Bindhoff, N. L., Willebrand, J., Artale, V. *et al.* (2007). Observations: oceanic climate change and sea level. In *Climate Change 2007: The Physical Science Basis. Contribution of Working Group I to the Fourth Assessment Report of the Intergovernmental Panel on Climate Change*, eds. S. Solomon, D. Qin, M. Manning *et al.* Cambridge: Cambridge University Press, pp. 385–432.

Brando, P. M., Nepstad, D. C., Davidson, E. A. *et al.* (2008). Drought effects on litterfall, wood production and belowground carbon cycling in an Amazon forest: results of a throughfall reduction experiment. *Philosophical Transactions of the Royal Society B*, **363**, 1839–1848.

Briffa, K. R., Shishov, V. V., Melvin, T. M. *et al.* (2008). Trends in recent temperature and radial tree growth spanning 2000 years across northwest Eurasia. *Philosophical Transactions of the Royal Society B*, **363**, 2269–2282.

Brondizio, E. S. and Moran, E. F. (2008). Human dimensions of climate change: the vulnerability of small farmers in the Amazon. *Philosophical Transactions of the Royal Society B*, **363**, 1803–1809.

Brooks, N., Brown, K. and Grist, N. (2009). Development futures in the context of climate change: challenging the present and learning from the past. *Development Policy Review*, **27**, 741–765.

Brovkin, V., Ganopolski, A. and Svirezhev, Y. (1997). A continuous climate-vegetation classification for use in climate-biosphere studies. *Ecological Modelling*, **101**, 251–261.

Butchart, S. H. M., Brooks, T. M. and Symes, A. (2010). Aves. In *Evolution Lost: Status and Trends of the World's Vertebrates*, eds. J. E. M. Baillie, J. Griffiths, S. T. Turvey, J. Loh and B. Collen. London: Zoological Society of London, pp. 47–53.

Challinor, A. J., Wheeler, T. R., Slingo, J. M., Craufurd, P. Q. and Grimes, D. I. F. (2004). Design and optimisation of a large-area process-based model for annual crops. *Agricultural and Forest Meteorology*, **124**, 99–120.

Chase, T. N., Pielke, R. A., Sr, Kittel, T. G. F., Nemani, R. R. and Running, S. W. (2000). Simulated impacts of historical land cover changes on global climate in northern winter. *Climate Dynamics*, **16**, 93–105.

Cook, B. I., Bonan, G. B., Levis, S. and Epstein, H. E. (2008). Rapid vegetation responses and feedbacks amplify climate model response to snow cover changes. *Climate Dynamics*, doi:/10.1007/s00382-007-0296-z.

Cornell, S. and Jackson, M. (in press). Social science perspectives on natural hazards risk and uncertainty. In *Natural Hazards Risk and Uncertainty*, eds. S. Sparks, J. Rougier and L. Hill. Cambridge: Cambridge University Press, ch. 17.

Cox, P. M. (2001). Description of the TRIFFID Dynamic Global Vegetation Model. Exeter: Met Office, Hadley Centre Technical Note 24.

Cox, P. M., Betts, R. A., Bunton, C. B. *et al.* (1999). The impact of new land surface physics on the GCM simulation of climate and climate sensitivity. *Climate Dynamics*, **15**, 183–203.

Cox, P. M., Betts, R. A., Jones, C. D., Spall, S. A. and Totterdell, I. J. (2000). Acceleration of global warming due to carbon-cycle feedbacks in a coupled climate model. *Nature*, **408**, 184–187.

Crutzen, P. J. (2004). New Directions: the growing urban heat and pollution island effect – impact on chemistry and climate. *Atmospheric Environment*, **38**, 3539–3540.

Dai, A. (2011). Drought under global warming: a review. *Wiley Interdisciplinary Reviews: Climate Change*, **2**, 45–65.

Dai, A. and Trenberth, K. E. (2002). Estimates of freshwater discharge from continents: latitudinal and seasonal variations. *Journal of Hydrometeorology*, **3**, 660–687.

Dai, A., Lin, X. and Hsu, K.-L. (2007). The frequency, intensity, and diurnal cycle of precipitation in surface and satellite observations over low- and mid-latitudes. *Climate Dynamics*, **29**, 727–744.

Dai, A., Qian, T., Trenberth, K. E. and Milliman, J. D. (2009). Changes in continental freshwater discharge from 1948 to 2004. *Journal of Climate*, **22**, 2773–2792.

D'Arrigo, R., Wilson, R., Liepert, B. and Cherubini, P. (2008). On the 'Divergence Problem' in northern forests: a review of the tree-ring evidence and possible causes. *Global and Planetary Change*, **60**, 289–305.

Delire, C., Behling, P., Coe, M. T. *et al.* (2001). Simulated response of the atmosphere–ocean system to deforestation in the Indonesian Archipelago. *Geophysical Research Letters*, **28**, 2081–2084.

Döll, P. and Siebert, S. (2002). Global modelling of irrigation water requirements. *Water Resources Research*, **38**, 8.1–8.10.

DTI – Department for Trade and Industry (2004). *Foresight: Future Flooding and Coastal Erosion.* London: Office of Science and Technology, DTI. See: www.foresight.gov.uk.

Environment Agency (2009). Managing flood risk through London and the Thames Estuary: the TE2100 Flood Risk Management Plan. London: TE2100, Environment Agency. See: www.environment-agency.gov.uk/research/library/consultations/106100.aspx.

Essery, R. L. H., Best, M. J., Betts, R. A., Cox, P. M. and Taylor, C. M. (2003). Explicit representation of subgrid heterogeneity in a GCM land surface scheme. *Journal of Hydrometeorology*, **4**, 530–543.

Feddema, J. J., Oleson, K. W., Bonan, G. B. *et al.* (2005). The importance of land-cover change in simulating future climates. *Science*, **310**, 1674–1678.

Field, R. D., van der Werf, G. R. and Shen, S. S. P. (2009). Human amplification of drought-induced biomass burning in Indonesia since 1960. *Nature Geoscience*, **2**, 185–188.

Fischlin, A., Midgley, G. F., Price, J. T. *et al.* (2007). Ecosystems, their properties, goods, and services. In *Climate Change 2007: Impacts, Adaptation and Vulnerability. Contribution of Working Group II to the Fourth Assessment Report of the Intergovernmental Panel on Climate Change*, eds. M. L. Parry, O. F. Canziani, J. P. Palutikof, P. J. van der Linden and C. E. Hanson. Cambridge: Cambridge University Press, pp. 211–272.

Foley, J. A., Prentice, I. C., Ramankutty, N. *et al.* (1996). An integrated biosphere model of land surface processes, terrestrial carbon balance, and vegetation dynamics. *Global Biogeochemical Cycles*, **10**, 603–628.

Forster, P., Ramaswamy, V., Artaxo, P. *et al.* (2007). Changes in atmospheric constituents and in radiative forcing. In *Climate Change 2007: The Physical Science Basis. Contribution of Working Group I to the Fourth Assessment Report of the Intergovernmental Panel on Climate Change*, eds. S. Solomon, D. Qin, M. Manning *et al.* Cambridge: Cambridge University Press.

Fraser, E. D. G. (2006). Food system vulnerability: using past famines to help understand how food systems may adapt to climate change. *Ecological Complexity*, **3**, 328–335.

Fraser, E. D. G. (2007). Travelling in antique lands: using past famines to develop an adaptability/resilience framework to identify food systems vulnerable to climate change. *Climatic Change*, **83**, 495–514.

Frei, T. and Gassner, E. (2008). Trends in prevalence of allergic rhinitis and correlation with pollen counts in Switzerland. *International Journal of Biometeorology*, **52**, 841–847.

Friedlingstein, P., Cox, P. M., Betts, R. A. *et al.* (2006). Climate–carbon cycle feedback analysis, results from the C^{4}MIP model intercomparison. *Journal of Climate*, **19**, 3337–3353.

Friend, A. D., Stevens, A. K., Knox, R. G. and Cannell, M. G. R. (1997). A process-based, terrestrial biosphere model of ecosystem dynamics (Hybrid v3.0). *Ecological Modelling*, **95**(2–3), 249–287.

Gedalof, Z. and Berg, A. A. (2010). Tree ring evidence for limited direct CO_2 fertilization of forests over the 20th Century. *Global Biogeochemical Cycles*, **24**, BG3027.

Gedney, N., Cox, P. M., Betts, R. A. *et al.* (2006). Detection of a direct carbon dioxide effect in continental river runoff records. *Nature*, **439**, 835–838.

Gerten, D., Schaphoff, S., Haberlandt, U., Lucht, W. and Sitch, S. (2004). Terrestrial vegetation and water balance – hydrological evaluation of a dynamic global vegetation model. *Journal of Hydrology*, **286**, 249–270.

Gerten, D., Rost, S., von Bloh, W. and Lucht, W. (2008). Causes of change in 20th century global river discharge. *Geophysical Research Letters*, **35**, L20405.

Gillett, N. P., Weaver, A. J., Zwiers, F. W. and Flannigan, M. D. (2004). Detecting the effect of climate change on Canadian forest fires. *Geophysical Research Letters*, **31**, L18211.

Global Humanitarian Forum (2009). *The Anatomy of a Silent Crisis.* Human Impact Report, Climate Change. Geneva: Global Humanitarian Forum.

Goetz, S., Steinberg, D., Dubayah, R. and Blair, B. (2007). Laser remote sensing of canopy habitat heterogeneity as a predictor of bird species richness in an eastern temperate forest, USA. *Remote Sensing of Environment*, **108**, 254–263.

Gohar, L. and Lowe, J. A. (2009). Summary of the emissions mitigation scenarios. AVOID Workstream 1/Deliverable 1/Report 2. Exeter: Hadley Centre, Met Office.

Golubiewski, N. and Hall-Beyer, M. (2008). Woody encroachment in the southwestern United States. In *Encyclopedia of Earth*, ed. C. J. Cleveland, Washington, DC: Environmental Information Coalition, National Council for Science and the Environment. See: www.eoearth.org.

Gordo, O. and Sanz, J. J. (2009). Long-term temporal changes of plant phenology in the Western Mediterranean. *Global Change Biology*, **15**, 1930–1948.

Gornall, J., Betts, R., Burke, E. *et al.* (2010). Implications of climate change for agricultural productivity in the early twenty-first century. *Philosophical Transactions of the Royal Society B*, **365**, 2973–2989.

Govindasamy, B., Duffy, P. B. and Caldeira, K. (2001). Land use changes and northern hemisphere cooling. *Geophysical Research Letters*, **28**, 291–294.

Gregory, P. J., Ingram, J. S. I. and Brklacich, M. (2005). Climate change and food security. *Philosophical Transactions of the Royal Society B*, **360**, 2139–2148.

Hegerl, G. C., Zwiers, F. W., Braconnot, P. *et al.* (2007). Understanding and attributing climate change. In *Climate Change 2007: The Physical Science Basis. Contribution of Working Group I to the Fourth Assessment Report of the Intergovernmental Panel on Climate Change*, eds. S. Solomon, D. Qin, M. Manning *et al.* Cambridge: Cambridge University Press.

Henderson-Sellers, A. (1993). Continental vegetation as a dynamic component of global climate models: a preliminary assessment. *Climatic Change*, **23**, 337–378.

Hill, P. W., Farrar, J., Roberts, P. *et al.* (2011). Vascular plant success in a warming Antarctic may be due to efficient nitrogen acquisition. *Nature Climate Change*, **1**, 50–53.

Holton, J. R., Curry, J. A. and Pyle, J. A. (2003). *Encyclopedia of Atmospheric Sciences.* New York, NY: Academic Press.

Hurtt, G. C., Chini, L. P., Frolking, S. *et al.* (2011). Harmonization of land-use scenarios for the period 1500–2100: 600 years of global gridded annual land-use transitions, wood harvest, and resulting secondary lands. *Climatic Change*, **109**, 117–161.

Ichinose, T., Shimodozono, K. and Hanaki, K. (1999). Impact of anthropogenic heat on urban climate in Tokyo. *Atmospheric Environment*, **33**, 3897–3909.

International Energy Agency (2008). World Energy Outlook 2008. Paris: International Energy Agency.

Jacob, D. J., Avissar, R., Bond, G. C. *et al.* (2005). *Radiative Forcing of Climate Change: Expanding the Concept and Addressing Uncertainties.* Washington, DC: The National Academies Press.

Jones, C. D., Hughes, J. K., Bellouin, N. *et al.* (2011). The HadGEM2-ES implementation of CMIP5 centennial simulations. *Geoscientific Model Development*, **4**, 543–570.

Jones, R. N. (2000). Managing uncertainty in climate change projections – issues for impact assessment. *Climatic Change*, **45**, 403–419.

Koutavas, A. (2007). Late 20th century growth acceleration in Greek firs *(Abies cephalonica)* from Cephalonia Island, Greece: a CO_2 fertilization effect? *Dendrochronologia*, **26**, 13–19.

Labat, D., Godderis, Y., Probst, J. L. and Guyot, J. L. (2004). Evidence for global runoff increase related to climate warming. *Advances in Water Resources*, **27**, 631–642.

Lapola, D. M., Priess, J. A. and Bondeau, A. (2009). Modeling the land requirements and potential productivity of sugarcane and jatropha in Brazil and India using the LPJmL dynamic global vegetation model. *Biomass Bioenergy*, **33**, 1087–1095.

Leech, D. I. and Crick, H. Q. P. (2007). Influence of climate change on the abundance, distribution and phenology of woodland bird species in temperate regions. *Ibis*, **149**(S2), 128–145.

Lloyd, S. J., Kovats, R. S. and Chalabi, Z. (2011). Climate change, crop yields, and undernutrition: development of a model to quantify the impact of climate scenarios on child undernutrition. *Environmental Health Perspectives*, **119**, 1817–1823.

Millennium Ecosystem Assessment (2005). *The Millennium Ecosystem Assessment*, eds. R. Hassan, R. J. Scholes and N. Ash. Washington, DC: Island Press.

Lo, A. W. and Mueller, M. T. (2010). Warning: physics envy may be hazardous to your wealth. Social Science Research Network working paper, See: http://ssrn.com/abstract=1563882.

Lobell, D. B. and Bonfils, C. (2008). The effect of irrigation on regional temperatures: a spatial and temporal analysis of trends in California, 1934–2002. *Journal of Climate*, **21**, 2063–2071.

Long, S. P., Ainsworth, E. A., Leakey, A. D. B., Nösberger, J. and Ort, D.R. (2006). Food for thought: lower-than-expected crop yield stimulation with rising CO_2 concentrations. *Science*, **312**, 1918–1921.

Lowe, J. A. and Gregory, J. M. (2006). Understanding projections of sea level rise in a Hadley Centre coupled climate model. *Journal of Geophysical Research*, **111**, C11014.

Lowe, J. A., Howard, T., Pardaens, A. *et al.* (2009). Marine and coastal projections, UK Climate Projections 2009 Report. Exeter: Hadley Centre, Met Office. See: http://ukclimateprojections.defra.gov.uk.

Magrath, J. (2007). From weather alert to climate alarm. Oxfam Briefing Paper 108. Oxfam International. See: www.oxfam.org/en/policy/bp108_climate_change_alarm_0711.

Maier, H. R., Ascough, J. C., Wattenbach, M. *et al.* (2008). Uncertainty in environmental decision making: issues, challenges and future directions. In *Environmental Modelling, Software and Decision Support: State of the Art and New Perspectives*, eds. A. J. Jakeman, A. A. Voinov, A. E. Rizzoli and S. H. Chen. Amsterdam: Elsevier, pp. 69–85.

Malhi, Y., Aragão, L. E. O. C., Galbraith, D. *et al.* (2009). Exploring the likelihood and mechanism of a climate-change-induced dieback of the Amazon rainforest. *Proceedings of the National Academy of Sciences of the United States of America*, **106**, 20610–20615.

Malhi, Y., Roberts, J. T., Betts, R. A. *et al.* (2008). Climate change, deforestation and the fate of the Amazon. *Science*, **319**, 169–172

McCarthy, M. P., Best, M. J. and Betts, R. A. (2010). Climate change in cities due to global warming and urban effects. *Geophysical Research Letters*, **37**, L09705.

Mearns, L. O. (2010). The drama of uncertainty. *Climatic Change*, **100**, 77–85.

Mearns, L. O., Easterling, W., Hays, C. and Marx, D. (2001). Comparison of agricultural impacts of climate change calculated from high and low resolution climate model scenarios: Part I. The uncertainty due to spatial scale. *Climatic Change*, **51**, 131–172.

Meijer, I. S. M., Hekkert, M. P., Faber, J. and Smits, R. E. H. M. (2006). Perceived uncertainties regarding socio-technological transformations: towards a framework. *International Journal of Foresight and Innovation Policy*, **2**, 214–240.

Met Office (2010). Can we avoid dangerous impacts? AVOID Briefing. 10/0202d. Exeter: Hadley Centre, Met Office. See: www.metoffice.gov.uk/media/pdf/a/o/avoid4.pdf.

Metz, B., Davidson, O. R., Bosch, P. R., Dave, R. and Meyer, L.A., eds. (2007). *Climate Change 2007: Mitigation. Contribution of Working Group III to the Fourth Assessment Report of the Intergovernmental Panel on Climate Change*. Cambridge: Cambridge University Press.

Mitchard, E. T. A., Saatchi, S. S., Woodhouse, I. H. *et al.* (2009). Using satellite radar backscatter to predict above-ground woody biomass: a consistent relationship across four different African landscapes. *Geophysical Research Letters*, **36**, L23401.

Moore, R. J. (2007). The PDM rainfall-runoff model. *Hydrology and Earth System Sciences*, **11**, 483–499.

Morgan, P. B., Ainsworth, E. A. and Long, S. P. (2003). How does elevated ozone impact soybean? A meta-analysis of photosynthesis, growth and yield. *Plant, Cell and Environment*, **26**, 1317–1328.

Moss, R. H. and Schneider, S. H. (2000). Uncertainties in the IPCC TAR: guidance to authors for more consistent assessment and reporting. In *Third Assessment Report: Cross-cutting Issues Guidance Papers, eds.* R. Pauchuri, T. Tanagichi and K. Tanaka. Geneva: World Meteorological Organisation.

Murphy, J. M., Sexton, D. M. H., Jenkins, G. J. *et al.* (2009). UK Climate Projections Science Report: Climate Change Projections. Exeter: Hadley Centre, Met Office.

Murray, S. J., Foster, P. N. and Prentice, I. C. (2011). Evaluation of global continental hydrology as simulated by the Land-surface Processes and eXchanges dynamic global vegetation model. *Hydrology and Earth System Sciences*, **15**, 91–105.

Myhre, G. and Myhre, A. (2003). Uncertainties in radiative forcing due to surface albedo changes caused by land use changes. *Journal of Climate*, **16**, 1511–1524.

Nemani, R. R., Keeling, C. D., Hashimoto, H. *et al.* (2003). Climate-driven increases in global terrestrial net primary production from 1982–1999. *Science*, **300**, 1560–1563.

Nicholls, R. J., Marinova, N., Lowe, J. A. *et al.* (2011). Sea-level rise and its possible impacts given a 'beyond 4 degree world' in the 21st century. *Philosophical Transactions of the Royal Society A*, **369**, 161–181.

Nohara, D., Kitoh, A., Hosaka, M. and Oki, T. (2006). Impact of climate change on river runoff. *Journal of Hydrometeorology*, **7**, 1076–1089.

Oki, T. and Sud, Y. C. (1998). Design of Total Runoff Integrating Pathways (TRIP) – a global river channel network. *Earth Interactions*, **2**, 1–37.

Pall, P., Aina, T., Stone, D. A. *et al.* (2011). Anthropogenic greenhouse gas contribution to flood risk in England and Wales in autumn 2000. *Nature*, **470**, 382–385.

Palmer, W. C. (1965). Meteorological drought. Office of Climatology Research Paper 45. Washington DC: US Weather Bureau. See: www.ncdc.noaa.gov/temp-and-precip/drought/docs/palmer.pdf.

Parmesan, C., Duarte, C., Poloczanska, E., Richardson, A. J. and Singer, M. C. (2011). Overstretching attribution. *Nature Climate Change*, **1**, 2–4.

Parry, M. L., Canziani, O. F., Palutikof, J. P., van der Linden, P. J. and Hanson, C. E., eds. (2007). *Climate Change 2007: Impacts, Adaptation and Vulnerability. Working Group II Contribution to the Intergovernmental Panel on Climate Change Fourth Assessment Report*. Cambridge: Cambridge University Press.

Pauli, H., Gottfried, M., Reier, K., Klettner, C. and Grabherr, G. (2007). Signals of range expansions and contractions of vascular plants in the high Alps: observations (1994–2004) at the GLORIA master site Schrankogel, Tyrol, Austria. *Global Change Biology*, **13**, 147–156.

Phillips, O. L., Malhi, Y., Higuchi, N. *et al.* (1998). Changes in the carbon balance of tropical forest: evidence from long-term plots. *Science*, **282**, 439–442.

Phillips, O. L., Aragão, L. E. O. C., Lewis, S. L. *et al.* (2009). Drought sensitivity of the Amazon rainforest. *Science*, **323**, 1344–1347.

Pielke, R. A., Marland, G., Betts, R. A. *et al.* (2002). The influence of land-use change and landscape dynamics on the climate system: relevance to climate change policy beyond the radiative effect of greenhouse gases. *Philosophical Transactions of the Royal Society London A*, **360**, 1705–1719.

Pielke, R. A., Pitman, A., Niyogi, D. *et al.* (2011). Land use/land cover changes and climate: modeling analysis and observational evidence. *Wiley Interdisciplinary Reviews: Climate Change*, **2**: 828–850. doi: 10.1002/wcc.144.

Porter, J. R. and Semenov, M. A. (2005). Crop responses to climatic variability. *Philosophical Transactions of the Royal Society B*, **360**, 2021–2035.

Potter, C. S., Klooster, S., Huete, A. and Genovese, V. (2007). Terrestrial carbon sinks for the United States predicted from MODIS satellite data and ecosystem modeling. *Earth Interactions*, **11**, 1–21.

Prentice, I. C., Cramer, W., Harrison, S. P. *et al.* (1992). A global biome model based on plant physiology and dominance, soil properties and climate. *Journal of Biogeography*, **19**, 117–134.

Prentice, I. C., Farquhar, G. D., Fasham, M. J. R. *et al.* (2001). The carbon cycle and atmospheric carbon dioxide. In *Climate Change 2001: The Scientific Basis. Contribution of Working Group I to the Third Assessment Report of the Intergovernmental Panel on Climate Change*, eds. J. T. Houghton, Y. Ding, D. J. Griggs *et al.* Cambridge: Cambridge University Press, pp. 184–238.

Puppi, G. (2007). Origin and development of phenology as a science. *Italian Journal of Agrometeorology*, **12**, 24–29.

Ramankutty, N., Foley, J. A., Norman, J. and McSweeney, K. (2002). The global distribution of cultivable lands: current patterns and sensitivity to possible climate change. *Global Ecology and Biogeography*, **11**, 377–392.

Randall, D. A., Wood, R. A., Bony, S. *et al.* (2007). Climate models and their evaluation. In *Climate Change 2007: The Physical Science Basis. Contribution of Working Group I to the Fourth Assessment Report of the Intergovernmental Panel on Climate Change*, eds. S. Solomon, D. Qin, M. Manning *et al.* Cambridge: Cambridge University Press.

Regan, H. M., Colyvan, M. and Burgman, M. A. (2002). A taxonomy and treatment of uncertainty for ecology and conservation biology. *Ecological Applications*, **12**, 618–628.

Ridley, J. K., Huybrechts, P., Gregory, J. M. and Lowe, J. A. (2005). Elimination of the Greenland ice sheet in a high CO_2 climate. *Journal of Climate*, **18**, 3409–3427.

Rosegrant, M., Cai, X. and Cline, S. (2002). *World Water and Food to 2025. Dealing with Scarcity.* Washington, DC: International Food Policy Research Institute (IFPRI).

Salazar, L. F., Nobre, C. A. and Oyama, M. D. (2007). Climate change consequences on the biome distribution in tropical South America. *Geophysical Research Letters*, **34**, L09708.

Scholze, M., Knorr, W., Arnell, N. W. and Prentice, I. C. (2006). A climate change risk analysis for world ecosystems. *Proceedings of the National Academy of Sciences USA*, **103**, 13116–13120.

Sheffield, J. and Wood, E. F. (2008). Global trends and variability in soil moisture and drought characteristics, 1950–2000, from observation-driven simulations of the terrestrial hydrologic cycle. *Journal of Climate*, **21**, 432–458.

Sitch, S., Smith, B., Prentice, I. C. *et al.* (2003). Evaluation of ecosystem dynamics, plant geography and terrestrial carbon cycling in the LPJ dynamic global vegetation model. *Global Change Biology*, **9**, 161–185.

Smith, P., Martino, D., Cai, Z. *et al.* (2007). Agriculture. In *Climate Change 2007: Mitigation. Contribution of Working Group III to the Fourth Assessment Report of the Intergovernmental Panel on Climate Change*, eds. B. Metz, O. R. Davidson, P. R. Bosch, R. Dave and L. A. Meyer. Cambridge: Cambridge University Press.

Soja, A. J., Tchebakova, N. M., French, N. H. F. *et al.* (2007). Climate-induced boreal forest change: predictions versus current observations. *Global and Planetary Change*, **56**, 274–296.

Solomon, S., Qin, D., Manning, M. *et al.*, eds. (2007). *Climate Change 2007: The Physical Science Basis. Contribution of Working Group I to the Fourth Assessment Report of the Intergovernmental Panel on Climate Change.* Cambridge: Cambridge University Press.

Sparks, T. H. and Yates, T. J. (1997). The effect of spring temperatures on the appearance dates of British butterflies 1883–1993. *Ecography*, **20**, 368–374.

Stern, N. (2007). *The Economics of Climate Change: The Stern Review.* Cambridge: Cambridge University Press.

Stott, P. A., Stone, D. A. and Allen, M. R. (2004). Human contribution to the European heatwave of 2003. *Nature*, **432**, 610–614.

Stott, P. A., Mitchell, J. F. B., Allen, M. R. *et al.* (2006). Observational constraints on past attributable warmings and predictions of future global warming. *Journal of Climate*, **19**, 3055–3069.

Tett, S. F. B., Jones, G. S., Stott, P. A. *et al.* (2002). Estimation of natural and anthropogenic contributions to twentieth century temperature change. *Journal of Geophysical Research*, **107**(D16), 4306.

Thorne, C., Evans, E. and Penning-Rowsell, E., eds. (2007). *Future Flooding and Coastal Erosion Risks.* London: Thomas Telford Ltd.

Toth, F. L., Cramer, W. and Hizsnyik, E. (2000). Climate impact response functions: an introduction. *Climatic Change*, **46**, 225–246.

UNFCCC (1992). United Nations Framework Convention on Climate Change, Article 2. UNEP/IUC/99/2, Information Unit for Conventions. Geneva: UNEP. See: www.unfccc.int/resource/convkp.html.

van Asselt, M. B. A. and Rotmans, J. (2002). Uncertainty in integrated assessment modelling: from positivism to pluralism. *Climatic Change*, **54**, 75–105.

van der Sluijs, J. P. (1996). Integrated Assessment Models and the management of uncertainties. IIASA Working Paper WP 96–119. Laxenburg, Austria: International Institute for Applied Systems Analysis.

Wilby, R. L. and Dessai, S. (2010). Robust adaptation to climate change. *Weather*, **65**, 180–185.

Woodworth, P. L. and Player, R. (2003). The Permanent Service for Mean Sea Level: an update to the 21st century. *Journal of Coastal Research*, **19**, 287–295.

Chapter 7

The role of the land biosphere in climate change mitigation

Joanna I. House, Jessica Bellarby, Hannes Böttcher, Matthew Brander, Nicole Kalas, Pete Smith, Richard Tipper and Jeremy Woods

Human interaction with the land biosphere has contributed to climate change. The land biosphere can play an important role in climate mitigation, through measures such as the management of forests and other carbon sinks, management of agricultural practices, and shifts from fossil-fuel energy to renewable forms of bioenergy. The potential for mitigation must be assessed with regard to the multiple demands for land and the services that ecosystems provide to human society.

7.1 Introduction: from human perturbation to biosphere management

Living organisms have co-evolved with the atmosphere, oceans and land surface, contributing to the climate that supports life on Earth today. The increasing human appropriation of the biosphere for food, energy and construction materials, which has brought enormous benefits, has also inadvertently contributed to a loss of biodiversity, widespread pollution, environmental degradation, and climate change.

Human activities have altered the balance of terrestrial greenhouse-gas sources and sinks. The replacement of forests and other natural ecosystems with crops, pastures and urban settlements has caused emissions of CO_2 due to losses from the carbon stock in vegetation and soils, and increased emissions of nitrous oxide (N_2O) and methane (CH_4) due to diverse agricultural practices.

Human land-use since 1750 accounts for about 10 to 30% of the anthropogenic radiative forcing by CO_2, and a significant (although less well quantified) proportion of the radiative forcing due to CH_4 and N_2O (House *et al.*, 2006). Anthropogenic land-based contributions to the radiative forcing of the atmosphere include the following:

- Land-use change is the term used to describe the conversion from one land cover type to another, such as forest to cropland. Land-use change accounted for around 20% of total CO_2 emissions over the last three decades (Denman *et al.*, 2007; Friedlingstein *et al.*, 2010), and about 12.5% from 2000 to 2010 (Friedlingstein *et al.*, 2010). Land-use-change emissions are currently dominated by the clearing of forests for agriculture in the tropics. Afforestation and reforestation, mainly in temperate countries, are causing smaller changes in the opposite direction, increasing the land uptake of CO_2.
- Around 50% of global anthropogenic CH_4 emissions in 2005 came from agricultural activities, notably rice cultivation and the husbandry of ruminant livestock (Smith *et al.*, 2007a; Denman *et al.*, 2007)
- Around 60% of global anthropogenic N_2O emissions in 2005 came from agriculture, primarily due to use of fertilizers (Smith *et al.*, 2007a). Biomass burning and bioenergy contributed around 10% in the 1990s (Denman *et al.*, 2007).

Land ecosystems also affect climate through their biophysical properties, including albedo and fluxes of water and energy. Thus human activities that affect vegetation cover also influence climate through these mechanisms. For example, planting trees in high-latitude areas that are frequently snow-covered reduces albedo, and thus causes localized warming that offsets

Understanding the Earth System: Global Change Science for Application, eds. Sarah E. Cornell, I. Colin Prentice, Joanna I. House and Catherine J. Downy. Published by Cambridge University Press

the cooling effects of their CO_2 uptake. Vegetation enhances regional rainfall through water recycling due to evapotranspiration, and is also a source of the chemical precursors to cloud condensation nuclei. These topics are discussed further in Chapter 4, Section 4.2 and Chapter 6.

Anthropogenic changes in climate and atmospheric composition in turn affect plant growth, feeding back on climate change both positively and negatively. Warming increases the decomposition of soil organic matter and the return of carbon to the atmosphere. Vegetation growth is enhanced by the fertilizing effects of rising CO_2 in the atmosphere, deposition of reactive nitrogen (although excessive amounts of nitrogen are damaging) and by climate change in cooler climates, removing carbon from the atmosphere. The balance of all these processes means that the land is currently a net sink for CO_2, despite deforestation. This net land sink helps to offset the effect of fossil-fuel emissions on the atmospheric concentration of CO_2 (see CO_2 budget in Chapter 2, Section 2.4, and this chapter, Section 7.3.1, Table 7.1).

Management of the biosphere also provides opportunities to mitigate some of the adverse effects of environmental change. Measures that could mitigate climate change through the management of carbon in terrestrial ecosystems can be classified as follows:

- **Conservation**: preventing emissions from existing carbon stocks, for example, through protecting forest regions from deforestation and degradation, and the protection of other high-carbon ecosystems, principally peatlands. The theoretical potential of this measure, assessed as an amount of carbon kept out of the atmosphere, equals the carbon stock that could be released from the biosphere.
- **Sequestration**: increasing stocks in existing carbon pools including the accumulation of carbon in plant litter and soil, for example through tree-planting and improved forest and agricultural soil management. The aim of these approaches is to increase carbon stores up to a theoretical maximum carbon-carrying capacity. The natural carbon-carrying capacity of land could potentially be enhanced by application of fertilizers, or exclusion of fire. Additionally, biologically sequestered carbon could be removed and 'locked away', for example, by transformation to biochar, which is a very stable form of carbon, or conversion to long-lived wood products. Significant sequestration is possible in areas where carbon stocks have been depleted.
- **Substitution**, either (a) of fossil-fuel energy with energy from biomass, or (b) of fossil-energy-intensive products with products based on regrowing resources (e.g. the substitution of aluminium window frames with timber). When biomass is used for energy, or when the wood-based product eventually decays, carbon is released. However, this carbon was taken up during growth, so the carbon saving through substitution arises from the avoided emissions that would have been associated with fossil-fuel use (minus whatever energy is used in producing, transporting and transforming the biomass energy or bio-product). Unlike sequestration, the effect of substitution as a mitigation measure accumulates over time with each subsequent harvest and product-use cycle. Thus, the theoretical potential is partly dependent on the timescale considered. It is also dependent on the source of biomass, the efficiency of the conversion process, the fossil fuel being displaced and, often critically, the change in carbon stocks due to land conversion. If bioenergy is used with CO_2 capture (through chemical or physical removal processes) and subsequent storage of the CO_2, this could potentially produce energy with net negative CO_2 emissions.

Conservation and sequestration aim at maximizing carbon stocks in the biosphere, while substitution aims at optimizing the carbon balance of energy and materials production compared to a fossil-fuel reference case. Biomass production for energy, food or materials might reduce or increase the carbon stocks of the land on which it is produced, yet still it relies on harvesting of the biomass, which is in tension with the objectives of conservation and sequestration strategies. Thus, conservation, sequestration and substitution strategies for climate mitigation not only compete with each other, but also compete for limited land resources with the production of food, timber and with other ecosystem services. Yet there is an increasing demand for all these products and services, and a need to meet them in a sustainable way. The competition between land-based sectors for available land, the environmental and social consequences of mitigation measures, their economic and technological feasibility, and the way we quantify and account for the full climate impacts

Box 7.1 Comparative measures of carbon and greenhouse gases

Carbon dioxide equivalent emissions, CO_2e:

For comparative purposes, different greenhouse gases have been weighted according to their potency in terms of radiative forcing. Greenhouse-gas emissions are often expressed in terms of what would be the equivalent emission of CO_2. The IPCC (IPCC, 2007) defines CO_2 equivalent emission as:

> '*The amount of CO_2 emission that would cause the same integrated radiative forcing, over a given time horizon, as an emitted amount of a well-mixed greenhouse gas or a mixture of well-mixed greenhouse gases. The equivalent CO_2 emission is obtained by multiplying the emission of a well-mixed greenhouse gas by its global warming potential for the given time horizon. … it does not imply exact equivalence of the corresponding climate change responses*'

The global-warming potential (GWP) is:

> '*An index based upon radiative properties of well-mixed greenhouse gases, measuring the radiative forcing of a unit mass of a given well-mixed greenhouse gas in the present-day atmosphere integrated over a chosen time horizon, relative to that of CO_2. The GWP represents the combined effect of the differing times these gases remain in the atmosphere and their relative effectiveness in absorbing outgoing thermal infrared radiation.*'

As these definitions indicate, the time horizon considered is a critical factor in using CO_2 equivalent emissions for different greenhouse gases. The global-warming potential of each greenhouse gas varies for given time horizons, depending on its biogeochemical behaviour, as the table below shows.

Global-warming potentials for a given time horizon (IPCC, 2007a):

	20-year	100-year	500-year
CO_2	1	1	1
CH_4	72	25	7.6
N_2O	289	298	153

The timeframe most commonly used (for example in the Kyoto Protocol and in all the scientific analyses described in this chapter) is 100 years. Over a shorter time horizon (20 years), methane (CH_4) has a strong radiative forcing effect. However, since the lifetime of CH_4 in the atmosphere (the time it takes for a given increase of emitted gas to be removed from the atmosphere) is only 12 years, on the 100-year time horizon its impact is much less.

Nitrous oxide (N_2O) has a lifetime of 114 years and therefore its global-warming potential is nearly the same over 20 or 100 years, but it is considerably less over 500 years.

Carbon dioxide does not have a single atmospheric lifetime as it is removed from the atmosphere by a variety of processes operating on different timescales. About half of it is removed on a timescale of 30 years, a further 30% takes a few centuries, and the remaining 20% remains in the atmosphere for many thousands of years. Thus, to have an impact on the climate over the next few decades, mitigation of CH_4 and N_2O is very effective, but in the long term, it is the CO_2 that counts.

Conversion factors for carbon and carbon dioxide:

The amount of carbon in the atmosphere can be expressed as parts per million (ppm) or in terms of the mass of C molecules or CO_2 molecules in the atmosphere. A rise of 1 ppm of atmospheric CO_2 corresponds to an increase of 2.13 Pg C (taking a total atmospheric mass of 5.137×10^{18} kg)

$$1 \text{ mole } CO_2 = 44.009 \text{ g } CO_2 = 12.011 \text{ g C}$$

$$1 \text{ g C} = 0.083 \text{ mole } CO_2 = 3.664 \text{ g } CO_2$$

In this chapter, carbon fluxes are presented in Pg C, for consistency of units with the carbon budgets presented in other chapters in this book, and also in Pg CO_2 to enable easier comparison with CO_2e values used across different land-based greenhouse-gas mitigation options.

of human activities on the land are critical science and policy issues for biosphere-based climate mitigation.

7.2 How big a mitigation effort is required?

A total of 194 nations[1] have signed up as parties to the United Nations Framework Convention on Climate Change (UNFCCC), which has the objective of stabilizing greenhouse-gas concentrations in the atmosphere at a level that would 'prevent dangerous anthropogenic interference with the climate system'.[2] Ongoing

[1] Number correct as of May 2011.

[2] http://unfccc.int/essential_background/convention/background/items/1353.php.

negotiations led in 2009 to the Copenhagen Accord,[3] which places a quantified target on the UNFCCC objective:

> the increase in global temperature should be below 2 degrees Celsius, on the basis of equity and in the context of sustainable development ... We agree that deep cuts in global emissions are required.

Conceptual difficulties, both with the framing of the climate change problem by the UNFCCC, and the adoption of a particular global-warming target, are noted in Chapters 6 and 8. Here we take a pragmatic view, accepting that the two-degree target has political traction and considering how biosphere management measures could contribute to reaching it.

Quantifying the greenhouse-gas emissions corresponding to a specified maximum warming, as in the UNFCCC target, involves the use of climate models. For comparability, the climate impacts of different greenhouse gases are often expressed in terms of CO_2 equivalent, typically over a 100-year time horizon (see Box 7.1). However, around 20% of the CO_2 emitted affects the atmosphere for many thousands of years and thus, in the long term, CO_2 dominates temperature impacts.

There is uncertainty in the relationship between emissions of CO_2 and the resulting change in concentration in the atmosphere. This is mainly due to uncertainty in the contribution of feedbacks between climate, CO_2 emissions and the uptake and loss of carbon by the biota (climate–carbon-cycle feedbacks), as discussed in Chapter 4. There is further uncertainty in the actual temperature change associated with a given change in CO_2 concentration (climate sensitivity),[4] as discussed in Chapters 2, 3 and 4.

Meinshausen *et al.* (2009) and Allen *et al.* (2009) analyzed the results of previously published climate model studies of emissions pathways and their temperature outcome that included a range of climate sensitivity and feedback strengths. Allen *et al.* (2009) found that total cumulative emissions of one trillion tonnes of carbon (1,000 Pg C or 3,670 Pg CO_2) results in a most likely peak CO_2-induced warming of 2 degrees above pre-industrial temperatures (with a 5 to 95% confidence interval of warming of 1.3 to 3.9 degrees). About half of this budget has already been emitted. Meinshausen *et al.* (2009) looked at future CO_2 emissions budgets in the medium term. They estimated that in order to have a 75% chance of keeping global mean temperature below two degrees of warming, cumulative CO_2 emissions over the period 2000 to 2050 would need to be limited to 273 Pg C (1,000 Pg CO_2 or a trillion tonnes of CO_2). Emissions of CO_2 from fossil fuels and land-use change from 2000 to 2010 were 322 Pg CO_2 (88 Pg C; Friedlingstein *et al.*, 2010). By this reckoning, a third of the 2000 to 2050 CO_2 budget is already spent.

The United Nations Environment Program (UNEP) Emissions Gap Report (UNEP, 2010) compiled a synthesis of estimates of emissions reductions that would be necessary by 2020 to limit climate warming to two degrees. To have a likely chance[5] (greater than 66%) of limiting the increase in global temperature to two degrees, emissions in 2020 need to be below 50 Pg CO_2e per year for all greenhouse-gas emissions, or below 44 Pg CO_2e per year excluding emissions from land use, land-use change and forestry (LULUCF;[6] Figure 7.1). This compares with business-as-usual estimated emissions in 2020 of 54 to 60 Pg CO_2e per year (fossil fuels only, excluding LULUCF).

The emissions reductions that would meet the two-degree target are calculated using integrated assessment models (IAMs). These models generate quantitative scenarios of possible futures based on human drivers such as population, economic indicators, technological capabilities and demand for energy and food. They can be used to explore the effects of different resource-use and energy policies, such as the pathways that could achieve stabilization at a given level of warming. Future emissions due both to fossil-fuel use and LULUCF are calculated. Since IAMs focus on human drivers, they often include quite simplified climate and land models. Analysis from complex climate models (GCMs) show a similar 'ceiling' of allowable emissions for limiting warming to two degrees as that calculated by the IAMs (UNEP, 2010).

---

3 http://unfccc.int/meetings/cop_15/copenhagen_accord/items/5262.php.

4 Climate sensitivity is defined as the equilibrium temperature change for a change in radiative forcing equivalent to a doubling of the CO_2 concentration (see Chapter 4).

5 Assessments made about the likelihood of a given outcome are described using the agreed terminology of the IPCC (Le Treut *et al.*, 2007).

6 LULUCF is a term commonly used in the UNFCCC and the climate science and policy communities to refer to emissions arising from changing land use or cover from one type to another (e.g. from forests to agriculture), and from the use or management of land without a change in use or cover (e.g. forest management and agricultural emissions of CH_4 and N_2O).

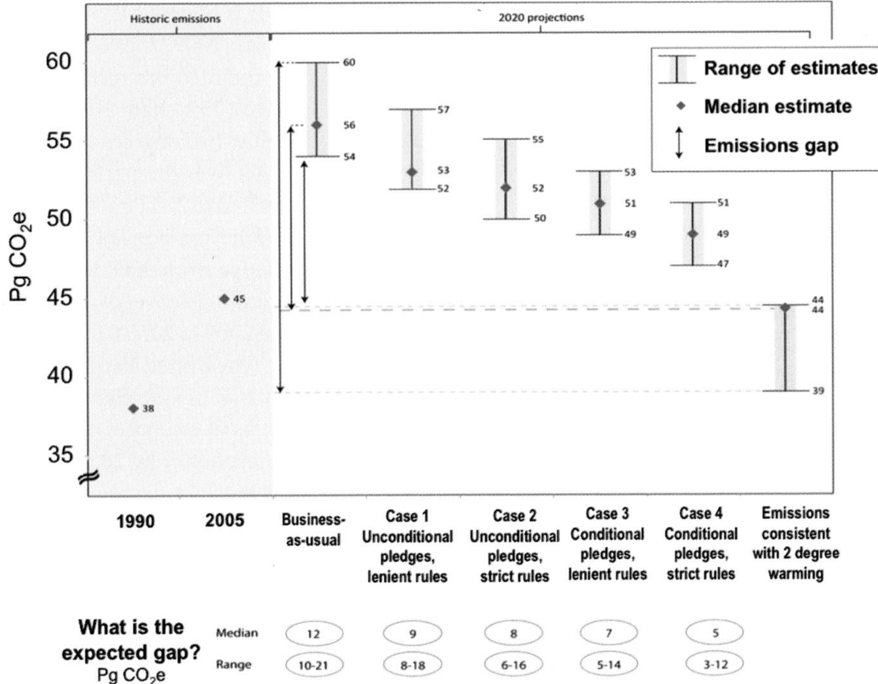

Figure 7.1 The gap between emissions reductions in 2020 pledged under the Copenhagen Accord and the emission levels consistent with a 'likely' chance of meeting the two-degree warming limit. The pale green shaded area shows modelled fossil-fuel emission levels in 2020 consistent with a 'likely' (66%) chance of meeting the two-degree policy limit. The blue bar within this area shows the median (purple diamond) and range (20th to 80th percentile) across estimates available from integrated assessment models in the published literature.

The blue bars above the shaded area show expected emissions in 2020 as an outcome of reduction pledges made under the Copenhagen Accord. Four mitgation cases are shown, along with a business-as-usual analysis. The range within each case is due to different assumptions and methods used by the groups who analyzed the pledges. Unconditional and conditional pledges can be interpreted as minimum and maximum emissions reductions pledged in each country. Maximum reductions are conditional on, for example, reaching a global agreement or being provided with financing (in developing countries). 'Lenient' rules imply that certain controversial emissions are credited, such as those due to 'forest management' under LULUCF accounting and 'surplus-emissions credits' (so-called 'hot-air reductions'). These occur for reasons other than deliberate mitigation activity, such as a fuel switch to gas, economic downturn, or regrowth in forests planted prior to 1990. If these 'lenient' emissions reductions are credited, it means a country's fossil-fuel emissions can be higher to meet the same pledged target, so the gap between necessary fossil-fuel-emissions reduction and deliberate activity for fossil-fuel-emissions reduction is greater. For each case, the magnitude of the gap between expected emissions and the two-degree level is given at the bottom of the diagram. Numbers are in Pg CO_2 equivalent. 1 Pg CO_2e = 1 Gt CO_2e = 0.27 Pg C. Reproduced with permission from UNEP, 2010.

7.2.1 What is being delivered by global policy?

Given that we have a fairly clear idea of what is necessary for mitigation, how close is global policy coming to achieving it? In 1997, the Kyoto Protocol[7] set binding emissions-reduction targets for 37 industrialized countries and the European community. An agreement of an average reduction compared to 1990 levels of about 5% during the 2008 to 2012 'commitment period' represented what could be achieved politically at the time. This target is set in relation to fossil-fuel and industrial emissions, but it is possible to count LULUCF sinks towards meeting the target.[8] Inclusion of LULUCF

Box 7.2 Summary description of LULUCF accounting rules under the Kyoto Protocol

Parties to the Kyoto Protocol are required to achieve target levels of emissions during the first five-year commitment period, which runs from 2008 to 2012. These target levels are expressed as a percentage of the emissions reported for a base year of 1990.

Accounting is **mandatory** for emissions and removals related to **afforestation, reforestation** and **deforestation** activities that have taken place **since 1990**.

Accounting is **optional** for emissions and removals related to **forest management** (i.e. management of forests in existence **before 1990**), **cropland management, grazing-land management** and **revegetation**.

A **cap** is applied to the removals due to **forest management** that can be claimed by a party. The caps on forest management are country-specific, but the limit to what may be counted is approximately 15% of the estimated sink.

See Section 7.7.3 for further details.

7 http://unfccc.int/kyoto_protocol/items/2830.php.

8 Emissions targets under the Kyoto Protocol are based on reductions in emissions from energy and industry, specifically excluding LULUCF emissions, although

under the Kyoto Protocol is summarized in Box 7.2, and explained in more detail in Section 7.7.3.

The Kyoto Protocol allows for International Emissions Trading of carbon credits. These are permits or certificates equivalent to 1 tonne of CO_2e. Countries with emissions below their target can sell surplus units to those that are exceeding their targets. The Kyoto Protocol also includes mechanisms for developed countries to claim credit for mitigation activities that they support or implement in other countries. The Clean Development Mechanism and Joint Implementation projects include LULUCF activities such as reducing agricultural emissions, afforestation, reforestation, forest management and bioenergy. These mechanisms represent a way of financing mitigation in developing countries, and can enable developed countries to find cost-effective means of obtaining carbon credits.

While the Kyoto Protocol marks a significant achievement in climate policy, it has had limited impact on reducing global emissions. Between 1990 and 2005, global emissions from fossil fuels had already increased by 18% to 45 Pg CO_2e per year (Figure 7.1; UNEP, 2010). Carbon dioxide emissions from fossil fuels plus land-use change were 22% higher in 2009 than in 1990 (data from the Global Carbon Project, www.globalcarbonproject.org/; Friedlingstein *et al.*, 2010).

In moving beyond the Kyoto Protocol to a new global 'climate deal,' the G8[9] nations issued a non-binding aim in 2007 '*to at least half global emissions of CO_2 by 2050*' (G8, 2008). Meinshausen *et al.* (2009) estimated that meeting this target would give a 55 to 88% probability of exceeding;[8] two degrees of warming. Meanwhile, UNFCCC negotiations are leading to stronger participation from developing countries and countries in transition, including specified mitigation commitments such as REDD (reducing emissions from deforestation and degradation). Both industrialized and developing countries have made pledges for reducing emissions under the 2009 Copenhagen Accord.[10] As of June 2011, pledges by industrialized countries represent a 12 to 19% reduction of emissions below 1990 levels by 2020. This range depends on the assumptions made about the details of the pledges, some of which are conditional on other countries also taking action. Examples of industrialized-country pledges include:

- **Norway**, which has pledged a 30 to 40% emissions reduction in 2020 compared to 1990. Reductions will come from energy efficiency, renewables (including bioenergy) and carbon capture and storage, including bioenergy for heat and a 2.5% increase of biofuels in the transport sector.
- **The EU27** have pledged a 20 to 30% reduction in 2020 compared to 1990. The aim is to increase renewables to a 20% share of total energy, increase energy efficiency by 20%, and increase renewable fuels (mainly biofuels) for transport to 10%. Details in the individual EU member states' National Renewable Energy Action Plans slightly exceed these targets and show the importance of bioenergy. Out of a total projected demand for renewable energy by 2020 of over 9.2 EJ, or 220 million tonnes of oil-equivalent (Mtoe), it is planned that bioenergy will provide around 50%, requiring the provision of about 250 million tonnes of dry biomass.
- **USA** has pledged a 3% reduction in 2020 compared to 1990, including 55 Tg CO_2 from bioenergy and 31 Tg CO_2 from the capture of waste CH_4 (landfill gas).

Several non-Annex I (developing) countries have also made pledges. These are conditional on actions taken by Annex I countries or on the provision of funding. The first three pledges in the list below represent significant-sized contributions from LULUCF:

- **Brazil** has pledged to reduce emissions by 36% to 39% compared to forecast business-as-usual emissions in 2020. The pledge outlined specific actions related to forestry, energy and agriculture,

LULUCF credits such as through Clean Development Mechanism and Joint Implementation projects can be used towards meeting targets. Similarly, pledges by industrialized countries under the Copenhagen Accord are based on emission from energy and industry excluding LULUCF. Bioenergy offsets are counted in the energy sector rather than the LULUCF sector. Emissions accounting and reporting are discussed in more detail in Section 7.7.

[9] The Group of Eight (G8) is an international forum for the governments of Canada, France, Germany, Italy, Japan, Russia, the United Kingdom, the United States and the European Union.

[10] Annex I country pledges: http://unfccc.int/resource/docs/2011/sb/eng/inf01r01.pdf.
For non-Annex I Nationally Appropriate Mitigation Action Plans see http://maindb.unfccc.int/library/view_pdf.pl?url=http://unfccc.int/resource/docs/2011/awglca14/eng/inf01.pdf.

including a reduction of deforestation in the Amazon region by 80% between 2005 and 2020.

- **China**'s pledge includes afforestation of 40 million hectares of land, and increasing forest stock by 1.3 billion cubic metres by 2020.
- **Indonesia** has pledged to cut emissions by 26% by 2020 compared to business-as-usual. About 80% of Indonesia's current business-as-usual emissions come from deforestation and peatland draining and burning.
- **The Maldives** and **Bhutan** have pledged to be carbon neutral by 2020. In Bhutan this includes an increase in forest area and a reduction in slash-and-burn agriculture.

When the contributions from all countries are added up, a gap still remains between what countries have pledged to do under the Copenhagen Accord and what the models indicate is necessary to avoid exceeding the two-degree target. This gap has been estimated to be between 5 and 12 Pg CO_2e per year, depending on what is included in the pledge analysis (Figure 7.1). Biospheric mitigation is included in a number of different ways that contribute to the range in the 'gap'.

The inclusion of REDD measures raises complex and controversial policy issues about financing. International financing is a key precondition to many developing-country pledges, as reflected in the difference between conditional and unconditional pledge cases shown in Figure 7.1: LULUCF initiatives in developing countries only contribute to the 'conditional' cases of emissions reductions.

The inclusion of LULUCF carbon sinks such as 'forest management' in industrialized countries is responsible for 0.8 Pg CO_2e of the difference between the 'lenient' and 'strict' rules cases in Figure 7.1. The forest-management sink accounted for under the Kyoto Protocol is typically based on measurements of increasing carbon stock in land classed as forest in 1990 (the baseline year). However, much of this carbon uptake is likely due to normal growth in areas that were afforested or reforested some time prior to 1990. Some is also due to enhanced forest growth under CO_2 and nitrogen fertilization effects and climate change. Since the measured carbon sink is not entirely due to specific management activities focused on climate mitigation, the amount of carbon that can be credited through this measure under Kyoto accounting is capped (Box 7.2; Section 7.7.3). There is still considerable debate about the level of the cap, and whether forest management should be included or not. Thus the 'lenient' case in Figure 7.1 allows for a capped level of forest management in Annex I countries (0.8 Pg CO_2e), while the strict case does not allow for forest management.

The potential contribution of LULUCF to mitigation is significant, yet the estimation of this contribution needs to be treated in a more systematic way in order to inform more robust policy decision-making. There are inconsistencies between what is included in the pledges and the IAM analyses, particularly regarding LULUCF and enhanced land sinks (House, 2012).

Bioenergy is included in both the IAM scenarios and the pledges. Again, there is some controversy around the full accounting of the impacts of bioenergy on the atmosphere (discussed in Sections 7.6 and 7.7). The contribution of bioenergy to mitigation is not very significant in the near term (2020) compared to emissions from forestry and agriculture, but it becomes very important in long-term mitigation scenarios. Most IAM scenarios that limit warming to two degrees assume a large contribution of biomass energy, with a substantial share coupled to carbon capture and storage. Bioenergy with carbon capture and storage (BECCS) is not part of the pledge cases of either Annex I or non-Annex I countries by 2020, as carbon capture and storage is not yet a proven large-scale technology for bioenergy, nor for conventional fossil fuels. In these IAM scenarios, however, BECCS becomes important later in the century, and is accounted as 'negative emissions' in the energy sector. It remains unclear whether BECCS is feasible to the degree required within the IAM scenarios to meet the strong long-term mitigation targets, either from a technological or land-availability perspective, nor what the associated land-use, environmental and social impacts (both positive and negative) would be.

7.2.2 Theoretical potential for biospheric mitigation

Simple calculations can place bounds on the ability of interventions in the biosphere to influence climate change via the carbon cycle. To take an extreme hypothetical case, if all of the carbon so far released by land-use change could be restored to the terrestrial biosphere over the timeframe of a century (implying the conversion of a great deal of agricultural land back to forest), the atmosphere would contain 310 to 550 Pg CO_2 (90 to 150 Pg C) less than would be the case if no such intervention had occurred, after accounting for the compensatory response of land and ocean sinks.[11] This is the sequestration potential. Conversely, complete deforestation of the remaining global forest over the same timeframe would increase atmospheric CO_2 by 1,020 to 2,260 Pg CO_2 (280 to 620 Pg C) (House *et al.*, 2002). This is the conservation potential.

In comparison, the projected increase in atmospheric CO_2 in 2100, under the SRES emissions scenarios developed for the IPCC, is 950 to 6870 Pg CO_2 (260 to 1,870 Pg C). Thus, our capacity to increase CO_2 in the atmosphere by fossil-fuel burning far exceeds the maximum possible CO_2 reduction that could be produced by restoring biospheric carbon stocks. These figures, although approximate, are very robust. Although they indicate that there is an intrinsic limit to the use of carbon stocks as a tool for climate mitigation, the role of biospheric carbon stocks is not negligible. In particular, the remaining stock of carbon in forests has substantial mitigation 'value' in the sense that the amount of carbon that *could* theoretically be returned to the atmosphere by continued deforestation is much larger than the amount that could be sequestered by planting trees.

Biosphere management can make a further contribution to climate change mitigation, beyond these intrinsic physical limits, by ongoing substitution of sustainable biomass energy sources for fossil fuels, or through carbon burial. While it is possible to determine what is theoretically possible (i.e. the biological and technological potential capacity), it is critical to recognize that what is actually achievable in practice will depend on the assumptions and choices made, the economic and technological context, and the practical constraints faced (see Sections 7.4 to 7.6 and discussions in the closing sections).

7.3 How has the biosphere influenced climate change in the recent past?

7.3.1 Land-use change and carbon fluxes

Land vegetation has been estimated to store about 650 Pg C and non-permafrost soils about 1,500 Pg C (Prentice *et al.*, 2001), compared with around 800 Pg C currently in the atmosphere. Land-based carbon fluxes arise from converting high-carbon systems to low-carbon systems such as forests and peatlands to agricultural land, and natural grasslands to pasture. Cumulative carbon losses to the atmosphere due to land-use change during the past one to two centuries are estimated at 180 to 200 Pg C (660 to 730 Pg CO_2; DeFries *et al.*, 1999) compared to cumulative fossil-fuel emissions up to the year 2000 of 280 Pg C (1,030 Pg CO_2; Marland *et al.*, 2008). At the same time that ecosystems are losing carbon due to land-use change, natural processes in terrestrial ecosystems are taking up carbon from the atmosphere. This 'natural' ecosystem uptake has been a sink for about a third of total anthropogenic (i.e. energy plus land-use) historical CO_2 emissions (House *et al.*, 2002). The ocean has taken up a further third (see more detailed discussions of the carbon cycle in Chapter 2, Section 2.4)

The land was a *net* source of CO_2 until around the middle of the last century, owing to agricultural conversion. After this it became a *net* sink, despite the high rates of tropical deforestation. In some areas such as Europe, North America, India and China, deliberate afforestation and reforestation in recent years is creating carbon sinks (Figure 7.2), albeit at a much smaller scale than deforestation losses, so this only partially accounts for the current net sink. The global terrestrial sink has been enhanced in recent years due to the indirect effects of human-induced environmental change on vegetation and soil fluxes.

Rising atmospheric CO_2 concentration and reactive nitrogen deposition (from combustion, industry, urbanization and agriculture) have a fertilizing effect on plant growth. The enhancement of uptake of carbon in plants and soils due to CO_2 fertilization effects represents a negative feedback in the climate system. Climatic changes, such as warmer temperatures in the northern extratropics, have extended the length and range of the growing season for many plant species, enhancing uptake (also a negative feedback). Increases in forest biomass of mature forests have been measured in recent years in both tropical (Phillips *et al.*, 1998; Lewis *et al.*, 2009) and temperate (Harrison *et al.*, 2008; Luyssaert *et al.*, 2008) regions, showing a combination of fertilization and climate effects.

Climate change is also expected to enhance CO_2 emissions from land through positive feedbacks involving processes such as faster decomposition of organic matter at warmer temperatures, and increased

[11] Cumulative land-use-change emissions over the last two centuries were about 200 Pg C, but some of these emissions have been taken up by land and ocean sinks. If that carbon was restored to the land through revegetation, there would be a compensatory loss of CO_2 from the land and ocean carbon stores (House *et al.*, 2002).

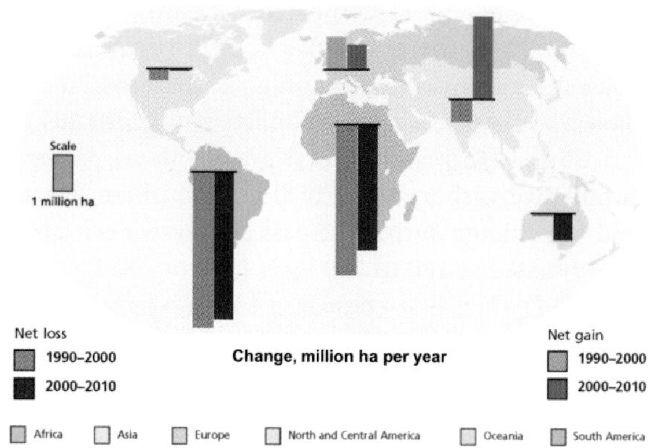

Figure 7.2 Annual change in forest area by region 1990 to 2010 (FAO, 2010a). 1 million hectares per year = 10,000 $km^2\ a^{-1}$.

forest fires and droughts. Model results indicate the net climate–carbon-cycle feedback in the future will be positive, as discussed in Chapter 4.

The importance of land-based fluxes of CO_2 compared to fossil-fuel emissions and ocean uptake in recent decades can be seen in estimates of the global CO_2 budgets published by the IPCC and the Global Carbon Project (e.g. Denman *et al.*, 2007; Friedlingstein *et al.*, 2010; Table 7.1). These results show that despite tropical deforestation, the land was a net sink of ~1 Pg C a^{-1} during the 1990s and 2000s. The principles involved in the construction of CO_2 budgets are summarized in Chapter 2. The atmospheric increase in CO_2 is known from direct measurements of its concentration. Using these measurements together with high-precision measurements of atmospheric O_2:N_2 ratios enables us to separate out fluxes involving the land biosphere from those involving the ocean. This is because the biological processes of photosynthesis and respiration involve simultaneous exchange of both O_2 and CO_2, whereas the ocean surface takes up CO_2 through dissolution without involving O_2. The atmosphere 'sees' all changes that happen to land carbon, thus atmospheric measurements give the total net flux of carbon from all land sources and sinks due to both anthropogenic and natural processes.

Models are used to calculate the impact of land use on the land–atmosphere CO_2 flux. Data on the area of land-use change come from country statistics, and from satellite data in more recent years (Figure 7.2). These are combined with data on the carbon density in vegetation and soils before and after the land-use change, the fate of the biomass, rates of decay and regrowth, and so on.

The difference between the total net land flux (from atmospheric measurements) and the land-use-change flux (calculated by models) is the residual terrestrial sink, representing the natural responses of ecosystems to environmental change. For some time, it was considered a 'missing sink' for the known anthropogenic emissions of CO_2 to the atmosphere, but this terminology is obsolete: a sink of the right magnitude has been found in forest inventory data (Pan *et al.*, 2011) and is consistently calculated by terrestrial biosphere models (le Quéré *et al.*, 2009; see also Chapter 2).

The total net land flux is the most uncertain term in the carbon budget. Breaking it down into components of anthropogenic LULUCF activity and the natural response of ecosystems adds further uncertainty. Recent estimates summarized in House (2009, 2012) and Houghton *et al.* (2012) range between 0.75 and 1.5 Pg C a^{-1} for the 1990s with a mean of 1.13 Pg C a^{-1} (Houghton *et al.*, 2012). A selection of these estimates is shown for comparative purposes in Figure 7.3. The differences between estimates are due to:

- different data sets, for example on land-use-change area and carbon density,
- different modelling approaches, e.g. bookkeeping models (in which estimates of each type of land-use change are combined with prescribed response curves for decay and regrowth) and dynamic global vegetation models (that explicitly model the response of vegetation and soils to change), and
- the inclusion (or not) of different processes such as the fertilizing effect of CO_2 on vegetation, shifting cultivation and management, and the fate of the biomass (e.g. instantaneous loss of carbon such as through burning, or slow decay on site or through the product chain).

The main source of uncertainty in estimates of the total net land-use-change flux of CO_2 is in the area of deforestation, but this uncertainty is declining with the availability of satellite data. Satellite observations have caused the FAO (2010a) to substantially downscale their recent estimates of deforested area. This is one of the reasons why the IPCC's estimate of the land-use-change flux during the 1980s and 1990s (shown as shaded boxes in Figure 7.3) are larger than more recent estimates. The IPCC range was based on Houghton (2003) and DeFries *et al.* (2002). Houghton's values at the higher bound of the IPCC range have recently been revised downwards (Friedlingstein *et al.*, 2010) based on new FAO data that takes account of satellite data on

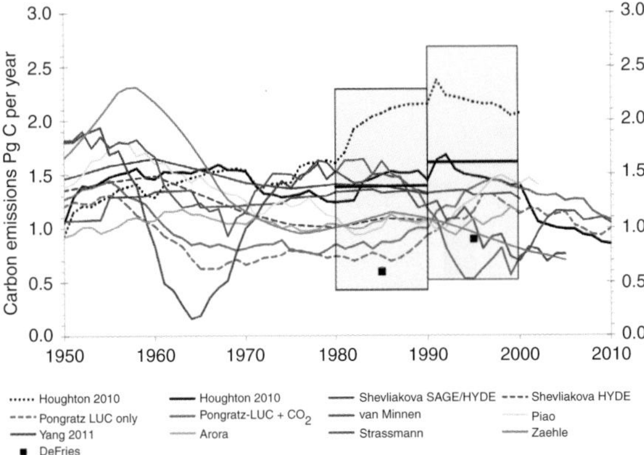

Figure 7.3 Annual net flux of carbon to the atmosphere due to land use, land-use change and forestry (LULUCF). (Updated from House, 2009, 2012 and Houghton *et al.*, 2012 to include an illustrative rather than comprehensive selection of land-use-change flux estimates). The IPCC-estimated mean and range (thick black line within orange shaded boxes) for the 1980s and 1990s (Table 7.1; Denman *et al.*, 2007). The IPCC estimate was based on the range across deFries et al., 2002 (black square shows the decadal mean) and Houghton 2003 (black dashed line). Houghton (2003) used FAO statistics and other data to create his own continuous data set of land-use area change, while deFries *et al.* used satellite estimates of deforestation rates to look at the resultant emissions due to change during the 1980s and 1990s only. Both used Houghton's book keeping model. Houghton 2010 (black line) is the estimate published in the Global Carbon Project analysis by Friedlingstein *et al.* (2010). This estimate is significantly revised down from the Houghton (2003) result used for the IPCC assessment, primarily due to greater use of satellite data in FAO (2010a) estimates of area change.

Shevliakova *et al.* (2009) included management, CO_2 and climate effects in their vegetation model. Results are shown for their analysis using just the HYDE land-use data base (dashed blue line) and a combination of SAGE and HYDE data (solid blue line). Pongratz *et al.* (2009, updated) used their own land-use change data set with a fully coupled vegetation and climate model allowing feedbacks between land, climate and CO_2, but included no management. Results are shown for analyses with and without the CO_2 fertilization feedback effect included (solid and dashed green lines). Van Minnen *et al.* (2009) used HYDE data and included management, CO_2 and climate effects. Piao *et al.* (2009) used SAGE and HYDE, including crop management, climate and CO_2 effects. The Yang *et al.* (2010, updated) vegetation model includes climate, CO_2 and nitrogen fertilization effects with various land-use change data sets; the version shown is based on SAGE with wood harvest from Hurtt *et al.* (2006). Arora *et al.* (2010) used a fully coupled climate and vegetation model, with Hurtt *et al.* (2011) prepared for the upcoming IPCC Fifth Assessment Report; the line shown here is from the uncoupled model run (without interactive CO_2 and climate effects) due to uncertainties in interpreting the fully coupled-model results. Strassmann *et al.* (2008, updated) used HYDE and include CO_2 and climate effects. Zaehle *et al.* (2011) use a vegetation model that includes nitrogen fertilization effects, with the Hurtt *et al.* (2006) data set.

forest-cover change (FAO, 2010a). The DeFries *et al.* (2002) values (at the lower bound of the IPCC range) are likely underestimates as they do not account for emissions due to land-use change prior to 1980 (i.e. they only considered emissions due to land-use change that happened from 1980 to 1990, but deforestation before that date would still be causing emissions from decay of soil, carbon plant biomass and products).

The second largest source of uncertainty is in the carbon density of vegetation and soils, which is highly variable in space. Soil and vegetation inventories (ground measurements) in the tropics are increasing but coverage is still very poor in most areas. Remote-sensing methods using three-dimensional imaging, radar and lasers are capable of providing additional information about vegetation height and density. These may lead to improved global land-use-change flux estimates in the next decade or so.

Few studies give an estimate of uncertainty in land-use-change flux, and no systematic analysis has been carried out. Recent scientific discussions estimate the uncertainty due to data and incomplete understanding of all processes affecting the carbon flux to be ± 0.5 Pg C (Houghton *et al.*, 2012). In the near term, this land-use-change uncertainty is greater than the uncertainty in climate feedbacks on the carbon cycle, which we estimate to be 0.4 Pg C a^{-1} in 2020 (House, 2009). If land-use-change emissions are at the high end of the range of estimates, it will be more difficult than expected to meet targets for global emissions reduction. On the other hand, the effectiveness of REDD measures could be even greater. Higher land-use-change emissions, for the same estimated total net flux from the land, also imply a greater land sink must be operating.

Pan *et al.* (2011) used a combination of inventory data and modelling to calculate the forest carbon flux in different regions (Figure 7.4). In temperate areas, their net 'forest carbon flux' was based on measured changes in carbon stocks in forests. When measuring carbon stock changes, it is not possible to separate what is due to regrowth following harvest or disturbance from the enhanced sink capacity. Therefore implicitly includes the flux due to land-use change, natural processes and the enhanced sink due to CO_2 and nitrogen fertilization effects and climate. In tropical areas, Pan *et al.* split the flux into three components. The gross deforestation flux and the forest-regrowth flux together make up the net land-use-change flux. This was calculated using the Houghton model (as in the IPCC and GCP analyses) based on FAO (2010a)

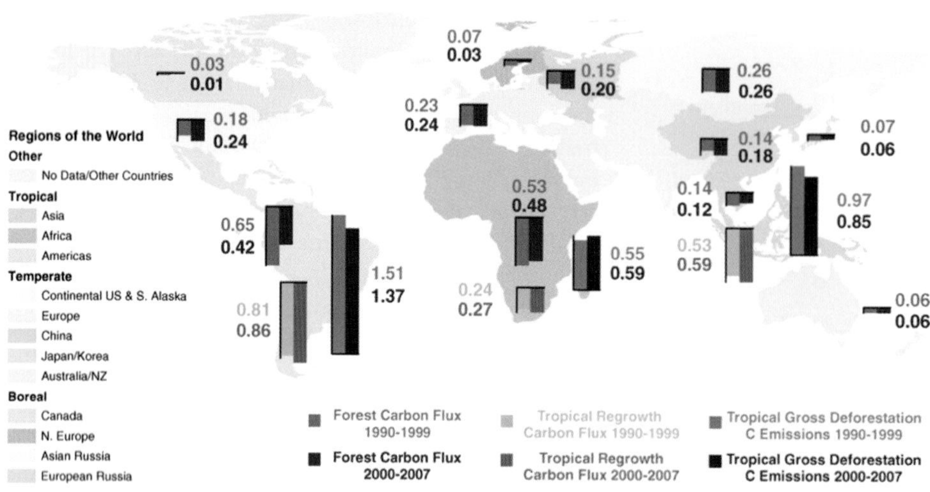

Figure 7.4 Sources and sinks of carbon in the world's forests (Pan *et al.*, 2011). Bars in the downward direction represent sinks, and up direction represent sources of carbon to the atmosphere. In tropical areas, purple colours represent fluxes in established (intact) forests estimated from inventory measurements, yellow/green and brown/orange colours represent modelled fluxes due to land-use change separated into gross deforestation and regrowth fluxes, respectively. For temperate areas, the purple-coloured bars represent the net flux from established (intact) forests plus fluxes due to afforestation and deforestation and land-use change, based on forest inventory measurements of changes in forest stock.

data on forest-area change. The tropical 'forest carbon flux' shown in Figure 7.4 is actually the 'enhanced' residual sink in intact forests as measured in forest inventory data, which is sparse in many tropical countries. It declined between the 1990s and 2007 due to a decrease in the area of intact forest, and a severe Amazon drought in 2005. The total net forest carbon flux in the tropics should be the combination of all three forest fluxes.

7.3.2 Peatlands

Peatlands (ecosystems on deep, highly carbon-rich soils) cover ~ 4 million square kilometres globally. Russia, Canada, Indonesia and the United States have the largest peatland areas, totalling just under 3 million square kilometres (Joosten, 2010). Roughly 20% of the world's soil carbon is stored in boreal and subarctic peatlands (Maltby and Immirzi, 1993). Low decomposition rates due to wet and/or cool conditions cause accumulation of carbon in soils. The total peatland carbon stock in 2008 was estimated to be ~ 450 Pg C (Joosten, 2010); however, a more recent estimate indicates that carbon stocks in northern peatlands alone are as high as 550 Pg C (Yu *et al.*, 2010). Peatlands continuously accumulate carbon and thus constitute a small but persistent sink, about 0.1 Pg C a^{-1} (0.37 Pg CO_2 a^{-1}). The rate of carbon accumulation globally over the last 6–8,000 years has been estimated to be between 20 and 30 g C m^{-2} a^{-1} (73.4 to 110 g CO_2 m^{-2} a^{-1}; Gorham, 1991; Vitt *et al.*, 2000; Turunen *et al.*, 2002, 2004). Northern peatlands are a small sink for CO_2 but an important source of CH_4 and dissolved and particulate organic carbon, which are eventually oxidized to CO_2 (Billett *et al.*, 2004; Blodau *et al.*, 2004). Tropical peatlands are less well studied, but are highly productive, leading to fast carbon accumulation rates and deep deposits of several metres in some areas of Amazonia and southeast Asia (Page *et al.*, 2002; Lähteenoja *et al.*, 2009).

Land-use change, draining of peatland soils, and fires can result in rapid decomposition and loss of carbon from peatlands. Loss of carbon from highly organic soils can make a significant contribution to national greenhouse-gas emissions. For example, approximately 15% of Scotland's total greenhouse-gas emissions come from land-use changes on Scotland's carbon-rich soils (Smith *et al.*, 2007b). Estimating the global loss of carbon from peatland soils is highly uncertain, but global peatland emissions from drained soils have been estimated to account for 1.3 Pg CO_2e a^{-1} while peat fires in south-east Asia have been estimated to contribute > 0.4 Pg CO_2e a^{-1} (Joosten, 2010). Warming and drying of peatlands due to climate change in the future would increase decomposition rates, causing positive feedback as discussed in Chapter 4, Section 4.5.

Peatland emissions are rarely included in estimates of land-use-change flux such as those presented in Figure 7.3 and Table 7.1 or those reported to the UNFCCC. This is primarily due to a lack of observational data. However, there is an urgent need to include such fluxes, from both the scientific perspective of correctly calculating the emissions budget and understanding processes, and from the policy perspective of setting robust global targets and identifying activities

Table 7.1 The global carbon budget (data shown are Pg C a^{-1})

	IPCC Fourth Assessment Report[a]			Global Carbon Project[b]	
	1980s	**1990s**	**2000–2005**	**1990s**	**2000–2009**
Emissions (fossil fuels + cement)	5.4 ± 0.3	6.4 ± 0.4	7.2 ± 0.3	6.4 ± 0.4	7.7 ± 0.5
Atmospheric increase	3.3 ± 0.1	3.2 ± 0.1	4.1 ± 0.1	3.1 ± 0.2	4.1 ± 0.2
Ocean–atmosphere flux	−1.8 ± 0.8	−2.2 ± 0.4	−2.2 ± 0.5	−2.2 ± 0.5	−2.3 ± 0.5
Land–atmosphere flux, partitioned as follows:	−0.3 ± 0.9	−1.0 ± 0.6	−0.9 ± 0.6	−1.1	−1.3
Land-use-change flux	1.4 (0.4 to 2.3)	1.6 (0.5 to 2.7)	n.a.	1.5 ± 0.7	1.1 ± 0.7
Residual terrestrial sink	−1.7 (−3.4 to 0.2)	−2.6 (−4.3 to −0.9)	n.a.	−2.6	−2.4

[a] IPCC Fourth Assessment Report (Denman *et al.*, 2007): Emissions from fossil fuels and cement are derived from energy and production statistics. The atmospheric increase and net land and ocean fluxes are derived from atmospheric measurements of CO_2 and O_2:N_2 ratios. The land-use-change flux was calculated using models, based on the mean values and ranges reported in Houghton (2003) and deFries *et al.* (2002). The residual terrestrial sink was calculated as the difference between the net land–atmosphere flux and the land-use-change flux.

[b] Global Carbon Project data are available from www.globalcarbonproject.org, published in Friedlingstein *et al.*, (2010). The ocean flux is derived from models. The land-use-change flux is derived using Houghton's bookkeeping model. The residual terrestrial sink is calculated as the difference between the other terms.

Positive fluxes denote net emissions to the atmosphere, negative fluxes denote a sink.

'n.a.' = data not available to make an assessment.

that will reduce emissions, such as protection and restoration of peatlands and wetlands.

7.3.3 Fire

Fires are a significant source of CO_2 and other reactive gases (CO, CH_4) and aerosols. The controls of fire regimes are discussed further in Chapters 2 and 4. While some fires are triggered naturally (commonly by lightning), fires are also set deliberately to clear forests to establish pastures and cropland. Land that is not allowed to return to its pre-disturbance state will experience a net loss of carbon.

Emissions due to fire are highly variable from year to year due to the state of the vegetation, the cause of the fire and climate variability (van Leeuwen and van der Werf, 2011). Average global fire emissions were around 2.1 Pg C a^{-1} during 2002 to 2007, declining to 1.7 Pg C a^{-1} in 2008 and 1.5 Pg C a^{-1} in 2009, partly due to a decline in deforestation fires. Over this period, tropical deforestation and degradation fires contributed about 20 % of all fire emissions, and peatland fires about 3%, giving a total of around 0.5 Pg C a^{-1} (van der Werf *et al.*, 2010). Loss of carbon due to deforestation fires is accounted for in land-use-change estimates (as in Figure 7.3), but losses arising from peat fires typically are not.

7.3.4 Agriculture

Agriculture accounted for estimated emissions of 5.1 to 6.1 Pg CO_2e a^{-1} in 2005, amounting to 10–12% of total global anthropogenic emissions of greenhouse gases. Despite large annual exchanges of CO_2 between the atmosphere and agricultural lands, the net flux is approximately balanced, with net CO_2 emissions around 0.04 Pg CO_2 a^{-1} only, as harvesting is balanced by regrowth (Smith *et al.*, 2007a).

Thus, the significance of agriculture – aside from the CO_2 emissions involved in the initial conversion of land – lies in its CH_4 and N_2O emissions. Agricultural CH_4 contributes 3.3 Pg CO_2e a^{-1}, and N_2O emissions amount to 2.8 Pg CO_2e a^{-1}. Of global anthropogenic emissions in 2005, agriculture accounted for about 60% of N_2O emission and about 50% of CH_4 emission. Globally, agricultural CH_4 and N_2O emissions have increased by nearly 17% from 1990 to 2005, an average annual emission increase of about 60 Tg CO_2e a^{-1}. During that period, non-Annex I countries showed a 32% increase, and were, by 2005, responsible for about three-quarters of total agricultural emissions. Annex I countries collectively showed a decrease of 12% in the emissions of these gases (Smith *et al.*, 2007a, 2008).

7.4 Mitigation potential in the forest sector

7.4.1 Reducing emissions from deforestation and degradation (REDD)

Of all the options for biospheric climate mitigation, reducing emissions from deforestation is recognized as one of the most urgent. This led to continuing intense negotiations under the UNFCCC about activities for protecting (and enhancing) carbon stocks by avoiding deforestation and forest degradation, especially in countries not currently committed in the Kyoto Protocol. The theoretical potential for mitigation of a REDD scheme is as high as the currently observed emissions from deforestation and degradation in the targeted tropical countries, i.e. 2.82 Pg C a^{-1} or 10.3 Pg CO_2e a^{-1} (the estimated tropical gross deforestation emissions for period 2000–2007; Pan *et al.*, 2011).

The Bali Action Plan,[12] adopted by UNFCCC signatories in 2007, extended REDD to include forest conservation, sustainable forest management and enhancement of forest carbon stocks under a REDD+ scheme. Further proposals may additionally allow for the accounting of afforestation under REDD+. A number of issues remain to be resolved in negotiating a workable international architecture addressing deforestation and degradation in a post-Kyoto agreement (Angelsen *et al.*, 2009). These issues are often interdependent, and include:

- providing sufficient financial resources to implement REDD schemes;
- procedures for setting reference levels against which mitigation efforts can be benchmarked;
- methodologies for monitoring, reporting and verification (known as MRV); and
- processes to promote the participation of indigenous peoples and local communities living in forest regions.

The financial needs for REDD readiness and implementation have been estimated to be in the range of US $15–35 billion per year for a 50% global reduction in forest emissions, while funds currently available are around US $2 billion (Hoare *et al.*, 2008). These costs depend crucially on how national baseline emissions from deforestation are determined (i.e. the level of deforestation to which future activity is compared). Griscom *et al.* (2009) found that, by applying different suggested approaches to the calculation of baselines, the total *credited* emissions avoided could range over two orders of magnitude for the same quantity of *actual* emission reductions. The costs also depend on the design and implementation of methods and the requirements for support for capacity building (training and technology transfer). The financial costs of monitoring, reporting and verification of REDD activities can be kept relatively low, or become prohibitive to implementation, depending on policy decisions about these factors (Böttcher *et al.*, 2009). Existing assessments of the economic costs of avoided deforestation tend to focus on direct costs without accounting for broader macroeconomic impacts such as the costs of structural transition, and food and land-price inflation, which are more difficult to quantify. Implementation of REDD could have multiple benefits in addition to offsetting carbon emissions, such as preserving biodiversity, maintaining regional hydrological cycles (avoiding floods and droughts), and protecting indigenous use of forest lands. These additional social benefits and ecosystem services are not well captured in economic assessments, and are therefore the focus of considerable research and policy attention. Box 7.3 describes an avoided deforestation project that is seeking to meet the multiple challenges of preventing environmental degradation, alleviating poverty and reducing carbon emissions.

Box 7.3 Avoiding deforestation: the Bale Mountains National Park Project

The Bale Mountains National Park Project is a proposed avoided deforestation project in Ethiopia, covering approximately 100,000 hectares of montane forest within the national park. It builds on an on going conservation project within the National Park (balemountains.org; Frankfurt Zoological Society, 2007). Its experiences to date illustrate some of the benefits, barriers, and constraints of trying to establish such projects. These are similar to those documented for an established REDD project outside of the National Park

[12] http://unfccc.int/meetings/bali_dec_2007/meeting/6319.php.

(Tadesse, 2010; www.pfmp-farmsos.org/BALE%20 HOME.html).

As a result of human disturbance, settlement and agricultural expansion, the forest area had been decreasing for several decades at an average rate of 374 hectares per year, rising to 1,505 hectares per year during 2001 to 2007. The mitigation project intends to generate revenue to invest in avoided deforestation through the sale of voluntary offset credits. The planned lifetime of the project is 20–30 years, with the carbon credits guaranteed for 100 years. The project will need to create sustainable sources of employment and provide alternative sources of fuel to replace the unsustainable firewood extraction. If these alternatives are not created the pressures to deforest will remain.

The project aims to reduce the rate of deforestation by:

- promoting and facilitating alternative fuel options;
- creating opportunities for diversification of livelihoods;
- promoting community protection of resources;
- developing sustainable natural-resource agreements; and
- introducing management plans for fire, grazing and other anthropogenic factors affecting biomass and soil conservation.

However, there are several operational constraints that need to be overcome. At present, there is a lack of a formal system for issuing logging rights and penalizing illegal logging. The legal status of the national park is not formally established, and there is a lack of capacity for land registration. Furthermore, there is a dynamic social context that needs to be considered with care. Migration rates in the region are very high. Incomers moving into the national park may increase deforestation pressures, and weaken the coherence of project activities.

In the Bale Mountains Project, the presence of an independent, qualified and funded project proponent (in this case a non-governmental organization) has catalyzed the initiative, but in order for the avoided deforestation project to succeed, there are still several necessary conditions to be met. A market needs to be created for the carbon offset credits. The active co-operation of local communities is also required, whose day-to-day practices and livelihoods need to change. The migration pressure also needs to be managed and contained. Several environment and development charities and NGOs are supporting the communities through training in alternative livelihoods and the participatory development of forest management plans, recognizing that 'bottom-up' measures are a vital part of the sustainability of mitigation and conservation efforts of this kind (Tesfaye, 2011).

7.4.2 Enhancing carbon stocks in forests

Options for carbon-sink enhancement include: creating new forest areas; changing the management of existing forests by changing the rotation length (the number of years between harvests), the amount of removals between harvests (thinning), and the treatment of residues from harvesting (such as leaving slash on the ground); enhancing biomass growth by measures such as irrigation or fertilizer application; changing the choice of tree species; and protection from disturbances (Böttcher and Lindner, 2010). Through these measures, forest management can introduce new sinks and enhance existing sinks, affecting the above- and below-ground biomass, litter, dead wood, and soil organic matter (e.g. Jandl *et al.*, 2007; Don *et al.*, (2009).

Carbon sequestration in the forestry sector is to a large degree a non-permanent strategy. Tree plantations that are established, harvested but not re-established do not contribute to carbon sequestration. The sequestration phase is finite, lasting for some decades during the active growing stage of the trees. The gained carbon stocks (mature forest) need to be protected thereafter, to keep the carbon withdrawn from the atmosphere. Sequestration therefore always needs to be guarded by conservation measures to ensure effective mitigation.

Future model projections of the forest carbon sink are particularly dependent on scenarios of future human land use, demand for wood products and environmental restrictions (e.g. REDD, protected areas, sustainability criteria). Estimates of the global area suitable and available for afforestation range from 3 to 10 million square kilometres. The potentials vary with assumptions on land quality, demand for agricultural land, policy interventions such as subsidies and commodity prices, and other factors (Campbell *et al.*, 2008; Thomson *et al.*, 2008; Zomer *et al.*, 2008). Figure 7.5 shows one example, produced for the IPCC Fourth Assessment Report, of the estimated cumulative potential from REDD and afforestation assuming a carbon[13] price of US $2.7 per tonne of CO_2 with a 5% annual carbon-price increment (Sathaye *et al.*, 2006). Box 7.4 outlines some of the practical issues arising in a long-standing, large-scale afforestation and reforestation programme in China.

Box 7.4 Afforestation for multiple benefits: China's Sloping Land Conversion Programme

China's Sloping Land Conversion Programme is the largest land retirement/reforestation programme in the developing world. Its scope and institutional dimensions have been documented by Bennett (2008) and Xu *et al.* (2010). The aim of the US $40 billion programme, initiated in 1999, was to convert nearly 15 million hectares of cropland to forests by 2010, with an additional 'soft' goal of converting a similar area of wasteland to forestry. The programme involves direct payment to millions of farmers and households undertaking afforestation or reforestation activities on suitable land. By the end of its first phase, in 2003, over 7.2 million hectares of cropland had been retired (Bennett, 2008) and 4.9 million hectares of barren land had been afforested (Xu *et al.*, 2006).

The principal motivations for the programme were soil stabilization and flood prevention in response to China's serious environmental problems, but by providing substantial and evidenced additionality in terms of creating carbon stocks, the programme also has had carbon sequestration benefits. Moberg and Persson (2011) estimated the carbon sequestered during the first decade of the programme as 222 to 468 Tg C, depending on the method used for calculation. This amounts to about 14% of China's CO_2 emissions due to fossil-fuel combustion and cement production over the same period. In addition to direct carbon sequestration, the programme will increase the domestic supply of timber, reducing China's demand for timber on the international markets, and consequently reduce pressure on the world's forests (Sun *et al.*, 2004).

Several studies have reflected on the programme's performance, focusing on its institutional dimensions and its approach to payment for ecosystem services (Bennett, 2008; Xu *et al.*, 2010; Li *et al.*, 2011). These studies note that despite the large overall budget, there were shortfalls and some misdirection of funds in the payments and tax breaks delivered to the farmers taking part in the initiative. The sheer numbers of participants, over such a large area, meant that local implementation costs were much higher than budgeted. The balance of top-down control and local volunteerism in making land-use decisions was unclear; consequently the scheme did not benefit fully from the theoretical efficiency of a market-based payment scheme relative to other institutional means of driving the desired environmental changes. Livelihood changes away from farming did not materialize as anticipated by the government, in part because there was insufficient technical support for implementation of forestry income streams.

These studies of the Sloping Land Conversion Programme give insight into the necessary conditions for implementing major land-based mitigation programmes. These include:

- well-defined land-tenure and land-use rights;
- participants' access to technical support, with regard both to the transformation of their lands and the adjustments to their livelihoods;
- governance, payment and inspection systems (together with effective institutions for implementing these); and
- significant up-front capital, in order to compensate participants for their short-term opportunity costs before longer-term benefits accrue.

It is unlikely that similar programmes would succeed if these necessary conditions were not met. Learning from the constraints and outcomes of this programme can help in achieving higher levels of success in mitigation, in a sustained way.

Although a sink enhanced by forest management is effectively removing carbon from the atmosphere, it is a challenging task to include forest sink management in a balanced carbon-accounting scheme (Böttcher *et al.*, 2008). Because the effects are relatively small compared to the large land-use-change effects, they are harder to monitor. Also, as already discussed, many processes leading to carbon accumulation in existing forests, such as CO_2 fertilization, are not directly human-induced (Maguani *et al.*, 2007; Ciais *et al.*, 2008).

To make carbon sequestration in ecosystems more effective globally, it is not just current environmental constraints but also their expected changes in the future that should be taken into account. There have been suggestions of a framework for identifying forests

[13] The carbon price reflects the value of a tonne of CO_2e that has not been emitted to the atmosphere. It can be fixed (e.g. by carbon taxes on emissions), or determined by the international market in trading of carbon credits. The higher the carbon price, the larger amount of mitigation that is cost effective.

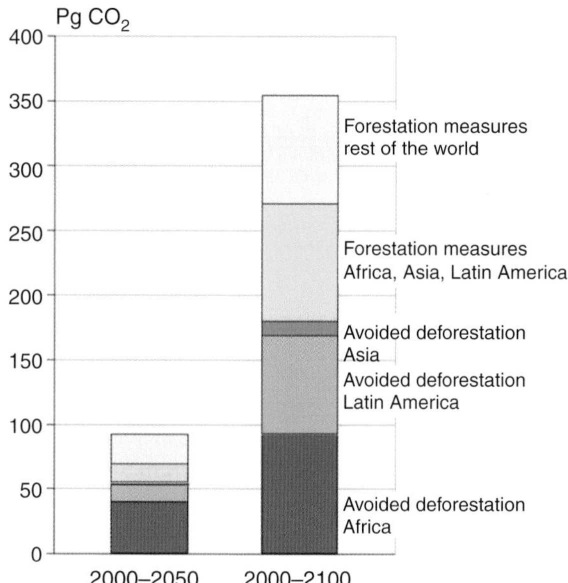

Figure 7.5 Cumulative mitigation potential in global forests from forestation and avoided deforestation at a carbon price of US $2.7 per tonne of CO_2 with a +5% annual carbon price increment. Data from Sathaye *et al.*, 2006, figure reproduced from Nabuurs *et al.*, 2007.

of high potential for carbon storage using criteria like relatively cool temperatures and moderately high precipitation, which would produce rates of fast growth but slow decomposition (Keith *et al.*, 2009). The risk of natural disturbance could be added to such a framework.

7.4.3 Management of harvested wood biomass

Wood products and biomaterials form a carbon pool that can be managed as part of a climate mitigation strategy. The use of biomass products (e.g. wood in construction) can in addition lead to the substitution of more greenhouse-gas- and energy-intensive materials (the production of wood specifically for bioenergy is dealt with in Section 7.6). When biomass products reach the end of their lifetime there are two options for end use: landfilling or combustion (incineration). Both options result in emissions of all or part of the biogenic carbon, and both options can include energy recovery (biogas or electricity and heat), but they differ in the timing of the release of carbon and energy (Christensen *et al.*, 2009; Manfredi *et al.*, 2009). Improving the overall greenhouse-gas balance of biomass use involves efficiency throughout the whole processing chain, from the use of by-products through to recycling and energy recovery in an integrated biomass management framework (Bahor *et al.*, 2009).

The best carbon strategies in forestry in the long term assume maximizing biomass yield, and optimizing utilization in the processing of the resulting wood in long-lived products and energy generation, in combination (e.g. using wood residues for energy) or in cascade use, recycling wood products for alternative use or for energy (Böttcher, 2008; Werner *et al.*, 2010).

Proposals have also been made for wood burial, a more direct way of transferring the living biomass in forests to long-term storage, by-passing the product and recycling chain (Zeng, 2008). Burial of carbon captured through photosynthesis has potential to create negative carbon emissions (Obersteiner *et al.*, 2010 in a similar way to burial of charcoal after energy recovery (biochar Section 7.6.1).

The trade of harvested wood products causes a displacement of carbon emissions. For example, timber cut in New Zealand or south-east Asia may be exported to Europe or China for paper production, making furniture, or use in construction. Emissions reporting under the UNFCCC currently determines emissions at the point of harvest. The same is true for most scientific analyses (including those shown in Figures 7.3 and 7.4). Determining the overall emissions at the point of actual use can lead to very different emissions calculations for importing and exporting countries, and hence different perspectives on the mitigation benefit of this measure, compared to those based on assumptions of emissions at the point of harvest (Kohlmaier *et al.*, 2007; Hashimoto, 2008).

7.5 Mitigation potential in the agricultural sector

The most prominent options for mitigation of greenhouse-gas emissions in agriculture are: improved management of cropland and grazing land (e.g. improved agronomic practices such as controlling the timing and amount of fertilizer application, reduced tillage or ploughing, and the management of unwanted biomass residues, such as leaving them on the ground or using for energy rather than burning on site); restoration of organic soils that have been drained for crop production; and restoration of degraded lands. Lower but still significant mitigation is possible with improved water management (especially in rice fields), set-asides, land-use change (e.g. conversion of cropland to grassland), agro-forestry, and improved livestock and manure management. Greenhouse-gas emissions could also be reduced by substituting fossil fuels with energy produced from agricultural feedstocks (e.g. crop residues, dung, and energy crops. For the purposes of carbon offsets and trading schemes, only bioenergy used within the agricultural sector can be accounted towards greenhouse gas reductions in this sector. The wider use of agricultural feedstocks for bioenergy is discussed in the next section.

Enhancement of the sinks through soil-carbon sequestration is the mechanism responsible for most of the mitigation potential, with an estimated contribution of 89%. Mitigation of CH_4 and N_2O emissions from soils accounts for 9% and 2%, respectively, of the total mitigation potential. The principal sources of uncertainties inherent in these mitigation potential estimates include: (a) the future level of adoption of mitigation measures; (b) the effectiveness of adopted measures in enhancing carbon sinks or reducing N_2O and CH_4 emissions; and (c) the persistence of mitigation, as influenced by future climatic trends, economic conditions and social behaviour (Smith *et al.*, 2007a).

Many agricultural mitigation opportunities use current technologies and can be implemented immediately, but technological development will be a key driver ensuring the efficacy of additional mitigation measures in the future (Smith *et al.*, 2007a, 2007c, 2008, 2011). Technological advances, such as increased crop yields, can reduce the need for expansion of agricultural lands and offset some of the problems of land degradation. However, while technological innovation has been shown to spare some land from conversion in the past, the type of land that has been preserved may not be that of highest value for biodiversity and other ecosystem services. Finding ways of obtaining a more integrated perspective on land use is becoming a prominent concern in mitigation research and policy (Section 7.7).

The role of alternative strategies changes across the range of prices for carbon. At low prices, dominant strategies are those consistent with existing production such as changes in tillage, fertilizer application, livestock diet formulation, and manure management. Higher prices elicit land-use changes that displace existing production, such as biofuels, and allow for use of costly animal-feed-based mitigation options. A practice effective in reducing emissions at one site may be less effective or even counterproductive elsewhere. Consequently, there is no universally applicable list of mitigation practices; practices need to be evaluated for individual agricultural systems based on their local climate, soil and landscape characteristics, the social setting and historical patterns of land use and management.

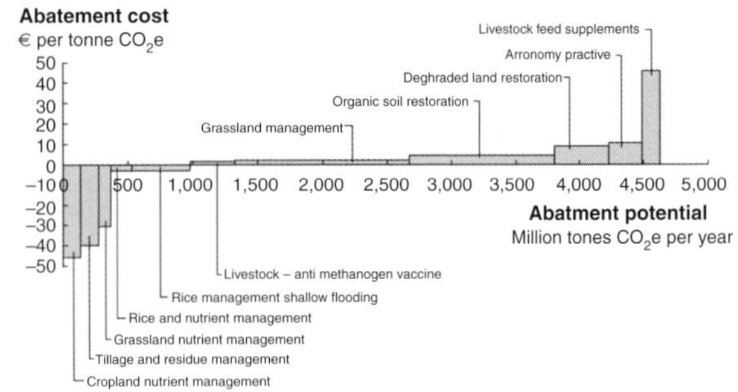

Figure 7.6 Global mitigation marginal abatement cost curve for agriculture for 2030 (McKinsey and Company, 2009)

Nevertheless, agricultural greenhouse-gas mitigation options are cost-competitive with non-agricultural options such as changes in energy, transportation and forestry in achieving long-term climate objectives. Figure 7.6 shows the global mitigation potential and costs of a range of options. Some have very low or even negative costs. Considering all greenhouse gases, the global technical mitigation potential from agriculture by 2030, excluding fossil-fuel offsets from biomass, is estimated to be ~ 5,500 to 6,000 Tg $CO_2e\ a^{-1}$. Economic potentials are estimated to be 1,500 to 1,600, 2,500 to 2,700, and 4,000 to 4,300 Tg $CO_2e\ a^{-1}$ at carbon prices of US $ 20, 50 and 100 per tonne of CO_2e, respectively. About 70% of this potential lies in non-OECD countries, 20% in OECD countries, and 10% in economies in transition (Smith *et al.*, 2007a). Global long-term modelling suggests that non-CO_2 crop and livestock abatement options could cost-effectively contribute 270 to 1,520 Tg $CO_2e\ a^{-1}$ globally in 2030 with carbon prices up to US $20 per tonne of CO_2e, and 640 to 1,870 Tg $CO_2e\ a^{-1}$ with carbon prices up to US $50 per tonne of CO_2e. Soil-carbon management options are not currently considered in long-term mitigation modelling (Smith *et al.*, 2007a, 2008).

The contribution to the mitigation potential within the agricultural sector made by using bioenergy for agriculture depends on relative prices of the fuels and the balance of supply and demand. Using top-down models that include assumptions on this balance, the economic mitigation potential for bioenergy use in the agricultural sector in 2030 is estimated to be 70 to 1,260 Tg $CO_2e\ a^{-1}$ at US $20 per tonne of CO_2e, and 560 to 2,320 Tg $CO_2e\ a^{-1}$ at up to US $50 per tonne. There are no modelled estimates for the additional potential at US $100 per tonne of CO_2e, but an estimate for the mitigation potential at prices above US $100 per tonne is 2,720 Tg $CO_2e\ a^{-1}$.

These are very large ranges, representing 5 to 80% (at carbon prices of up to US $20 per tonne), and 20 to 90% (up to US $50 per tonne) of all other agricultural mitigation measures combined. An additional mitigation of 770 Tg $CO_2e\ a^{-1}$ could be achieved by 2030 by improved energy efficiency in agriculture, although the mitigation potential is counted mainly in the buildings and transport sectors (Smith *et al.*, 2007a).

7.6 Mitigation in the bioenergy sector

In 2008, bioenergy provided approximately 50 EJ, or 10% of global primary energy (IEA, 2009a). Of this total bioenergy, traditional biomass energy use (wood, charcoal, animal manure and other agricultural residues used for cooking, heating and lighting) accounted for about 30 EJ (REN21, 2011; Edenhofer *et al.*, 2011). The remaining 4% of global primary energy was provided by 'modern' bioenergy, including transport fuels, electricity, and direct heating and cooling in domestic, agricultural, forestry and industrial applications.

It seems probable that bioenergy will play an increasingly important role in the provision of energy worldwide. Over the last decade, the use of transport biofuels in particular has grown rapidly, more than tripling from less than 1 EJ in 2005 to 2.7 EJ in 2010. This rapid growth was mainly as a result of policies aimed at decarbonizing the transport system, enhancing the security of energy supply, and economic development (IEA, 2011). The rising price of oil and its products, and of natural gas, have also led to biofuels becoming increasingly cost-competitive (Murphy *et al.*, 2011). Production of bioenergy coupled to carbon capture and storage is being actively evaluated using quantitative models in a range of climate-mitigation scenarios (van Vuuren *et al.*, 2007; Strapasson *et al.*, 2011; Metz *et al.*, 2005; Edenhofer *et al.*, 2011).

The combustion of biomass only releases as much carbon as was captured during its growth cycle. However, the production of bioenergy feedstocks, and their processing and distribution to end-users, accrues greenhouse-gas emissions due to use of fossil energy, fertilizers, land conversion, and so on. In order to assess the mitigation effect of bioenergy, these inputs and resulting emissions need to be taken into account in full life-cycle analysis. The greenhouse-gas savings of the bioenergy are then calculated compared to the fossil fuels they are assumed to replace. In the next section, we briefly describe the various forms of bioenergy, before discussing their significance in carbon offsetting.

7.6.1 Bioenergy options

Bioenergy can be obtained either from biomass grown directly for energy, or from recovered or recycled biomass wastes and by-products. Biomass can be converted through conventional bioenergy production pathways (first-generation) and 'advanced' bioenergy (second-generation) options that use state-of-the-art conversion technologies and novel feedstocks. Both first- and second-generation options often result in co-products such as biochemicals, animal feeds and materials (Figure 7.7).

Heating and cooling (industrial, domestic and traditional). The burning of biomass for heat is the oldest and most common way of converting solid biomass to energy. Heat energy from biomass can be generated by using the simplest of technologies, such as the three-stone fire, through to the advanced combustion of gases derived from the gasification or anaerobic digestion of the primary biomass. Cooling may also emerge as an important new market through the development of adsorption chilling technologies.

Bioelectricity. Biomass electricity, in contrast to other renewable energy technologies, provides a non-intermittent supply of electricity and can therefore be used for baseload and peakload generation. Biomass can be used in dedicated power plants or co-fired alongside coal in existing large-scale power plants. Fuel types include wood and agricultural residues, and the biogenic fraction of municipal solid waste. These organic wastes and residues can also be converted to secondary energy carriers such as biogas through anaerobic digestion (creating mostly methane, as in landfill gas), or syngas (a mixture of hydrogen and carbon monoxide) by gasification. These products can then be burned in conventional engines or turbines to produce electricity.

Biofuels for transport. Biofuels is the collective term for a variety of liquid and gaseous fuels derived from biomass for use in the transport sector. While electrification is a viable alternative to direct use of fossil fuels in the automotive transport fleet (IEAb, 2009), liquid fuels are currently the only viable fossil-fuel alternative for aviation, shipping and other applications. Bioethanol (produced from sugar and starch crops) and biodiesel (from oil crops and animal fats) are increasingly blended with fossil fuels, driven by

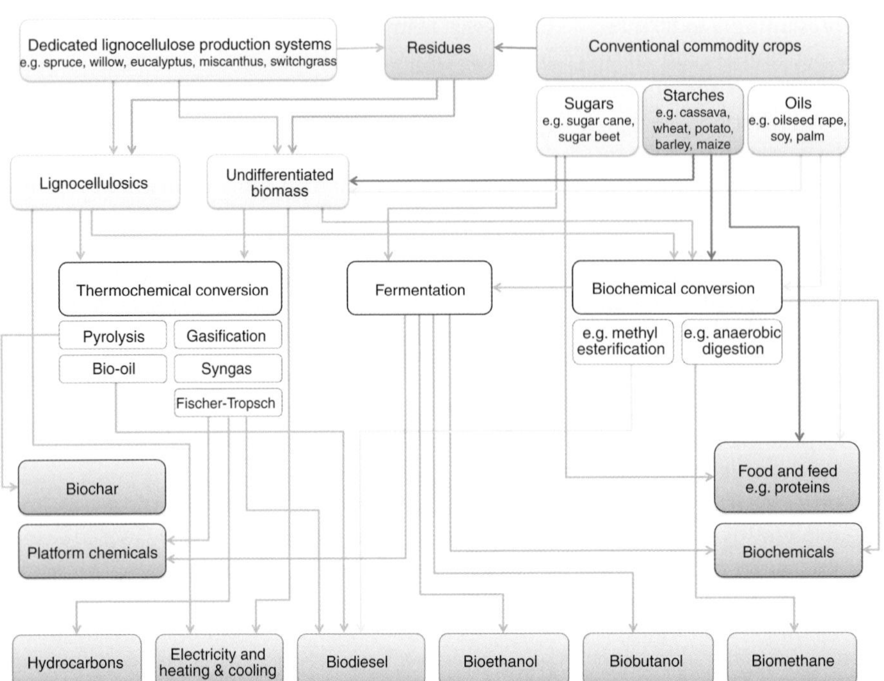

Figure 7.7 Bioenergy options and co-products (adapted from Black *et al.*, 2010). Different classes of biomass (denoted by colour) can be transformed through a variety of different processes (grey boxes) into a range of products (red boxes) with energy carriers shown along the bottom (red boxes with red borders) and by-products at the side (red boxes with black borders). Processes using conventional commodity crops ('first-generation' bioenergy processing) are already widely in use. Production systems using advanced processing of lignocellulosic crops ('second-generation' bioenergy) are under development.

mandated targets in the EU, USA, Brazil and elsewhere. In 2007, biofuels were approximately 2.5% of total European transport fuel consumption (Eurostat, 2011). Advanced biofuels technologies are the subject of intensive research, the objectives being to develop cost-effective and environmentally friendly substitutes, not only for fossil energy sources, but also for current biofuels, which are primarily derived from commodity crops (e.g. food, palm oil), or whose production may compete with arable lands and the production of animal feed (The Carbon Trust, 2010). The use of conventional biofuels is expected to dominate until around 2030, when advanced fuels will begin to prevail (Murphy *et al.*, 2011).

Integrated bioenergy production systems (combined heat and power systems and biorefineries). The integration of bioenergy production systems with other processes can significantly improve the greenhouse-gas balance and reduce the environmental impact of the individual products. Combined heat and power systems utilize 'waste' heat from the electricity generation for industrial processes or residential heating. Biorefineries integrate the processing of biomass so that it can be efficiently fractionated into a variety of outputs such as fuels, heat, power, biomaterials and chemicals (Clark *et al.*, 2006; Ragauskas *et al.*, 2006; NNFCC, 2009). The concept is analogous to that of a petroleum refinery, which in addition to fuels also produces an array of chemical feedstocks for the pharmaceutical, food and agricultural industries. A biorefinery can thus maximize the value of its production output by co-producing high-value, low-volume chemicals and low-value, high-volume fuels.

Biomass energy with carbon capture and storage. Carbon is absorbed from the atmosphere during the growth of the bioenergy crops. If CO_2 emissions are then captured at the point of energy production and stored in underground reservoirs and other geological formations, the result will be net negative emissions (as it is assumed that a proportion of this CO_2 will never return to the atmosphere). The same, or similar, carbon capture and storage technologies can be used for bioenergy as those being trialled for coal-fired power plants. Many climate mitigation scenarios rely on the increased implementation of bioenergy with carbon capture and storage technologies. For example, the IEA's BLUE Map 450 ppm stabilization scenario invokes global emissions of −2 Pg C a^{-1} by 2050 (IPCC, 2007a). It should be noted that while numerous carbon capture and storage demonstration and pilot projects are under way, geological carbon storage is not yet a proven technology at scale (Strapasson *et al.*, 2011).

Biochar. Bioenergy can also be carbon negative when biochar (charcoal) is produced as a co-product, and subsequently sequestered. In the biomass pyrolysis process, biomass is heated to high temperatures in the absence of oxygen. The resulting products are bio-oil, syngas and charcoal (McKendry, 2002, NNFCC, 2009). In contrast to conventional pyrolysis where the aim is to maximize bio-oil and syngas production, biochar production occurs at lower temperatures. This char is highly resistant to decomposition. When added to soil, it improves soil condition and the carbon can be sequestered for hundreds to thousands of years (Gathorne-Hardy, 2011). Overall, Lehmann (2006) estimates that pyrolysis produces 3–9 times more energy than the process itself requires, and that approximately 50% of the carbon in the feedstock can be sequestered in soil. Biochar in soil may also decrease emissions of nitrous oxide and methane (Lehmann, 2006), although this is likely to be variable (Gathorne-Hardy, 2011).

7.6.2 Carbon offsets and life-cycle analysis

The overall mitigation effects of bioenergy are highly dependent on how the measures are implemented. Greenhouse-gas fluxes can occur at every stage of the process from converting land and growing the crop, through to transporting the biomass and converting it to energy, and ultimately to the final energy use. An assessment of fluxes at every stage in the life cycle creates a level playing field when comparing various options with each other, and indeed with the full-life-cycle emissions from extraction, production and use of fossil fuels. Figure 7.8 shows the potential range of emission reductions of conventional and advanced transportation biofuels. Emissions associated with displacement activities, such as indirect land-use change (discussed later in this chapter), are not reflected in this analysis.

Technology has an effect: the production of advanced biofuels is generally associated with lower life-cycle greenhouse-gas emissions than that of conventional biofuels (IEA, 2011). Co-products arising from main-crop processing to biofuels are highly significant differentiating factors for all bioenergy crops when evaluating their emissions, energy and land requirements (Murphy *et al.*, 2011). The greenhouse-gas balance of some biofuels can be improved when a portion of the emissions is proportionally allocated to one or more co-products.

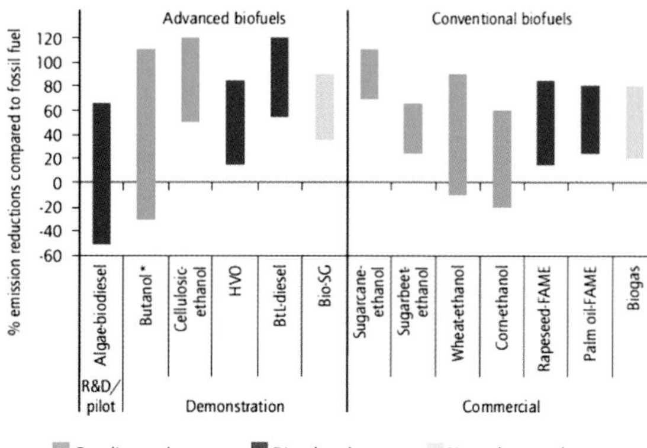

Figure 7.8 Life-cycle greenhouse-gas balance of different conventional and advanced biofuels, and current state of technology (IEA, 2011). Based on a review by the International Energy Agency (IEA) and UNEP of 60 life-cycle analysis studies. Note: the assessments exclude emissions from indirect land-use change. Emissions savings of more than 100% are possible through use of co-products. Bio-SG = bio-synthetic gas; FAME = fatty acid methyl ester; HVO = hydrolysed vegetable oil.

Life-cycle greenhouse-gas emissions from fossil-fuel-derived electricity generation can range between 100 to 300 g CO_2e per MJ of electricity, with coal-fired power production typically emitting between 200 and 300 g CO_2e per MJ of electricity, and gas-fired power production between 100 and 200 g CO_2e per MJ of electricity (RFA, 2010; Woods *et al.*, 2006). For gasoline and diesel, life-cycle emissions range between 85 to 90 g CO_2e per MJ of fuel (RFA, 2010). Estimating the direct emissions associated with bioenergy use is more complex due to the multiple feedstocks and pathways, and the temporal and spatial variability of production and use. Assessing the net impacts of the supply of biological products remains highly controversial with uncertainty arising from two main components:

(1) The length and technical complexity of the supply chain: how these products are produced (particularly with regard to carbon-stock changes and emissions if land-use change occurs), and how they are then stored, treated and transported materially affects their environmental and social impacts.
(2) The knock-on or 'consequential' impacts of increasing the demand and therefore the supply of one product in terms of competition for alternative products or for the resources needed to produce the products, e.g. land, water, capital, labour, etc. These consequential impacts are assessed through integrated regional and global systems modelling.

Table 7.2 Typical total carbon-stock changes, (above and below ground) in 10^6 kg C km^{-2} when converting to croplands. Data from Hamelinck *et al.* (2008).

	To:						
From:	**Wheat**	**Sugar beet**	**Sugar cane**	**Maize**	**Palm oil**	**Rape-seed**	**Soy bean**
Set-aside	−0.9	−0.9		−0.9		−0.9	−0.9
Temperate grassland	−0.9	−0.9		−0.9		−0.9	
Temperate forest	−4.8	−4.8		−4.8		−4.8	
Tropical grassland					**+6.1**		
Tropical forest			−15.1		−6.1		−15.1

Calculating the life-cycle emissions from bioelectricity production is subject to lower uncertainty than for biofuels (Figure 7.8) due to restricted feedstocks, simpler and fewer supply chains, and few co-products. For co-firing of biomass with coal in the UK, Woods *et al.* (2006) reported emissions associated with the mining and transport of coal of 48 g CO_2e per MJ of fuel, compared to a range of 19 to 34 g CO_2e per MJ of fuel for indigenously produced miscanthus and willow. This represents a carbon saving of 29 to 60%.

One of the chief concerns around the promotion of bioenergy at large scales is the need to expand biomass production into non-agricultural land, or the displacement by biomass of current agricultural production into non-agricultural land. In particular, the appropriation of high carbon-stock lands can lead to initial losses of carbon (as seen in Table 7.2) that can result in an adverse carbon balance in the short term; for example, when a natural forest is replaced by an annual bioenergy cropping system. However, the longer the timeframe considered in the analysis, i.e. with more harvest cycles and bioenergy offsets of fossil-fuel emissions, the better the carbon balance becomes. The accounting system must consider the resulting emissions over time (see for example BSI, 2008).

Not all land conversions result in the loss of carbon stocks, as can be seen from biofuel data from Brazil (Table 7.3). The expansion of sugarcane production in Brazil, primarily to produce ethanol as a transport fuel, involved land use being transferred between different types of cropland, but predominantly from degraded pastureland (Nassar *et al.*, 2011). Establishing sugarcane on cerrado (natural savannas) can result in carbon losses of 300 to 1,800 × 10^3 kg C km^{-2}, depending on whether the residues are burned or left on site to decay, while replacement of degraded pasturelands or corn/cotton croplands can result in an increase of 1,000 to 2,900 × 10^3 kg C km^{-2} (based on Macedo, 2010, and Amaral *et al.*, 2008 for above-ground biomass and soil carbon down to 20 cm).

The type of bioenergy crop regime established affects both the net change in land carbon (Tables 7.2 and 7.3) and the annual offset savings from replacing fossil fuels (Table 7.4). Different regimes therefore have different payback times (Table 7.3). Payback times are the number of years it takes for the fuel offsets to neutralize losses of carbon on the land. This can range from zero years (where there is a net gain in carbon from land conversion) to estimates of more than 700 years when tropical forest is replaced with palm-oil plantations for biodiesel. Fully accounting for such losses is important in order to avoid 'perverse mitigation' efforts that actually result in a net emission of carbon to the atmosphere. However, the methodologies and mechanisms to carry out this accounting and the tools to monitor carbon-stock changes remain controversial (see for example, EEA, 2011 and further discussion in Section 7.7).

7.6.3 Projections of future bioenergy use

Projections of bioenergy in the future are derived for a variety of different purposes, such as meeting energy demand or meeting a specific climate mitigation target. They are affected by assumptions on fossil-fuel prices, policies concerning energy and climate, the demand and supply of food and feed, rural development, assumptions about crop yield, short- and long-term carbon stocks, land availability, and innovation in conversion technologies. In particular, understanding the share of new energy production that is provided by feedstock-yield increases, versus land-use intensification (e.g. double or triple cropping), versus access to 'new' land, taking into account the likely yields on

Table 7.3 Change in carbon stocks from land-use-change conversions, and estimated payback periods when land is converted to bioenergy (fuel ethanol and biodiesel)

	Carbon stocks[a] 10^6 kg CO_2 km^{-2}		Change of land use[b] 10^6 kg CO_2 km^{-2}			Payback time[c] Years	
	Above ground	Below ground	To degraded pasture	To unburned sugarcane	To soybean cropland	Cane ethanol	Soy biodiesel
Tropical forest	16.6	60.9	−45.4	−35.6	−40.8	51	728
Degraded pastureland	4.2	15.5	–	**9.7**	**4.6**	−14	−82
Cultivated pastureland	5.9	21.5	−6.0	**3.8**	**1.4**	−5	−24
Soybean cropland	5.5	20.1	−4.6	**5.1**	–	−7	–
Corn cropland	4.4	16.1	−0.6	**9.1**	**4.0**	−13	−71
Cotton cropland	4.0	14.7	**0.8**	**10.5**	**5.4**	−15	−96
Cerrado	7.2	26.2	−10.7	−1.0	−6.1	1	109
Campo limpo	8.0	29.5	−14.0	−4.3	−9.4	6	168
Cerradão	8.7	31.7	−16.2	−6.5	−11.6	9	208
Burned sugarcane	5.4	19.7	−4.2	**5.5**	**0.4**	−8	−7
Unburned sugarcane	6.9	25.2	−9.7	–	−5.1	–	92

[a] Carbon stocks for 'tropical forest' from Stickler *et al.* (2007). All other carbon stocks from Amaral *et al.*, (2008) (all 0–20 cm below-ground depths).

[b] Calculated change in carbon stock in vegetation and soils when converting from the existing land cover to the stipulated land cover. Negative values denote a loss of stock on the land or emissions to the atmosphere. Carbon gain on the land or net uptake from the atmosphere is shown in bold.

[c] Payback times for bioenergy conversions are calculated based on typical emission and savings factors for the two different bioenergy conversion pathways, fuel ethanol and biodiesel (Table 7.4) and change in carbon stocks as shown. A negative value for the payback period results from an increase in carbon stocks when the biofuel crop replaces the existing vegetation land type (for example, unburned cane replacing 'degraded pastureland'). A positive value for the payback period (an adverse outcome) results when carbon stocks are reduced when the biofuel crop replaces the existing vegetation or land type.

Table 7.4 Typical emissions and savings factors for sugarcane bioethanol and soy biodiesel (assuming production in Brazil)

Annual offset	Sugarcane ethanol	Soy biodiesel	Unit
Biofuel production	12,000,000	1,600,000	MJ km^{-2}
Default emission factor: gasoline/diesel	85	85	g CO_2e MJ^{-1} fuel
Default direct emission reduction (EU RED, 2009)	71%	40%	
Avoided emissions	60	34	g CO_2e MJ^{-1} fuel
Avoided emissions per hectare of land	700	56	10^3 kg CO_2e km^{-2}

that new land and its geographic location, requires the coupling of agricultural and forestry market models to spatial land-use models. Yet confidence remains low that economic parameters can accurately predict the spatial location and scale of future land demand, and therefore the impacts on carbon stocking due to the land-use change (Nassar *et al.*, 2011; Oladosu *et al.*, 2011).

The availability of land to meet the demand for bioenergy is a key issue. Estimates of land potentially available for bioenergy production range from 3.2 to 37 million square kilometres (see Table 7.5). The wide range is explained by the variation in the assumptions about the 'availability' of currently non-productive land underlying the estimates. At the lower end, more realistic assessments of societal limiting factors, such as sustainability and food security considerations, are taken into account than

Table 7.5 Estimates of bioenergy potential on 'available' land

Land available for bioenergy (million km²)	Bioenergy potential (EJ)	Source
3.2 to 7.0	61–143 (net energy gain)	Cai *et al.* (2011)
6.7	38–101	WWF (2011)
–	40–170	Dornburg *et al.* (2010)
4.4	120–300	van Vuuren *et al.* (2009)
7.3	59–215	Smeets *et al.* (2007)
3.8 to 4.7	32–41	Campbell *et al.* (2008)
26 to 37	694–988	Hoogwijk *et al.* (2005)

in the earliest studies, as well as other environmental assumptions such as water availability, suitability of land, and competing uses. Similarly, estimates of the bioenergy potential in these studies vary depending on future feedstock yields, the potential range in energy-conversion efficiencies between the different feedstocks for bioenergy carriers, types and scale of end-uses, consideration of by-products, and displacement factors.

The land issue can also be looked at from the opposite perspective starting with energy demand. Murphy *et al.* (2011) estimate that the net land demand for biofuels alone could range between 1 to 6.4 million square kilometres by 2050 to meet a projected biofuel demand of 48 EJ a^{-1}, 30% of total transport energy demand. This land demand is based on estimates ranging from 7.5 to 15 TJ km^{-2}. For comparison, current yields of ethanol from wheat (UK), maize (US) and sugarcane (Brazil) are 6, 6 and 12 TJ km^{-2} respectively. For biodiesel produced from rapeseed (UK), soybean (US) and oil palm (Malaysia), biofuel productivities are currently 1.6, 3.5 and 15 TJ km^{-2}, respectively. These crops all produce energy and non-energy co-products, making precise calculations of net land demand difficult. Dedicated lignocellulosic pathways, such as SRC willow (UK), eucalyptus (Brazil) and sugarcane (integrated conventional and lignocellulosic biofuel production) are likely to have biofuel productivities of 10, 11 and 19 TJ km^{-2}.

The inherent complexity of various bioenergy supply chains and interactions with other land uses makes it difficult to determine its technical potential. Estimates of the theoretical potential range from zero to 1,500 EJ (Chum *et al.*, 2011; Figure 7.9). In their integrated literature and model assessments, Dornburg, *et al.* (2010) found the upper bound of the technical global biomass potential to lie around 500 EJ a^{-1}. This projection is based on a policy framework that ensures food security, sustainable land use, improvements in agricultural management, and safeguards for the protection of biodiversity water, and soils. Of that 500 EJ, 40–170 EJ a^{-1} are derived from residues from forestry, agriculture and the biodegradable fraction of municipal solid waste. Surplus forestry products (excluding residues) could provide another 60 to 100 EJ a^{-1}, while agricultural intensification and improved management could yield another 140 EJ a^{-1}. Energy crop production on surplus agricultural and pastureland amounts to 120 EJ a^{-1}, whereas biomass production on degraded or abandoned land is estimated to be 70 EJ a^{-1}. Biomass energy deployment levels possible by 2050 are less than the technical potential (Fischedick *et al.*, 2011; Chum *et al.*, 2011). The most likely range, according to projections, is 80 to 190 EJ a^{-1}, potentially reaching around 300 EJ a^{-1}. The 190 EJ includes the appropriation of 8 million square kilometres (800 Mha) of marginal and surplus land (Chum *et al.*, 2011).

The IEA (2008a) BLUE Map 450 ppm stabilization scenario adopt a challenging 50% reduction of global energy-related CO_2 emissions between 2005 and 2050. In this scenario, emissions peak in the coming decade and then decrease steadily. By 2100, the economy would be virtually decarbonized. But the scenario cannot be met with current technologies and relies heavily on low-carbon technologies that are still under development, such as carbon capture and storage, or that are not yet economically competitive.

In the power sector, 19% of savings in this scenario are from carbon capture and storage, accompanied by a switch to renewable energy technologies. By 2050, 46% of global power is generated by renewables, resulting in CO_2 emissions savings of 21%; a further 6% of power is nuclear. In the transport sector, low-carbon biofuels contribute 30 EJ by 2050. Overall, biomass is expected to contribute 150 EJ a^{-1}, which equates to approximately 8 billion tonnes of biomass per year. The marginal costs associated with the BLUE scenarios are estimated to be up to US \$200 per tonne of CO_2 (up to US \$500 per tonne in the transport sector), or approximately 1.1% of GDP per year until 2050 (IEA, 2008).

Based on the updated BLUE Map scenario in IEA (2010), Chum *et al.* (2011) use a biofuel demand of

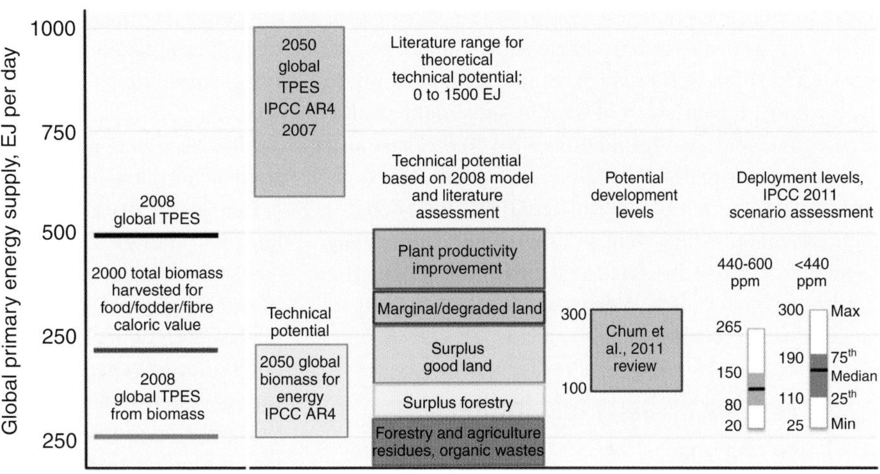

Figure 7.9 Synthesis of projections of bioenergy potential from the scientific literature (Chum *et al.*, 2011; IPCC, 2011). Lines on the left-hand side indicate the 2008 total primary energy supply (TPES) from all sources, the equivalent energy within the biomass that was harvested for food, fodder and fibre in year 2000, and the TPES that was provided by biomass fuels in 2008. A summary of major global projections to 2050 of TPES from biomass is shown from left to right: (1) Global estimates for TPES (grey box) and technical potential for primary biomass for energy (blue box) from the IPCC's Fourth Assessment Report (Metz *et al.*, 2007); (2) Theoretical primary biomass potential for energy, and the upper bound of biomass technical potential based on global assessments and model studies for five resource categories; models included criteria on biodiversity protection, water limitations, and soil degradation, assuming policy frameworks that secure good governance of land use (Dornburg *et al.*, 2010); (3) Potential deployment levels of terrestrial biomass for energy by 2050, from expert review of scientific literature by Chum *et al.* (2011); and (4) Deployment levels of biomass for energy from two long-term mitigation scenarios (CO_2 concentrations by 2100 of 440–600 ppm (orange) or < 440 ppm (blue) (Fischedick *et al.*, 2011). Figure adapted from chum *et al.* 2011 including data from that report.

12 EJ (at 2030) to estimate a 0.1 Pg C abatement. Assuming the conservative biofuel productivity above, a gross land demand of 1.1 million square kilometres would occur. Estimating gross or net land demand for heat and electricity production is complicated by needing to estimate the share of feedstock energy that would be provided from residues and wastes versus dedicated cropping. Assuming a biomass productivity of 15 TJ km^{-2} and a biomass demand for electricity of 5 EJ by 2020 (Chum *et al.*, 2011, based on IEA, 2010), a gross land demand of 333,000 km^2 is calculated.

A different analysis by the IEA (2009a) looked at rising demand for energy based on trend analysis. Compound annual growth rates of 5% for biofuels, bioelectricity and bioheat and 1.2% for traditional biomass energy led to a bioenergy demand of around 110 EJ by 2030 and 250 EJ by 2050 (25% of total primary energy demand). This equates to a demand of 14 billion tonnes of dry biomass annually, or about 10% of global terrestrial net primary production. When coupled to projected increases in demand for food production, this implies that the majority of the biomass must come from enhanced photosynthesis and better use of existing (and future) biomass flows rather than an expansion of biomass production area.

7.7 Critical issues in land-based mitigation

The analysis outlined in the earlier sections indicates that large amounts of bioenergy, together with substantial efforts to reduce deforestation and increase forest area and forest biomass density, are required to make significant contributions to climate mitigation through land use. Numerous environmental and socio-economic implications of these land-based mitigation measures have been documented, resulting in a complex decision-making landscape shaped by issues of land suitability, environmental integrity, social equity, economic costs and technology issues.

These issues must be considered as part of the debate on how we can mitigate climate change, while expanding global food production to meet the world's needs in 2050, providing shelter and energy, and preserving our natural-resource base. What measures do we need to take into account to ensure that mitigation is sustainable, i.e. that the unintended consequences of policies and incentive programmes do not result in unsustainable practices, or even lead to overall increases in greenhouse-gas emissions?

The major source of controversy around land-based mitigation stems from the multiple demands for services provided by the land itself. Debates around large-scale bioenergy implementation, and the assessment of the full range of ecosystem services provided by land, have led to shifts in how we view and wish to evaluate the human appropriation of land. A secondary source of controversy arises around our ability to measure and account for mitigation accurately, along with its associated impacts, both positive and negative. These debates do not remain in the science domain only, but are the subject of intense international negotiations for the setting of global policy. We discuss them more fully in the following sections.

7.7.1 Land availability and indirect land-use change

Land is a finite resource that provides a variety of 'services' to human society besides the provision of food and materials. These include biodiversity, rainfall recycling, control of flooding, nutrient cycling, amenity and indeed climate mitigation through uptake of CO_2. Converting land currently under natural ecosystems may provide additional food, fuel or fodder, but it can result in the loss of a number of these other ecosystem services. The development of more sustainable policy and practice will depend on better accounting for the changes in services provided by land, and better planning for multiple uses and benefits.

Changes in a variety of drivers and pressures impact on competition for land and prices of land-based goods such as food. Population increase is a prominent concern, but demand for land is strongly influenced by other factors such as crop yields, changes in diet and the prevalence of pollution impacts on agricultural production. Competing uses and pressures on land is a theme that runs through all discussions of land-based mitigation. With the exception of agricultural measures aimed at non-CO_2 emission reductions, all of the terrestrial mitigation measures increase competition for land, so trade-offs are needed between different mitigation measures, but these are rarely assessed in combination.

Analyses of the potential for land-based mitigation tend to rely on relatively large estimates of the area of land that is already degraded or set aside and that could be 'available' for mitigation (see Table 7.5). However, even this land is being used in some way, or is providing some kind of ecosystem function or habitat, so there will be a trade-off of services. Modelling studies show that future policy decisions in the agriculture, forestry, energy and conservation sectors could have profound effects on land use, with the different demands for land to supply multiple ecosystem services usually intensifying competition for land in the future (Smith *et al.*, 2010).

Indirect land-use change occurs when the production of biomass feedstocks in one area displaces the existing land-use activities to other areas. The resulting land-use change can have negative impacts on carbon stocks and thereby potentially increase the bioenergy life-cycle greenhouse-gas emissions (Fargione *et al.*, 2008; Gibbs *et al.*, 2008; RFA, 2008; Searchinger *et al.*, 2008). For instance, if the demand for palm oil for biofuel production is supplied from existing plantations that have previously supplied the food market, it displaces food production from existing farmland to non-agricultural land.

There are widely differing views on the extent to which biofuels are currently causing indirect land-use change, the scale of the resulting greenhouse-gas emissions and the extent to which the impacts might decline or increase in the future. Research indicates that the relative contribution of biofuels to indirect land-use change is smaller than that of other drivers of change (FAO, 2010b). A recent empirically based statistical evaluation, by Kim and Dale (2011), could not find conclusive evidence that the world's largest and most aggressive biofuel programme, the US Corn Ethanol Program, had caused indirect land-use change. Other research suggests that enough suitable land exists to accommodate biofuels without necessarily causing displacement effects (Perlack *et al.*, 2005; EEA, 2007; Smeets *et al.*, 2007). However, most analyses depend on models that are not yet advanced enough to predict indirect effects with confidence or in detail, nor to differentiate reliably among feedstocks (Woods *et al.*, 2010).

Scenarios of future land-use change are generally derived using IAMs. Those that explicitly model LULUCF typically estimate demand for land for bioenergy, timber and food. They model changes in crop yields and the degree of intensification of crop and livestock production. They then allocate the required changes in land use to the available land following a set of allocation rules. Most assume broadly similar population growth and changes in consumption patterns, such as a link with increased consumption of livestock products as countries develop economically. Yet there is considerable uncertainty in the projections of intensity of competition for land in the future, and the regional distribution of this competition. Probably

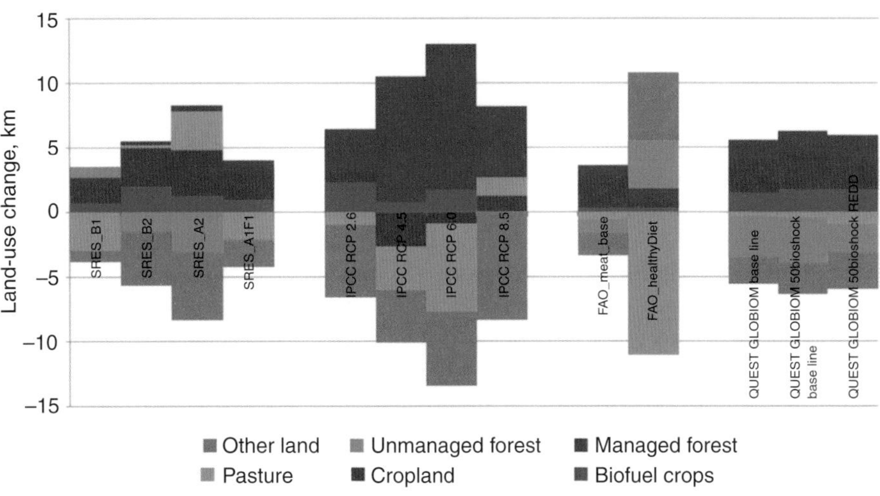

Figure 7.10 Land-use-change between 2000 and 2030 based on different scenarios. SRES scenarios were produced for the IPCC using several integrated assessment models (Nakicenovic and Swart, 2000) and outline different global storylines of environmental change without specific mitigation policy. IPCC RCP scenarios explore mitigation policy (Moss *et al.*, 2010; http://luh.sr.unh.edu/). Data available gives no 'unmanaged forest' category, so loss of 'unmanaged forest' is seen in 'other lands', and gain of unmanaged forest is included in 'managed forest' IPCC RCP 8.5 does not give 'bioenergy' but includes 'wood harvest,' which is therefore categorized under 'managed forest.' Stehfest *et al.*, (2009) used the FAO (2006b) trends of meat consumption as a baseline (FAO-meat-base) in comparison to a 'healthy diet' (FAO healthy-diet) variant. GLOBIOM model scenarios were produced, under the UK research programme QUEST to explore the integrated effects of different mitigation policies, for example, increasing amounts of bioenergy by 50% (50BIOSHOCK) above expected trends (QUEST GLOBIOM baseline) and limiting the environmental impact via policies such as REDD. See: http://quest.bris.ac.uk/research/themes/QUATERMASS.html.

the greatest uncertainty lies in the relationship between increased demand for biological products and the supply response, in terms of the share of biomass provided through land expansion compared to increased yield (productivity per unit area).

Figure 7.10 shows a small selection of scenarios and how they affect land use. The IPCC SRES business-as-usual storylines and RCP mitigation scenarios include a range of varying factors including yield change, diet change and bioenergy use change. The study based on FAO food scenarios focuses on the effects of changing meat consumption (Stehfest *et al.*, 2009), which is discussed in more detail in the next sectiones. The GLOBIOM study focuses on the effects of land-based mitigation, in particular bioenergy, and imposing sustainability criteria.

The widely used IPCC SRES scenarios (Nakicenovic and Swart, 2000) explored different possible futures without any explicit climate change mitigation policy. Rapid economic development and technological advances assumed in the A1FI and B1 storylines tend to lead to agricultural intensification, reducing demand on land, and also to the modernization of energy, reducing demand for traditional biofuels. These two factors lead to a reduction in CO_2 emissions from deforestation, and in some cases net uptake due to afforestation. The cumulative abandoned agricultural area ranges from 7.3 and 9.4 million square kilometres in 2100. Using this land for mitigation means potentially sequestering 116 to 146 Pg C (426 to 536 Pg CO_2) over the century (van Vuuren *et al.*, 2007).

The scenarios being developed for the IPCC Fifth Assessment Report do explore mitigation options. The Representative Concentration Pathways (Moss *et al.* 2010; see Chapter 5) are scenarios selected from the published literature to represent a broad range of radiative forcing outcomes by the end of the century. These scenarios provide projections of emissions concentrations and land use/cover to the Earth system models that produce the IPCC's climate projections. Each of the four pathways is supplied by a different IAM (IMAGE, GCAM, AIM and MESSAGE), and represents just one of many scenarios that would lead to the specified radiative forcing.

The RCP 2.6 W m^{-2} scenario, provided by IMAGE, is associated with a stabilization of CO_2 concentration at 450 ppm and a global temperature rise remaining below two degrees. It is also one of the key mitigation scenarios used in the UNEP GAP report (UNEP, 2010; see Figure 7.1). A global bioenergy use of 200 EJ a^{-1} by 2100 was invoked (van Vuuren *et al.*, 2007). This represents a theoretical potential, taking geographical, technical and economic considerations into account. Subsequent modelling that imposes restrictions in terms of environmental and social safeguards has limited the bioenergy implementation potential to 65–115 EJ a^{-1} by 2050 (van Vuuren *et al.*, 2009). Within the RCP 2.6 scenario, it is assumed that both livestock and cropping increase in yield and intensity and therefore free up land. The increased demand for land for food and bioenergy is met by abandoned crop and pastureland, and by natural grassland, but no natural forest is used to meet demand for food or bioenergy. Plantations are established on abandoned agricultural land to meet increasing demand for wood. If this is not sufficient, then natural forests are converted to plantation forests with associated loss of carbon. Therefore, in this scenario, bioenergy production may displace plantation forestry into natural forests. Another important feature of this scenario is that its vast carbon saving from bioenergy is tied to the assumption of large-scale use of carbon capture and storage technology in the second half of the century.

The GCAM RCP 4.5 scenario assumes that carbon in natural vegetation will be valued as part of global climate policy. In this scenario, there is an overall expansion in forested area throughout the twenty-first century. The use of cropland and grassland decreases, following considerable yield increases and dietary changes. The AIM RCP 6 scenario initially shows a decline in cropland by 2030, but by the end of the century total cropland area increases by 26% compared to 2005 due to increasing food demand. There is a decline in pasture throughout the century as the increase in production of animal products is met through a shift from extensive to more intensive animal husbandry.

The RCP 8.5 scenario (MESSAGE) assumes a business-as-usual pathway with continuously rising emissions. The area of both cropland and pasture increases, mostly driven by an increasing global population. By 2100, cropland increases by 300 million ha, 16% above 2005 levels. Forest cover declines by 300 million ha from 2000 to 2050, and another 150 million ha from 2050 to 2100.

Research carried out in the UK research programme QUEST using the GLOBIOM[14] sectoral land-use model explored the interacting effects on land use of different mitigation policies including bioenergy, REDD and afforestation, with the implementation of different sustainability criteria (e.g. restricting use of natural forest with implied large losses of carbon on conversion). The baseline case followed the POLES scenario (Russ *et al.*, 2009), which includes an increase of bioenergy use. The Bioshock scenario follows a more aggressive mitigation policy, increasing bioenergy production by an additional 50%. It leads to an additional loss of natural forest. Implementing a REDD policy safeguards the loss of natural forest compared to both the baseline bioenergy provision and the additional Bioshock. In this latter scenario, the same total land area is used for bioenergy crops as in the Bioshock, but instead of natural forest being used, the demand is met by using 'other land' (such as natural grassland).

7.7.2 Land-use competition and interaction with the food system

One of the greatest concerns about land-based mitigation is the competition with food production for land, water and other resources, and the effects that large-scale mitigation activities will have on food prices. The strong linkages between terrestrial climate mitigation and food and energy security are receiving growing attention in many countries.

The rapid increase in the use of food crops for bioenergy, driven by direct economic competitiveness with oil, could put pressure on food stocks, particularly cereals and vegetable oils, exacerbating price volatility in food markets. Early assessments (e.g. Mitchell, 2008; Searchinger *et al.*, 2008) of the scale of the impact on global food prices often ascribed a significant share of the increase in global food prices to biofuels. For example, Mitchell (2008) controversially estimated that biofuels caused 75% of the increase in internationally traded food prices between 2002 and 2008. Subsequent evaluations (including a more recent analysis by Mitchell (2011)), some using economic models of agricultural markets that are resolved on finer regional scales, show a more complex picture. Many interacting factors have affected food price, for example, severe drought in some important crop-producing countries towards the end of this period, and growth in demand for meat from China.

In contrast, a model analysis of the impact of existing biofuels policy in Europe (10% of transport fuels to be met by renewables by 2020) by the International Food Policy Research Institute (IFPRI, 2010) found only marginal impacts on food price and indirect land-use change. With biofuels contributing 5.6% of transport fuels by 2020, most biodiesel was projected to be produced domestically while most bioethanol was imported from Brazil. Over a 20-year time horizon

[14] www.iiasa.ac.at/Research/FOR/globiom.html.

from 2020, this led to a projected real increase in income in Brazil (+0.06%), and a lesser increase for the EU, while food prices went up by a maximum of +0.5% in Brazil and +0.14% in Europe. All other regions showed a marginal decrease in income due to factors such as lower export oil prices or higher food prices. The global area of cropland increased by 0.07%, indicating a small indirect land-use change effect associated with the EU biofuels policy. Direct emission savings were estimated at 18 Tg CO_2, and additional emissions from indirect land-use change were 5.3 Tg CO_2 (mostly in Brazil), resulting in a global net balance of nearly 13 Tg CO_2 savings over 20 years.

Economic markets, demand and international trade are notoriously complex and difficult to model, leading to a lack of clear consensus in the literature about the relationship between food prices and biofuels. The underlying methodologies and results – in both economic and climate-mitigation terms – of such studies remain controversial (e.g. the detailed analysis of the US Corn Ethanol Program by Kim and Dale, 2011; Oladosu *et al.*, 2011).

It cannot be assumed that the expanding role of for bioenergy is automatically a threat to food security or a cause of geopolitical conflict. In a follow-up to his 2008 report, Mitchell (2011) explored possible co-benefits of bioenergy production with food for Africa and concluded that, despite potential problems such as conflicts with food production, biofuels offer a development opportunity for African countries. In fact, bioenergy already plays a role in the provision of energy services to agriculture and the rural poor, supporting crop production, harvesting, storage and transport (Lynd and Woods, 2011). Policies addressing the primary drivers of competition for land (e.g. population growth and migration, dietary preferences, environmental conservation and protected areas, forestry policy, and infrastructure development) are more fundamental to food provision than bioenergy policy and could have significant impacts in reducing tension about land use for bioenergy.

In many regions, especially the less-developed ones, the gap between potential and actual yields of biological production is still very wide, implying that it should be possible to increase production on existing agricultural land to meet increasing demand – without converting additional land. For example, actual cereal production yields in 2000 were 0.7 kg m^{-2} in Western Europe, 0.5 kg m^{-2} in the USA and 0.2 kg m^{-2} in eastern Europe compared to a possible 0.8 kg m^{-2} with advanced farming. In many developing regions, yields were only 0.1 to 0.3 kg m^{-2} compared to a potential between 0.5 and 0.7 kg m^{-2} (IIASA, 2011). The projected 70% increase in demand for food production between 2000 and 2050 implies an annual increase in crop yields of 1.4%. However, as Bruinsma (2011) points out, despite a historical global growth rate of 1.7% per year, projected increases slow to around 0.7% per year by 2050 without substantial intervention. Land-based mitigation therefore needs to be developed in ways that support overall increases in productivity and will therefore need to be predominantly through increased cropping intensity as opposed to increased cropland area.

It remains a challenge to deliver these increased levels of production in a way that does not damage the environment and compromise other ecosystem services (Smith *et al.*, 2010), but there is much scope for improvement in current practices. Unwanted agricultural or forestry surpluses and wastes can be used for energy, keeping crop and timber prices stable. Many crops are suitable for the simultaneous production of both food and energy, with different parts of the plant providing different needs. Table 7.6 summarizes the main types of co-products from transportation biofuel production and their substitution. Palm oil can provide animal feed, oil for biodiesel or for other consumer products, solid biomass for burning and biogas from different parts of the plant. Cereal grains or rape meal, when used as animal feed, reduce the need for soy production and thereby reduce net land-use requirements (Murphy *et al.*, 2011). Given that approximately one-third of arable land globally is dedicated to the production of animal feed, the positive land-use implications of avoided feed production can be significant (FAO, 2011).

To explore the effects of changing human diet, Stehfest *et al.* (2009) ran some simple scenarios reducing the amount of meat being consumed compared to FAO (2006a) projections of trends in meat consumption (the land-use change associated with the 'healthy diet' scenario is shown in Figure 7.10). Livestock husbandry currently occupies 80% of anthropogenic land use. Consequently, the scenarios reducing meat consumption had a profound impact on projected land use, with a substantial area of land being freed from livestock production for conservation or other uses. If all of this additional land were used for bioenergy, the potential for energy from woody biomass would be 450 EJ by 2050. (note that Figure 7.10 shows the land use changes associated with the bioenergy use in the

Table 7.6 Selected co-products from use of crops for transportation biofuels and their uses (© Elsevier, reproduced with permission from Murphy *et al.*, 2011)

Biofuel crop	Co-product	Use	Substitution
Cereals (e.g. maize, wheat)	Distillers grains	High-protein animal feed, solid fuel	Soy meal, other protein feeds, electricity, heat
Sugarbeet	Sugarbeet pulp	High-calorie animal feed	Wheat, other feeds
Rapeseed	Rape meal	Animal feed	Soy meal, other feeds
Oil palm	Empty fruit bunches	Animal feed, solid fuel	Other animal feed, electricity, heat
	Palm-kernel expeller	Solid fuel	Electricity, heat
	Palm-oil mill effluent	Anaerobic digestion	Electricity, heat
Sugarcane	Bagasse	Solid fuel,	Electricity, heat
Lignocellulose crops (e.g. willow, switchgrass)	Lignin	Solid fuel, chemicals	Electricity, heat, petrochemicals

'healthy diet' scenario, 170 EJ, not this total potential.) There is also a substantial difference in land use and energy intensity of different forms of meat production, with both being higher in red-meat production than in poultry and pig rearing. Thus a switch in type of meat consumed would further reduce land demand.

A wide range of options exist for producing food, animal feed, timber and bioenergy more sustainably in integrated land-management systems. Sustainable intercropping, co-production and maximizing the use of co-products and waste can increase the productivity of land, recycle nutrients, improve water resources and reduce the need to convert land for the dedicated production of bioenergy (FAO, 2011; Porter *et al.*, 2009; Dale *et al.*, 2010). On a large scale, this integration has the potential to improve national food and energy security and maintain or enhance ecosystem services, while reducing greenhouse-gas emissions (FAO, 2011; Valentine *et al.*, 2012). Constructively managing the interaction between food and conventional biofuels, therefore, represents a real opportunity for both food security as well as for energy supplies (Woods *et al.*, 2010; Rosillo-Calle and Johnson, 2010; Valentine *et al.*, 2012).

7.7.3 Monitoring, reporting and verification of biospheric mitigation

One of the most important goals of the UNFCCC is to reduce greenhouse gas sources and enhance their sinks. Effective climate mitigation policy and action requires criteria, instruments and methods for monitoring, reporting and verification of emissions and sinks. These should follow the UNFCCC principles of transparency, consistency, comparability, completeness and accuracy (Grassi *et al.*, 2008). Accounting rules should therefore measure as accurately as possible the atmospheric outcome of deliberate activity. This implies the necessity for a transparent and simple accounting of the atmospheric effects of various intended activities and actions to provide incentives for good practice. Methods need to minimize loopholes that would allow perverse incentives. These might include claiming carbon credits for something that would have happened anyway without additional activity, or a failure to account for all sources and sinks of greenhouse gases such that the total actual mitigation is at best less than the sink accounted, or at worst a net source.

The UNFCCC rules and guideline methods for reporting greenhouse-gas emissions and sinks (e.g. IPCC, 2003) have been painstakingly negotiated over several years, and are still the subject of much scientific and political controversy. They are continually being revised alongside the development of new principles and methodologies that go beyond the Kyoto Protocol, such as REDD in developing countries. Methodologies need to reflect the level of accuracy required for verification. On the other hand, mitigation policy and accounting rules cannot be formulated independently of the currently available monitoring techniques because they need to be applicable and verifiable.

Annex I (developed) countries that are signatories to the UNFCCC are required annually to report greenhouse-gas emissions and sinks due to changes in land cover and land management. This includes sources and sinks of greenhouse gases in the LULUCF sector. Countries use methodologies appropriate to their data availability and capabilities, so methods vary considerably between countries. For example, some use default values for carbon density and account for changes in vegetation carbon stock only, while others use detailed carbon-inventory measurements and modelling approaches for different land-cover types. Many non-Annex I countries have also reported emissions data but rarely on an annual basis, and reports are often based on sparse data and lower-level methodology.

Under the Kyoto Protocol, Annex B countries (essentially the same as UNFCCC Annex I developed countries) have committed to reduce their emissions during the commitment period (2008 to 2012) relative to a baseline year, typically 1990. The target is based on emissions from fossil fuels and industry, but LULUCF within a country can be counted towards targets. Accounting methods basically follow that of UNFCCC accounting with some critical differences, described below.

Emissions due to land-use change are reported to the UNFCCC based on annual *net* change in area. This can lead to the perverse outcome that a country could cut old-growth (mature) forest in one area and plant new forest in another and not have to report emissions, even though the loss of carbon from the old-growth forest would be far greater than that taken up in the newly planted forest. Under Kyoto Protocol accounting, therefore, countries must report *gross* change in forest area (accounting for deforestation and afforestation separately), reporting all changes since 1990.

For UNFCCC reporting, countries elect the area of forests considered 'managed' on which they report emissions and sinks (termed 'Forest Land Remaining Forest Land,' or FLRFL). Countries do not report emissions in 'unmanaged' or 'natural' forests. This was intended as a way to approximate accounting only for direct human activity contributing to carbon sinks. However, the definition of what is 'managed' varies considerably between countries. For example, across EU countries, the area classed as 'managed' varies between 5 and 100% of the total national forest area. The flux calculations are typically based on data of changes in forest carbon stock (carbon in above-ground biomass) measured in forest inventories. Thus, the carbon sink in the FLRFL category is not necessarily due to direct human management of the forest. Some of it results from recovery or growth of vegetation due to past changes or disturbance (Magnani *et al.*, 2007; Ciais *et al.*, 2008). Most of it is likely due to the effects of environmental change (climate, increased CO_2, and nitrogen deposition), depending on the data and method of calculation (Harrison *et al.*, 2008; Luyssaert *et al.*, 2008). For this reason, under the Kyoto Protocol, the amount of carbon sink that can be accounted from forest management is capped at around 15%, if countries wish to account for it at all. Accounting for forest management, cropland management and grazing-land management are all optional under the Kyoto Protocol.

There are scientific methods to model and factor out the effects of direct human activity against the background of natural and indirect effects (Vetter *et al.*, 2005; Böttcher *et al.*, 2008). These analyses involve first the determination of a baseline of business-as-usual activity. The changes relative to that baseline can then be assessed. For example, management activities such as changing the amount of tree removals in forest thinning, changing the length of time between establishment and harvesting of trees, or changing the tree species can all result in a change in carbon stock that can be measured and modelled, and their mitigation effect calculated in order for credits to be claimed. This accounting for specific activities would allow for better market integration of mitigation efforts, by increasing the incentives for improved greenhouse-gas management and by reducing the variability of fluxes that are accounted towards the overall national emission target.

The activity-based accounting is only a partial accounting of the total carbon flux that the atmosphere 'sees', or that is measurable from changes in vegetation and soil-carbon stock. It serves a different purpose from scientific efforts to estimate the total global land carbon sources and sinks (as described in Section 7.3) to better understand processes, system responses and feedbacks, and build more realistic Earth system models.

Implementing REDD raises some specific scientific challenges around establishing baselines for forest condition, forest cover and human perturbation of forest ecosystems, against which REDD measures can be judged, as well as operational concerns about the need for improvement in monitoring, accounting and verification in deforesting countries. A further controversial issue in the context of current international emissions trading schemes and the future development of REDD (with implementation often dependent on international financing) is who can claim the credits for a mitigation activity: the country where the activity is implemented, or the country paying for it.

Capacity building in developing countries for monitoring of mitigation measures is essential for the effective implementation of bio-mitigation measures. The assessment of emissions from land use and land-use change requires both data on changes in land cover and estimates of carbon-stock changes associated with the transition between land-use types (IPCC, 2003). Traditional (national) forest inventories provide data of the growing stock timber volume per unit area, obtained by measuring tree diameter or age classes and species composition. To estimate changes in growing stocks, repeated measurements at permanent sample plots are carried out. However, only a few developing countries are conducting their national forest inventories on a regular basis (GOFC-GOLD, 2008).

Remote sensing has been identified as a key technology for the successful implementation and monitoring of a future REDD mechanism (Herold and Johns, 2007). Estimations of above-ground biomass combined with remotely sensed imagery from airplanes or satellites can produce coherent maps of change (Goetz *et al.*, 2009; Baccini *et al.*, 2012). Satellite sensors collect data routinely and data are often freely available (at the global scale). The quality of global products derived from those sensors depends upon many factors, such as conditions at the time of data capture (e.g. cloud cover), the period of time and area covered, and the availability of ancillary data for validation. Optical remote-sensor measurements in the visible and infrared wavebands are correlated with vegetation cover, but short-term and small-scale (few-kilometre) changes such as shifting cultivation (burning small patches of forests to grow crops, followed by abandonment and re-establishment of forests) are often hard to detect. Airborne optical sensors such as three-dimensional digital infrared cameras can provide much more detailed information at the sub-metre scale, and can also provide information about tree height. However, there is still a limited ability to develop accurate biomass estimation models, especially for the complex canopy structures of tropical forests, based on remotely sensed optical data (Goetz *et al.*, 2009).

A number of new and innovative technologies have recently approached operational feasibility for better-resolved monitoring of land-cover change and land-based mitigation efforts. Synthetic aperture radar remote sensors have the advantage of being able to penetrate clouds and, to a limited extent, the forest canopy (dependent upon wavelength), whilst LiDAR uses laser light to estimate forest height and vertical structure. It can accurately measure the spatial variability of forest carbon stocks, but requires extensive calibration data from ground measurements (Patenaude *et al.*, 2004; Magnusson *et al.*, 2007; McRoberts and Tomppo, 2007; Simard *et al.*, 2011). A 'hierarchical nested approach' combines high- and coarse-resolution optical, radar and/or LiDAR data (DeFries *et al.*, 2002; van Laake and Sánchez-Azofeifa, 2004; Morton *et al.*, 2005; DeFries *et al.*, 2006; Goete *et al.*, 2009; Baccini *et al.*, 2012). With the help of such an approach, monitoring systems at national levels in developing countries can also benefit from pan-tropical and regional observations, mainly by identifying hot spots of change and prioritizing areas for on-the-ground monitoring at finer spatial scales (Achard *et al.*, 2007). However, finding an adequate sampling method that is dense enough and well designed to capture single land-use-change events remains a scientific and technological challenge that will depend on the accuracy and precision requirements from the policy process.

Reporting and accounting for mitigation through bioenergy has its own unique and controversial set of issues. Bioenergy use is counted as carbon-neutral in the energy sector as it is assumed that the carbon emitted is taken up during growth. However, as discussed in Section 7.6, a full life-cycle analysis shows that there are emissions associated with bioenergy production and use. Emissions arising from the transporting of biomass should be reported in the transport sector; emissions due to fertilizer use and land-use change should be reported in the LULUCF sector. However, not all countries fully report emissions in all sectors. This means that the total emissions from bioenergy are often underestimated, or are not estimated at all. This creates perverse incentives that encourage bioenergy with small or even net negative climate effects. This is a particularly problematic issue where a developed country claims reduced emissions or carbon credits for bioenergy end-use in their own country's accounting, when the biomass was produced in a developing country with poor reporting of emissions in the LULUCF and other sectors. For example, palm oil could be produced in Brazil from cleared tropical forest areas and transported to the USA. The USA could claim credit for the reduced fossil-fuel emissions, but the LULUCF emission would not have been reported. National-level accounting under the UNFCCC does not account for any carbon-stock changes or emissions outside the national border, even for the production of goods used within the country.

Even if there were full accounting of the separate emissions in each different sector in each country, this would still not provide all the essential information for decision-makers. In order to ensure a 'level playing field' when comparing bioenergy to other renewable-energy options or other land-based mitigation options, it is necessary to count all the fluxes in a full life-cycle analysis for each bioenergy production chain *and* for each of the other mitigation options (Obersteiner *et al.*, 2010). The European Union has attempted to take account of full life-cycle greenhouse-gas emissions in its Renewable Transport Fuels Obligation as part of its sustainability criteria. Emissions due to direct land conversion and processing should be accounted for when considering emissions due to bioenergy used for transportation biofuels. However, even in this information-rich region, full information is often unavailable or not applied, and default factors are very often used, so there is still some way to go in ensuring full account is made of land-conversion emissions. Also, although effort is being made to assess the life-cycle impacts of biofuels, there is no such control currently on biomass used for electricity generation. Thus issues remain with the multiple accounting systems for carbon emissions and offsets that are not really compatible with each other. Until there is universal standardized accounting, these kinds of issues will continue to present challenges.

7.7.4 Sustainability

Human development is inextricably linked to our ability to exploit the biosphere. Economic development and population growth, with their associated increasing demand for food, material goods and energy, place ever-greater pressure on Earth's finite and renewable resources. The exploitation of land, water and soil resources beyond their replenishment capacity, as well as the overtaxing of the ecosystem's mediating and restorative facilities, result in resource shortages and degradation of ecosystem services such as soil and water quality, regulation of nutrient fluxes and disease, and loss of biodiversity and amenity (Foley *et al.*, 2005; Millennium Ecosystem Assessment, 2005; West *et al.*, 2010).

Mitigation efforts potentially add an additional burden onto an already pressured system. There is particular concern about bioenergy production systems. Environmentally, the unregulated production of bioenergy feedstocks can lead to deforestation and other land-use change with detrimental effects on carbon stocks, biodiversity, water and other ecosystem resources (Howarth *et al.*, 2009; Smeets *et al.*, 2008; UNEP, 2009; Rice, 2010). Indirect land-use changes can also negatively affect biodiversity, water and other resources (Tait *et al.*, 2011). On the positive side, if carefully managed, feedstock production may enhance carbon stocks and biodiversity, reduce soil erosion and have a positive impact on other ecosystem services (UNEP, 2009; FAO, 2011; Chum *et al.*, 2011). The adverse socio-economic impacts that can arise from bioenergy production include threats to food security, infringements on workers' rights and land-rights conflicts (Smeets, 2008; Sawyer, 2008; Marti, 2008; Rosillo-Calle and Johnson, 2010; Rice, 2010). Again, positive socio-economic impacts are also possible, with the production of biofuels generating new employment opportunities, revenue streams and improved livelihoods, education and health compared to other comparable agricultural activities (Diaz-Chavez *et al.*, 2010; Moraes, 2010). Better understanding and quantification of the full range of impacts and opportunities of implementation of bioenergy and other mitigation measures are urgently needed, to support more informed decision-making, and the development of policies and guidelines to avoid negative impacts on food, forestry and people and to ensure co-benefits and sustainability.

A number of climate-mitigation policies and programmes that involve ecosystem management have introduced sustainability criteria as a means of minimizing their non-greenhouse-gas environmental impacts. For example, the EU's Renewable Energy Directive (European Commission, 2009) requires that biofuels and bioliquids which qualify for incentives under the Directive must not originate from biodiverse areas or protected areas. However, to date, these sustainability criteria address some of the direct environmental impacts of bioenergy feedstock cultivation, but do not address potential indirect impacts, whereby feedstock cultivation displaces other land-use activities to biodiverse areas, or areas designated for natural protection. The issue of possible indirect environmental impacts may be addressed through the development of commodity-wide sustainability standards, such as the Roundtable on Sustainable Palm Oil standard and the Round Table on Responsible Soy standard, which can be applied to all production and not only the proportion used for bioenergy. However, indirect or displacement effects will still be possible unless all, or nearly all, land-use activities adhere to sustainability standards. This does not appear likely in the near future.

For avoided deforestation, reforestation and afforestation projects and policies there appears to be a greater focus on social sustainability, such as the protection of indigenous peoples' rights and the promotion of sustainable livelihoods. At the project level, there are a number of sustainability standards that address both environmental and social sustainability, including the Climate, Community and Biodiversity (CCB) standard and the Gold Standard. At the global policy level, the text adopted at the Conference of the Parties at Cancún (UNFCCC, 2011) includes guidance and safeguards for policy approaches and incentives relating to reducing emissions from deforestation and forest degradation, including respect for the rights of indigenous peoples and local communities and the protection of natural forests and biodiversity.

7.7.5 Potential versus actual capacity

Sustainability criteria, limits on land availability and biological, technical and socio-economic factors all limit the amount of mitigation that can be achieved compared to the theoretical maximum. Estimating the climate-mitigation potential of terrestrial ecosystem management involves overlaying different layers of potential capacity (Cannell, 2003; Hoogwijk *et al.*, 2005):

- biological or theoretical potential;
- technological potential;
- economic potential;
- ecological potential; and
- realistic potential for implementation.

The climate-change mitigation potential becomes progressively smaller as each successive layer is introduced. In order to accurately estimate the realistic potential, it is necessary to include all the significant determining factors and their effects. Typically, models work with quantifiable parameters and relationships. However, many important factors – including some that may be determinants of success for the implementation of a mitigation measure – are not readily quantifiable, and are therefore difficult to model. These factors include security, governance, land tenure, institutional capacity and the local availability of technical support as illustrated by the case studies. It is important for policy-makers to understand that there are constraints that are not accounted for in modelled estimates, so that the level of mitigation that is realistically available is not overestimated.

Different real-world constraints may have different implications for mitigation potential. Some constraints may increase the cost of mitigation, where there are resource implications associated with overcoming the constraints (e.g. creating new governance arrangements where there is a lack of institutional framework). Other constraints may be better characterized as a reduction in the total mitigation potential available. The schematic marginal abatement cost curve in Figure 7.11 illustrates these two effects. The effects of constraints may either increase the cost of mitigation, shifting the marginal-abatement cost curve (MACC) from MACC 1 to MACC 2, or they may reduce the level of mitigation that is realistically available, shifting the total abatement available from point A to point B. There may also be a combination of these effects, resulting both in more expensive measures and less total mitigation.

The availability of financing is a critical factor in constraining or facilitating mitigation, as indicated in the IPCC (2007b) analysis of mitigation potential across sectors at a variety of carbon prices (Figure 7.12). Applying the analysis of the effects of other real-world constraints on the MACC as in Figure 7.11, it is likely that the amount of mitigation which is realistically achievable at each carbon price is lower than the theoretical estimates of what is economically viable shown in Figure 7.12.

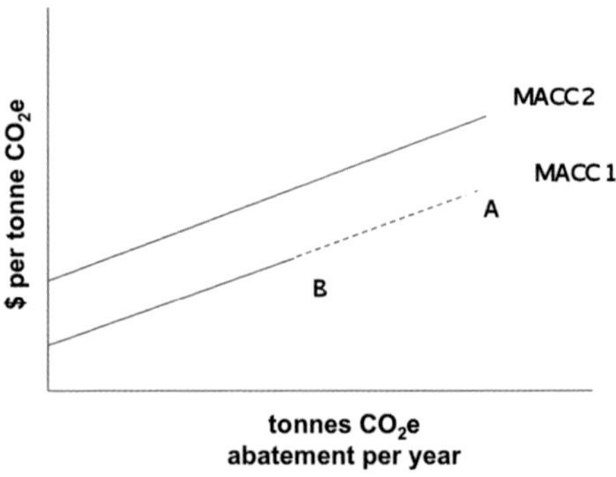

Figure 7.11 Effects of constraints on mitigation potential and costs illustrated using marginal-abatement cost curves (MACCs)

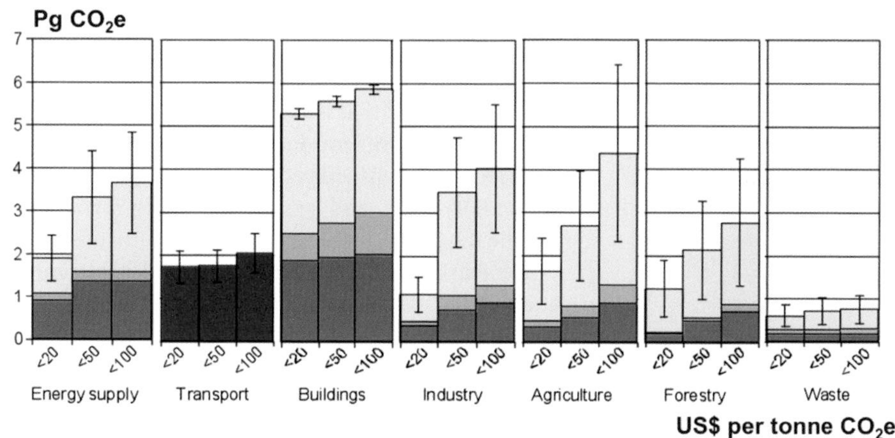

Figure 7.12 Global economic mitigation potentials for all sectors at three carbon prices (< 20, < 50 and < 100 US dollar per tonne CO_2e). Darkest blue (for transport only) denotes world totals where disaggregation by region was not possible. Light blue is for developing countries (non-OECD), mid blue is for countries with economy in transition (EIT) and dark blue is for developed countries (OECD). Reproduced with permission from Metz *et al.*, 2007.

As yet, Earth's biophysical limits are not seen as a fixed constraint, but there is growing awareness of the increasing pressures placed on the biosphere by human activities. One indicator to measure the scale of human impact on the biosphere is the 'human appropriation of net primary production' or HANPP (see e.g. Haberl *et al.*, 2007). According to calculations, humankind already appropriates a substantial proportion of global net primary production, over half of which is dedicated to crop production. Rojstaczer *et al.* (2001) considered the anthropogenic impact in terms of human appropriation of photosynthesis products and estimated the range to be between 10 and 55%. Foley *et al.* (2005) estimated global HANPP to be 30–50%, while Imhoff (2004) suggested that the HANPP for Europe may be as high as 72%. The rapidly expanding demands for food, timber and bioenergy imply we will need to increase the amount of production we use both through yield increases, and through the conversion of natural lands to productive cropland, pastureland or forestry.

7.8 Opportunities and priorities for action

This chapter has highlighted the mitigation potential in land-based sectors, with much of it available at medium, low or even negative costs (Metz *et al.*, 2007; McKinsey and Company, 2009).

The relative advantages of different approaches to land-based mitigation depend on the timeframe considered. For example, if a new forest is planted it will quickly take up carbon. If it replaced non-forest land, it will end up with a higher carbon density than the original vegetation type and will have removed carbon from the atmosphere. This carbon uptake will stop once the forest reaches maturity, and then this carbon store will need to be protected so that it is 'permanent'. It is important to note that if the plantation forest replaces natural forest, it is likely to have lower carbon density and will therefore represent a net loss of carbon to the atmosphere. The same is true if it is planted on drained peatlands, as the soil carbon is lost rapidly as peatlands dry out.

If the new planting is for bioenergy or to produce timber for long-lived products, then over the long term the carbon benefit can be greater because the same area of land can continue to contribute to offsetting fossil fuels. However, in the short term, carbon will be lost during conversion of the land and it may take several harvest cycles until the system shows net carbon benefit (Marland *et al.*, 1997; Figure 7.13).

Mitigation action is needed in the short term as the later or higher global emissions peak, the more technologically difficult and more expensive it will be to get back on track towards climate stabilization (Meinshausen *et al.*, 2009). Long-term mitigation options and permanent sinks are also needed to sustain low emission levels into the future. Different options are possible in the short term compared to the longer term, where technological advances may be realized. It can take decades to change over infrastructure such as power stations, or to realize growth of mature trees. It can also take many years to put the right policies and laws in place. Therefore rapid implementation of short-term options and advanced planning for longer-

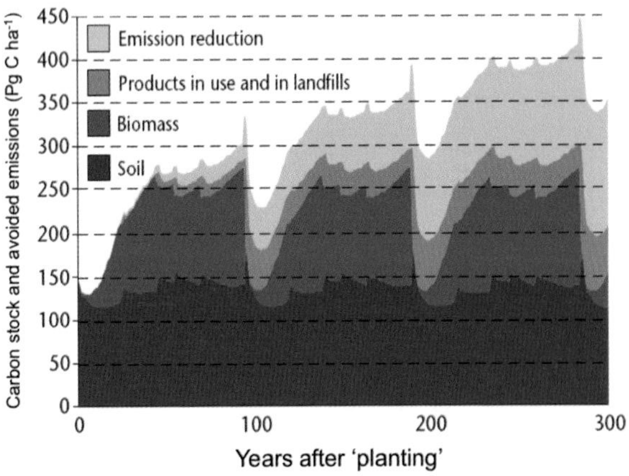

Figure 7.13 'Accumulated' benefit from forest management for production of wood products, including credit for the emissions reduction from avoided fossil-resource consumption. Reproduced with permission. Image © Forestry Commission, produced using model output from G. J. Nabuurs.

term options are both critical, and need to be addressed simultaneously.

Options that are relatively cheap and easy to implement in the short term tend to be close to current practice, such as improved cropland and grazing land management, energy from wastes, energy from conventional agricultural crops with co-production of animal feed, avoided deforestation, afforestation and peatland protection, rewetting and re-vegetation (Smith, 2012). Longer-term options are those requiring more research and development or a change of infrastructure, such as second-generation bioenergy from woody crops. Some options will depend on the choices that people make such as fuel switching and dietary preferences.

The most important and immediate actions we can take in terms of land-based mitigation are to reduce deforestation and protect peatlands. This action protects a massive store of carbon along with biodiversity and other ecosystem services. On high-productivity land, after accounting for provision of food, bioenergy crops are probably the most effective land-based mitigation option, particularly if they are replacing cropland or pasture. Increasing agricultural yields and reducing (or intensifying) meat production could increase the availability of high-productivity land for bioenergy, especially where there is co-production of bioenergy, food and animal feed. However, establishing bioenergy crops is not a suitable option on lands with currently high carbon stocks in vegetation and soils, such as tropical forests and peatlands, due to the large net losses of carbon that would occur. On low-productivity land, afforestation is a cost-effective and easy option for quick uptake of CO_2 (again providing this is not on land that is currently forest or peatland). In the longer term, production of woody crops for advanced bioenergy will become more feasible on low-productivity land, and will be especially favourable with the range of possible co-products.

The feasibility and impacts of all land-based mitigation options strongly depend on their interactions with other uses of land and natural ecosystems. There is a risk of mitigation activities adversely affecting provision of food or other ecosystems services. However, there are also many under-exploited opportunities for co-production of food, fuel and timber, and for win–win options that have environmental or social benefits. The concept of 'available land' is a misnomer: all land is providing some kind of function for ecosystems and service for humanity. However, the 'food-versus-fuel' dichotomy is simplistic and counterproductive as we have shown above. There is a limited land resource base, and we have many competing and increasing demands on the land. The real issue is, rather, how we assure the provision of multiple services in the most efficient, sustainable and equitable way.

The lack of consistent and more comprehensive accounting and reporting of land-based emissions, either within or across national borders and sectors (LULUCF, energy) can lead to perverse incentives for mitigation that are sub-optimally effective or have negative outcomes (e.g. when tropical forests or peatlands are replaced with palm oil for bioenergy). Knowing the full life-cycle greenhouse-gas emissions of all land-based and other mitigation options on different time-frames is necessary to create a level playing field for informed decision-making. There has been a significant improvement in scientific knowledge and tools to carry out full life-cycle analyses. Yet there is still a long way to go in implementing such accounting, leading to many misconceptions among decision-makers and the public. Bioenergy only accounts for 1 to 2% of total global crop production, while most land-based emissions are due to land conversion for food production, and the growing and processing of foods. This highlights the need for all types of land use to be assessed and accounted for in the same way. Land accounting in an integrated way will lead to better understanding of the land and its interactions with the Earth system, and therefore better decision-making regarding what we want to get out of the land and how we do it.

References

Achard, F., DeFries, R., Eva, H. *et al.* (2007). Pan-tropical monitoring of deforestation. *Environmental Research Letters*, **2**, 045022.

Allen, M. R., Frame, D. J., Huntigford, C. *et al.* (2009). Warming caused by cumulative carbon emissions towards the trillionth tonne, *Nature*, **458**, 1163–1166.

Amaral, W. A. N., Marinho, J. P., Tarasantchi, R., Beber, A. and Guiliani, E. (2008). Environmental sustainability of sugarcane ethanol in Brazil. In *Sugarcane Ethanol: Contributions to Climate Change Mitigation and the Environment*, eds. P. Zuurbier and J. van de Vooren. Wageningen: Wageningen Academic Publishers, pp. 113–138.

Angelsen, A., Brown, S., Loisel, C. *et al.* (2009). *Reducing Emissions from Deforestation and Forest Degradation (REDD): An Options Assessment Report*. Washington, DC: Meridian Institute.

Arora, V. K. and Boer, G. J. (2010). Uncertainties in the 20th century carbon budget associated with land use change. *Global Change Biology*, **16**, 3327–3348.

Baccini, A., Goetz, S. J., Walker, W. S. *et al.* (2012). Estimated carbon dioxide emissions from tropical deforestation improved by carbon-density maps. *Nature Climate Change*, **2**, 182–185.

Bahor, B., Van Brunt, M., Stovall, J. and Blue, K. (2009). Integrated waste management as a climate change stabilization wedge. *Waste Management and Research*, **27**, 839–849.

Bennett, M. T. (2008). China's Sloping Land Conversion Program: institutional innovation or business as usual. *Ecological Economics*, **65**, 699–711.

Beurskens, L. W. M. and Hekkenberg, M. (2011). *Renewable Energy Projections as Published in the National Renewable Energy Action Plans of the European Member States*. Energy Research Centre of the Netherlands. ECN-E-10-069. See: www.ecn.nl/nreap.

Billett, M. F., Palmer, S. M., Hope, D. *et al.* (2004). Linking land–atmosphere–stream carbon fluxes in a lowland peatland system. *Global Biogeochemical Cycles*, **18**, GB1024.

Black, M. J., Whittaker, C., Housseini, S. A. *et al.* (2010). Life cycle assessment and sustainability methodologies for assessing industrial crops, processes and end products. *Journal of Industrial Crops and Products*, **34**, 1332–1339.

Blodau, C., Basiliko, N. and Moore, T. R. (2004). Carbon turnover in peatlands mesocosms exposed to different water table levels. *Biogeochemistry*, **67**, 331–351.

Böttcher, H. (2008). Forest management for climate change mitigation – modeling of forestry options, their impact on the regional carbon balance and implications for a future climate protocol. Doctoral Thesis, Fakultät für Forst- und Umweltwissenschaften, Albert-Ludwigs-Universität, Freiburg.

Böttcher, H. and Lindner, M. (2010). Managing forest plantations for carbon sequestration today and in the future. In *Ecosystem Goods and Services from Plantation Forests*, eds. J. Bauhus, P. van der Meer and M. Kanninen. Oxford: Earthscan, pp. 164–170.

Böttcher, H., Kurz, W. A., and Freibauer, A. (2008). Accounting of forest carbon sinks and sources under a future climate protocol/factoring out past disturbance and management effects on age-class structure. *Environmental Science & Policy*, **11**, 669–686.

Böttcher, H., Eisbrenner, K., Fritz, S. *et al.* (2009). An assessment of monitoring requirements and costs of 'Reduced Emissions from Deforestation and Degradation'. *Carbon Balance and Management*, **4**, 7.

Bruinsma, J. (2011). The resource outlook to 2050: by how much do land, water and crop yields need to increase by 2050? Presented at the Expert Meeting on How to Feed the World in 2050, Rome, 24–26 June 2009. Food and Agriculture Organization of the United Nations (FAO), Economic and Social Development Department. See: ftp://ftp.fao.org/docrep/fao/012/ak971e/ak971e00.pdf.

BSI – British Standards Institution (2011). Specification for the assessment of the life cycle greenhouse gas emissions of goods and services. Publically Available Specification PAS 2050:2011. London: BSI. See: www.bsigroup.com/Standards-and-Publications/How-we-can-help-you/Professional-Standards-Service/PAS-2050.

Cai, X., Zhang, X. and Wang, D. (2011). Land availability for biofuel production. *Environmental Science and Technology*, **45**, 334–339.

Campbell, J. E., Lobell, D. B., Genova, R. C. and Field, C. B. (2008). The global potential of bioenergy on abandoned agriculture lands. *Environmental Science and Technology*, **42**, 5791–5794.

Cannell, M. G. R. (2003). Carbon sequestration and biomass energy offset: theoretical, potential and achievable capacities globally, in Europe and the UK. *Biomass and Bioenergy*, **24**, 97–116.

Christensen, T. H., Simion, F., Tonini, D. and Møller, J. (2009). Global warming factors modelled for 40 generic municipal waste management scenarios. *Waste Management and Research*, **27**, 871–884.

Chum, H., Faaij, A., Moreira, J. *et al.* (2011). Bioenergy. In *IPCC Special Report on Renewable Energy Sources and Climate Change Mitigation*, eds. O. Edenhofer, R. Pichs-Madruga, Y. Sokona *et al.* Cambridge: Cambridge University Press, ch. 2.

Ciais, P., Schelhaas, M. J., Zaehle, S. *et al.* (2008). Carbon accumulation in European forests. *Nature Geoscience*, **1**, 425–429.

Clark, J. H., Budarin, V., Deswarte, F. E. I. *et.al.* (2006). Green chemistry and the biorefinery: a partnership for a sustainable future. *Green Chemistry*, **10**, 853–860.

Dale, B., Bals, B. D., Kim, S. and Eranki, P. (2010). Biofuels done right: land efficient animal feeds enable large environmental and energy benefits. *Environmental Science and Technology*, **44**, 8385–8389.

DeFries, R., Field, C. B., Fung, I., Collatz, G. and Bounoua, L. (1999). Combining satellite data and biogeochemical models to estimate global effects of human-induced land cover change on carbon emissions and primary productivity. *Global Biogeochemical Cycles*, **13**, 803–815.

DeFries, R. S., Houghton, R. A., Hansen, M. C. *et al.* (2002). Carbon emissions from tropical deforestation and regrowth based on satellite observations for the 1980s and 1990s. *Proceedings of the National Academy of Sciences*, **99**, 14256–14261.

DeFries, R. S., Archard, F., Brown, S. *et al.* (2006). Reducing greenhouse gas emissions from deforestation in developing countries: considerations for monitoring and measuring. GTOS Report 46, GOFC-GOLD Report 26. Rome: Global Terrestrial Observing System. See: http://nofc.cfs.nrcan.gc.ca/gofc-gold/Report%20Series/GOLD_26.pdf.

Denman, K. L., Brasseur, G., Chidthaisong, A. *et al.* (2007). Couplings between changes in the climate system and biogeochemistry. In *Climate Change 2007: The Physical Science Basis. Contribution of Working Group I to the Fourth Assessment Report of the Intergovernmental Panel on Climate Change*, eds. S. Solomon, D. Qin, M. Manning *et al.* Cambridge: Cambridge University Press.

Diaz-Chavez, R., Mutimba, S., Watson, H., Rodriguez-Sanchez, S. and Nguer, M. (2010). Mapping food and bioenergy in Africa. A report prepared on behalf of FARA. Ghana, Forum for Agricultural Research in Africa (FARA).

Don, A., Rebmann, C., Kolle, O., Scherer-Lorenzen, M. and Schulze, E. D. (2009). Impact of afforestation-associated management changes on the carbon balance of grassland. *Global Change Biology*, **15**, 1990–2002.

Dornburg, V., van Vuuren, D., van de Ven, G. *et al.* (2010). Bioenergy revisited: key factors in global potentials of bioenergy. *Energy and Environmental Science*, **3**, 258–267.

Edenhofer, O., Pichs-Madruga, R., Sokona, Y. *et al.*, eds. (2011). *IPCC Special Report on Renewable Energy Sources and Climate Change Mitigation.* Cambridge: Cambridge University Press.

EEA – European Environment Agency (2007). Land-use scenarios for Europe: qualitative and quantitative analysis on a European scale. EEA Technical report No 9/2007. Copenhagen: European Environment Agency. www.eea.europa.eu/publications/technical_report_2007_9/at_download/file.

EEA Scientific Committee (2011). Opinion of the EEA Scientific Committee on greenhouse gas accounting in relation to bioenergy. Copenhagen: European Environment Agency. See: www.eea.europa.eu/about-us/governance/scientific-committee/sc-opinions/opinions-on-scientific-issues/sc-opinion-on-greenhouse-gas.

European Commission (2009). Directive 2009/28/EC of the European Parliament and of the Council of 23 April 2009 on the promotion of the use of energy from renewable sources. See: http://eur-lex.europa.eu/LexUriServ/LexUriServ.do?uri=OJ:L:2009:140:0016:0062:en:PDF

Eurostat (2011). *Energy, Transport and Environment Indicators, 2010 edition*. Luxembourg: Publications Office of the European Union. See: http://epp.eurostat.ec.europa.eu/cache/ITY_OFFPUB/KS-DK-10-001/EN/KS-DK-10-001-EN.PDF.

FAO (2006a). *Livestock's Long Shadow*. Rome: Food and Agriculture Organization of the United Nations.

FAO (2006b). *World Agriculture: Towards 2030/2050. Prospects for Food, Nutrition, Agriculture and Major Commodity Groups*. Rome: Food and Agriculture Organization of the United Nations, Global Perspectives Studies Unit.

FAO (2010a). *Global Forest Resources Assessment 2010*. Rome: Food and Agriculture Organization of the United Nations.

FAO (2010b). *Making Integrated Food-Energy Systems Work for People and Climate. An Overview.* Rome: Food and Agriculture Organization of the United Nations.

FAO (2011). *Bioenergy and Food Security. The BEFS Analytical Framework. The Bioenergy and Food Security Project.* Rome: Food and Agriculture Organization of the United Nations.

Fargione, J. (2008) Land clearing and the biofuel carbon debt. *Science*, **319**, 1235.

Fischedick, M., Schaeffer, R., Adedoyin, A. *et al.* (2011). Mitigation potential and costs. In *IPCC Special Report on Renewable Energy Sources and Climate Change Mitigation*, eds. O. Edenhofer, R. Pichs-Madruga, Y. Sokona *et al.* Cambridge: Cambridge University Press, ch. 10.

Foley, J., DeFries, R. S., Asner G. P. *et al.* (2005). Global consequences of land use. *Science*, **309**, 570–574.

Frankfurt Zoological Society (2007). Bale Mountains National Park, Ethiopia: General Management Plan 2007–2017. Frankfurt Zoological Society (Bale Mountains Conservation Project) and Institute of Biodiversity Conservation. See: www.zgf.de/download/166/BMNP_GMP_2007.pdf.

Friedlingstein, P., Houghton, R. A., Marland, G. *et al.* (2010). Update on CO_2 emissions. *Nature Geoscience*, **3**, 811–812.

Friends of the Earth (2008). Losing Ground: The human rights impacts of oil palm plantation expansion in Indonesia. Friends of the Earth, LifeMosaic and Sawit Watch. See: www.foe.co.uk/resource/reports/losingground.pdf.

G8 (2008). Hokkaido Toyako Summit Leaders Declaration. See: www.mofa.go.jp/policy/economy/summit/2008/doc/.

Gathorne-Hardy, A. (2011). The role of biochar in English agriculture – agronomy, biodiversity, economics and climate change. PhD Thesis, Imperial College London.

Gibbs, H., Johnston, M., Foley, J. A. *et al.* (2008). Carbon payback times for crop-based biofuel expansion in the tropics: the effects of changing yield and technology. *Environmental Research Letters*, **3**, 034001.

Goetz, S. J., Baccini, A., Laporte, N.T. *et al.* (2009). Mapping and monitoring carbon stocks with satellite observations: a comparison of methods. *Carbon Balance and Management*, **4**, 2.

GOFC-GOLD (2008). Reducing greenhouse gas emissions from deforestation and degradation in developing countries: a sourcebook of methods and procedures for monitoring, measuring and reporting. GOFC-GOLD report version COP13-2. Calgary: GOFC-GOLD Project Office, Natural Resources Canada.

Gorham, E. (1991). Northern peatlands: role in the carbon budget and probable responses to global warming. *Ecological Applications*, **1**, 182–195.

Grassi, G., Monni, S., Federici, S., Archard, F. and Mollicone, D. (2008). Applying the conservativeness principle to REDD to deal with the uncertainties of the estimates. *Environmental Research Letters*, **3**, 035005.

Griscom, B., Shoch, D., Stanley, B., Cortez, R. and Virgilio, N. (2009). Sensitivity of amounts and distribution of tropical forest carbon credits depending on baseline rules. *Environmental Science and Policy*, **12**, 897–911.

Haberl, H., Erb, K. H., Krausmann, F. *et al.* (2007). Quantifying and mapping the human appropriation of net primary production in Earth's terrestrial ecosystems. *Proceedings of the National Academy of Sciences*, **104**, 12942–12947.

Hamelinck, C., Koop, K., Croezen, H. *et al.* (2008). Technical specification: greenhouse gas calculator for biofuel. Report GAVE-08-01. Utrecht: SenterNovem, Ecofys.

Harrison, R. G., Jones, C. D. and Hughes, J. K. (2008). Competing roles of rising CO_2 and climate change in the contemporary European carbon balance. *Biogeosciences*, **5**, 1–10.

Hashimoto, S. (2008). Different accounting approaches to harvested wood products in national greenhouse gas inventories: their incentives to achievement of major policy goals. *Environmental Science and Policy*, **11**, 756–771.

Herold, M. and Johns, T. (2007). Linking requirements with capabilities for deforestation monitoring in the context of the UNFCCC-REDD process. *Environmental Research Letters*, **2**, doi:10.1088/1748-9326/2/4/045025.

Hoare, A., Saunders, J., Nussbaum, R. and Legge, T. (2008). Estimating the cost of building capacity in rainforest nations to allow them to participate in a global REDD mechanism. Report produced for the Eliasch Review by Chatham House and ProForest with input from the Overseas Development Institute and EcoSecurities.

Hoogwijk, M., Faaij, A., Eickhout, B., de Vries, B. and Turkenburg, W. (2005). Potential of biomass energy out to 2100, for four IPCC SRES land-use scenarios. *Biomass and Bioenergy*, **29**, 225–257.

Houghton, J. T., Ding, Y., Griggs, D. J., Noguer, M. and van der Linden, P. J., eds. (2001). *Climate Change 2001: The Scientific Basis. Contribution of Working Group I to the Third Assessment Report of the Intergovernmental Panel on Climate Change.* Cambridge: Cambridge University Press.

Houghton, R. A. (2003). Revised estimates of the annual net flux of carbon to the atmosphere from changes in land use and land management 1850–2000. *Tellus B*, **55**, 378–390.

Houghton, R. A., van der Werf, G. R., DeFries, R. S. *et al.* (2012). Carbon emissions from land use and land-cover change, *Biogeoscences Discussions*, **9**, 835–878.

House, J. I. (2009). Emissions from LULUCF (land use, land use change and forestry): implications of uncertainty for policy targets. Workstream 2, Report 2 of the AVOID Programme [AV/WS2/D1/R02].

House, J. I. (2012). Reconciling LULUCF accounting with modelled mitigation targets. Report to UK Department of Energy and Climate Change. AVOID AV/WS2/D1/R29 See: www.metoffice.gov.uk/avoid/files/resources-researchers/AVOID_WS2_D1_29.pdf.

House, J. I., Prentice, I. C. and Le Quéré, C. (2002). Maximum impacts of future reforestation or deforestation on atmospheric CO_2. *Global Change Biology*, **8**, 1047–1052.

House, J. I., Brovkin, V., Betts, R. *et al.* (2006). Climate and air quality. In *The Millennium Ecosystem Assessment*, eds. R. Hassan, R. J. Scholes and N. Ash. Washington, DC: Island Press, pp. 355–390.

Howarth, R. W., Bringezu, S., Martinelli, L. A. *et al.* (2009). Introduction: biofuels and the environment in the 21st century. In *Biofuels: Environmental Consequences and Interactions with Changing Land Use*, eds. R.W.

Howarth and S. Bringezu. Proceedings of the Scientific Committee on Problems of the Environment (SCOPE) International Biofuels Project Rapid Assessment, 22–25 September 2008. Gummersbach and Ithaca: Cornell University, pp. 15–36.

Hurtt, G. C., Frolking, S., Fearon, M. G. *et al.* (2006). The underpinnings of land-use history: three centuries of global gridded land-use transitions, wood-harvest activity, and resulting secondary lands, *Global Change Biology*, **12**, 1208–1229.

Hurtt, G. C., Chini, L. P., Frolking, S. *et al.* (2011). Harmonization of land-use scenarios for the period 1500–2100: 600 years of global gridded annual land-use transitions, wood harvest, and resulting secondary lands. *Climatic Change*, **109**, 117–161.

IEA (2008). *World Energy Outlook, 2008.* Paris: International Energy Agency.

IEA (2009a). *World Energy Outlook, 2009.* Paris: International Energy Agency.

IEA (2009b). *Bioenergy – A sustainable and Reliable Energy Source. A Review of Status and Prospects.* Paris: International Energy Agency.

IEA (2010). *World Energy Outlook, 2010.* Paris: International Energy Agency.

IEA (2011). *Bioenergy, Land Use Change and Climate Change Mitigation – Background Technical Report.* IEA Bioenergy:ExCo:2011:04. Paris: International Energy Agency.

IFPRI (2010). *Global Trade and Environmental Impact Study of the EU Biofuels Mandate.* Washington, DC: International Food Policy Research Institute.

IIASA (2011). Options. IIASA GAEZ MODEL. Political will is the only way to end hunger. Vienna: International Institute for Applied Systems Analysis.

Imhoff, M. L. (2004). Global patterns in human consumption of net primary production. *Nature*, **429**, 870–873.

IPCC (2003). *Good Practice Guidance for Land Use, Land-Use Change and Forestry.* Hayama, Japan: IPCC/OECD/IEA/IGES.

IPCC (2007a). *Climate Change 2007: Synthesis Report*, eds. R. K. Pachauri and A. Reisinger. Geneva: IPCC.

IPCC (2007b). Summary for Policymakers. In *Climate Change 2007: Mitigation. Contribution of Working Group III to the Fourth Assessment Report of the Intergovernmental Panel on Climate Change*, eds. B. Metz, O. R. Davidson, P. R. Bosch, R. Dave and L. A. Meyer. Cambridge: Cambridge University Press, pp. 1–23.

Jandl, R., Lindner, M., Vesterdal, L. *et al.* (2007). How strongly can forest management influence soil carbon sequestration? *Geoderma*, **137**, 253–268.

Joosten, H. (2010). The global peatland CO_2 picture: peatland status and drainage related emissions in all countries of the world. Wetlands International. See: www.wetlands.org/WatchRead/tabid/56/mod/1570/articleType/ArticleView/articleId/2418/The-Global-Peatland-CO2-Picture.aspx.

Keith, H., Mackey, B. G. and Lindenmayer, D. B. (2009). Re-evaluation of forest biomass carbon stocks and lessons from the world's most carbon-dense forests. *Proceedings of the National Academy of Sciences*, **106**, 11635–11640.

Kim, S. and Dale, B. E. (2011). Indirect land use change for biofuels: testing predictions and improving analytical methodologies. *Biomass and Bioenergy*, **35**, 3235–3240.

Kohlmaier, G., Kohlmaier, L., Fries, E. and Jaeschke, W. (2007). Application of the stock change and the production approach to Harvested Wood Products in the EU-15 countries: a comparative analysis. *European Journal of Forest Research*, **126**, 209–223.

Lähteenoja, O., Ruokolainen, K., Schulman, L. and Oinonen, M. (2009). Amazonian peatlands: an ignored C sink and potential source. *Global Change Biology*, **15**, 2311–2320.

Le Quéré, C., Raupach, M. R., Canadell, J. G. *et al.* (2009). Trends in the sources and sinks of carbon dioxide. *Nature Geoscience*, **2**, 831–836.

Le Treut, H., Somerville, R., Cubasch, U., *et al.* (2007). Historical overview of climate change. In *Climate Change 2007: The Physical Science Basis. Contribution of Working Group I to the Fourth Assessment Report of the Intergovernmental Panel on Climate Change*, eds. S. Solomon, D. Qin, M. Manning *et al.* Cambridge: Cambridge University Press.

Lehmann, J. (2006). Bio-char sequestration in terrestrial ecosystems – a review. *Mitigation and Adaptation Strategies for Global Change*, **11**, 395–419.

Lewis, S. L., Lopez-Gonzalez, G., Sonké, B. *et al.* (2009). Increasing carbon storage in intact African tropical forests. *Nature*, **457**, 1003–1006.

Li, J., Feldman, M. W., Li, S. and Daily, G. (2011). Rural household income and inequality under the Sloping Land Conversion Program in western China. *Proceedings of the National Academy of Sciences*, **108**, 7721–7726

Luyssaert, S., Schulze, E. D., Borner, A. *et al.* (2008). Old-growth forests as global carbon sinks. *Nature*, **455**, 213–215.

Lynd, L. R. and Woods, J. (2011). Perspective: a new hope for Africa. *Nature*, **474**, S20–S21.

Macedo, I. C. (2010). Biofuels and sustainability: Brazilian developments, international expansion. Contribution to 'Sustainable biofuels: recent developments, international opportunities', 16 November 2010, a Chatham House seminar in collaboration with the Embassy of Brazil and the Imperial College Seminar. London: Chatham House.

Magnani, F., Mencuccini, M., Borghetti, M. *et al.* (2007). The human footprint in the carbon cycle of temperate and boreal forests. *Nature*, **447**, 849–851.

Magnusson, M., Fransson, J. E. S. and Holmgren, J. (2007). Effects on estimation accuracy of forest variables using different pulse density of laser data. *Forest Science*, **53**, 619–626.

Maltby, E. and Immirzi, P. (1993). Carbon dynamics in peatlands and other wetland soils: regional and global perspectives. *Chemosphere*, **27**, 999–1023.

Manfredi, S., Tonini, D., Christensen, T. H. and Scharff, H. (2009). Landfilling of waste: accounting of greenhouse gases and global warming contributions. *Waste Management and Research*, **27**, 825–836.

Marland, G., Schlamadinger, B. and Leiby, P. (1997). Forest/biomass based mitigation strategies: does the timing of carbon reductions matter? *Critical Reviews in Environmental Science and Technology*, **27**, 213–226.

Marland, G., Boden, T. A. and Andres, R. J. (2008). *Global, Regional, and National Fossil Fuel CO_2 Emissions.* Oak Ridge, TN: Carbon Dioxide Information Analysis Center, Oak Ridge National Laboratory, US Department of Energy.

Marti, S. (2008) Losing ground: the human rights impacts of oil palm plantation expansion in Indonesia. UK and Indonesia: Friends of the Earth, LifeMosaic and Sawit Watch. See: www.foe.co.uk/resource/reports/losingground.pdf.

McKendry, P. (2002). Energy production from biomass (Part 2): conversion technologies. *Bioresource Technology*, **83**, 47–54.

McKinsey and Company (2009). Pathways to a low-carbon economy: Version 2 of the Global Greenhouse Gas Abatement Cost Curve. See: https://solutions.mckinsey.com/ClimateDesk/default.aspx.

McRoberts, R. E. and Tompp. E. O. (2007). Remote sensing support for national forest inventories. *Remote Sensing of Environment*, **110**, 412–419.

Meinshausen, M., Meinshausen, N., Hare, W. *et al.* (2009). Greenhouse-gas emission targets for limiting global warming to 2 degrees C. *Nature*, **458**, 1158–1162.

Metz, B., Davidson, O., de Coninck, H.C., Loos, M. and Meyer, L.A., eds. (2005). *IPCC Special Report on Carbon Dioxide Capture and Storage*. Cambridge: Cambridge University Press.

Metz, B., Davidson, O. R., Bosch, P. R., Dave, R., Meyer, L. A., eds. (2007). *Climate Change 2007: Mitigation. Contribution of Working Group III to the Fourth Assessment Report of the Intergovernmental Panel on Climate Change*. Cambridge: Cambridge University Press.

Millennium Ecosystem Assessment (2005). *The Millennium Ecosystem Assessment*, eds. R. Hassan, R. J. Scholes and N. Ash. Washington, DC: Island Press.

Mitchell, D. (2008). A note on rising food prices. Policy Research Working Paper WPS 4682. Washington, DC: World Bank.

Mitchell, D. (2011). Biofuels in Africa: opportunities, prospects, and challenges. Report Number 58438. Washington, DC: World Bank.

Moberg, J. and Persson, M. (2011). The Chinese Grain for Green Program – assessment of the land reform's carbon mitigation potential. Master of Science Thesis, Sustainable Energy Systems, Chalmers University, Sweden.

Moraes, M. A. F. D. (2011). Socio-economic indicators and determinants of the income of workers in sugar cane plantations and in the sugar and ethanol industries in the north, north-east and centre-south regions of Brazil. In *Energy, Bio Fuels and Development: Comparing Brazil and the United States*, eds. E. Amann, W. Baer and D. Coes. Abingdon: Routledge.

Morton, D. C., DeFries, R. S., Shimabukuro, Y. E. *et al.* (2005). Rapid assessment of annual deforestation in the Brazilian Amazon using MODIS data. *Earth Interactions* **9**, 1–22.

Moss, R. H., Edmonds, J. A., Hibbard, K. A. *et al.* (2010). The next generation of scenarios for climate change research and assessment. *Nature*, **463**, 747–756.

Murphy, R. J., Woods, J., Black, M. J. and McManus, M. (2011). Global developments in the competition for land from bio-fuels. *Food Policy*, **36**, S52–S61.

Nabuurs, G. J., Andrasko, K., Benitez-Ponce, P. *et al.* (2007). Forestry. In *Climate Change 2007: Mitigation. Contribution of Working Group III to the Fourth Assessment Report of the Intergovernmental Panel on Climate Change*, eds. B. Metz, O. R. Davidson, P. R. Bosch, R. Dave and L.A. Meyer. Cambridge: Cambridge University Press, pp. 541–584.

Nakicenovic, N. and Swart, R. (2000). *IPCC Special Report on Emissions Scenarios*. Cambridge: Cambridge University Press.

Nassar, A. M., Harfuch, L., Moreira, M. M., Bachion, L. and Antoniazzi, L. B. (2011). Simulating land use and agriculture expansion in Brazil: food, energy, agro-industrial and environmental impacts. São Paulo: Instituição Sede, Instituto de Estudos do Comércio e Negociações Internacionais (ICONE).

NNFCC – National Non-Food Crops Centre (2009). Biorefineries. www.nnfcc.co.uk/metadot/index.pl?id=2185;isa=Category;op=show.

Obersteiner, M., Böttcher, H. and Yamagata, Y. (2010). Terrestrial ecosystem management for climate change mitigation. *Current Opinion in Environmental Sustainability*, **2**, 271–276.

Oladosu, G., Kline, K., Uria-Martinez, R. and Eaton, L. (2011). Sources of corn for ethanol production in the United States: a decomposition analysis of the empirical data. *Biofuels, Bioproducts and Biorefining*, **5**, 640–653.

Page, S. E., Siegert, F., Rieley, J. O. *et al.* (2002). The amount of carbon released from peat and forest fires in Indonesia during 1997. *Nature*, **420**, 61–65.

Pan, Y., Birdsey, R. A., Fang, J. *et al.* (2011). A large and persistent carbon sink in the world's forests. *Science*, **333**, 988–993.

Patenaude, G., Hill, R. A., Milne, R. *et al.* (2004). Quantifying forest above ground carbon content using LiDAR remote sensing. *Remote Sensing of Environment*, **93**, 368–380.

Perlack, R. D., Wright, L. L., Turhollow, A. F. *et al.* (2005). Biomass as a feedstock for a bioenergy and bioproducts industry: the technical feasibility of a billion-ton annual supply. Tennessee: US Department of Energy. See: www.osti.gov/bridge.

Phillips, O. L., Malhi, Y., Higuchi, N. *et al.* (1998). Changes in the carbon balance of tropical forests: evidence from long-term plots. *Science*, **282**, 439–442.

Piao, S., Ciais, P., Friedlingstein, P. *et al.* (2009). Spatiotemporal patterns of terrestrial carbon cycle during the 20th century. *Global Biogeochemical Cycles*, **23**, GB4026.

Pongratz, J., Reick, C. H., Raddatz, T. and Claussen, M. (2009). Effects of anthropogenic land cover change on the carbon cycle of the last millennium. *Global Biogeochemical Cycles*, **23**, GB4001.

Porter, J., Costanza, R., Sandhu, H., Sigsaard, L. and Wratten, S. (2009). The value of producing food, energy, and ecosystem services within an agro-ecosystem. *Ambio*, **38**, 186–193.

Prentice, I. C., Farquhar, G. D., Fasham, M. J. R. *et al.* (2001). The carbon cycle and atmospheric carbon dioxide. In *Climate Change 2001:The Scientific Basis. Contribution of Working Group I to the Third Assessment Report of the Intergovernmental Panel on Climate Change*, eds. J. T. Houghton, Y. Ding, D. J. Griggs, M. Noguer and P. J. van der Linden. Cambridge: Cambridge University Press, pp. 183–237.

Ragauskas, A. J., Williams, C. K., Davison, B. H. *et al.* (2006). The path forward for biofuels and biomaterials. *Science*, **311**, 484–489.

REN21 (2011). Renewables 2011 – Global Status Report. Paris: Renewable Energy Policy Network for the 21st Century. See: www.ren21.net/REN21Activities/Publications/GlobalStatusReport/GSR2011/tabid/56142/Default.aspx.

RFA (2008). The Gallagher review of the indirect effects of biofuels production. London: UK Renewable Fuels Agency. See: www.dft.gov.uk/topics/sustainable/biofuels/.

RFA (2010). Carbon and sustainability reporting within the Renewable Transport Fuel Obligation – Technical guidance part one (Version 3.1). London: UK Renewable Fuels Agency. See: http://assets.dft.gov.uk/publications/carbon-and-sustainability-technical-guidance/technical-guidance-part1.pdf.

Rice, T. (2010). Meals per gallon: the impact of industrial biofuels on people and global hunger. London: ActionAid. See: www.actionaid.org.uk/doc_lib/meals_per_gallon_final.pdf.

Rojstaczer, S., Sterling, S. M. and Moore, N. J. (2001). Human appropriation of photosynthesis products. *Science*, **294**, 2549–2552.

Rosillo-Calle, F. and Johnson F. X., eds. (2010). *Food versus Fuel. An Informed Introduction to Biofuels*. London: Zed Books.

Russ, P., Wiesenthal, T., van Regemorter, D. and Ciscar, J. C. (2009). Economic assessment of post-2012 global climate policies – analysis of greenhouse gas emission reduction scenarios with the POLES and GEM-E3 models. Ispra: European Commission Joint Research Centre Institute for Prospective Technological Studies. See: http://ftp.jrc.es/EURdoc/JRC50307.pdf

Sathaye, J. A., Makundi, W., Dale, L., Chan, P. and Andasko, K. (2006). GHG mitigation potential, costs and benefits in global forests: a dynamic partial equilibrium approach. *Energy Journal*, **SI 3**, 127–172.

Sawyer, D. (2008). Climate change, biofuels and eco-social impacts in the Brazilian Amazon and Cerrado. *Philosophical Transactions of the Royal Society B*, **363**, 1747–1752.

Searchinger, T., Heimlich, R., Houghton, R. A. *et al.* (2008). Use of US croplands for biofuels increases greenhouse gases through emissions from land-use change. *Science*, **319**, 1238–1240.

Shevliakova, E., Pacala, S. W., Malyshev, S. *et al.* (2009). Carbon cycling under 300 years of land use change: importance of the secondary vegetation sink. *Global Biogeochemical Cycles*, **23**, GB2022.

Simard, M., Pinto, N., Fisher, J. B. and Baccini, A. (2011). Mapping forest canopy height globally with spaceborne lidar. *Journal of Geophysical Research*, **116**, G04021.

Smeets, E. M. W. (2008). Possibilities and limitations for sustainable bioenergy production systems. Doctoral Dissertation, Utrecht University.

Smeets, E. M. W., Faaij, A. P. C., Lewandowski, I. M. and Turkenburg, W. C. (2007). A bottom-up assessment and review of global bio-energy potentials to 2050. *Progress in Energy and Combustion Science*, **33**, 56–106.

Smith, P. (2012). Agricultural greenhouse gas mitigation potential globally, in Europe and in the UK: what have we learned in the last 20 years? *Global Change Biology*, **18**, 35–43.

Smith, P., Martino, D., Cai, Z. *et al.* (2007a). Agriculture. In *Climate Change 2007: Mitigation. Contribution of Working Group III to the Fourth Assessment Report of the Intergovernmental Panel on Climate Change*, eds. B. Metz, O. R. Davidson, P. R. Bosch, R. Dave and L. A. Meyer. Cambridge: Cambridge University Press, pp. 499–540.

Smith, P., Smith, J., Flynn, H. *et al.* (2007b). *ECOSSE – Estimating Carbon in Organic Soils Sequestration and Emission*. Edinburgh: The Scottish Executive.

Smith, P., Martino, D., Cai, Z. *et al.* (2007c). Policy and technological constraints to implementation of greenhouse gas mitigation options in agriculture. *Agriculture, Ecosystems and Environment*, **118**, 6–28.

Smith, P., Martino, D., Cai, Z. *et al.* (2008). Greenhouse gas mitigation in agriculture. *Philosophical Transactions of the Royal Society B*. **363**, 789–813.

Smith, P., Gregory, P. J., van Vuuren, D. *et al.* (2010). Competition for land. *Philosophical Transactions of the Royal Society B*, **365**, 2941–2957.

Stehfest, E., Bouwman, L., van Vuuren, D. P. *et al.* (2009). Climate benefits of changing diet. *Climatic Change*, **95**, 83–102.

Stickler, C., Coe, M., Nepstad, D., Fiske, G. and Lefebvre, P. (2007). Readiness for REDD: a preliminary global assessment of tropical forested land suitable for agriculture. Falmouth, MA: The Woods Hole Research Center. See: www.whrc.org/resources/publications/pdf/WHRC_REDD_crop_suitability.pdf.

Strapasson, A., Dean, C., Shah, N. and Workman, M. (2011). Negative CO_2 emissions technology database and recommendations. Draft Report to UK Department of Energy and Climate Change (DECC). London: Imperial College London.

Strassmann, K. M., Joos, F. and Fischer, G. (2008). Simulating effects of land use changes on carbon fluxes: past contributions to atmospheric CO_2 increases and future commitments due to losses of terrestrial sink capacity. *Tellus B*, **60**, 583–603.

Sun, X., Katsigris, E. and White, A. (2004). Meeting China's demand for forest products: an overview of important trends, ports of entry, and supplying countries, with emphasis on the Asia–Pacific Region. *International Forestry Review*, **6**, 227–236.

Tadesse, T. (2010). Pursuing REDD+ through PFM: early experiences from the Bale Mountain's REDD Project in Ethiopia. In *Pathways for Implementing REDD+: Experiences from Carbon Markets and Communities*, eds. X. Zhu, L. R. Møller, T. De Lopez and M. Z. Romero. Denmark: UNEP Risø Centre, pp. 113–126.

Tait, J., Adcock, M., Barker, G. C. *et al.* (2011). *Biofuels: Ethical Issues*. London: Nuffield Council on Bioethics.

Tesfaye, Y. (2011). Participatory forest management for sustainable livelihoods in the Bale Mountains, Southern Ethiopia. Doctoral Thesis, Sveriges lantbruksuniversitet, Uppsala (*Acta Universitatis agriculturae Sueciae*, **64**).

The Carbon Trust (2010). Bioenergy. See: www.carbontrust.co.uk/emerging-technologies/technology-directory/bioenergy/pages/bioenergy.aspx.

Thomson, A. M., Izaurralde, R. C., Smith, S. J. and Clarke, L. E. (2008). Integrated estimates of global terrestrial carbon sequestration. *Global Environmental Change*, **18**, 192–203.

Turunen, J., Tomppo, E., Tolonen, K. and Reinikainen, A. (2002). Estimating carbon accumulation rates of undrained mires in Finland – application to boreal and subarctic regions. *Holocene*, **12**, 69–80.

Turunen, J., Roulet, N. T., Moore, T. R. and Richard, P. J. H. (2004). Nitrogen deposition and increased carbon accumulation in ombrotrophic peatlands in eastern Canada. *Global Biogeochemical Cycles*, **18**, GB3002.

UNEP (2009). Towards sustainable production and use of resources: assessing biofuels. Paris: United Nations Environment Program.

UNEP (2010). Are the Copenhagen Accord pledges sufficient to limit global warming to 2 °C or 1.5 °C? A preliminary assessment. Nairobi, Kenya: United Nations Environment Program.

UNFCCC (2011). Report of the Conference of the Parties on its sixteenth session, held in Cancun from 29 November to 10 December 2010. United Nations Framework Convention on Climate Change. See: http://unfccc.int/resource/docs/2010/cop16/eng/07a01.pdf.

Valentine, J., Clifton-Brown, J., Hastings, A. *et al.* (2012). Food vs. fuel: the use of land for lignocellulosic "next generation" energy crops that minimize competition with primary food production. *Global Change Biology Bioenergy*, **4**, 1–19.

van der Werf, G. R., Randerson, J. T. Giglio, L. *et al.* (2010). Global fire emissions and the contribution of deforestation, savanna, forest, agricultural, and peat fires (1997–2009). *Atmospheric Chemistry and Physics*, **10**, 11707–11735.

van Laake, P. E. and sánchez-Azofeifa, G. A. (2004). Focus on deforestation: zooming in on hot spots in highly fragmented ecosystems in Costa Rica. *Agriculture, Ecosystems and Environment*, **102**, 3–15.

van Leeuwen, T. T. and van der Werf, G. R. (2011). Spatial and temporal variability in the ratio of trace gases emitted from biomass burning. *Atmospheric Chemistry and Physics*, **11**, 3611–3629.

van Minnen, J. G., Klein Goldewijk, K. Stehtest, E. *et al.* (2009). The importance of three centuries of land-use change for the global and regional terrestrial carbon cycle. *Climatic Change*, **97**, 123–144.

van Vuuren, D. P., den Elzen, M. G. J., Lucas, P. L. *et al.* (2007). Stabilizing greenhouse gas concentrations at low levels: an assessment of reduction strategies and costs. *Climatic Change*, **81**, 119–159.

van Vuuren, D.P., van Vliet, J. and Stehfest, E. (2009). Future bio-energy potential under various natural constraints. *Energy Policy*, **37**, 4220–4230.

Vetter, M., Wirth, C., Böttcher, H. *et al.* (2005). Partitioning direct and indirect human-induced effects on carbon sequestration of managed coniferous forests using model simulations and forest inventories. *Global Change Biology*, **11**, 810–827.

Vitt, D. H., Halsey, L. A., Bauer, I. E. and Campbell, C. (2000). Spatial and temporal trends in carbon storage of peatlands of continental western Canada through the Holocene. *Canadian Journal of Earth Sciences*, **37**, 683–693.

Werner, F., Taverna, R., Hofer, P., Thürig, E. and Kaufmann, E. (2010). National and global greenhouse gas dynamics of different forest management and wood use scenarios: a model-based assessment. *Environmental Science and Policy*, **13**, 72–85.

West, P. C., Gibbs H. K., Monfreda, C. *et al.* (2010). Trading carbon for food: global comparison of carbon stocks vs. crop yields on agricultural land. *Proceedings of the National Academy of Sciences*, **107**, 19645–19648.

Woods, J., Tipper, R., Brown, G. *et al.* (2006). *Evaluating the Sustainability of Co-firing in the UK*. London: Department for Trade and Industry (DTI).

Woods, J., Williams, A., Hughes, J. K., Black, M. J. and Murphy, R. J. (2010). Energy and the food system. *Philosophical Transactions of the Royal Society B*, **365**, 2991–3006.

WWF (2011). *The Energy Report – 100% Renewable Energy by 2050*. Gland: WWF International.

Xu, J. T., Tao, R., Xu, Z. G. and Bennett, M. T. (2010). China's Sloping Land Conversion Program: does expansion equal success? *Land Economics*, **86**, 219–244.

Xu, Z., Xu, J., Deng, X. *et al.* (2006). Grain for green versus grain: conflict between food security and conservation set-aside in China. *World Development*, **34**, 130–148.

Yang, X., Richardson, T. K. and Jain, A. K. (2010). Contributions of secondary forest and nitrogen dynamics to terrestrial carbon uptake, *Biogeosciences*, **7**, 3041–3050.

Yu, Z., Loisel, J., Brosseau, D. P., Beilman, D. W. and Hunt, S. J. (2010). Global peatland dynamics since the last glacial maximum. *Geophysical Research Letters*, **37**, L13402.

Zaehle, S., Ciais, P., Friend, A.D. and Prieur, V. (2011). Carbon benefits of anthropogenic reactive nitrogen offset by nitrous oxide emissions, *Nature Geoscience*, **4**, 601–605.

Zeng, N. (2008). Carbon sequestration via wood burial. *Carbon Balance and Management*, **3**, 1.

Zomer, R. J., Trabucco, A., Bossio, D. A. and Verchot, L. V. (2008). Climate change mitigation: a spatial analysis of global land suitability for clean development mechanism afforestation and reforestation. *Agriculture, Ecosystems and Environment*, **126**, 67–80.

Society's responses and knowledge gaps

Sarah E. Cornell and I. Colin Prentice

Society's needs for the knowledge that Earth system science can provide are urgent, but the challenges of knowledge integration and application are substantial. This closing chapter explores some of the issues that arise with the move towards an increasingly 'applied' Earth system science.

8.1 Introduction

At the start of this book, we traced the development of Earth system science from its early foundations, including its evolving interfaces with other academic disciplines and policy. Here we take an exploratory forward look, with a particular focus on some of the more contentious issues that currently surround climate science. We draw attention to unaddressed knowledge gaps and unstated conceptual problems, which we believe are making it harder than it need be to establish an effective communication and accommodation between policy-making and science. We argue that it is important to recognize what science can and cannot achieve, and what scholarship *could* achieve in the service of good policy-making, if the right questions were addressed and methodologies developed.

Nothing we say should be interpreted as diminishing the value of Earth system science as a fundamental investigation of the interacting biological and physical/chemical processes that have sustained life on Earth. However, much of the funding that has supported the rapid advances in this field of research during the past two decades has been made available by governments with a more focused agenda, keen to assess the likely magnitude and consequences of anthropogenic climate change and (increasingly) the options for mitigation of, and necessities for adaptation to, climate change on a policy-relevant timescale. Like it or not, scientists have thereby become embroiled in debates and controversies for which they were not well prepared.

Under these unusual circumstances, we suggest that Earth system scientists have an exceptional responsibility to try seriously to answer the questions posed by policy-makers about climate change. In doing so, they may well discover fruitful lines of fundamental enquiry about human–environment interactions that break new ground by transcending disciplinary constraints.

8.2 Some unresolved issues

8.2.1 The politicization and media profile of climate science

There are many areas, including energy policy, public health and environmental protection, where scientific evidence is recognized as essential input to the policy process. But policy-making is never just a conversation between policy-makers and scientists. In the area of climate policy, the process has come to include innumerable debates involving a much wider public. There are many reasons why this is so, some of them excellent, others more worrying, such as the ease with which disinformation can be disseminated via internet-based media. However, most commentators agree that it is good to have an engaged and informed populace. And with the extraordinary, worldwide variety of social actors and conflicting interests that are potentially impacted both by climate change and by climate policy, it should be no surprise that globally concerted action is hard to achieve.

Understanding the Earth System: Global Change Science for Application, eds. Sarah E. Cornell, I. Colin Prentice, Joanna I. House and Catherine J. Downy. Published by Cambridge University Press

Widespread public scrutiny of 'applicable' science can be regarded as a positive good in that it promotes public engagement with and participation in the end-to-end process through which knowledge is produced and policies made. However, it also opens the door to the cynical manipulation of opinion. Disinformation – including high-profile presentations of deliberately falsified data, as in the case of the polemical documentary *The Great Global Warming Swindle* (WagTV/Channel 4, 2007; Ofcom, 2008) – continues to be a major problem for the wider understanding of climate science. It is firmly established that the same mechanisms and organizations once used by the tobacco industry to cast doubt on the dangers of smoking, and by the chemical industry to cast doubt on the dangers of ozone-depleting compounds, have been deployed again in the interests of self-preservation by sections of the fossil-fuel industry to cast doubt on the evidence for anthropogenic climate change (Oreskes and Conway, 2010). Demeritt (2001, p. 328) noted that:

> In the highly adversarial contexts of [the climate change] debate, the values of openness, disinterestedness, and good faith upon which science depends are quickly suspended. Personal motives are subjected to corrosive scrutiny and expert judgments discounted out-of-hand as competing groups seek to advance their interests by deconstructing opposing claims in the media or on the witness stand. Fortunately, interested skepticism about the science of climate change – paid for by the fossil fuel industry – is largely restricted to the political culture of the United States. It is difficult to see how this very cynical and deeply interested campaign to discredit the science of climate change is either reflexive or enlightened.

However, as Demeritt (2001) went on to observe, the reaction by scientists has not necessarily been the most effective one. Specifically, there has been a tendency to emphasize consensus, and yet (as climate change 'sceptics' often like to point out, quite correctly in this case) science actually advances by controversy, not consensus (Pielke, 2001; Sarewitz, 2011). Most fields of science have nothing like the IPCC process that periodically reviews and summarizes the entire state of knowledge. This collaborative effort is actually a huge and invaluable resource for researchers and educators. The secret of the IPCC reports is that they are '*written by scientists, for scientists*', as climate physicist Reto Knutti has pointed out (Haag, 2007). But the IPCC is constantly portrayed in skeptical media, absurdly, as a clique intent on presenting a biased interpretation of climate change science. Although this view has nothing to do with reality, it seems to be rather easy for people (not just interest groups) to imagine that by presenting the areas where there is scientific consensus, climate scientists are covering up disagreements. Continuing in this vein, Demeritt (2001, p. 328) suggested that

> ... efforts to win public trust by basing policy on scientific certainty can actually increase public skepticism and make the resulting policy decisions more politically uncertain. They invite political opponents to conduct politics by waging war on the underlying science (and scientists!), which in turn breeds scientific defensiveness rather than reflexive engagement in the face of criticism and debate.

It is important for scientists to state, and where necessary to repeatedly re-state, those scientific findings that are relevant to policy and for which there is strong evidence. We suggest it is also important to insist on the unique value of science in providing testable (and continually re-tested) descriptions of how the world works. But to 'close ranks' and downplay controversies may be counterproductive; it excludes the public from insights into how science really advances, and the controversies that go on behind the scenes. We are *not* suggesting that the causes of contemporary climate change are open to serious contention. Chapter 2 summarizes the evidence that contemporary climate change is unequivocal, and its causes well established. But the same cannot be said, for instance, of current trends in net primary production (NPP), the causes of abrupt climate changes in the past, the numbers of species at risk of extinction due to climate change, the role of ancient humans in the demise of mammoths and giant marsupials, or a host of other hotly disputed topics.

8.2.2 The problem of defining 'dangerous' climate change

With the benefit of hindsight, the framing of the climate change issue by the UNFCCC has created a problematic conundrum (discussed by Pielke, 2005). Specifically, the convention binds its signatories to develop policies striving to avoid '*dangerous anthropogenic interference with the climate system*' (Article 2; UNFCCC, 1992). The difficulty is that this definition carries unstated, but potentially counterproductive, implications: (a) that we live in a world today where climate is benign (the reality is very different; see UNISDR, 2009: enormous damage in terms of life and livelihood occurs episodically due to weather-related disasters such as tropical cyclones, landslides, droughts and insect infestations, whose frequency and severity have remained fairly constant although the economic costs of natural disaster have risen conspicuously in recent decades); (b) that there is a definable level of change that is 'dangerous' (impossible, as discussed in Chapters 2 and 6); and (c) that the primary purpose of climate policy should be the mitigation of anthropogenic climate change, rather than improving the adaptation of the human population to the climate we already have and any changes that might occur soon, whether anthropogenic or not.

It could perhaps be argued that the UNFCCC definition of the climate change problem is now immaterial. It is certainly true that adaptation and 'climate-proofing' have entered the mainstream of policy-making by many countries and international organizations (e.g. Swart *et al.*, 2009; ADB, 2005). Equally, there is no need for policy-making to be constrained by the wording of just one of the global environmental treaties. However, the hidden assumptions of the UNFCCC may have deflected research priorities, casting the spotlight on an artificially narrow subset of the greater issue of climate and its relation to environment and development. These assumptions may also have encouraged both scientists and policy-makers to focus on the perceived imperative of limiting climate change to within certain bounds, rather than on development goals that are more fundamental from a humanitarian perspective. At the same time, policy targets that have been framed in terms of development and human rights, notably the Millennium Development Goals (UN, 2000), have limited scope for leverage and integration with climate policy.

There is still a huge gap in our knowledge as to what measures are available for different societies to adapt better, both to their present climate and to climate change. While society can in principle make decisions that will influence the extent of future global warming, it will need to adapt in any case to the component of climate change that is inevitable regardless of how quickly emissions are abated. This lack of knowledge is problematic – as discussed in Chapter 6 – because whereas we can use the toolkit of Earth system science to assess the impacts of climate change, many of these are 'potential impacts', in the sense that we have much more limited information about the nature and costs, and thus the current and future affordability, of adaptation measures. Much of the growing literature on adaptation is strictly place-based, resulting in a scale mismatch with the global impacts assessments. It has been argued that this situation is intractable, because all adaptation, and the socio-economic background against which adaptation decisions are made, is intrinsically local. We suggest that this is an unnecessarily despairing view, and that many aspects of adaptation are generic and should be included in action-oriented analyses of climate change. It should be possible, for example, to assess the feasibility and cost of different measures to conserve water or increase year-round water supplies in regions where per-capita freshwater resources are declining; or to quantify the viability of different options for enhanced coastal defence in areas where sea-level rise presents threats to habitation. Indeed, we suspect that the magnitude of these adaptation costs for 'high-end' warming scenarios may provide the most important and persuasive arguments for climate-change mitigation.

Climate change is deeply interconnected with the other challenges facing humanity. The framing of climate change as if it were isolated from these other challenges is unhelpful, both in terms of advancing our fundamental understanding of global change processes, and in terms of informing the decisions by which society steers itself towards its desired goals. In particular, the idea that climate change as a physical phenomenon is unique and distinct, or uniquely 'dangerous' in its impacts on human activities, is both counterproductive and wrong.

8.2.3 Beyond prediction?

Earth system science offers predictive power. In the early stages of this field, this was 'small-p prediction', in the narrow scientific sense of generating testable hypotheses about the system being studied. In the context of climate change, the combination of models that incorporate multiple interlinked and interdependent processes with the compilation of detailed global pictures of environmental and biogeochemical changes in the past, and Earth observation that allows present-day real-time changes to be monitored, now marks a shift to 'proper prediction'. These tools offer a valuable glimpse into Earth's future. The same toolkit that provides insights into the risks of global environmental change also needs to be deployed in the identification and evaluation of options for coping in the changed world of the future.

The future will not be an extrapolation of the present. The previous chapters of this book have described several reasons why this is the case. For example, from the social perspective, Malthus was radically wrong in extending his logic about demographic trends because he did not anticipate transformations in the way that people interact with their environment. From the physical perspective, a rich body of evidence shows that complex feedbacks and thresholds come into play with a perturbation to climate. Model predictions differ fundamentally from other foresight efforts – the predicted changes are not a matter of anyone's subjective opinion. Many of the predicted changes are not intuitive (such as the further drying of already-dry areas discussed in Chapter 2), and their consequences are hard to envisage because there is often no contemporary analogue of the projected outcome. This is why process-based models that link climate, chemistry and the biosphere are needed. Nevertheless, the power and realism of Earth system models have their limits, and there are some areas that we argue that scientists should avoid.

The first and most obvious of these is the quantification of practically unquantifiable things, such as the 'value of biodiversity'. Another is attempting to predict phenomena that lack any useful predictive framework. For example, major geopolitical events or technological breakthroughs may well have causal influence on a future outcome, but there is no known mechanism that shapes their dynamics. A further category is situations where prediction would be unhelpful, even if it were possible. For example, using models to try to predict which policies *will* be adopted is a futile direction if what is needed is to use the model output to inform a decision-making process. Where there are questions of human agency, a more useful approach is deliberately to retain that part of the process outside the model 'black box'.

Predictions of climate and its potential impacts can in principle be made up to ~ 30 years ahead, because of the combined inertia in climate, the carbon cycle and the world's energy systems. Driven by the desire to inform climate adaptation, the growth in research on seasonal to decadal prediction has been very rapid in recent years (e.g. Royal Meteorological Society, 2011). No doubt there is scope for considerable economic value in an improved ability to predict; for example, knowledge of next year's growing conditions in different regions could potentially provide information useful to farmers (crop selection and planting dates). Of course, there is a *But* – or several. The skill and reliability of seasonal to decadal forecasting are still very limited (Meehl *et al.*, 2009; Palmer, 2011) and the potential to greatly improve this skill is not established. From a modelling perspective, the availability of data on the internal variability of the climate system at these timeframes is a major constraint, compounded by the need to assimilate this initialization data into the forecasting model in ways that yield information about the climate quantities that actually matter for adaptation. From an economic point of view, the expense of developing forecasting on this scale would be high, as would the costs of rapid responses in some climate-impacted sectors. Improved prediction would need to be justified in terms of net economic benefits. It has even been argued (Dessai *et al.*, 2009) that major investments in this area reflect a misplaced technocratic worldview. 'Optimizing' human activities on a global scale on the basis of quantitative model predictions might not be a desirable pathway even if it were a feasible one. It risks disregarding the rich cultural dimensions of human–environment interactions, and if taken to its logical extreme could imply a need for an extraordinary level of social control.

Even in the natural sciences, when studying complex systems (including ecosystems and the biosphere), *the most important conclusions from quantitative modelling are often qualitative*. Quantification is necessary to establish, for example, which effects matter and which are trivial. This limitation of quantification applies *a fortiori* to human systems. Rademaker (1982) argued that a process of model 'dequantification' is needed after the analysis stage, to reduce the hazards of modelling complex (social) systems, rather than hiding conjectures away – because there will always be a need to make some conjectures in modelling, instead of simply ignoring processes known to be important. This argument from a social science perspective is, we suggest, equally valid for modelling complex natural systems, especially when the biota and biodiversity are included in what is modelled. Dawson *et al.* (2011) argued that the most popular frameworks for modelling the impact of climate change on biodiversity are both spuriously precise and potentially misleading because (a) they do not take into account all of the available evidence, much of it qualitative or semi-quantitative, about how species have actually responded to climate changes in the past, which would show that their assumptions are flawed; and (b) although they presume to inform conservation policy, they start in the wrong place – with a mechanistic prediction – instead of first posing the important practical question (what conservation measures are likely to be effective in a changing climate?) and then asking what types of information exist, or are required, to answer it.

In short, prediction and quantification are important pursuits; they can be tremendously valuable; they are not 'the answer'. Much relevant understanding is qualitative by nature. Quantitative models are essential routes to the qualitative understanding of complex systems, but their quantitative nature should not be mistaken for precision or accuracy.

8.2.4 Problems in defining expertise

Why should society listen to scientists when they make statements about climate change? Embedded in this question is a morass of concerns about what counts as expertise and who can claim to have it. A large part of the authority of science comes from its proven success in explaining the natural world. Scientists trained in the skills and methods that apply to their field can contribute to the development of that field, over time building up expertise and the capability to speak authoritatively on their subject. In climate change research, the intrinsic requirement for engagement across what are often highly specialized scientific disciplines presents non-trivial challenges. Collins and Evans (2007), in their 'periodic table of expertise', drew attention to the essential role of interaction across disciplines in many contemporary fields of research and practice. Interactional expertise, they suggested, should be recognized in its own right. It involves the ability to engage in credible high-level dialogue with the contributory experts in a field of study, which in turn must involve a high degree of intellectual engagement with the concepts and issues. Interactional expertise is also an essential component in the effective communication of science beyond the specialist field. However, when it comes to the nuanced and tacit knowledge that all specialist fields contain (and which often come to the fore in areas of contention and debate), there is no substitute for the extended, profound intellectual immersion that underpins contributory expertise.

There are occasions when 'experts' should not have a platform, or at least not a privileged one. Scientists from other fields opining about climate science is one such situation. At best, they cloud the picture of the state of understanding, and on some occasions they have unhelpfully provided fodder for the tendentious arguments of climate 'skeptics'. Again, the approach of Collins and Evans (2007) for differentiating the types and contexts of expertise helps deal with this misuse of expertise. There is an important place in research for open and inclusive discussion, not least because exposing science (including its models and methods) to a broad range of critique at worst cannot harm the scientific enterprise and at best may help scientists to answer new questions, or old ones in new ways. But the critique needs to meet on the level – scientists from whatever field need to have contributory expertise on the issue in question. Equally problematic is the situation where climate scientists use their platform to unhelpfully state poorly thought-out positions that are beyond *their* realm of expertise.

Expertise should be valued. To avoid devaluing it, experts should take care to emphasize what expertise they have – and also to be clear about what they don't have.

8.2.5 The role of science and scientists in decision-making

Why should scientists make statements about climate change to society, and how? The short answer to the first question is because scientists' knowledge about the present state of the environment and the prospects for change matters to society. Scientific statements are not specious futurology, which is precisely why attention needs to be given to the way in which science links with society's decision-making processes. Pielke (2007) noted that a broad possibility space exists for engagement at the science–policy interface. This space is marked out by different views on what experts should provide to the process of decision-making, and on how this happens. With regard to the former, experts may either provide scientific evidence for or against an argument, or provide information about a suite of alternatives on an issue. Similarly there is a range of views on how science informs policy. On one hand, 'pure' science can be seen as the input to a linear pipeline, where knowledge is progressively translated or developed until it reaches the decision-maker at the other end, whose job it then is to apply the knowledge. On the other hand, scientists can be viewed as one of many groups of stakeholders in a contested field of knowledge inputs that informs decision-making. Within this possibility space, Pielke's framework yields four roles for scientists. A 'pure scientist' operates detached from the stakeholder fray, building evidence on a narrow issue. A 'science arbiter' (a consultant, for instance) similarly operates under the linear model of science in society, but presents information on a range of issues. An 'issue advocate' gathers selected arguments to present in the multi-stakeholder context. An 'honest broker' seeks to present the available alternatives in the contested stakeholder space.

In the context of global environmental change, both the need for and the value of the 'honest broker' role are evident to us, but apparently under-appreciated by the scientific community at large. Fulfilling this role would entail doing and communicating research explicitly about the effectiveness and consequences (including possible perverse side-effects) of alternative policies, with respect to the goals they are supposed to promote. This is a highly interdisciplinary area, as yet seriously understudied largely because it does not fit comfortably into any of our current academic discipline boxes. In not doing this research, we suggest that science is failing policy-makers.

Scientists may play any or all of the four roles. Scientists may legitimately advocate a particular policy, from the point of view of a well-informed insider. But it is important to be transparent about where one's science stops and advocacy begins. It is not uncommon for scientists to blur this boundary (sometimes apparently unwittingly), but this is one of the most counterproductive tactics of all. Climate science risks being discredited by the perceived association between the communication of well-established science on climate change with the advocacy by some of particular positions, for instance on population, aviation, economic growth, energy policy or geoengineering, that do not actually follow from science.

In improving the engagement of science with decision-makers and wider society, there are several things science can already provide:

- observations of what is happening;
- models seeking to explain what is happening, together with the analysis of their consistency with observations; and
- 'what if' answers (though many questions remain), including estimating the consequences of policies on multiple scales, to inform decisions about their overall effectiveness.

There are also some things that could be done better. Earth system scientists have a responsibility to make a serious attempt to answer stakeholders' questions, not only academic questions formulated within the various disciplines and 'interdisciplines' that make up the field. In many parts of the world, research funders are promoting and incentivizing this kind of engagement, because it raises the value and the impact of the research they invest in. We need to move beyond the interdisciplines that underpin today's climate and Earth system science to a deeper interdisciplinarity that entrains the much broader range of knowledge and skills needed for application and action in a complex human context. Orienting the research community towards addressing questions that are articulated by decision-makers, or that are co-developed through dialogue with them, could drive new areas of enquiry with serious intellectual content. But many communication barriers exist, and must be overcome if this vision is to be realized.

8.3 Envisioning the future

Science has useful things to say to society about the future. Scientists are recognizing their responsibility to share this knowledge, policy-makers are keen to incorporate it in their decision-making, and society has more access to information and is more literate than ever before. But given the intricacies and interconnections of the Earth system, a major challenge still remains in envisioning the future and channelling this information.

8.3.1 Scenarios: strengths and limitations

Scenarios of the future underpin much of the science outlined in this book. The emissions pathways used as inputs for climate modelling and impacts assessment are based on quantitative scenarios that cover a wide range of possible environmental, social and economic conditions. The first IPCC synthesis assessment had just one scenario (SA90) describing the baseline world with no climate policy interventions. The six IS92 baseline scenarios developed for the Second Assessment Report aimed to embody a much wider array of assumptions affecting how future greenhouse-gas emissions could evolve, to better represent the associated range in the ultimate stabilization levels. The IPCC's Third and Fourth Assessments used the series of SRES scenarios (Nakicenovic and Swart, 2000), which are all 'non-intervention' scenarios, describing the evolution of emissions assuming no explicit mitigation policy (although they do allow for the potential adoption of policies driven by other concerns than climate). The process of developing scenarios for climate studies has remained broadly the same throughout this period: assumptions are taken from published forecasts and expert analysis of factors including energy use, demographics and trade, and the resulting storylines are subject to external expert review before the emissions pathways over time are calculated using integrated assessment models. Girod *et al.* (2009) describe the evolution of the IPCC's 'official' scenarios, arguing that this process has involved ebbs and flows in saliency, credibility and legitimacy in response to scientific criticisms, policy demands and the political processes of the IPCC itself.

The IPCC scenarios are just one of many sources of quantitative emission scenarios. Wigley *et al.* (1996) developed the WRE scenarios that redistribute the carbon budget for stabilization over a different timeframe from the IS92 scenarios, arguing that it would be both more realistic and more cost-effective to allow a longer time for the necessary economic transitions, even though this would delay the emissions reductions and subsequently require steeper cuts. It was partly in response to this kind of criticism about the realism of the embedded technological and economic assumptions that the SRES scenarios were developed. But a new criticism emerged: while the technical aspects of the SRES scenario generation have been documented in rich detail, involving efforts from several modelling teams in a more open process than previously, the process of constructing the initial storylines is much less transparent, and the full set of 40 scenarios in the series has rarely been used as intended.

In the Millennium Ecosystem Assessment, Cork *et al.* (2005) took particular care in explaining both the process of storyline description and the relationship between the qualitative aspects of the scenarios and the modelling activities, in its development of a set of four scenarios to explore the implications of global futures on ecosystems. They argued that models, which must restrict themselves to processes for which there is quantitative evidence, should be used together with the scenario storylines, which can explore a wider range of plausible changes. The Representative Concentration Pathways (RCPs) developed by the climate science community for the IPCC's Fifth Assessment (Moss *et al.*, 2010) represent a conscious effort to allow for improved interaction between the models and the scenario storylines in the climate context. Unlike the IPCC's previous scenarios based on (non-intervention) emission pathways, the RCP scenarios are defined in terms of greenhouse-gas concentrations, for which a range of socio-economic storylines can be devised, including an explicit consideration of intervention options for adaptation and mitigation of climate change.

Despite several decades of scenario use in climate modelling and global change research, misunderstandings about their design and purpose are frequently evident. These include:

- **believing that every scenario must be a plausible future** – In most management contexts, where scenarios are used as visioning tools to scope possible futures, the plausibility and internal coherence of the storyline are essential criteria. For this kind of application, there is no point in including situations that lie beyond the realms of possibility. But in climate science, there are often very good reasons to develop quantitative scenarios that are not viable in reality, or that do not seem plausible from our current perspectives. There is an important place for thought experiments in Earth system science (e.g. Smith, 2011), both for fundamental inquiry and for education and communication, and scenarios are an important part of that.
- **believing that the scenario line can be followed in reality** – Despite all the warnings in scenario literature that the ubiquitous coloured lines on scenario graphs are merely indicative of possible or conceivable futures, there is a very widespread inclination to interpret them as actual pathways for society. People talk of the desirability of following, say, the SRES B2 trajectory without reflecting on the concomitant requirement to live in the (constructed and simplified) B2 world. The translation from emissions scenarios via greenhouse-gas concentration trajectories to the line of temperature projections is also imbued with a concreteness that does not exist, given the remaining uncertainties about climate sensitivity and feedbacks as discussed in Chapter 4.
- **believing that things could be 'worse' than the 'worst case'** – The SRES scenarios were designed to cover a wide range of possible futures in terms of global greenhouse-gas emissions. The fact that emissions in recent years have tracked or exceeded the trajectories of the higher emission scenarios has led to concerns that the 'worst case' scenarios for the coming century are not severe enough. But the scenarios were never intended to be accurate predictions of what might happen over a particular period of just a few years. The cumulative emissions by 2100 of the 'high-end' SRES scenarios (in the A1 family) are large, and

constrained by the maximum plausible rates of population and economic growth.

- **mistaking scenario differences for 'uncertainty'** – This misunderstanding is nearly universal in media reports, and there is often ambiguity about the relationship between scenario analysis and uncertainty in scientific contexts too. The range delineated by scenario lines is intended to capture much of the 'possibility space' for the future, but it is not a measure of the uncertainty in climate modelling, nor an indication of the uncertainty about Earth's future socio-environmental trajectories. The former requires careful statistical approaches to uncertainty characterization and assessment (discussed in Chapter 5), and the latter cannot be predicted quantitatively by scenario approaches. It is in part a matter of human agency, best dealt with through deliberation rather than abstraction into models.

A key role of scenarios is to fill gaps where prediction is impossible. Emissions and concentration scenarios designed to be inputs to climate models are necessarily quantitative but, as Cork *et al.* (2005) explain, scenarios do not need to be quantitative to be of great value to global change science. In particular, research, policy and practice for adaptation and the reduction of vulnerability depend on the development and use of qualitative scenarios, often with the narratives constructed through dialogue with the stakeholders in the issue. As in many other areas of global change science, however, distinct research communities are engaged in quantitative and narrative scenario approaches. This pronounced divide contributes to the persistent misunderstandings about climate change scenarios that we describe above.

As climate science becomes more deeply interdisciplinary, it will need to continue to develop the scenario methodologies used in the Millennium Ecosystem Assessment and the RCP scenarios for parallel but interacting storyline and modelling processes. There is also considerable scope for embedding scientific insights better into many aspects of society's future planning. Foresight initiatives are used extensively for strategic planning of communities, sectors and nations, entraining considerable expertise to take a longer-term perspective than in normal policy processes. Foresight studies differ from scenario analysis in that they generally actively seek a desirable future outcome, rather than merely mapping out the possibility space for the future. Better scientific engagement with these processes can avoid their making speculative or subjective treatments of environmental change, as if it were as unpredictable or uncertain as future social and economic directions.

Scenarios fulfil a good and useful role in climate science, as long as they are not over-interpreted. Given that we cannot foresee the future, by imagining and elaborating descriptions of what could happen in ways that take into account the likely impacts of climate change and other drivers, we stand a better chance of preparing ourselves to face the challenges ahead.

8.3.2 Decision-making frameworks

Scientific research into global change has practical value when it addresses issues that matter to decision-makers and engages with them effectively. Throughout this book, we have explored areas of fundamental research that are valuable for informing decision-making, giving insights about how global change might affect future plans, and also how society's decisions might affect global change.

Economic assessments are an important part of the toolkit routinely used by industry and governments to assess the feasibility, benefits and costs of possible courses of action. Climate change and energy policy are not exceptions; so Earth system scientists need to engage with and be informed about the underlying principles of economic assessments, their value and their limits.

It is a rather big stretch to apply cost–benefit analysis to the global question of balancing investments in decarbonizing the energy system against climate change impact costs over timescales of a century and beyond – but it has been done. The high-profile conclusions of both the Stern (2007) and Garnaut (2008, 2011) reviews are that immediate major investment in climate change mitigation is justified economically, in terms of avoided damages in the future. Stern took into account the possibility of 'high-end' damage costs, 40% greater than the basic assessment that used the SRES scenarios, for a scenario based on the supposed 'fat tail' of climate sensitivity estimates (although his qualitative conclusion does not depend on this assessment). Weitzman (2009) also emphasized the uncertainty in climate sensitivity, drawing drastic conclusions about the damage costs associated with climatic disasters and mass extinctions if very high temperatures were to result from the true climate sensitivity lying in the fat tail. Since these economic studies were made, palaeoclimate modelling and observations have been used to provide constraints on climate sensitivity: see Chapter 3 for more on this topic. Effectively, climate sensitivities lying in the fat tail of the distribution are ruled out by observations of ice-age greenhouse gases and regional climates.

The Stern Review has been the subject of several critiques. Some of these have tried to discredit the underlying science; we shall ignore these. (See Chapter 2 for a summary of the relevant scientific evidence.) However, others have pointed out deficiencies from an economic perspective. Although full consideration of these extensive economic critiques is outside our scope, we wish to draw attention to one caveat that is quite fundamental: the issue of discount rates. The net economic benefit of a given degree of mitigation must, inevitably, depend on the economic discount rates applied to the damage costs. This dependence becomes very strong indeed when the time horizon extends more than two centuries into the future.

All decision-making involves discounting the future. Discounting is involved when we assume, for example, that contemporary policy-makers need not be concerned with events 1000 or 10,000 years hence, even if such timescales are a mere instant in the history of the Earth. A form of discounting is implicit whenever global-warming potentials with a specific (commonly 100-year) time horizon are used to describe the climate impact of greenhouse gases, and to establish 'equivalency' among greenhouse gases with different lifetimes.

In economics, future values are discounted using an exponential decay function with a rate constant r. The standard theoretical formulation of the 'social discount rate' (r) is based on the Ramsey equation:

$$r = \rho + \eta g \tag{8.1}$$

where ρ is the pure rate of time preference, η is the elasticity of the marginal utility of consumption, and g is the growth rate of consumption, which in the long run depends on technological development (Nordhaus, 2007). The terms contributing to r lack the certainty of physical constants, but they are nevertheless constrained by observations. The term r itself is constrained by the observed rate of return on capital. The term g is reasonably well determined by historical data and is commonly taken to be around 2% per year. The term η functions as an intertemporal weighting factor for growth. It expresses the fact that additional costs become more affordable with increasing wealth. Values of around 1.5 have been derived from observations. The term ρ expresses people's preference for spending now, as opposed to saving for the future. This is the term whose values are the most controversial, but this too is constrained by observations of the savings ratio. It is usually taken to be around 3%.

Stern, however, set η at the conservative value of 1, and ρ at the exceptionally low value of 0.1% for the base-case analysis, resulting in a value of r much lower than that normally used in the assessment of capital projects (for example by the UK Treasury) and, indeed, lower than that implicit in the Stern Review's own assessment of mitigation costs (Mendelsohn, 2007). Stern argued that the rates used for typical capital projects are unlikely to be appropriate for a sensible treatment of risks and uncertainties that extend beyond the investors' lifetimes to more than 100 years into the future. The choice of an unusually low discount rate has been described as 'ethical' (apparently on the grounds that it seems to be more in favour of future generations than conventional economic choices). But this notion is problematic, as it seeks to impose a particular ethical preference that is at variance with people's actual preferences, as revealed in economic observations (Nordhaus, 2007). Furthermore, its 'ethical' nature is arguably spurious: it does no service (actually a disservice) to future generations to apply less stringent criteria to assess the cost-effectiveness of climate change mitigation than are applied to other investments. We suggest that this may be another example of the fallacy of assigning a unique status to climate change in relation to other issues deserving of investment.

Tol and Yohe (2006) drew attention to a further limitation of Stern's analysis: the dichotomous choice between stabilizing at 550 ppm CO_2 or doing nothing at all. They suggested that the jury is still out on the economics of mitigation, in part because the various alternatives (including intermediates between these positions) have not been quantified. But they also emphasized a qualitative consensus, that there is *no economic case for doing nothing* to mitigate climate change; and that there are many good reasons to tackle climate change immediately, many of which are not well represented within the framework of cost–benefit analysis.

These debates in climate economics bring theory and models right up against moral concerns. It is vital to apply expert knowledge to key societal problems and laudable to aim for the greater good in doing so. But using climate policy to address intra- or intergenerational income inequality, for example, is as problematic as presenting politically defined global-warming targets as science. Where decisions are made about what components to include in expert analysis, the rationale behind those choices needs to be explicit and value judgements made transparent. This requires a much greater degree of self-reflection – for both science and economics – and the way in which it can be done and the way that decision-makers handle this more nuanced expert input both present new operational challenges for the study of global change.

Of course, climate policy decisions are not made solely on the basis of economic assessments. These are not only inherently uncertain when dealing with long-time-scale concerns, but they also manifestly fail to cover many aspects of the world that people care about. This problem is not adequately solved by trying to place monetary values on things that are not bought and sold in the marketplace, as Stern attempted to do. As mentioned in Chapter 1, the results of such valuations for non-marketed goods and services vary hugely according to methodology, and the direct applicability of these monetizations to economic analysis in the first instance, and beyond that to actual decision-making, is moot. Nevertheless, these efforts at valuation of ecosystems and 'natural capital' have served an important role in drawing attention to the sheer magnitude of our dependence on the natural world, and on goods and services for which there is no market.

In decision-making, many different concerns are always in play. Institutions such as the OECD, the World Bank and the UN have long had a set of defined criteria for 'good governance' for economic development in a globalized world: transparency, accountability, responsiveness, compliance with the rule of law, efficiency, effectiveness and equity. These remain an excellent foundation for a globalized world undergoing climate change. With the unique perspective and predictive power that Earth system science brings, attention to these issues is needed in parallel to the scientific advances – and even more so as the field entrains more of the social and economic dimensions of global change. The UK sustainable development strategy (Defra, 2005) incorporates a framework that is useful for those seeking the right balance of expert knowledge in decision-making. It frames its top-level goals in terms of the interacting social and environmental systems ('*ensuring a strong, healthy and just society*' while '*living within environmental limits*'). The mediators of the dynamic balance between society and environment are the economic and governance sub-systems, based on sound science. All three of these dimensions are needed for sustainability, and they act as 'correctives' to each other.

8.4 Concluding remarks

The feasibility of seriously warming the world is undeniable. There is no physical obstacle to melting all the ice sheets (raising sea level by ~ 70 m) and driving the planet to a state somewhat resembling the Eocene, some 50 million years ago, even though there are both physical and economic constraints on how rapidly such a transformation could occur. But there is probably an easy consensus that we do not want to carry on making the world warmer and warmer without limit, even if the speed, extent and nature of required mitigation actions are not agreed upon by all actors.

We presume that the pace of responses to climate change will ultimately be set by what is politically feasible and, to some extent, seen as desirable on other grounds (win–win situations). The technical feasibility of decarbonization is not in doubt, and it is no longer disputed that a transformation of the global energy system (albeit with questions about the rate) is economically feasible as well. What is clear, however, is that the pace will not be set by scientists defining targets, like the ubiquitous 'two degrees'; especially as such targets lack a clear scientific basis.

Despite well-intentioned statements about targets, the policy process designed to achieve internationally binding agreements has been sluggish, riven with stalemates. But the slow progress to date in international negotiations between countries should not draw attention away from the reality of actions being taken at the domestic and regional levels. China and India are adopting tough climate targets, while their 'bottom line' in negotiations remains their social and economic development. India '*needs to grow at 8–10% for 20–30 years*' (Parikh, 2010), and this will entail major increases in energy use. China similarly will not compromise on energy use for development, even as it makes extraordinary efforts to increase energy efficiency and reduce the carbon intensity of its energy supply. In the USA, there are major efforts involving individual states that collectively account for a substantial fraction of the US economy, as well as collaborations involving Canadian provinces and even other nations. In Australia, where climate change is still a highly contentious and politicized topic, the state of New South Wales at least has made strong climate mitigation commitments (e.g. Anon., 2002). Internationally, the World Bank is sponsoring many projects involving alternative energy and promoting energy efficiency. These are actions that, if pursued and extended, will eventually lead to climate stabilization. The actors involved have, in effect, decided that they need to 'learn by doing' and this is seen as a politically more palatable option than signing up to treaties whose feasibility of implementation is unknown. In this context, drivers other than climate change are important: nations are increasingly aware of energy security concerns, and this awareness is likely to drive a move away from fossil fuels. Denmark moved away, more than 30 years ago, for reasons unconnected with climate change, using tax structure as the vehicle for the transition to renewable energy sources (Hadjilambrinos, 2000; Mendonça *et al.*, 2009). There were no adverse effects on the economy or the people of Denmark, and the country, although small, is now a leading exporter of renewable-energy technology. Climate stabilization is one major driver for an energy system transition, but it is far from being the only one.

The human context of Earth system science, in a world that is 'wired' (observed systematically) and globalized (therefore interconnected), defines a huge field of endeavour linked to a potential transition to a more sustainable mode of energy production and use. But where is the next generation being trained in the required skills for addressing problems that span the physical, biological and human sciences? The late Stephen Schneider, who pioneered so much in climate research and its responsible engagement with society, commented 20 years ago on the narrow focus of higher education. He characterized the discipline-bound focus of most university departments and faculties in the context of the global change challenge as '*intellectually derelict and socially irresponsible*' (Schneider, 1992). Has anything changed?

References

ADB – Asian Development Bank (2005). Climate proofing: a risk-based approach to adaptation. Manila: Asian Development Bank. See: www.adb.org/Documents/Reports/Climate-Proofing.

Anon. (2002). Electricity Supply Amendment (Greenhouse Gas Emission Reduction) Act 2002. Sydney: Government of New South Wales.

Collins, H. and Evans, R. (2007). *Rethinking Expertise*. Chicago, IL: University of Chicago Press.

Cork, S., Peterson, G., Petschel-Held, G. *et al.* (2005). Four scenarios. In *Millennium Ecosystem Assessment – Scenarios Assessment*, eds. S. Carpenter and P. Pingali. Washington, DC: Island Press, pp. 223–294.

Dawson, T. P., Jackson, S. T., House, J. I., Prentice, I. C. and Mace, G. M. (2011). Beyond predictions: biodiversity conservation in a changing climate. *Science*, **332**, 53–58.

Defra (2005). *Securing the Future: the UK Sustainable Development Strategy*. London: The Stationery Office.

Demeritt, D. (2001). The construction of global warming and the politics of science. *Annals of the Association of American Geographers*, **91**, 307–337.

Dessai, S., Hulme, M., Lempert, R. and Pielke, R., Jr (2009). Do we need better predictions to adapt to a changing climate? *Eos*, **90**, 111–112.

Garnaut (2008). *The Garnaut Climate Change Review. Final Report*. Cambridge: Cambridge University Press.

Garnaut (2011). *The Garnaut Review 2011. Australia in the Global Response to Climate Change*. Cambridge: Cambridge University Press.

Girod, B., Wiek, A., Mieg, H. and Hulme, M. (2009). The evolution of the IPCC's emission scenarios – changes, causes and critical aspects. *Environmental Science and Policy*, **12**, 103–118

Haag, A. L. (2007). What's next for the IPCC? *Nature Reports Climate Change*. doi:10.1038/climate.2007.73.

Hadjilambrinos, C. (2000). Understanding technology choices in electricity industries: a comparative study of France and Denmark. *Energy Policy*, **28**, 1111–1126.

Meehl, G.A., Goddard, L., Murphy, J. *et al.* (2009). Decadal prediction: can it be skillful? *Bulletin of the American Meteorological Society*, **90**, 1467–1485.

Mendelsohn, R. O. (2007). A critique of the Stern report. *Regulation*, **29**, 42–46.

Mendonça, M., Lacey, S. and Hvelplund, F. (2009). Stability, participation and transparency in renewable energy policy: lessons from Denmark and the United States. *Policy and SocietyPolicy and Society*, **27**, 379–398.

Moss, R. H., Edmonds, J. A., Hibbard, K. A. *et al.* (2010). The next generation of scenarios for climate change research and assessment. *Nature*, **463**, 747–756.

Nakicenovic, N. and Swart, R. (2000). *IPCC Special Report on Emissions Scenarios*. Cambridge: Cambridge University Press.

Nordhaus, W. (2007). A review of *The Stern Review on the Economics of Climate Change*. *Journal of Economic Literature*, **45**, 686–702.

Ofcom (2008). *Ofcom Broadcast Bulletin*, July 21, **114**, 6–22 and 36–80.

Oreskes, N. and Conway, E. M. (2010). *Merchants of Doubt: How a Handful of Scientists Obscured the Truth on Issues from Tobacco Smoke to Global Warming.* London: Bloomsbury Press.

Palmer, T. (2011). What's needed to produce reliable decadal predictions? Presentation at RMetS National Meeting, Reading, 16 February. See: www.rmets.org/pdf/presentation/20110216-palmer.pdf and www.rmets.org/pdf/presentation/20110216-palmer.mp3.

Parikh, K. (2010). Climate change and sustainable development: view from the developing world. Keynote presentation at the Worldwatch Institute Workshop, Strengthening India's low carbon growth strategy, New Delhi, 13 September.

Pielke, R. A., Jr (2001). Room for doubt. *Nature*, **410**, 151.

Pielke, R. A., Jr (2005). Misdefining "climate change": consequences for science and action. *Environmental Science and Policy*, **8**, 548–561.

Pielke, R. A., Jr (2007). *The Honest Broker: Making Sense of Science in Policy and Politics.* Cambridge: Cambridge University Press.

Rademaker, O. (1982). On modelling ill-known social phenomena. In *Groping in the Dark: the First Decade of Global Modeling*, eds. D. Meadows, J. Richardson and G. Bruckmann. Chichester: John Wiley and Sons, pp. 206–208.

Royal Meteorological Society (2011). Decadal forecasting: where are we now, where are we going, and how do we get there? RMetS National Meeting, Reading. 16 February. See: www.rmets.org/events/detail.php?ID=4521.

Sarewitz, D. (2011). The voice of science: let's agree to disagree. *Nature*, **478**, 7.

Schneider, S. (1992). The role of the university in interdisciplinary global change research: structural constraints and the potential for change. *Climatic Change*, **20**, vii–x.

Smith, R. (2011). Would our Earth look the same with 0 degree tilt? *BBC 23 Degrees blog.* www.bbc.co.uk/blogs/23degrees/2011/03/what_if_the_earth_had_no_tilt.html.

Stern, N. (2007). *The Economics of Climate Change: The Stern Review.* Cambridge: Cambridge University Press.

Swart, R., Biesbroek, R., Binnerup, S. *et al.* (2009). Europe adapts to climate change: comparing National Adaptation Strategies. PEER Report No 1. Helsinki: Partnership for European Environmental Research.

Tol, R. S. J. and Yohe, G. W. (2006). A review of the Stern review. *World Economics*, 7, 233–250.

UN (2000). UN Millennium Declaration. Resolution 55/2 of the UN General Assembly, 18 September 2000. A/RES/55/2. New York, NY: United Nations. See: www.un.org/millenniumgoals.

UNFCCC (1992). United Nations Framework Convention on Climate Change – Article 2. UNEP/IUC/99/2, Information Unit for Conventions. Geneva: UNEP. See: www.unfccc.int/resource/convkp.html.

UNISDR (2009). Global assessment report on disaster risk reduction: risk and poverty in a changing climate. United Nations International Strategy for Disaster Risk Reduction. Geneva: United Nations. See: www.preventionweb.net/gar09.

Wigley, T. M. L., Richels, R. G. and Edmonds, J. A. (1996). Economic and environmental choices in the stabilization of atmospheric CO_2 concentrations. *Nature*, **379**, 240–243.

Weitzman, M. L. (2009). On modeling and interpreting the economics of catastrophic climate change. *Review of Economics and Statistics*, **91**, 1–19.

Acronyms

AABW	Antarctic Bottom Water
AOGCM	coupled atmosphere–ocean general circulation model
AIM	Asian–Pacific Integrated Model
AIMES	Analysis, Integration and Modeling of the Earth System, an IGBP project established in 2003
AMOC	Atlantic Meridional Overturning Circulation
AR4	The IPCC's Fourth Assessment Report, published in 2007
AVHRR	Advanced Very High Resolution Radiometer
AVOID	Avoiding Dangerous Climate Change, a UK research and policy programme
BECCS	bioenergy with carbon capture and storage
BVOC	biogenic volatile organic compounds
C^4MIP	Coupled Carbon Cycle–Climate Model Intercomparison Project
CCDAS	Carbon Cycle Data Assimilation System
CCMAP	Climate–Carbon Modelling, Assimilation and Prediction, a QUEST project
CIESIN	Consortium for International Earth Science Information Network
CLIMAP	Climate: Long range Investigation, Mapping, and Prediction
CMIP	Coupled Model Intercomparison Project
COHMAP	Cooperative Holocene Mapping Project
D–O	Dansgaard-Oeschger events
Defra	UK Government Department for Environment, Food and Rural Affairs
DESIRE	Dynamics of the Earth System and the Ice core Record, a QUEST project
DGVM	dynamic global vegetation model
Diversitas	An International Programme of Biodiversity Science
DJF	December–January–February, a shorthand description of the northern-hemisphere winter season and the southern-hemisphere summer
DPSIR	Driver–Pressure–State–Impact–Response, a framework for describing social-environmental changes
ECHAM	Max Planck Institute for Meteorology's global climate model
EEA	European Environment Agency
EMICs	Earth system models of intermediate complexity
ENSO	El Niño Southern Oscillation
EOF	empirical orthogonal function, used in a multivariate statistical technique
EPICA	European Project for Ice Coring in Antarctica

ERBE	Earth Radiation Budget Experiment
erSSTv3	NOAA global sea-surface temperature series
ESM	Earth system model
ESSP	Earth System Science Partnership, involving the Global Change Projects IGBP, IHDP, WCRP, and Diversitas
EU	European Union
FACE	Free Air Carbon Dioxide Enrichment
FAMOUS	Fast Met Office/UK Universities Simulator, a lower-resolution UK Earth system model
FAO	Food and Agriculture Organization of the United Nations
fAPAR	fraction of absorbed photosynthetically active radiation
FLUXNET	network of regional networks making eddy-covariance flux measurements of land–atmosphere exchanges
GARP	Global Atmospheric Research Program
GCAM	Global Change Assessment Model
GCM	general circulation model
GCP	Global Carbon Project
GDP	gross domestic product
GENIE	Grid-enabled Integrated Earth System Model (a fast model)
GFDL	Geophysical Fluid Dynamics Laboratory (and also their climate model)
GFED	Global Fire Emissions Database
GI	Greenland interstadial, the abrupt warming and gradual cooling periods of Dansgaard-Oeschger events
GISS	NASA Goddard Institute for Space Studies
GLACE	Global Land–Atmosphere Coupling Experiment
GLOBIO	global biodiversity model, developed by the PBL Netherlands Environmental Assessment Agency, UNEP–World Conservation Monitoring Centre and UNEP GRID–Arendal.
GLOBIOM	IIASA's global dynamic partial equilibrium model integrating agricultural, bioenergy and forestry sectors
GPP	gross primary production
GS	Greenland stadial, the steep cooling and cold periods of Dansgaard–Oeschger events
GSWP	Global Soil Wetness Project, an international project
HadCM	family of Hadley Centre climate models
HadGEM	Hadley Centre global environmental model
HadISST1	Hadley Centre global sea-surface temperature series
HANPP	human appropriation of net primary production
HE	Heinrich events, major episodes of iceberg discharges in the North Atlantic Ocean
HYBRID	a dynamic global vegetation model
IAM	integrated assessment model (typically linking two or more of economics, land use, energy and climate)
IBIS	Integrated Biosphere Simulator, a dynamic global vegetation model
ICSU	International Council of Science Unions

IEA	International Energy Agency
IGBP	International Geosphere Biosphere Programme
IHDP	International Human Dimensions Programme on Global Environmental Change
IHOPE	Integrated History and future of People on Earth
IIASA	The International Institute for Applied Systems Analysis
IMAGE	Integrated Model to Assess the Global Environment
IPAT	the IPAT equation frames environmental impact *I* as a function of population *P*, affluence *A*, and technology *T*
IPCC	Intergovernmental Panel on Climate Change
ISSC	International Social Sciences Council
JJA	June–July–August
JULES	UK's Joint Land Surface model
LGM	Last Glacial Maximum
LIG	last interglacial
LPJ	Lund–Potsdam–Jena dynamic global vegetation model, now a family of models developed collaboratively
LUE	light-use efficiency
LULUCF	land use, land-use change and forestry
MA	Millennium Ecosystem Assessment
MACC	marginal-abatement cost curve
MacPDM	Macro-scale Probability-Distributed Moisture, a global hydrological model
MARGO	Multiproxy Approach for the Reconstruction of the Glacial Ocean, an international collaborative project
MEGAN	Model of Emissions of Gases and Aerosols from Nature
MESSAGE	Model for Energy Supply Strategy Alternatives and their General Environmental Impact
MODIS	Moderate Resolution Imaging Spectroradiometer
MOSES	Met Office Surface Exchange Scheme, one of the UK's global land-surface models
NADW	North Atlantic Deep Water
NASA	National Aeronautic and Space Association
NCAR	US National Center for Atmospheric Research
NCAS	National Centre for Atmospheric Sciences
NCEP	US National Centers for Environmental Prediction
NDVI	normalized difference vegetation index
NERC	Natural Environment Research Council
NPP	Net Primary Production
NRC	US National Research Council
NREAPs	National Renewable Energy Action Plans
OECD	Organization for Economic Cooperation and Development
ORCHIDEE	Organizing Carbon and Hydrology in Dynamic EcosystEms, the French global land surface model
PAR	photosynthetically active radiation

PDSI	Palmer Drought Severity Index
PETM	Paleocene–Eocene Thermal Maximum (~56 Ma before present)
PFT	plant functional type (in dynamic global vegetation models); plankton functional type (in dynamic marine biogeochemistry models)
PMIP	Palaeoclimate Modelling Intercomparison Project (there has now been a series of PMIPs)
POLES	Prospective Outlook on Long-term Energy Systems, an econometric model of the world energy system
QESM	QUEST Earth System Model (and the project that developed it)
RAINFOR	Amazon Forest Inventory Network
RCP	Representative Concentration Pathway, the scenarios used for IPCC's Fifth Assessment Report.
REDD	reducing emissions from deforestation and forest degradation
RESCUE	Responses to Environmental and Societal Challenges for our Unstable Earth, a European Science Foundation Forward Look assessing research needs for the 'Anthropocene'
SPI	Standard Precipitation Index
SRES	The IPCC Special Report on Emissions Scenarios, and the scenarios themselves
SSRC	Social Science Research Council (originally US, now international)
SST	sea-surface temperature
TEEB	The Economics of Ecosystems and Biodiversity, an international assessment of the benefits of well-functioning ecosystems to human society
TRIFFID	Top-down Representation of Interactive Foliage and Flora Including Dynamics, a dynamic global vegetation model
UKCA	UK Chemistry and Aerosols model
UN	United Nations
UNDP	United Nations Development Programme
UNECE	United Nations Economic Commission for Europe
UNEP	United Nations Environment Programme
UNESCO	United Nations Educational, Scientific and Cultural Organization
UNFCCC	United Nations Framework Convention on Climate Change
UNISDR	United Nations International Strategy for Disaster Reduction
VECODE	Vegetation Continuous Description, a dynamic global vegetation model
WCRP	World Climate Research Programme
WMO	World Meteorological Organization
WRE	Wigley, Richels and Edmonds climate scenarios (named after their authors)
WUE	water-use efficiency

Glossary of terms

Anthropocene	The period in which global changes due to human activities have been of a magnitude comparable with the natural events that define geological epochs. The Industrial Revolution is usually regarded as the start of the Anthropocene.
Atmospheric lifetime	In the calculation of global-warming potential for long-lived trace gases, the atmospheric lifetime is defined as the time taken for 63% of the additional amount added to be removed from the atmosphere by chemical, biological and physical processes.
Benchmarking	Quantitative evaluation of a model's ability to match observations to a specified degree of accuracy. For Earth system models, this involves the selection of observational data that capture key features of Earth system processes and feedbacks, not just of first-order changes such as temperature or greenhouse-gas concentration.
Biosphere	The part of the Earth system that comprises living organisms (in both terrestrial and marine ecosystems) and their immediate environment, including organic matter derived from living organisms.
Carbon dioxide equivalent	A measure for comparing concentrations and fluxes of various greenhouse gases based upon their radiative forcing (relative to that of carbon dioxide) over a specified timeframe.
Carbon dioxide fertilization	The enhanced growth of plants as a result of increased atmospheric CO_2 concentration.
Clathrate	A chemical lattice structure, where one molecule is held in a 'cage' made of molecules of another compound. Methane clathrates, also called methane hydrates, consist of methane in a water molecule lattice, and occur in permafrost regions and in deep-sea formations.
Climate sensitivity	The equilibrium change in the annual mean global surface temperature following a doubling of the atmospheric carbon dioxide concentration.
Coupling (of models)	Coupled climate models include numerical models of ocean and atmospheric dynamics, and take into account the effects of each component on the other. Earth system modelling involves linking additional components, such as the dynamics of land vegetation and ocean biogeochemistry, and representing their interactions.
Ecosystem services	The benefits to human society arising from well-functioning ecosystems.
Feedback	A change (in the climate system) that occurs in response to a change in forcing, and that in turn affects the initial forcing. It can be *positive*, leading to an amplification of the original signal, or *negative*, resulting in a tendency to dampen the effects of the perturbation.

Global-warming potential	An index of the cumulative radiative forcing effect of a greenhouse gas over a specified time period (typically 100 years) resulting from the emission of a unit mass of the gas, relative to that of carbon dioxide.
Land-use change	The human modification of Earth's land surface. This can involve changes to the land cover (e.g. forest, grassland, bare soil), modified land-surface processes, and shifts in the social and economic contexts of the land use.
Mitigation	Action taken to reduce the consequences of global warming. In climate policy, mitigation refers primarily to action to decrease radiative forcing by reducing the net emissions of greenhouse gases.
Non-linear responses	Abrupt changes in response to a smooth forcing. In climate science, examples include the Dansgaard–Oeschger events and the alternations in the states of the Atlantic circulation.
Prediction	In this book, we use this word in its scientific sense, as a statement of what will ensue from a given set of circumstances, based on current knowledge. It is not used to refer to general statements about what the future will hold, nor more narrowly to projections of future climate made using models.
Radiative forcing	The change in the downward longwave (thermal) radiation flux that results from a given change in atmospheric composition, relative to a defined baseline, typically the pre-industrial atmosphere.
Reanalysis data	Global data sets for several climate variables that are not directly observable, produced using previously observed climate data. Observations of temperature, wind speed, and pressure are interpolated onto a global system of grids, and used as inputs to an atmospheric forecasting model which outputs simulated data sets of variables such as sea-surface temperature, precipitation, and soil moisture.
Scenario	A description of the world. Quantitative scenarios are used to define the inputs to a model, and can include historical observations as well as descriptions of possible futures. Narrative scenarios are storylines characterizing possible futures, which can be translated into quantitative scenarios using data and simplifying assumptions about demographic, political, socio-economic and technological trends.
Transdisciplinarity	An approach to research that draws on and integrates knowledge from multiple disciplines, and that includes input from the stakeholders of the research in the knowledge development process.

Index